RESISTIVITY AND INDUCED POLARIZATION

Resistivity and induced polarization methods are used for a wide range of near-surface applications, including hydrogeology, civil engineering and archaeology, as well as emerging applications in the agricultural and plant sciences. This comprehensive reference text covers both the theory and practice of resistivity and induced polarization methods, demonstrating how to measure, model and interpret data in both the laboratory and the field. Marking the 100-year anniversary of the seminal work of Conrad Schlumberger (1920), the book covers the historical development of electrical geophysics, electrical properties of geological materials, instrumentation, acquisition and modelling, and includes case studies that capture applications to societally relevant problems. The book is also supported by a full suite of forward and inverse modelling tools, allowing the reader to apply the techniques to a wide range of applications using digital datasets provided online. This is a valuable reference for graduate students, researchers and practitioners interested in near-surface geophysics.

ANDREW BINLEY is Professor of Hydrogeophysics at Lancaster University. His research focuses on the use of near-surface electrical geophysics for hydrogeological characterization. He is the developer of widely used geoelectrical modelling computer codes. In 2012, he was awarded the Frank Frischknecht Leadership Award for his long-term contributions to the field of near-surface geophysics, and in particular hydrogeophysics. This award is jointly presented by the Society of Exploration Geophysicists and the Environmental and Engineering Geophysical Society. He was elected Fellow of the American Geophysical Union in 2013 for pioneering work on uncertainty modelling and hydrogeophysics.

LEE SLATER is Distinguished Professor and Henry Rutgers Professor of Geophysics at Rutgers University. His research focuses on near-surface geophysics, and he has performed extensive laboratory and field studies with resistivity and induced polarization. In 2013, he was awarded the Harold B. Mooney Award for long-term contributions in education and professional outreach in near-surface geophysics by the Society of Exploration Geophysicists and the Environmental and Engineering Geophysical Society. He was elected Fellow of the American Geophysical Union in 2018 for visionary experimentation in near-surface geophysics.

'This is without doubt the most comprehensive and thorough treatment of electrical geophysics anywhere in the literature. It is a brilliantly written book, covering theory and practice, with numerous real-world examples of the use of resistivity and induced polarization. It will certainly be first on my recommended reading list for students, researchers and practitioners working in the field of geo-electrics and near-surface geophysics.'

Jonathan Chambers, British Geological Survey

'Binley and Slater are two of the best electrical geophysicists in the world, and together have written a comprehensive, accessible textbook for anyone interested in electrical methods. By including a history of the methods, open-source software, and sections on theory, instrumentation, forward and inverse modelling, and applications, they've produced a "one-stop shop" for all things electrical. This book starts with a primer on the most fundamental mathematics and builds up from there to topics outlining the state of the science, including helpful figures and sidebar information along the way. I strongly recommend this book to any student or practitioner interested in learning more about how to apply electrical geophysical techniques to shallow-Earth problems, and look forward to sharing it with my research students.'

Kamini Singha, Colorado School of Mines

'Andrew Binley and Lee Slater, two experienced scientists in the field of near-surface geophysics, have compiled a modern textbook that describes the development and the state of the art of resistivity and IP technology. The book provides deep insight into the theoretical fundamentals, and presents the breadth of application of these geophysical methods. Considering the wealth of information and the clearly arranged presentation, the textbook will be useful both for academic education and as a reference work for researchers and practitioners. This book will certainly inspire further research work and practical application of resistivity and IP methods.'

Andreas Weller, Technische Universität Clausthal

RESISTIVITY AND INDUCED POLARIZATION

Theory and Applications to the Near-Surface Earth

ANDREW BINLEY
Lancaster University

LEE SLATER
Rutgers University Newark

CAMBRIDGE
UNIVERSITY PRESS

CAMBRIDGE
UNIVERSITY PRESS

University Printing House, Cambridge CB2 8BS, United Kingdom

One Liberty Plaza, 20th Floor, New York, NY 10006, USA

477 Williamstown Road, Port Melbourne, VIC 3207, Australia

314–321, 3rd Floor, Plot 3, Splendor Forum, Jasola District Centre,
New Delhi – 110025, India

79 Anson Road, #06–04/06, Singapore 079906

Cambridge University Press is part of the University of Cambridge.

It furthers the University's mission by disseminating knowledge in the pursuit of education, learning, and research at the highest international levels of excellence.

www.cambridge.org
Information on this title: www.cambridge.org/9781108492744
DOI: 10.1017/9781108685955

First published 2020

A catalogue record for this publication is available from the British Library.

Library of Congress Cataloging-in-Publication Data
Names: Binley, Andrew, 1961– author. | Slater, Lee, 1969– author.
Title: Resistivity and induced polarization : theory and applications to
the near-surface earth / Andrew Binley, Lancaster University, Lee Slater, Rutgers University
Newark, New Jersey.
Description: Cambridge, UK ; New York, NY : Cambridge University Press,
2020. | Includes bibliographical references and index.
Identifiers: LCCN 2020030992 (print) | LCCN 2020030993 (ebook) | ISBN
9781108492744 (hardback) | ISBN 9781108685955 (ebook)
Subjects: LCSH: Earth resistance (Geophysics) | Induced polarization. |
Geophysics – Technique. | Geophysics – Methodology. | Geophysics – Data processing.
Classification: LCC QC809.E15 B56 2020 (print) | LCC QC809.E15 (ebook) |
DDC 551.1/4–dc23
LC record available at https://lccn.loc.gov/2020030992
LC ebook record available at https://lccn.loc.gov/2020030993

ISBN 978-1-108-49274-4 Hardback

Additional resources for this publication at www.cambridge.org/binley

Contents

Colour plates are to be found between pp. 268 and 269

Preface

The year 2020 marks the 100-year anniversary of Conrad Schlumberger's publication on electrical methods, a document that serves as the foundation of modern-day resistivity and induced polarization methods. These methods are now two of the most widely used geophysical techniques for probing the near-surface (upper few hundred metres of the Earth's crust). Although originally developed for hydrocarbon and mineral exploration, their popularity in near-surface applications grew rapidly from the widespread recognition that subsurface electrical properties are often well correlated to physical and chemical properties of fluids within the pore space (e.g. saturation, salinity) and lithological characteristics (e.g. porosity, clay content) needed to understand a wide range of near-surface properties and processes. This has opened up numerous areas of application, including hydrology, civil engineering, agronomy, forensic science and archaeology. Furthermore, the theoretical concepts are well established, and field measurement techniques are highly scalable, allowing investigations from cm to km scale. Measurements can be made in a wide variety of configurations, allowing simple mapping through to 3D time-lapse imaging. Compared to many other near-surface methods, instrumentation is relatively low cost and straightforward to operate. Added to this, many interpretation (modelling) tools are now widely available and continue to evolve.

A number of texts on resistivity and induced polarization have been written, but none provide a comprehensive account of the fundamental electrical properties of natural Earth materials along with the theoretical basis of modelling approaches used to interpret measurements. Furthermore, most reference texts have focussed only on exploration applications of resistivity and induced polarization and thus do not cover many recent advancements in the techniques specifically aimed at near-surface applications. The literature covering these methods is vast, and the sheer volume of material can be daunting for some researchers and practitioners. Building on 25 years of continuous collaboration, we set out to produce a single text that explains how electrical properties are related to other characteristics of Earth materials, how we can make measurements of electrical properties in the laboratory and the field, and how we can analyse the signals in order to produce electrical models of the subsurface. Our target audience includes graduate students, researchers, advanced undergraduate students and practitioners. We include the historical context for much of the material discussed so that the reader can appreciate that several advances were made decades ago, even though full exploitation of concepts and observations may not have been realized early on. The concepts covered in the main text are illustrated throughout, with additional case studies later in the book to highlight the breadth of application and emerging areas. As we wanted to provide a single comprehensive text, we also include software to allow the reader to analyse example datasets using some of the modelling approaches covered in the book and, more importantly, apply these techniques to their own datasets. We hope that the text will inspire a new generation of geophysicists to further advance resistivity and induced polarization beyond the 100-year legacy of Schlumberger and other early adopters of these technologies for investigating the near-surface Earth.

Acknowledgements

Production of this book would not have been possible without the help of many people, to whom we are extremely grateful. Sina Saneiyan did a fantastic job preparing most of the figures for Chapters 2 and 3, along with a number of figures in Chapter 6. Guillaume Blanchy took the lead on developing ResIPy for the book, supported by Sina Saneiyan, Jimmy Boyd and Paul McLachlan. We are grateful to colleagues who provided input to some of the examples in the book, in particular Guillaume Blanchy, Jimmy Boyd, Jon Chambers, Baptiste Dafflon, Nikolaj Foged, Thomas Günther, Andreas Kemna, Kisa Mwakanyamale, Sina Saneiyan, Michael Tso, Florian Wagner, Ken Williams and Paul Wilkinson. Judy Robinson, Sina Saneiyan, Michael Tso, Chen Wang, Andreas Weller and Paul Wilkinson completed reviews of some early chapter drafts – their help on improving the text is much appreciated. We are also grateful to Konstantin Titov and Valeriya Hallbauer-Zadorozhnaya for providing some historical insight into early Russian developments in the methods. Within the book we have used many examples that originate from our own studies: work that was supported by past and present researchers working in our groups, especially (Lancaster) Siobhan Henry-Poulter, Ben Shaw, Roy Middleton, Peter Winship, Nigel Crook, Arash JafarGandomi, Lakam Mejus, Melanie Fukes, Heather Musgrave, Qinbo Cheng, Paul McLachlan, Michael Tso, Guillaume Blanchy, Jimmy Boyd; (Rutgers) Dimitrios Ntarlagiannis, Judy Robinson, Sina Saneiyan, Chen Wang, Kisa Mwakanyamale, Yuxin Wu, Jeff Heenan, Pauline Kessouri, Fardous Zarif and Alejandro Garcia. We have both been fortunate to have worked with many colleagues over the years. A number of these deserve particular recognition. AB is grateful to Maurice Beck for first introducing him to electrical tomography 30 years ago and to Fraser Dickin for early collaborations that built on this. Long collaborations between AB, Bill Daily, Abelardo Ramirez and Doug LaBrecque were both successful and enjoyable. AB and Andreas Kemna began working together on inversion when Andreas worked for a short period in AB's group in the mid-1990s – this partnership (and friendship) has remained and has helped mature many of the codes that AB has developed. LS is grateful to Fred Vine for introducing him to geophysics and to Stewart Sandberg for encouraging him to pursue experiments with induced polarization as a post-doctoral scientist. A productive and enjoyable collaboration with David Lesmes led to invaluable insights into electrical properties in the early 2000s. LS is also indebted to Andreas Weller for ten years of continuous collaboration and for generously sharing his unique understanding of induced polarization mechanisms. Finally, we should acknowledge the patience and understanding of friends and family during our obsession with writing this book.

Symbols

English Symbols

A	= positive current electrode label
AB	= distance between electrodes A and B
AM	= distance between electrodes A and M
AN	= distance between electrodes A and N
a	= electrode dipole spacing
B	= negative current electrode label
BM	= distance between electrodes B and M
BN	= distance between electrodes B and N
$\hat{B}$	= equivalent conductance of exchange ions appearing in the Waxman and Smits (1968) model
$C_{(c)}$	= concentration of charge carriers in mol m^{-3}
$C_{\boldsymbol{m}}$	= model covariance matrix
CEC	= cation exchange capacity (in meq g^{-1} of dry clay)
c	= Cole-Cole model exponent
c_p	= specific polarizability describing the role of EDL chemistry on σ''
$c_{(c)}$	= concentration of charge carriers in mol l^{-1}
D	= diffusion coefficient of charge carriers
D_+	= diffusion coefficient of ions in the Stern layer
d	= layer thickness or distance
$\boldsymbol{d}$	= data vector
$\boldsymbol{d}_{rat}$	= ratio dataset (time-lapse inversion)
d_0	= grain diameter
$\boldsymbol{E}$	= electric field
e	= elementary charge (1.6022×10^{-19} coulombs)
F	= formation factor describing σ_{el} (equal to Archie F when surface conductivity is zero)
F	= forward model operator
F_a	= measured (apparent) formation factor
F_s	= equivalent formation factor describing σ_{surf}
f	= frequency in Hz (s^{-1})

g	= gravitational acceleration
h	= layer thickness
i	$= \sqrt{-1}$
I	= electric current (amperes or A)
I_0	= ionic strength (mol m^{-3})
I_r	= saturation index from Archie's second law
I_s	= equivalent saturation index for surface conductivity
I	= identity matrix
J_1	= first-order Bessel function
$\boldsymbol{J}$	= current density
$\boldsymbol{J}$	= Jacobian matrix
$\boldsymbol{J_{diff}}$	= diffusion current density
$\boldsymbol{J_{mig}}$	= electromigration current density
$\boldsymbol{J_{reac}}$	= reaction current density
K	= geometric factor
K_h	= hydraulic conductivity
K_s	= Stefanesco kernel function
k	= permeability
k_w	= wavenumber or iteration number
k_b	= Boltzmann's constant $\left(1.3806 \times 10^{-23}\ \text{J K}^{-1}\right)$
$k_{i,j}$	= reflection coefficient for interface between layers i and j
L	= longest distance between electrodes
$L(\boldsymbol{m})$	= likelihood of parameter vector $\boldsymbol{m}$
l	= constant of proportionality defined as the ratio of σ'' to σ_{surf}
M	= positive potential electrode label
M	= size of parameter vector
M_a	= measured apparent chargeability
$M_{n(a)}$	= measured normalized apparent chargeability
MF	= metal factor
m	= cementation exponent from Archie's first law
$\tilde{m}$	= chargeability appearing in relaxation models (e.g. Cole-Cole)
$\hat{m}$	= intrinsic chargeability
m_n	= normalized chargeability (equivalent to $\sigma_\infty - \sigma_0$)
$\boldsymbol{m}$	= parameter vector
$\boldsymbol{m_0}$ (or $\boldsymbol{m_{hom}}$)	= reference parameter set
N	= number of measurements
N	= negative potential electrode label
N_A	= Avogadro's constant $\left(6.022 \times 10^{23}\ \text{mol}^{-1}\right)$
n	= saturation exponent from Archie's second law
n	= outward normal direction
n	= multiplier for electrode spacing or summation index
$\hat{n}$	= charge carrier density
n_r	= number of relaxation terms in Cole-Cole model
PFE	= percentage frequency effect
$P(\boldsymbol{A}\mid\boldsymbol{B})$	= the probability of event $\boldsymbol{A}$ given event $\boldsymbol{B}$

p	= saturation exponent for the surface conductivity
Q_v	= shaliness factor describing the excess charge involved in surface conduction
q	= electric charge
R	= resistance (ohm or Ω)
R_N and R_R	= normal and reciprocal measurement of resistance, respectively
$\boldsymbol{R}$	= roughness matrix
$\boldsymbol{R_m}$	= resolution matrix
$\boldsymbol{R_t}$	= component of roughness matrix in time dimension
$\boldsymbol{R_x}$ (or $\boldsymbol{R_z}$)	= component of roughness matrix in x (or z) direction
r	= distance between current and potential electrodes
r_i	= distance between imaginary current and potential electrodes
$\boldsymbol{S}$	= cumulative sensitivity matrix
S_{por}	= pore volume normalized surface area
S_w	= relative saturation (−)
s	= $AB/2$ (half of current electrode spacing)
T	= temperature
T_s	= Schlichter kernel function
T_e	= electrical tortuosity
T_h	= hydraulic tortuosity
T_p	= period of waveform
t	= time
V	= electric potential (volts)
V_a	= primary voltage field due to current source in homogenous medium
V_b	= secondary voltage field due to heterogeneity in the medium
V_p	= primary voltage during current on time
V_s	= secondary voltage after current shut off
V_{DC}	= voltage after application of infinitely long current pulse
v	= voltage in Fourier transformed space
$\hat{v}$	= volume fraction of electron conducting mineral
$\boldsymbol{W_d}$	= data weight matrix
x, y, z	= Cartesian coordinates
Z	= impedance (ohms)
$\hat{Z}$	= valence

Greek Symbols

α	= regularization scalar
α_L	= characteristic length of a finite element
α_s	= regularization scalar
α_t	= regularization scalar for time-lapse inversion
α_x	(or α_z) = regularization scalar in x (or z) direction
β	= mobility of a charge carrier, e.g. ions

β_+ = mobility of ions in Stern layer
β_P = apparent ionic mobility for real part of surface conductivity in POLARIS model
δ = Dirac delta
ε = dielectric permittivity
ε_0 = dielectric permittivity of free space (8.854×10^{-12} F m^{-1})
ε_R = error in resistance measurement
ε_M = modelling error
ε_{Ma} = error in chargeability measurement
ε_Z = error in magnitude of complex impedance measurement
ε_φ = error in phase angle measurement
η = dynamic viscosity
η_d = differential polarizability
θ = volumetric water content
θ = angle
κ = relative permittivity ($\varepsilon/\varepsilon_0$)
Λ = characteristic length scale (approximately twice the pore volume/pore surface area)
λ = integration variable
$\hat{\lambda}$ = equivalent imaginary conductance of exchange ions appearing in the Vinegar and Waxman (1984) model
λ_A = coefficient of anisotropy
λ_P = apparent ionic mobility for imaginary part of surface conductivity in POLARIS model
λ_{LM} = Levenberg-Marquardt damping factor
μ = Lagrange multiplier
ρ = resistivity (ohm m or Ωm)
ρ_a = apparent resistivity
ρ^* = complex resistivity
ρ_w = resistivity of fluid filling pores (e.g. groundwater)
ρ_g = fluid density
$\rho_{[ps]}$ = resistivity of a partially saturated material
$\rho_{\parallel}$ = resistivity parallel to bedding planes
$\rho_{[s]}$ = resistivity of a saturated material
$\rho_{\perp}$ = resistivity orthogonal to bedding planes
Σ = surface conductance (in siemens or S)
Σ^s = surface conductance of the Stern layer
Σ^d = surface conductance of the diffuse layer
σ = electrical conductivity (siemens m^{-1} or S m^{-1})
σ^* = complex electrical conductivity
σ' = real part of complex electrical conductivity
σ'' = imaginary part of complex electrical conductivity
$|\sigma|$ = magnitude of complex electrical conductivity
σ_{eff} = effective electrical conductivity from a volumetric mixing model
$\sigma_{[s]}$ = electrical conductivity of a saturated material

$\sigma_{[ps]}$	= electrical conductivity of a partially saturated material
σ_w	= electrical conductivity of fluid filling pores (e.g. groundwater)
σ_{el}	= electrolytic conductivity associated with the fluid-filled interconnected pore network
σ_{EDL}	= electrical conductivity of the electrical double layer (EDL)
σ_{surf}	= surface conductivity associated with ions in the electric double layer
σ^*_{surf}	= complex surface conductivity
σ^0_{surf}	= surface conductivity at the low-frequency limit
σ^∞_{surf}	= surface conductivity at the high-frequency limit
σ_0	= electrical conductivity at the low-frequency limit
σ_∞	= electrical conductivity at the high-frequency limit
τ	= characteristic relaxation time in spectral IP datasets
τ_{mean}	= mean relaxation time from a Debye decomposition
τ_p	= characteristic relaxation time defined from peak in phase spectrum
τ_0	= the time constant appearing in relaxation models (e.g. Cole-Cole)
$\tau(x, y, z)$	= cross gradient function
Φ_d	= data misfit
Φ_m	= model misfit
φ	= 'phase shift' or 'phase' of the complex conductivity or complex resistivity (positive for conductivity and negative for resistivity)
ϕ	= interconnected porosity (−)
ϕ_{eff}	= effective porosity probed by current flow lines (−)
ϕ_v	= volume fraction
χ_d	= Debye screening length
χ^2	= chi-squared statistic
ω	= angular frequency in radians s^{-1} ($\omega = 2\pi f$)

1
Introduction

1.1 Geophysical Investigation of the Subsurface

Geophysical methods were developed over 100 years ago to investigate the structure of the Earth from measurements made at the surface. The geophysical properties measured with these methods, such as density, seismic wave velocity and electrical resistivity (of interest here), are sensitive to changes in the physical and chemical properties of the solids and fluids that make up the Earth. Development of geophysical methods was foremost motivated by exploiting this sensitivity of the measurements in the search for large mineral and petroleum deposits located within the upper few kilometres of the Earth. Further technology development occurred during the Second World War, e.g. for the detection of unexploded ordnance (UXO) such as beach mines. Economic considerations drove the development of geophysical methods in mineral and petroleum exploration, foremost to reduce costs of drilling holes into the Earth. Techniques were needed to guide drilling to economically favourable locations. These locations were identified from anomalous geophysical measurements relative to more normal 'background' measurements in the region of study.

In the last ~50 years, geophysical techniques have been developed to investigate the near-surface Earth, which we define here as the upper few hundred metres of the Earth's crust. Much of this work falls within the evolving field of 'hydrogeophysics' (Binley et al., 2015), which refers to the use of geophysical techniques in water resources exploration and protection rather than petroleum and mineral exploration. Other areas of application include engineering, e.g. for assessing the integrity and performance of engineered subsurface structures, archaeology and forensic science. A number of these techniques focus on measuring the electrical resistivity of the Earth's near-surface, which, for reasons discussed later, can provide valuable information on a wide range of properties of Earth materials. Geophysical techniques that measure resistivity fall into two general categories: those that rely on galvanic (direct current) contact with the Earth and those that rely on the physics of electromagnetic (EM) induction. It is also possible to estimate variations in the electrical resistivity of the Earth from the attenuation of transmitted high-frequency (MHz range) electromagnetic waves using the ground-penetrating radar (GPR) geophysical method (Jol, 2008).

Electromagnetic instruments fall into two general categories: those using controlled sources and those that use a remote natural source (e.g. magnetotellurics). The controlled (localized) source methods are of most interest in near-surface investigations of the Earth. Frequency domain EM systems have primarily been developed for rapid mapping of the electrical conductivity of soils (McNeil, 1980) whereas time domain systems have primarily been developed for acquiring 1D models of the vertical electrical conductivity structure of the Earth (Kaufman et al., 2014). The major advantage of the EM methods over the galvanic methods is the non-contact nature of the measurement, which means that spatially extensive information on variations in conductivity is readily obtainable with these methods. This has led to the widespread development of mobile EM measurement platforms, deployed from land, water or air. The requirement for direct contact with the Earth is often a major impediment to comparable rapid data acquisition with galvanic methods. One exception is waterborne surveying, where continuous galvanic measurements are possible by pulling instrumentation in direct contact with the water column (Day-Lewis et al., 2006). Land-based mobile systems based on galvanic coupling tailored for agricultural applications have also proven effective (Gebbers et al., 2009). Some land-based mobile platforms have been developed using capacitive coupling concepts (Geometrics, 2001) but they remain relatively uncommon in use.

Despite their limitations, the galvanic electrical geophysical methods remain extremely popular. This is largely because they give the most direct and simplest (theoretically) measurements of the low-frequency electrical properties of the Earth. In contrast to EM methods, the concepts involved in understanding the measurements are relatively straightforward. Galvanic measurements are much simpler to implement for high-resolution, plot-scale characterization and monitoring and can be applied over a relatively broad range of scales. Importantly, information on the low-frequency charge storage properties of the ground, determined from induced polarization (IP) methods discussed in detail in this book, is straightforward to extract from galvanic measurements. Although EM datasets may contain IP signals (Flis et al., 1989), the interpretation of EM measurements in terms of IP targets is currently non-trivial.

Our focus is on the galvanic methods, although much of the material presented in the first part of Chapter 2 is relevant to the interpretation of EM measurements as well as resistivity estimated from attenuation of electromagnetic wave propagation recorded with GPR. Galvanic electrical resistivity methods remain one of the most widely used near-surface geophysical techniques, certainly for environmental and hydrological investigations, because of the sensitivity to a range of properties and states of subsurface materials and fluids. Although originally developed for hydrocarbon and mineral exploration, resistivity methods are increasingly used for water resources evaluation, including for the ever-important task of locating dwindling water supplies in arid parts of the world. Environmental applications of the techniques emerged in the 'green revolution' of the 1980s, including the mapping of contaminant plumes and assessment of groundwater clean-up technologies. More recent global issues such as food security, energy security, climate change and coastal resiliency demand more spatial and temporal (for process monitoring) information on the near-surface Earth. Mapping contrasts in resistivity and IP over multiple scales can provide information to help address many of these issues. Example case studies presented in Chapter 6 highlight this opportunity.

1.2 Importance of Electrical Properties

The broad popularity of the electrical geophysical methods for investigating the subsurface in large part stems from the dependence of electrical properties on a wide range of physical and chemical properties of interest to many fields of study. The controls of the subsurface structure and composition on electrical resistivity are discussed in detail in Chapter 2. Here, it is sufficient to note that resistivity varies over almost 17 orders of magnitude in the Earth in response to changes in the volume concentrations of the constituents (e.g. as defined by porosity and water content), texture (e.g. grain size distribution), fabric (e.g. the arrangement of particles and their connectivity), pore fluid composition, mineralogy and temperature. This wide range of controlling factors makes for a wide range of potential applications. However, it also highlights the potential 'Achilles Heel' of resistivity measurements, being a strong likelihood of misinterpretation or ambiguity of imaged structures in the absence of supporting information. A classic example from near-surface applications is distinguishing between the presence of saline groundwater in an aquifer and the presence of a fine-grained (e.g. clay) unit. Both will manifest as conductive anomalies that, without further information, cannot be discriminated. Fortunately, some of this ambiguity can be reduced through the additional information obtained from IP measurements when acquired with the same (or similar) instrumentation. IP was originally developed for mineral exploration due to its strong sensitivity to small-volume concentrations of electron conducting iron minerals (Bleil, 1953). However, relative to resistivity measurements, IP is also strongly sensitive to rock and soil texture and less sensitive to pore fluid composition when electron conducting minerals are absent. Consequently, electrical properties measured with IP methods can help reduce the inherent ambiguity in resistivity measurements. For example, IP measurements may provide an opportunity to distinguish clays from saline fluids in resistivity images (Slater and Lesmes, 2002a). The theoretical concepts involved in the understanding of IP methods are well established for mineral exploration where the signals are strongly controlled by the volume concentration (and size) of iron minerals (Pelton et al., 1978; Wong, 1979). The foundations for understanding IP effects in soils and rocks absent of electron conductors were laid early in the development of the methodology (Marshall and Madden, 1959). However, theoretical understanding of the IP response in the absence of electron conductors has solidified mostly in the last 25 years.

Another important reason for the broad popularity of galvanic electrical methods is that field measurement techniques are highly scalable, allowing investigations from centimetre to kilometre scale (in some instances using the same instrumentation). It is also straightforward to instrument experimental tanks, lysimeters and other such vessels with the sensors (electrodes) needed to perform electrical imaging of processes occurring within the object of study. Field-scale measurements can be made in a wide variety of configurations, allowing simple mapping with four-electrode probes (e.g. in archaeology), vertical profiling of formation properties using borehole logging sensors through to 4D (3D + time) imaging using permanent (or long-term) installations of electrode arrays. Compared to some other near-surface methods, instrumentation is relatively low cost and straightforward to operate. In fact, it is even possible to build a basic resistivity instrument from first principles for a cost less than a modern smartphone (Florsch and Muhlach, 2017). Added to these advantages, robust data inversion tools are now widely available and continue to evolve. Although it is important to recognize the scalable nature of these measurements, our focus in this book is on the development of the electrical resistivity and IP methods for studying the near-surface Earth from surface observations. We also consider the investigation of the electrical structure away from (or between) boreholes using electrodes placed in boreholes. Early examples of this approach include mapping the distribution of electric potentials at the surface due to current injected into a borehole drilled into a highly conductive body (historically known as the Mise-à-la-Masse method) (Beasley and Ward, 1986). More recently, cross-borehole electrical imaging has been developed to obtain higher-resolution information on the near-surface electrical structure relative to what can be obtained using surface measurements alone (Ramirez et al., 1993). We do not focus on borehole geophysical logging measurements in this book, but discuss them briefly in Chapter 4. Although this is an immensely popular use of the method that motivated early understanding of the relationship between electrical resistivity and rock properties described in Chapter 2, well logging is extensively treated in other contributions that focus on petroleum exploration (e.g. Ellis and Singer, 2007).

Despite the rich amount of information on the physical and chemical properties of the subsurface potentially extractable from electrical measurements, the inherent limitations of geophysical measurements must always be recognized and respected. As previously cautioned, these measurements are only ever proxies (at best) of the physical and chemical properties of the subsurface and therefore easily subject to misinterpretation. The reliability of such proxy measures will vary widely depending on how well the conceptual model for the subsurface is constrained, along with the strength of the relationship between the electrical measurements and the property of interest. In the classic case of electrical well logging, the porosity is often reliably estimated in clean (low clay) sandstone from Archie's empirical relation (Section 2.2.4.1) when the fluid salinity is known. At the other end of the spectrum, some geophysical contractors serving the environmental industry tasked with characterization and remediation of contaminated sites have misled this industry in terms of the information content extractable from the method, e.g. proposing that concentrations of dense non-aqueous phase liquids (DNAPLs) in groundwater can be estimated from resistivity measurements. In this case, the conceptual model is poorly developed (contaminated sites are often complex with variable geology and pore fluid chemistry) and the relationship between DNAPL concentration and resistivity poorly established. Such misuse of geophysical methods is unfortunate, as the reputation of the methods (and geophysics in general) becomes tarnished when they fail to meet (unrealistic) expectations.

1.3 Historical Development of Electrical Geophysics

1.3.1 DC Resistivity

The historical development of electrical methods often gets reported foremost from the perspective of petroleum and mineral exploration as this was the economic engine driving development. Van Nostrand and Cook (1966) provide a comprehensive account of western development of the resistivity method up until the mid-1960s, some of which we draw on here. Other historical reviews focusing on exploration include Barton (1927), Rust (1938) and Ward (1980). However, the historical development of the methods to where they stand in terms of implementation today actually spans a number of fields of study. Modern resistivity methods are widely assumed to originate from work in the early twentieth century by Conrad Schlumberger that focused on petroleum and mineral exploration (Schlumberger, 1920), although earlier studies (e.g. in Britain and USA), as discussed below, explored similar concepts.

As reported by Van Nostrand and Cook (1966), in 1720, British scientists Stephen Gray and Granville Wheeler first reported electrical properties of Earth (and other) materials. It was not until the early nineteenth century when Robert Fox (another British scientist) highlighted the possible value of geoelectrical measurements for mineral prospecting. Fox made many significant contributions to the study of geophysical mechanisms including the observation of natural electrical current flow between two electrodes associated with the presence of a mineral vein. This was the first documented record of what is today known as a self-potential. In 1830, Fox documented electrically conductive properties of a number of Earth materials (Van Nostrand and Cook, 1966). Carl Barus' invention of the non-polarizing electrode in 1880 (Rust, 1938) resulted in a significant improvement in the reliability of the technique.

Figure 1.1 illustrates key advances in the galvanic method throughout the first half of the twentieth century. The twin-electrode self-potential measurement became established for ore prospecting during the late nineteenth century. Towards the end of the century a rapid growth of new techniques evolved, many of which underpin modern-day methods. At the start of the twentieth century, Fred Brown and Augustus McClatchey (both from the USA) independently patented methods for detecting mineral ores using a twin (current) electrode device (Brown, 1900, 1901; McClatchey, 1901a, 1901b). Given our interest in the near-surface Earth, it is worthwhile noting that resistance methods were actually extensively explored by the United States Department of Agriculture (USDA) in the late nineteenth century for agricultural applications. Gardner (1897) illustrates how the resistance between two electrodes in the soil (with suitable calibration) can be used to infer soil moisture (although the influence of other factors, such as salinity and soil texture, ultimately limited the acceptance of this approach, which was effectively superseded by dielectric sensors in the 1980s).

These twin-electrode resistance measurements were limited in use because of the dominant effect of the resistance close to the electrodes (the contact resistance). In the late nineteenth century, British entrepreneurs Leo Daft and Alfred Williams developed a geoelectrical prospecting method based on driving current between two electrodes and mapping potential differences using two electrodes connected to a telephone earphone (telephone earphones were used in early instruments to allow the operator to assess the balance of a potentiometer audibly). In 1901, Daft and Williams set up The Electrical Ore-Finding Company Ltd. in England (Vernon, 2008) and in 1906 patented their methods as 'Improved Apparatus for Detecting and Localizing Underground Mineral Deposits' (Daft

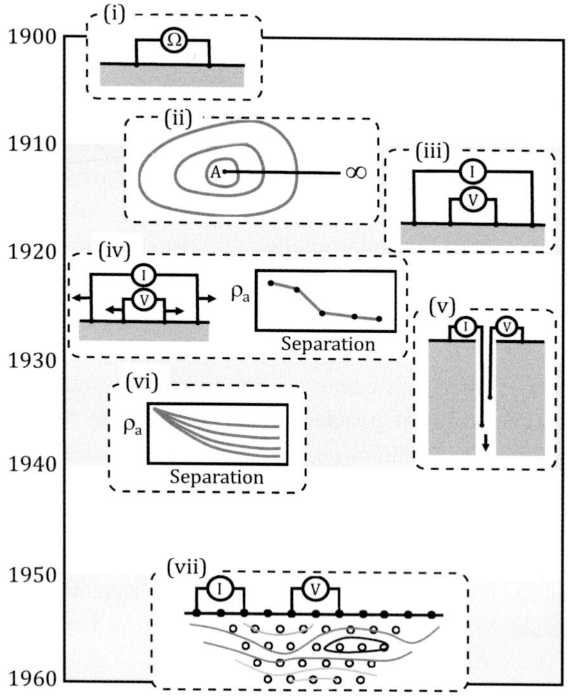

Figure 1.1 Evolution of the resistivity method in the first half of the twentieth century. (i) Twin-electrode measurement; (ii) potential mapping; (iii) four-electrode apparent resistivity measurement; (iv) vertical electrical sounding; (v) borehole logging; (vi) development of type curves for vertical soundings; (vii) pseudosections.

and Williams, 1906). It appears that the Williams and Daft method, despite demonstrations in Britain, Canada and Australia, had mixed successes. In 1905 their company went into receivership. Their method was, however, introduced to Sweden in the early twentieth century, where it was successfully deployed (Rust, 1938).

In 1912, Conrad Schlumberger field-tested equipment to map equipotential lines about a current source near his family estate in Caen, Normandy. He too was able to demonstrate that ore bodies influence the potential field, thus offering a means of prospecting. His method was patented in 1915 (Schlumberger, 1915). A few years later, Lundberg in Sweden applied a potential mapping system consisting of two 1 km long parallel current lines, 1 km apart, which, using regularly spaced potential dipoles, allowed identification of anomalies (from a parallel equipotential field that would be obtained under homogenous conditions) (Van Nostrand and Cook, 1966). The field implementation of such a method would, no doubt, have been challenging. Lundberg's method was, however, very successful, as evidenced by the discovery of several orefields (Barton, 1927). It is interesting to note that Lundberg adopted the same measurement approach for the first documented archaeological electrical geophysics survey, which was conducted in Williamsberg, Virginia, USA, in 1938 (Gaffney and Gater,

2003). Electrical prospecting in archaeology did not really take off, however, until 1946, through the work of Richard Atkinson in England (Atkinson, 1953).

Schlumberger recognized the impact of self-potentials in galvanic measurements and, like Frank Wenner in the USA around the same time (Wenner, 1912, 1915), the value of resistivity in discriminating Earth materials. The work on four-electrode measurements made by Wenner is often overlooked, but in fact provides a physical basis for the modern-day resistivity technique. Wenner is now widely known for the four-electrode configuration that bears his name, but in fact he also introduced the concept of an 'apparent resistivity' (the effective homogenous resistivity of the ground) and also detailed the concepts of reciprocity applied to four-electrode measurements (Wenner, 1912). Schlumberger adopted a four-electrode configuration similar to that proposed by Wenner, but with a shorter potential electrode spacing – a configuration still referred to as the Schlumberger array and widely used in vertical sounding because of its relative insensitivity to lateral variation in resistivity and efficiency of implementation.

In order to overcome problems with polarization of electrodes (experienced when using true DC sources), Schlumberger adopted a periodically reversed DC source. In 1920 Schlumberger applied the resistivity method in the iron-bearing region of May-Saint-Andié (Van Nostrand and Cook, 1966). In 1929 he detailed the resistivity method for locating oil-bearing formations (Schlumberger, 1926) and in 1932 a modification to the four-electrode configuration for well logging was patented (Schlumberger, 1933).

The electrical prospecting method became established and throughout the first half of the twentieth-century electrical methods remained relatively similar to the original approach. Van Nostrand and Cook (1966) document examples of early successes of the method for mineral exploration in Europe and North America. Measurements were typically made of apparent resistivity, potential fields from a current source using a 'potential-drop-ratio' method along transects radiating from the source, or by carrying out vertical soundings using progressively larger electrode spacing. Instrumentation advanced by building on Wenner's and Schlumberger's alternating current source design, most notably the double commutator method of Gish and Rooney (1925) (not dissimilar from that adopted in most field instruments today, although original designs used a hand-driven commutator).

A range of four-electrode configurations began to emerge. The 'Lee portioning' configuration (after F. W. Lee of the US Geological Survey) includes a third potential electrode in the centre of a Wenner configuration, allowing greater resolution. The dipole–dipole configuration is attributed by Seigel et al. (2007) to the work of Madden in 1954, but in fact West (1940) utilized the same configuration (naming it the 'Eltran' (electrical transients) array in line with electromagnetic methods using a separate transmitter and receiver). Ward (1980) points out that the Lee and Wenner arrays 'have not survived the test of time'. However, the Wenner array became a popular choice in the late twentieth century for near-surface applications partly due to its high signal-to-noise ratio, making it a good choice for relatively low cost, low power instrumentation for shallow investigations.

Early attempts at interpreting electrical resistivity data include Stefanesco et al. (1930), which underpinned much of the subsequent work on vertical soundings. Typically, type curves were provided to allow interpretation of electrical sounding data, e.g. the three-layer models of Wenner configuration measurements produced by Wetzel and McMurry (1937). Slichter (1933) proposed an alternative approach in which a vertical resistivity profile is computed directly from measured data. In the 1950s, interpretation methods began to evolve but were clearly constrained by computational demands. For two decades or more, many elegant, and often highly complex, analytical solutions to specific resistivity problems were developed (e.g. Keller and Frischknecht, 1966; Koefoed, 1970; Ghosh, 1971). Although these solutions have been largely superseded by more flexible numerical techniques, they still offer valuable approaches to modelling the resistivity structure of the Earth for specific situations. Chapter 4 discusses the application of these analytical approaches to a range of present-day near-surface problems.

In the 1970s, the digital computer permitted development and application of numerical modelling techniques applied to interpretation of resistivity data, opening up a wide range of extensions of the original approach (see Figure 1.2). Lytle and Dines (1978) (from Lawrence Livermore National Laboratory, USA) outlined their 'impedance camera' concept based on multiple configurations of four-electrode measurements that forms the foundation of modern-day electrical imaging systems. Field-scale, multiplexed instruments with automated switching capabilities first emerged in the 1980s, facilitating practical

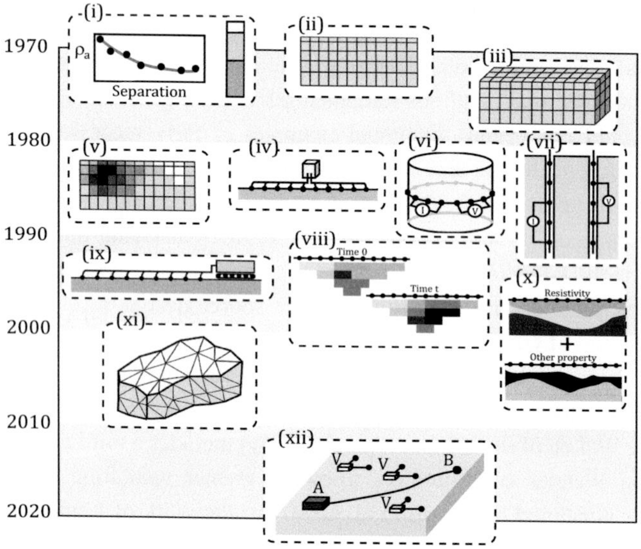

Figure 1.2 Further developments of the resistivity method. (i) Automated inversion of vertical sounding data; (ii) 2D forward modelling on structured grids; (iii) 3D forward modelling on structured grids; (iv) multi-electrode measurement systems; (v) automated inversion of 2D resistivity data; (vi) resistivity imaging of bounded regions; (vii) cross-borehole resistivity imaging; (viii) time-lapse resistivity imaging; (ix) land-based towed systems; (x) joint and coupled inversions; (xi) 3D inverse modelling on unstructured grids; (xii) distributed wireless resistivity measurements (A, B are current electrodes; V denotes distributed voltage gradient receivers).

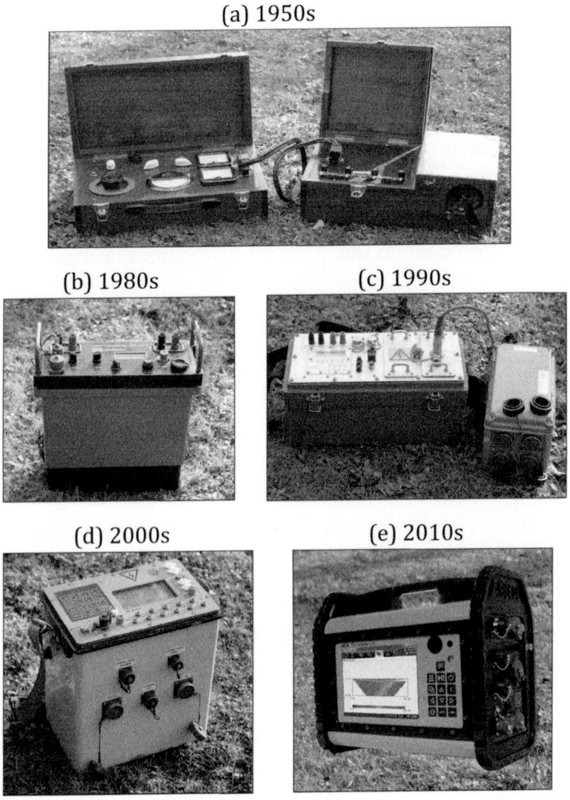

Figure 1.3 Examples of resistivity instrumentation, showing evolution over the past few decades. (a) Early analogue instrument with crank handle for current generation (box on the right); (b) ABEM SAS300B single-channel four-electrode instrument; (c) Campus Geopulse with separate multiplexer (to right of instrument) allowing 64 electrode connectivity (single channel) for resistivity and IP; (d) Iris Instruments Syscal Pro with integrated multiplexer (96 electrodes, 10 channels); (e) ABEM LS with multiplexer and 12 channels, with graphical display of pseudosection.

application of multi-electrode measurements (see e.g., Griffiths et al., 1990). Figure 1.3 illustrates some of the changes in resistivity instrumentation over the past few decades. Advances in measurement systems were coupled with the rapid growth and availability of digital computer technology in the 1970s and 1980s, which permitted the development of practical forward modelling tools. Madden (1972) proposed the transmission line method and Coggon (1971) the finite element method for solution of the 2D DC resistivity forward problem. Hohmann (1975) delivered the first 3D numerical solution of the forward problem. Dey and Morrison (1979) also developed a 3D numerical solution, which was utilized in numerous subsequent studies.

Inversion of field data (i.e. computation of a resistivity distribution that is consistent with a set of measured apparent resistivities) remained constrained by computational resources until the late 1980s and early 1990s. For example, Pelton et al. (1978) offer a method for inversion of resistivity and IP data that utilized a pre-calculated set of models stored on a computer. Such an approach has links to modern-day machine learning techniques.

Robust inversion tools followed and by the 1990s the widespread availability of personal computers meant that imaging data could be analysed without the need for access to remote mainframe computers. The proliferation of reasonably powerful laptop computers in the 2000s allowed computation of resistivity models in the field. Nowadays, modest 3D imaging can be carried out immediately after data acquisition. The growth of cloud-based computation will no doubt enhance the scale of problems that can be analysed in a field environment.

In parallel to geophysics, electrical imaging approaches, very similar to the geophysical methods, emerged in biomedical imaging. Barber and Brown (1984) coined the term 'applied potential tomography' for their Sheffield (UK) resistivity imaging system (the term 'electrical impedance tomography' (EIT) is now more widely used for such applications (e.g. Webster, 1990)). Similar approaches evolved in the field of process (chemical) engineering where the properties of fluids mixing in vessels are imaged. 2D and 3D imaging of resistivity using measurements from an array of electrodes are commonly referred to as electrical resistivity (or resistance) tomography (ERT). Some authors prefer the term 'electrical resistivity imaging (ERI)' as tomography is now synonymous with X-ray CT (computed tomography), in which measurements are made around the perimeter of an object. However, as the word tomography originates from the Greek words *tomos* for 'slice' and *graphe* for 'drawing' then ERT may be, in fact, an appropriate term, even if the measurement sensors (electrodes) do not bound the region of interest. However, Lionheart (2004) argues that 2D resistivity methods should not be considered tomographic because the measurements are sensitive to features off-plane. We recognize that ERI and ERT are both frequently used to represent a wide range of 2D and 3D applications of resistivity, so we retain both terms in this book.

Somewhat in tandem with the methodological and modelling efforts, critical advances in the understanding of the petrophysical relationships linking electrical resistivity to the geometry of a porous medium occurred over the last century. Sundberg (1932) was perhaps the first to identify that the resistivity of a rock will be proportional to the resistivity of the pore-filling fluid, recognizing that the proportionality constant should be a function of the interconnected porosity. However, G. E. Archie, an engineer at Shell Oil Company, defined the electrical formation factor to describe the increase in resistivity of a rock relative to the resistivity of the pore fluid due to the presence of a non-conducting rock matrix. His meticulous empirical observations on cores from well logs were published in a classic paper (Archie, 1942) that provided relationships linking the formation factor to interconnected porosity and degree of cementation (controlling tortuosity). His second relationship described the relationship between resistivity and the degree of saturation of the pore space of the electrically conducting fluid phase.

Archie's relations assumed that the only conduction path through a rock was via the pore-filling electrolyte. This caused problems in understanding the electrical conductivity of rocks containing clay minerals, which were found to contain excess conductivity and deviate from Archie's law. The next major advance in the understanding of electrical properties of soils and rocks occurred with the recognition that the ionic charges forming in the electrical double layer at mineral–fluid interfaces can also conduct current. Waxman and Smits (1968) introduced a model for shaly sandstones involving conduction pathways in both the electrolyte and in the double layer associated with clay minerals that were assumed to add in parallel. This second conduction pathway later was recognized to occur in all rocks and soils, where it was generally described as interfacial or surface conduction (Rink and Schopper, 1974; Revil and Glover, 1998). This parallel conduction model remains the foundation for interpreting electrical measurements and has been implemented in popular models used by soil scientists to estimate moisture content from resistivity (Rhoades et al., 1976). Extensive petrophysical research has established robust relationships between the surface conductivity and the cation exchange capacity (Waxman and Smits, 1968), surface area (Rink and Schopper, 1974) and grain size (Lesmes and Morgan, 2001) of rocks and soils. Figure 1.4 illustrates the evolution of electrical petrophysical models for both resistivity and IP.

1.3.2 Induced Polarization

The historical development of the IP method foremost is rooted in mineral exploration. The measurement of IP had a large impact on mineral prospecting as resistivity measurements alone were often incapable of detecting disseminated ore deposits. The discovery that measuring a reversible charge storage effect with the same basic instrumentation as resistivity could illuminate such deposits represented a major advance in minerals exploration with geophysics. Schlumberger (1920) is credited with making the first field observations of the IP method in the early twentieth century. Schlumberger observed the slow (rather than instantaneous) decay of the voltage following abrupt termination of the current pulse in the presence of metallic sulphides (Seigel et al., 2007) and referred to the phenomenon as 'polarisation provoquée' (Schlumberger, 1920). Schlumberger patented the method in 1912 (Schlumberger, 1912), and in 1924, Hermann Hunkel proposed the use of an alternating current and phase angle measurements for mineral prospecting (Hunkel, 1924). Given our focus on the near-surface Earth, it is important to recognize that Schlumberger also measured a storage of charge in rocks arising in the absence of mineral deposits, in addition to the enhanced polarization in the presence of mineral deposits. The mechanisms causing the background IP signal measured by Schlumberger would not be further investigated until almost 40 years later. According to Ward (1980), Schlumberger decided to pursue the self-potential method in preference to IP.

Extensive research in Russia during the 1930s, followed by studies in countries in Western Europe, along with parallel independent work by the US Navy focused on beach mine detection, led to development of the IP method. Weiss (1933) claimed the method to be capable of detecting electrochemical effects in the subsurface. In 1940, Gennady Potapenko patented the method (Potapenko, 1940) as a means of detecting oil-bearing sands based on laboratory measurements of the frequency dependence of impedance (what we now refer to as spectral induced polarization, SIP). According to Bertin and Loeb (1976), field trials of

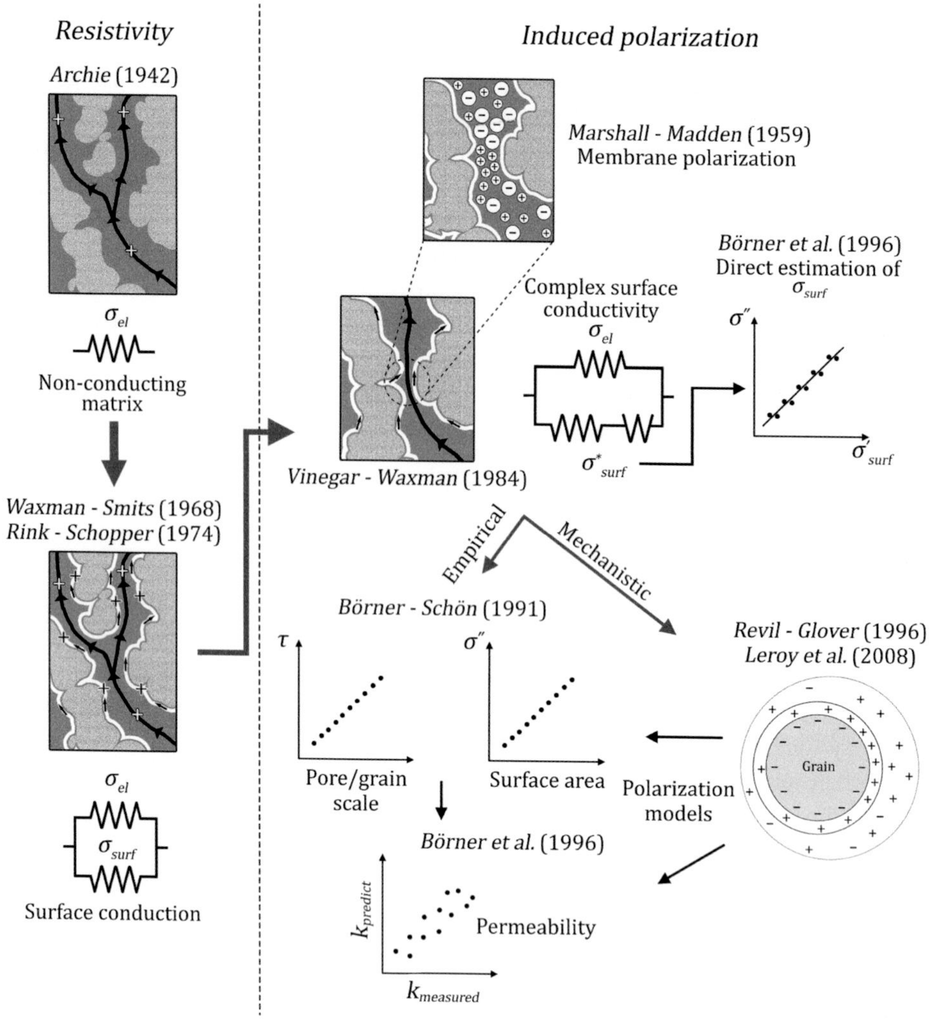

Figure 1.4 Evolution of the petrophysical modelling of resistivity and IP for rocks and soils devoid of electron conducting minerals. Some key publications highlighted.

Potapenko's method were not successful. Bleil (1953) is one of the first peer-reviewed journal publications on the possible implementation of IP for mineral discrimination that led to IP becoming an established geophysical technology for mineral exploration by the 1970s. In fact, Bleil (1953) is credited as first using the term 'induced polarization' to describe the method (Seigel et al., 2007).

Unlike most other geophysical methods, the IP signal results from an interfacial electrochemical phenomenon rather than a physical property of rocks and soils. The unique electrochemistry of the interface between an electron conductor and pore-filling electrolyte results in the largest IP signals and is the reason for the success of the method in mineral exploration. However, as noted above, Conrad Schlumberger identified the existence of a background polarization in the absence of electron conductors. These considerably smaller IP signals originate in the electrical double layer that develops to counteract the net surface charge of any mineral grain. In Russia, the pioneer in IP, Vladimir Komarov, showed that the IP signal in the presence of sulphide mineralization was always in excess of the background signal in the absence of mineralization (Komarov, 1980). By the 1960s, IP had become the primary electrical method applied to mineral exploration in Russia.

Interest in the background IP signals grew in the 1950s (Vacquier et al., 1957) as the method was identified as a candidate for mapping clays. Beginning in the 1970s, research began to focus on the link between IP signals and geometric properties of the interconnected pore space in soils and rocks, leading to the realization that IP might hold the answer to reliable geophysical estimation of the permeability of porous media. Vinegar and Waxman (1984) made an important petrophysics advancement by extending the basic parallel surface conduction model of Waxman and Smits (1968) to incorporate a complex surface conductivity to capture the IP phenomenon. As the imaginary part of the measured complex conductivity exclusively arises from the surface conductivity, a realization that IP measurements could resolve the inherent ambiguity in the interpretation of resistivity measurements (i.e. the separation of Archie-type electrolytic conduction from surface conduction) evolved (Börner 1992). In tandem, laboratory experiments established strong relationships between the imaginary conductivity and the surface area of soils and rocks (Börner and Schön, 1991). Other workers confirmed relationships between the relaxation times recorded with spectral (frequency dependent) IP measurements and the grain (Klein and Sill, 1982) or pore (Scott and Barker, 2003) size of soils and rocks. Given that surface area, grain size and pore size are key pore geometrical properties controlling permeability, the ability to predict permeability from IP measurements has been extensively explored (Börner et al., 1996; Slater and Lesmes, 2002b). The development of mechanistic models to explain the electrochemical controls on the surface conductivity (Revil and Glover, 1998) and its role in generating IP mechanisms (Leroy et al., 2008) has yielded fundamental insights into the information extractable from these measurements.

Despite the origins of the method in mineral exploration, these theoretical and experimental advances led to extensive investigations of the method within the field of hydrogeophysics (Binley et al., 2015). Example applications include: groundwater evaluation (Vacquier et al., 1957; Bodmer et al., 1968; Draskovits et al., 1990); delineation of saline and freshwater (Roy and Elliott, 1980); aquifer vulnerability (Draskovits and Fejes, 1994); groundwater contamination (Deceuster and Kaufmann, 2012); mine waste and landfill imaging (Yuval and Oldenburg, 1996; Gazoty et al., 2012; Power et al., 2018); and hydrocarbon exploration (Veeken et al., 2009). Today, the unique electrochemical interfacial information offered by IP is motivating many studies of the signals that arise from biogeochemical and microbiological processes that modify, or create, interfaces in soils and rocks (e.g. Zhang et al., 2014).

Like resistivity, the IP measurement is only ever a proxy of the property of interest. Similarly, the reliability of this proxy will depend on how well developed the conceptual model of the subsurface is along with the robustness of the relationship between IP measurement and the property of interest. In Chapter 2, we will see that IP is remarkably robust for estimating the concentration of disseminated electron conducting minerals in the subsurface. This is because the presence of electron conducting minerals usually enhances the IP signal by an order of magnitude or more over the 'background' signal arising in rocks devoid of conducting minerals. Furthermore, the relationship between the chargeability, a primary measure of the IP signal, and the volumetric concentration of electron conducting minerals is weakly controlled by the properties of the fluids filling the pores (relative saturation, ionic composition, temperature). However, many recent novel applications of the IP method have been proposed, based on exploiting the high sensitivity of the measurement to variations in the electrochemistry of the mineral–fluid interface. For example, the IP method was extensively investigated as a technology for locating DNAPL in the subsurface during the explosion of environmental geophysics in the 1980s (Olhoeft, 1985). This work led to high expectations about the field application of IP at complex, contaminated field sites. Unsurprisingly, these expectations were not met given the ambiguity of the IP signals originating in such a complex setting. More recently, some laboratory studies have shown that small IP signals are correlated with the volume concentrations of bacterial cells grown in synthetic soils (e.g. Ntarlagiannis et al., 2005a). Although intriguing and lending more insights into the nature of polarization mechanisms in soils, such observations do not currently form the basis for employing the methodology for quantifying microbes in natural soils.

1.4 Recent Methodological Developments

Electrical geophysical measurements have evolved from the early days of Conrad Schlumberger, where changes in resistivity of the subsurface were inferred from measuring voltages recorded along profiles, to modern data-acquisition technologies that exploit multi-dimensional imaging concepts similar to those used for medical tomography to determine high-resolution resistivity and IP models of the subsurface. Electrical geophysicists have harnessed increased computational power available from reduced core-memory costs and parallel computing platforms. This has facilitated the development of 3D inversion techniques, permitting computation of over a million parameter cell values using data from thousands of electrodes. Inversion methods continue to advance, e.g. allowing the computation of 4D images (e.g. for monitoring the effectiveness of groundwater remediation technology (e.g. Johnson et al., 2015)). Advances in data analysis include the integration of resistivity measurements into process-based (e.g. hydrological) models.

In contrast, resistivity and IP instrumentation has advanced at a somewhat slower pace, with limited technological developments occurring in the last 20 years. The technology is inherently limited by the need to establish a direct electrical contact with the Earth, so the reliance on electrodes, wiring and insulation prevents dramatic technological alterations to the basic technique (some developments with capacitively coupled electrodes have been made but they are limited to a narrow range of configurations). The development of multi-channel receivers in the 1970s resulted from the need for quicker data-acquisition rates in mineral exploration. The 1980s saw the introduction of the first multiplexing units, although the approach mostly remains foremost based on dated mechanical switches for field-based systems (high speed electronic switching has been used in biomedical imaging and process tomography since the 1980s, but rarely exploited in near-surface geophysics applications – but see Binley et al., 1996a). Some new technologies are emerging for field instruments, including faster and more scalable instruments, further expanding the potential application of this relatively simple but immensely successful geophysical technique. One promising development for fully 3D surveys is the distributed system, using time-synchronized recording of full transmitter and receiver waveforms (Truffert et al., 2019). This removes the reliance on resistivity cables connecting electrodes along a line or on a grid. That said, many applications do not require such sophisticated treatment and the widespread measurement of resistivity from simple configurations (e.g. vertical sounding, 2D imaging) is likely to remain for some time. On the other hand, many of the biogeochemical and geochemical processes detected in highly controlled laboratory studies using IP measurements are unlikely to be resolvable in the field without substantial new developments in instrumentation. The most promising near-surface applications of field-scale IP outside of mineral exploration likely pertain to (1) improving the interpretation of lithology from electrical resistivity measurements by resolving surface conductivity, and (2) characterization and monitoring of environmental processes that involve transformations of electron conducting minerals.

1.5 Outline of the Book

This book exclusively deals with the galvanic DC resistivity and IP geophysical methods. Although the material presented in Chapter 2 is also relevant to the interpretation of geophysical techniques that rely on the principles of EM induction, these methods are not included in this book. The interested reader can find excellent descriptions of the EM techniques published elsewhere (e.g. Kaufman et al., 2014). Similarly, this book does not consider the self-potential electrical method, which is sometimes confused with the resistivity and IP techniques. The self-potential method relies on measuring the small voltages that arise from natural electric currents flowing in the Earth due to the coupling of moving fluids, heat, ionic constituents and electric charge. Self-potential is a simple yet fascinating geophysical sensing technique that relies on measurements of voltage differences between a pair of electrodes from which the flow of fluids, heat and ions is inferred. The interested reader is referred to the relatively recent text by Revil and Jardani (2013) that provides a comprehensive evaluation of the self-potential geophysical method.

This book is broken into chapters that address the major elements of the galvanic electrical methods. Recognizing that some readers may care about the popular resistivity method more than the specialized IP technique, each chapter is broken into two major parts. The first part of each chapter introduces content from the perspective of DC resistivity alone. The second part of each chapter extends that content to consider IP. This approach represents the impression of the authors that IP is ultimately an extension of the resistivity method, where some additional information on the behaviour of charge in response to an applied electric field is obtained. As will be shown later, IP is often referred to as ‘complex resistivity’, the reference being to a complex variable that is characterized by two components rather than just one (in the case of a ‘real’ number). In essence, IP gives two (potentially more from frequency domain data) pieces of information whereas resistivity just gives one piece of information.

Following this introductory chapter, the book is divided into chapters that consider the electrical properties of soils and rocks, instrumentation, field data acquisition, modelling electrical datasets, and case studies that capture both established and emerging applications of the methods. The focus is on the development of the technology in the last 25 years, which has been largely driven by near-surface environmental and engineering characterization needs. A number of impressive, detailed and theoretically rigorous treatments of the galvanic electrical techniques exist (Telford et al., 1990), but these texts date from the 1990s or earlier. Such texts contain a wealth of information on fundamental aspects of the techniques that remain relevant to this day. However, the advent of electrical imaging technologies, which has revolutionized this geophysical technique, is not captured in these texts. The existing books that focus on electrical techniques (e.g. Keller and Frischknecht, 1966) contain numerous elaborate analytical solutions of geometries foremost representing either a layer-cake Earth (1D models) or simple 2D and 3D geometric models that capture the basic characteristics of important geological scenarios. As demonstrated in Chapter 4, some of these analytical solutions remain valuable to this day. As a result of their age, the existing texts contain only very cursory treatments of numerical methods for modelling complex 2D and 3D structures, as well as the inverse techniques that have been developed to automatically image resistivity and IP structures from field measurements. In this book, the focus is foremost on the numerical approaches to resistivity modelling and the ‘art’ of resistivity inversion. In addition to modelling static distributions of electrical properties, modelling and inversion of monitoring (4D time-lapse) datasets is considered. This book pays special attention to the mechanics of the inversion methods at the heart of the electrical imaging method, particularly the effects of data quality on image results. Emerging methods for interpretation, such as Bayesian modelling and joint inversion, are also discussed in Chapter 5, along with uncertainty estimation and other aspects of image appraisal.

Another major difference of this book relative to preceding texts relates to the context of the application of the methods. The existing texts draw foremost on efforts to apply the methods to image geological structures, reflecting development of the methods for mineral exploration and large-scale water resources characterization. Yet, since the 1990s, a parallel development of electrical imaging occurred in the geoscience, medical and process tomography fields. Although this book foremost considers applications to imaging the near-surface Earth, content developed in the medical and process tomography fields is incorporated because of its clear relevance and overlap. The geoscience applications considered in this book are more diverse than reported in prior texts, highlighting the rapid development of the methods in the last 25 years for shallow environmental and engineering applications, and include the use of cross-borehole electrical imaging and time-lapse investigations.

An important aspect of this book is the approach to the subject of IP, which is handled differently to earlier texts on the method. Previous dedicated texts on IP (e.g. Sumner, 1976) have been written primarily from the perspective of mineral exploration. These texts foremost describe the state of understanding of the subject from the perspective of the mineral exploration community based on the developments that occurred from the 1960s through to the 1980s. However, a substantial shift in understanding of the IP phenomenon occurred in the 1990s, in part driven by efforts in the well logging community to further exploit the 'background polarization' defined by the likes of Schlumberger and Komarov for estimating physical properties of the pore space controlling fluid flow and transport. This motivated a re-evaluation of IP mechanisms in soils and rocks devoid of electron conducting minerals, with substantial theoretical model developments occurring in the last 20 years. Near-surface environmental applications of IP have advanced, including efforts to use the measurements for estimation of permeability (Börner et al., 1996). Others have explored the small but intriguing IP signals that result from mineral transformations (e.g. precipitation and dissolution) and some microbiological processes (Ntarlagiannis et al., 2005b). In the last decade, environmental applications of IP have led to re-evaluation of well-established models for IP in the presence of electron conducting minerals, with new understanding of the sensitivity of the measurement to iron mineral concentration and size distribution (Revil et al., 2015). This book describes IP foremost from the perspective of the interpretational framework that has developed in the last 25 years, although the relationship to earlier ground-breaking studies is emphasized wherever possible.

The text is supported by a resistivity modelling and inversion software package built around the *R2*, *cR2*, *R3t*, *cR3t* codes (see Appendix A) developed by the first author. A graphical user interface developed in Python and known as *ResIPy* (see Appendix A), along with example datasets hosted in Jupyter notebooks, is also available to simplify learning of key concepts covered in the text. *ResIPy* has been designed to be intuitive in its use and yet give the user access to a comprehensive range of modelling options, including a means of assessing data quality. The code has been produced as open source to allow the interested user a means of developing and customizing the routines according to their own needs. Example datasets are provided to give the user an opportunity to acquaint themselves with the inversion of resistivity and IP datasets and investigate further several examples included in the text.

2
Electrical Properties of the Near-Surface Earth

2.1 Introduction

The electrical properties of porous media, specifically of interest here being soils and rocks, are controlled by the electrical properties of the constituents (solids, liquids and gasses), the volume concentrations of these constituents and how they are arranged. We will see later that the interfaces between these constituents also play a strong role in determining induced polarization (IP) signals. Decades of experimental observations, backed up by theoretical developments, have advanced our ability to interpret resistivity and IP measurements in terms of the physicochemical properties of soils and rocks. These developments have been largely driven by the need to quantify the primary physical properties of soils and rocks controlling the movement of fluids, along with constituents that are transported by the fluids. An additional important motivation for developing understanding of the electrical properties of rocks was the need to locate economically valuable minerals. More recently, our understanding of the electrical properties of soils and sediments subject to biogeochemical processes has been improved in response to the need to address contaminants in the subsurface and resulting contaminant transformations (Atekwana and Atekwana, 2009).

Some of the earliest work in electrical properties was performed by the United States Department of Agriculture (USDA). Scientists aimed at developing quantifiable relationships between electrical resistivity and both moisture content and salinity of agricultural soils (Gardner, 1897, 1898; Whitney et al., 1897). However, major breakthroughs in the understanding of how electrical resistivity depends upon rock properties came largely in response to the development of the resistivity logging method in oil exploration. Conrad Schlumberger introduced the IP technique to the former Soviet Union as a method for borehole logging of hydrocarbon reservoirs (Seigel et al., 2007). The investigation of oil reservoirs and the location of economically extractable reserves were critically dependent on the development of tools that could provide information on the porosity and permeability of an oil reservoir, along with the relative concentration of oil in the pore-filling fluids of the reservoir. Gustave Archie (1950) introduced the term *petrophysics* to describe the study of the physics of rocks, particularly with respect to the fluids they contain. At the centre of his petrophysical system of analysis was the pore-size distribution of a rock, which he recognized controlled both the fluid transport properties and the geophysical properties.

From the perspective of electrical properties, petrophysics refers to the laborious measurements of electrical resistivity as a function of the rock pore geometry and pore-filling fluid properties needed to interpret the borehole resistivity logs. The most impactful work in this area was that of Archie (1907–1978), who, through his extensive experimental research at Shell Oil Company, laid the foundations for the interpretation of electrical resistivity measurements on rocks (Thomas, 1992). Interpretations of electrical geophysical datasets routinely continue to make use of the foundational Archie's empirical law (Archie, 1942) and are supported by theoretical proofs of the empirical equation (e.g., Sen et al., 1981). Careful experimental research in petrophysics continued throughout the 1950s to 1960s, leading to important developments in the understanding of how conduction at the mineral–pore fluid interface not considered by Archie exerts a strong control on the electrical properties of fine-grained soils and rocks, particularly when pores are filled with relatively low electrical conductivity fluids. Waxman and Smits (1968) presented a widely adopted model, which couples Archie's law with an additional conductivity term that is controlled by the physical and chemical properties of the mineral–pore fluid interface.

Although the petroleum exploration industry has foremost determined the evolution in our thinking about electrical properties, parallel developments in advancing the use of electrical geophysics in soil science have also been important, particularly since the 1970s. Rhoades et al. (1976) introduced an empirical model that has been extensively used (and modified) to describe the electrical conductivity of soils. The model is parameterized in terms of the properties of most interest to a soil scientist rather than the properties of primary interest to a petroleum engineer.

In the same way that resistivity measurements were advanced by service industries supporting the growth of the oil industry, the understanding of how the electrical properties of rocks control IP measurements was foremost developed from the need to locate ore deposits. Schlumberger (1920) is credited with the first observations of the low-frequency polarizability of Earth materials resulting from the injection of an electric current, with his initial observations being estimated as early as 1913 (Seigel et al., 2007). The Second World War accelerated development of the IP method as potential military applications (e.g. location of mines buried in sea water saturated beach sediments) were identified. Seigel (1949) and Bleil (1953) represent the earliest publications in English of the use of IP for subsurface exploration. Disseminated mineral deposits were detectable at electron conducting mineral concentrations of 3% by weight (Bleil, 1953). Such work stimulated much interest in the IP method as a tool for mineral exploration and led to a new branch of petrophysics focusing on how the presence of ore minerals in rocks controls electrical properties. Decades of careful experimental research revealed how IP measurements are related to the concentration of ores in a rock as well as the size of the ore minerals. Some of the most significant work was led by William Pelton of Phoenix Geophysics and Stan Ward of the University of Utah. The complexity of the IP effect in ores limited the parallel development of theoretical models to describe this phenomenon. One exception is the groundbreaking work of Joe Wong, a physicist at the University of Toronto, who developed from first principles an electrochemical model to describe the IP response of electron conducting particles (Wong, 1979). Relatively little further theoretical development in the polarization of electron conductors (also commonly known as electrode polarization) occurred until recently, when the challenge was revisited partly for environmental applications of the IP method (Revil et al., 2015a).

Schlumberger (1920) also observed a much smaller background polarization of soils and rocks that occurs in the absence of electron conducting particles. There was relatively little interest in this effect until the 1990s when careful petrophysical research led by Frank Börner and Jürgen Schön (both at Bergakademie Freiberg, Institute of Geophysics and Geoinformatics, Germany) identified the strong relationship between this polarization and measurable properties of the total surface area of the insulating mineral–fluid interface in a rock (Börner and Schön, 1991). These relationships strengthened the link between electrical properties and permeability, being a key rock property needed to understand flow and transport in the subsurface. Prior attempts to link permeability to resistivity alone were largely unsuccessful as resistivity decreases with both increasing porosity and increasing surface area, whereas permeability increases with increasing porosity but decreases with increasing surface area (Purvance and Andricevic, 2000). The additional measurement of IP appeared to provide a means to track surface area independently, paving the way for efforts to directly estimate permeability from combined resistivity and IP measurements (e.g. Börner et al., 1996).

In this chapter, we explore the electrical properties of soils and rocks. We first examine the fundamentals of charge transport and define the intrinsic property of electrical resistivity. We then consider conduction processes in porous media, starting first with the controls on the electrical conductivity of a fluid. In order to understand the conductivity of a soil or rock, we examine the role of the interconnected pore space, the interconnected surface area and the presence of electron conducting minerals on electrical conduction. We pay attention to the electrical double layer (EDL), as it causes surface conduction and is also the source of the IP effect. We outline the commonly accepted conduction model that is used to explain the electrical conductivity of porous media in terms of parallel conduction pathways in the interconnected pores and along mineral–fluid interfaces via ions in the EDL. In the second half of the chapter, we turn our attention to IP and show how the parallel conduction model can be expanded to incorporate a complex surface conductivity term that accounts for both surface conduction and the surface polarization behind the IP effect. We also explore the frequency dependence of IP measurements and the relationship between the shape of the spectra and the grain/pore size of soils and rocks. We discuss empirical, phenomenological and mechanistic models for the IP response. We also investigate how IP measurements may hold the key to a reliable estimation of the permeability of soils and rocks.

2.2 DC Resistivity

The resistivity method is based on the direct current (DC) theory of electrical conduction. In an analogy to an electrical circuit, the porous material is assumed to behave purely as a resistor. This is often a reasonable assumption because the method is commonly implemented with low-frequency alternating electrical fields where displacement currents are

Box 2.1

Electric field and electric potential energy

An electrically charged body exerts a force ($\boldsymbol{F}$) on another charge (q), which is away from the body when the charge has the same polarity as the body and towards the body when the charge has the opposite polarity. The strength of this force is described by the electric field ($\boldsymbol{E}$),

$$\boldsymbol{E} = \frac{\boldsymbol{F}}{q}.$$

The direction of the field at a point is defined as the direction that a unit positive charge would move at that point.

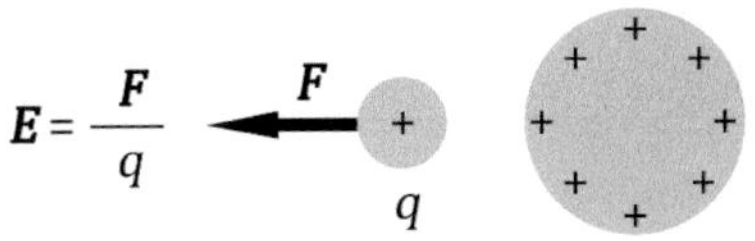

The electric field is determined by the force exerted on a unit positive charge at a point relative to a charged body.

The electric field created by the charged body exists everywhere in the space around it. The work done (W) by the electric field in moving a charge between two points ($a \rightarrow b$) is defined by the difference in the potential energy (U) of the field between the two points,

$$W_{a \rightarrow b} = U_a - U_b.$$

The potential energy of the field at a point is normalized by the unit charge (q) and reported as the electric potential (V), with units of volts (equal to joules/coulomb),

$$V = \frac{U}{q}.$$

Equipotential lines (in 2D) or surfaces (in 3D) connect points in space with the same electric potential. They are used to visualize the electric field associated with the charged body.

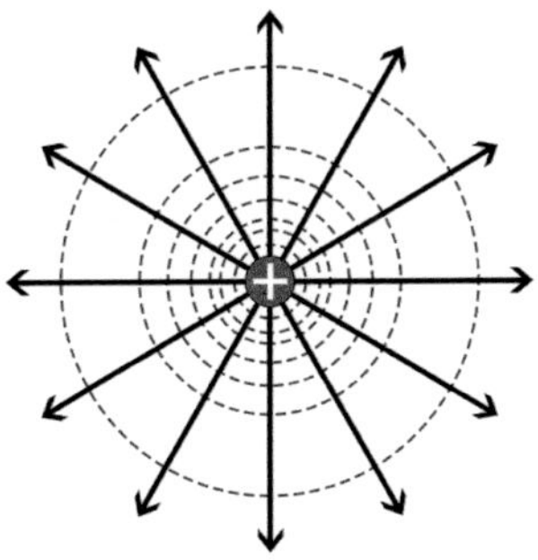

Equipotential lines describe the 2D spatial distribution of the electric field where the direction of the field is indicated by the arrows.

neglected. However, the temporary reversible storage of charge in a porous medium will be considered under the extension to the IP method in the second part of this chapter.

2.2.1 Electrical Conduction

The DC resistivity method is fundamentally about measuring the resistance of a porous medium to the transport of electrical charge. This charge transport in response to an electric field ($\boldsymbol{E}$) is termed *electromigration*. In order to determine resistivity, the forces acting on a charge and the resulting amount of charge moving in response to the force must be measured. An electrically charged body exerts a force on another charge (Box 2.1). The magnitude of this force is determined by the electric field, which equals the force at a point divided by the unit charge (q). The strength of the field is measured by the difference in electric potential (physical SI unit of volts).

Conduction (or electromigration) is the term used to describe the transport of electric charges, which defines an electric current (I with physical SI unit ampere [A], being the flow of charge through a unit surface at a rate equal to 1 coulomb per second [s]). The coulomb is the SI unit of charge (1 C = 1 As), being the charge (q) transported by a constant current of one ampere in one second. This charge is carried by moving electrons in a metal, but charge is also carried by ions in an electrolyte, as in saline water. The moving charged particles that form an electric current are referred to as the charge carriers (Box 2.2).

2.2.2 Resistivity/Conductivity Definitions

Ohm's law is the constitutive equation describing how the electric current (I in units of amperes) flowing between two points in a conductor is directly proportional to the electric potential (or voltage) difference (ΔV in units of volts) between the two points,

$$I = \frac{\Delta V}{R}, \tag{2.1}$$

where R is the constant of proportionality known as the resistance (units of ohms, Ω). Ohm's law implies that R is independent of the current.

Electric current density is defined from the current I flowing through a small surface A oriented orthogonal to the direction of motion of charges,

$$J = \lim_{A \to 0} \frac{I(A)}{A}. \tag{2.2}$$

The current density vector $\boldsymbol{J}$ has a magnitude J and is oriented in the direction of the movement of charges. It is generally assumed to be proportional to the electric field strength, which is equal to the gradient of the electric potential ($\boldsymbol{E} = -\nabla V$),

$$\boldsymbol{J} = \sigma \boldsymbol{E} = \frac{1}{\rho} \boldsymbol{E}. \tag{2.3}$$

Box 2.2

Electric current and current density

An electric current (I) exists due to the transport of electric charge (q) that results from an applied electric field ($\boldsymbol{E}$). This charge transport is referred to as *electromigration*. The current is equal to the net charge (ΔQ) flowing per unit time (Δt),

$$I = \frac{\Delta Q}{\Delta t} = \frac{\hat{n} q v A \Delta t}{\Delta t} = \hat{n} q v A,$$

where $\hat{n}$ is the charge carrier density (the number of charged particles per volume), q is the charge of the individual carrier, v is the drift velocity of the charge carriers and A is the cross-sectional area of the conductive medium.

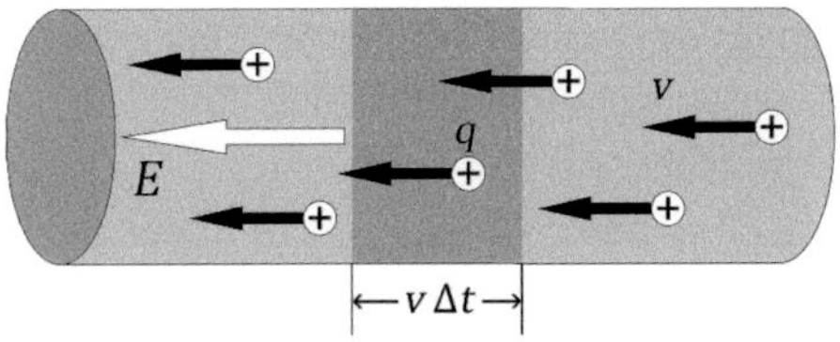

Transport of charge in response to an electric field.

The current density (J) with the physical unit A/m^2 is defined as the current per unit cross-sectional area,

$$J = \frac{I}{A} = \hat{n} q v.$$

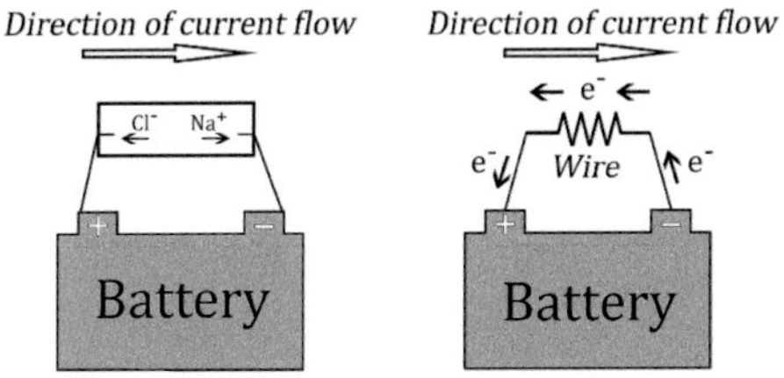

Electron conduction in a wire versus electrolytic conduction in an ionic fluid.

The charge carriers can be electrons as in the case for current flowing down a wire. Electrons can be charge carriers in the subsurface, e.g. in the presence of high concentrations of ore minerals or buried objects (e.g. steel drums). However, the charge carriers can also be ions, which is the common case for charge transport in fluids in the near-surface Earth.

The constant of proportionality σ is the electrical conductivity of the material (units of siemens/metre = S/m), and the electrical resistivity (ρ in units of ohm m = Ωm) is the reciprocal of this conductivity,

$$\rho = \frac{1}{\sigma}. \tag{2.4}$$

The electrical conductivity and electrical resistivity are intrinsic properties of the material through which current is flowing. They quantify the ability of a material to conduct electric charge. The conductivity of a porous medium depends on the temperature-dependent intrinsic electrical properties of the three phases (solid, liquid and gas) making up the medium, as well as the volumetric concentration and spatial arrangement of the different phases. It follows that the intrinsic electrical properties of porous media can be described in terms of either the conductivity or the resistivity. Traditionally, electrical resistivity has been the preferred property shown in equations describing electrical conduction in geophysics-focused publications. We prefer to use electrical conductivity as (1) it is directly related to the equivalent properties used to describe transport of fluids and heat, and (2) it simplifies the integration of properties controlling IP measurements with the properties controlling resistivity measurements.

2.2.3 Conduction Processes in Earth Materials

The electrical conductivity of the minerals making up the soil and rock of the subsurface varies over a huge range (almost 18 orders of magnitude) and more than any other physical property (Van Nostrand and Cook, 1966). The most commonly occurring minerals (i.e. the silicates, carbonates and sulphates) are effectively insulators with a resistivity ranging from $\sim 10^7$ Ωm for kaolinite to $\sim 10^{14}$ Ωm for quartz (Schön, 2011). The oxides and sulphates exhibit a wide range of resistivity, with some being insulators (e.g. sphalerite) and others being semiconductors, including the iron sulphide mineral pyrite with a resistivity of 10^{-3} Ωm and the iron oxide mineral magnetite with a resistivity of 10^{-4} Ωm. These electron conductive minerals will be important in the discussion of IP mechanisms appearing later. Schön (2011, Table 8.1) provides a comprehensive summary of the resistivity of the most common minerals.

Ignoring the presence of electron conducting minerals, charge is primarily transported by the movement of ions in the near-surface Earth. Two mechanisms contribute to this ionic transport of charge: (1) conduction via the ions dissolved within the pore fluid filling the interconnected pore space and (2) conduction via the ions in the EDL that forms at the mineral–fluid interface. The first mechanism is more intuitive, and it was Barus (1882) who was possibly the first person to conclude '*that the conductivity of the rock is largely, if not wholly, due to the presence of moisture in its pores, and is therefore electrolytic*' (Van Nostrand and Cook, 1966). The second mechanism, commonly referred to as 'surface' or 'interfacial' conduction, was not fully understood until the middle of the twentieth century (Hill and Milburn, 1956). The relative importance of ionic conduction in the pore fluid versus ionic conduction in the EDL is foremost determined by a trade-off between the salinity (aka ionic concentration) of the pore fluid versus the amount of mineral–fluid surface area. Ionic conduction in the pore fluid is promoted by high-salinity fluids in low surface area soils/rocks. Ionic conduction in the EDL is promoted by low-salinity pore fluids and high interfacial surface areas.

The conductivity of any material depends upon both the number of charges and the mobility of the charge carriers. For a single charge carrier *i*,

$$\sigma_i = \hat{n}_i \hat{Z}_i e \beta_i, \tag{2.5}$$

where $\hat{n}_i$ is the charge carrier density, $\hat{Z}_i$ is the valence of the charge carrier, e is the elementary charge (1.6022×10^{-19} C) and β_i is the mobility of the charge carrier (in $m^2\ s^{-1}\ V^{-1}$). The direct proportionality between conductivity and these three properties of the charge carriers holds for all conduction processes encountered in the Earth.

2.2.3.1 Ionic Conduction in a Fluid

The presence of water in the shallow subsurface means that ionic conduction is the most important charge transport process determining conductivity (this changes in the deeper subsurface as pressure closes connected pore spaces and temperatures increase). Ionic conduction involves the transport of ions when an electric field is present. For example, in an aqueous solution of NaCl, the Na^+ and Cl^- ions are responsible for the charge transport.

The mobility of ionic charges plays a critical role in determining the conductivity of a fluid. The mobility of a single ion is proportional to $\hat{Z}_i$, the diffusion coefficient of the charged species (D_i, units of $m^2\ s^{-1}$) in that fluid, and inversely proportional to the temperature (T, in kelvin (K)),

$$\beta_i = \frac{\hat{Z}_i e D_i}{k_b T}, \tag{2.6}$$

where k_b is Boltzmann's constant ($1.3806 \times 10^{-23}\ m^2\ kg\ s^{-1}\ K^{-1}$). The ionic mobility can also be related to the viscosity (η in Pa s) of the fluid,

$$\beta_i = \frac{\hat{Z}_i e}{6\pi\eta r_i}, \tag{2.7}$$

where r_i is the radius of the hydrated ion. Equation 2.7 dictates that anions are less mobile than the smaller cations, which will be important in some IP mechanisms discussed later.

The electrical conductivity of an ion in solution is described by the Nernst–Einstein relationship,

$$\sigma_i = \frac{D_i \hat{Z}_i^2 e^2 N_A c_{(c)i}}{k_b T}, \tag{2.8}$$

where N_A is the Avogadro constant ($6.022 \times 10^{23}\ mol^{-1}$) and $c_{(c)i}$ is the concentration of that charge carrier in mol L^{-1}. At low temperatures (e.g. below 200°C) conductivity increases with temperature (an apparent contradiction on a cursory look at Equation 2.8) because the dependence of D_i on temperature is more significant than the direct inverse dependence of conductivity on temperature shown in Equation 2.8 (Glover, 2015).

An alternative expression for the conductivity of an ionic solution comes from inserting Equation 2.7 into Equation 2.5 (Glover, 2015),

$$\sigma_i = \frac{\hat{n}_i \hat{Z}_i^2 e^2}{6\pi\eta r_i}. \tag{2.9}$$

This highlights how the conductivity of a solution is proportional to the charge concentration, the square of the charge, and indirectly proportional to both the fluid viscosity and hydrated radius of the ion (Glover, 2015). This equation links the electrical conductivity of the pore fluid in a soil/rock sensed with the resistivity method to the aqueous chemistry of this fluid.

Equation 2.9 could be used to determine the total conductivity of a natural groundwater (σ_w) by accounting for the relative contributions from the individual ionic constituents. However, groundwater will typically have a complex and often poorly defined ionic composition such that the properties appearing in Equation 2.9 are not known. Consequently, empirical formulae are normally used to estimate how conductivity of a certain solution (commonly NaCl) varies with two easily measurable properties: salinity and temperature (e.g. Hilchie, 1984; Sen and Goode, 1992). For example, Bigelow (1992) gives the resistivity of a NaCl brine as

$$\rho_{NaCl} = \left(0.0123 + \frac{3467.5}{\left(C_{NaCl}\right)^{0.96}}\right)\frac{81.77}{1.8T + 38.77}, \tag{2.10}$$

where C_{NaCl} is the NaCl concentration (in ppm) and temperature is in °C.

Such empirical equations are based on a single aqueous solution (NaCl in Equation 2.10) and must be modified for a groundwater composed of a complex ionic mixture. This can be done by defining an equivalent NaCl solution that gives a conductivity equal to σ_w. Dunlap coefficients are multipliers for each ion other than Na^+ and Cl^- in a solution that can be applied to account for the differences in the conductivity dependence on salinity for these ions relative to that for Na^+ and Cl^- (Dunlap and Hawthorne, 1951). However, the approach is an approximation as it does not consider ion interactions or how the coefficients might vary with temperature (Glover, 2015).

Within the expected ranges of salinity and temperature encountered for groundwaters, the ionic concentration typically exerts the overwhelming control on σ_w, although the effect of temperature is important in geothermal applications, some environmental remediation monitoring (e.g. active heating) and also in monitoring groundwater–surface water exchange. At the low salinities encountered for natural groundwaters, increases in salinity (ionic concentration) increase σ_w. At high salinities encountered in brines, the rate of increase of σ_w with increasing salinity is reduced due to decreasing ionic mobility of the charge carriers. The dependence of σ_w on salinity varies substantially with major electrolyte composition and should be considered in the interpretation of electrical measurements.

The dependence of σ_w on temperature over the temperature range encountered in natural groundwaters is foremost linked to the viscosity. Viscosity decreases with temperature, leading to an increase in σ_w with temperature (Equation 2.9) that must be considered in hydrogeophysical applications of electrical methods. Over a limited range of low temperatures (T < 100°C), $\sigma_w[T]$ can be corrected to a reference temperature (T_{ref}) using the approximation

$$\sigma_{w[T_{ref}]} = \frac{\sigma_{w[T]}}{1 + \alpha_T \left(T - T_{ref}\right)}, \tag{2.11}$$

where α_T is an empirical coefficient with a typical value between 0.02 and 0.025 K^{-1}(Keller and Frischknecht, 1966; Bairlein et al., 2016). Such empirical approximations are used to correct electrical data for temperature variations in order to better relate σ_w to salinity variations. At high temperatures (T > 400°C) the viscosity increases with temperature and causes σ_w to decrease with increasing temperature. This effect must be considered in geothermal applications of electrical methods.

2.2.3.2 The Electric Double Layer (EDL)

The charged mineral surface of the interconnected pore space attracts and absorbs charged ions from the pore fluid, forming the EDL. The EDL can be described with specific reference to quartz, being the most abundant mineral in near-surface unconsolidated sediments (Figure 2.1). Reactions at the quartz surface result in positively charged $> SiOH^{2+}$ sites when $pH < pH_{pzc}$ (> represents the surface), i.e. the pH is less than the pH for the point of zero charge (pzc), and negatively charged $> SiO^-$ sites when $pH > pH_{pzc}$. These fixed (in space), charged sites form adjacent to the mineral surface. Both charges exist in this layer, with $[SiOH^{2+}] > [SiO^-]$ at $pH<pH_{pzc}$ and $[SiO^-] > [SiOH^{2+}]$ at $pH > pH_{pzc}$. The condition $[SiOH^{2+}] = [SiO^-]$ is met at pH_{pzc}, which is approximately pH = 3 for quartz minerals.

These silica sites react with charged ions in the solution. The exact reactions depend on the ionic composition of the solution. For the case of a 1:1 electrolyte (each ion singly charged) such as NaCl, the following surface reactions occur in the pH range of natural groundwaters (pH 6–8) (Davis et al., 1978; Revil and Glover, 1997; Glover, 2015),

$$> SiOH^0 \overset{K_-}{\rightleftarrows} > SiO^- + H^+, \tag{2.12}$$

$$> SiOH^0 + Me^+ \overset{K_{Me}}{\rightleftarrows} > SiOMe^0 + H^+, \tag{2.13}$$

where Me^+ denotes a metal cation, H^+ a proton, and K_- and K_{Me} are constants describing the equilibrium positions of the two reactions. The $>SiOH^{2+}$ site concentrations are very low for pH > 6 (Glover, 2015). The surface therefore consists of two neutral sites ($SiOH^0$ and $SiOMe^0$) and one negatively charged site (SiO^-). The Stern layer is composed of those metal cations engaging in these surface adsorption reactions.

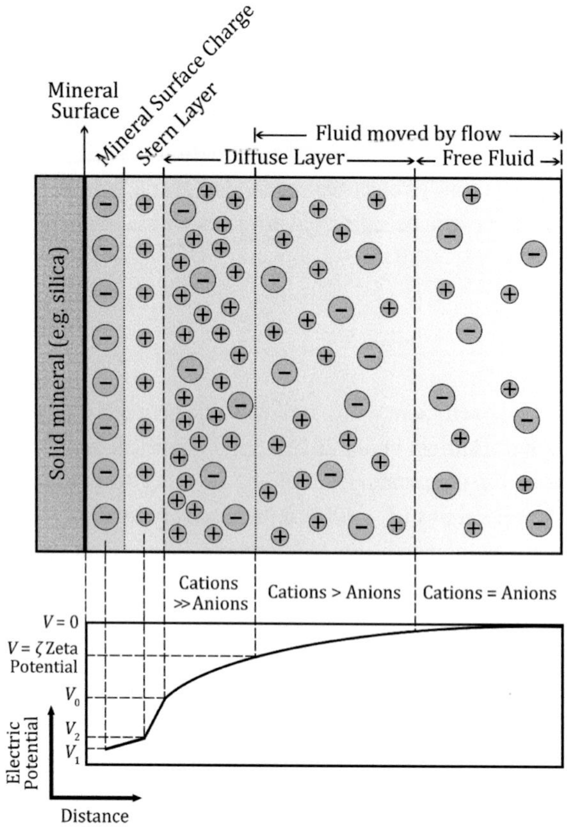

Figure 2.1 The EDL forming at a net negatively charged mineral surface (the most common case when groundwater pH is greater than the point of zero charge) showing the distribution of potential F with distance from the surface (modified from Glover, 2015).

The total surface site density represents the sum of the surface site densities. In the common case for groundwater where $pH > pH_{pzc}$, the mineral surface has a net negative charge, attracting cations from the bulk electrolyte, resulting in the formation of the diffuse part of the double layer that is depleted in anions. This depletion is greatest close to the plane of the Stern layer and it decays exponentially with distance from this plane and out to the bulk electrolyte (Glover, 2015). The formation of an EDL happens for all rock minerals, although more elaborate triple-layer models are required to fully describe the electrical properties of the surface of clay minerals (e.g. Leroy and Revil, 2004).

The Stern layer thickness is on the order of a hydrated metal ion (about 10^{-10} m (Glover, 2015)). The thickness of the diffuse part of the EDL is typically assumed to be equal to twice the Debye screening length (χ_d),

$$\chi_d = \sqrt{\frac{\varepsilon k_b T}{2N_A e^2 I_0}}, \tag{2.14}$$

where ionic strength (I_0) in mol m^{-3} is given by

$$I_0 = 0.5 \sum_i^n \hat{Z}_i^2 C_{(c)i}\,, \tag{2.15}$$

where $C_{(c)i}$ is the concentration (mol m^{-3}) of each ionic species (i) in solution and ε is the dielectric permittivity. The diffuse layer is thicker at low salinities than at high salinities. This results from the lower number of cations per unit volume in a low-salinity fluid relative to a high-salinity fluid. Consequently, a greater volume (i.e. thickness) of low-salinity fluid is needed to supply the cations to compensate the negative charge of the mineral surface (Glover, 2015). The EDL thickness is postulated to play an important role in generating a pore throat polarization mechanism known as membrane polarization (Bücker and Hördt, 2013a), which is described later (see Section 2.3.5.2). In low-salinity fluids with narrow pore throats, the width of the two diffuse layers on either side of the throat is assumed to become equal to or greater than the pore throat thickness. The diffuse layer contains a net charge (usually positive under conditions associated with natural groundwater), acting as a cation-selective zone.

2.2.3.3 Electron Conduction

Electron conduction in the Earth becomes significant in the presence of high concentrations of ore minerals. Although all materials contain electrons, these electrons are only mobile in good electrical 'conductors' such as metals where they occupy partially filled energy bands (the energy bands in insulators are fully occupied). An external electric field drives movement in these partially filled bands and/or jumps to higher energy bands. The electron conductivity depends on the distance between energy bands and the temperature, as these both control the number of electrons that are mobile and can contribute to conduction. The conductivity of metals increases strongly with temperature as the number of electrons available to conduction increases. The electron mobility depends on time between collisions between electrons. The number of collisions increases as temperature increases, with a resulting decrease in the electron mobility. Electron conduction also occurs in semiconductors (along with charge transport by 'holes'). The presence of conducting and semiconducting minerals in the Earth will be important in the discussion of IP in Section 2.3 of this chapter.

2.2.4 Conduction in a Porous Medium

The solids of a porous medium are most commonly insulators, e.g. silicate and carbonate minerals have resistivities exceeding 10^9 Ωm (Schön, 2011). However, the solids can be semiconductors when significant concentrations of sulphide or oxide minerals are present.

These semiconducting minerals are normally only present in low concentrations, such that the solids generally impede the flow of electric current. In the case that all solids are insulators and we ignore the presence of the EDL, the electric current is restricted to ionic conduction through the electrolyte filling the pore network. The component of the porous medium conductivity associated with this network is referred to here as the electrolytic conductivity (σ_{el}) and is partly controlled by σ_w. It is therefore dependent on the properties (ionic concentration, temperature) that control σ_w as discussed earlier.

A second ionic conduction mechanism emerges in a porous medium as a result of the previously described EDL at the mineral–fluid interface. Ions in this EDL transport charge along this interface when an electric field is applied. The total conductivity of a porous medium is commonly assumed to equal the sum of the conductivity associated with the electrolytic mechanism (σ_{el}) and that associated with the mineral–fluid interface, referred to here as the surface conductivity (σ_{surf}),

$$\sigma = \sigma_{el} + \sigma_{surf}. \tag{2.16}$$

Equation 2.16 assumes that the conduction paths through the pore-filling electrolyte and along the mineral surface add in parallel (Waxman and Smits, 1968; Rink et al., 1974), a concept first proposed to account for the presence of electron conductive minerals (as opposed to ionically conductive interfaces) in rocks (Wyllie and Southwick, 1954; Marshall and Madden, 1959). This parallel addition of two conducting phases is conceptually reasonable and often assumed, although Revil (2013) emphasizes that it may not always represent the correct geometric arrangement of conducting phases.

The resistivity of Earth materials varies over a wide range, depending on the relative controls of pore fluid conductivity, saturation, pore geometry and mineralogy on electrolytic and surface conduction (Table 2.1). The addition of water to geologic materials is responsible for much of the variation in the electrical properties of rocks. This effect is illustrated in Table 2.1 by the direction of the arrow.

2.2.4.1 Electrolytic Conduction and Archie's Laws

As the solids of a porous medium are generally excellent insulators, it follows that the conductivity of a saturated porous medium must be reduced relative to the conductivity of the pore-filling electrolyte σ_w if we assume that surface conduction is insignificant ($\sigma_{surf} \approx 0$). This conductivity reduction is quantified by the electrical formation factor (Archie, 1942),

$$F = \frac{\sigma_w}{\sigma_{el[s]}} = \frac{\rho_{el[s]}}{\rho_w}, \tag{2.17}$$

where the incorporation of [s] in the subscript is to remind the reader throughout this book that this refers to a fully saturated rock porous medium.

Table 2.1 Typical range of resistivity for major rock and sediment types. Arrow indicates direction of effect of presence of water (modified from Schön, 2011)

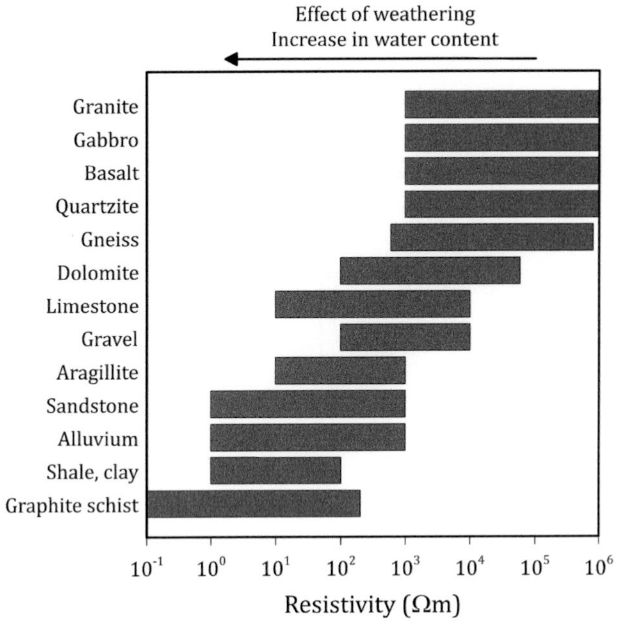

Box 2.3

Archie's laws

With the need to develop the world's oil reserves, geophysical exploration methods advanced rapidly from the 1930s. New geophysical well logging techniques were being developed, with tools that could measure the electrical resistivity of a formation local to a borehole. The search for petrophysical relationships that could reliably transform these geophysical measurements into estimates of the rock properties controlling the economic value of oil reservoirs was then a major field of petroleum geology. This research was performed by correlating geophysical properties (e.g. electrical resistivity) with important rock properties (e.g. porosity and permeability) using careful laboratory measurements on rock cores obtained from extraction boreholes. By far the most important petrophysical relationship that resulted from this meticulous laboratory work is known as Archie's law. G.E. Archie (1907–1978) worked for Shell Oil Company where he demonstrated, for the first time, how well logging data could be used to identify 'pay zones', a petroleum geology term for those rocks where oil and gas occur in exploitable quantities. Archie presented his findings at the 1941 meeting of the American Institute of Mining, Metallurgical, and Petroleum Engineers and the work was subsequently published in the transactions of this meeting (Archie, 1942). As of the time of writing, this work had been cited 8,085 times (Google scholar), making it the most highly cited piece of applied/ exploration geophysics literature.

Box 2.3 (cont.)

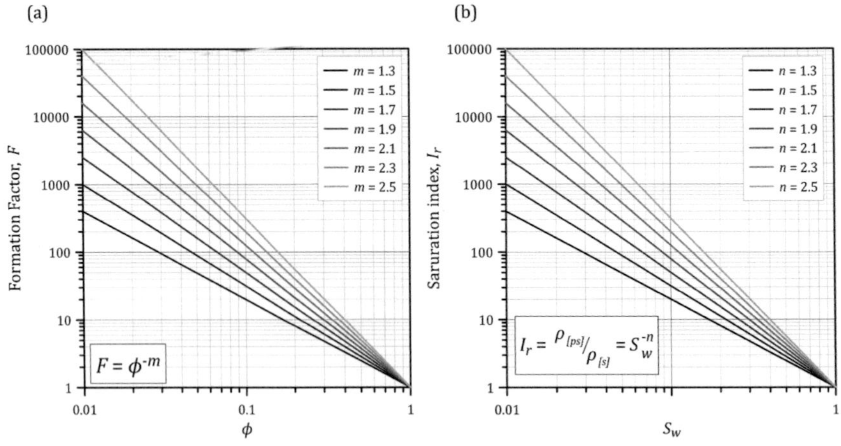

Graphical representation of Archie's laws. Left: First Archie Law relating formation factor (F) to porosity (ϕ). Right: Second Archie Law relating saturation index (I_r) to saturation (S_w).

Archie presented a set of carefully performed measurements on rock cores saturated with a brine (an electrolyte with a salinity ranging from 20 to 100 g of NaCl per litre). His first law relates the electrical formation factor (F) to the porosity through a power law quantified by a cementation exponent (m). His second law relates a resistivity saturation index (I_r) to the degree of saturation (S_w) through a power law quantified by a saturation exponent (n). Archie's laws apply to a rock saturated with a saline brine such that only electrolytic conduction via the connected, fluid-filled network occurs. These laws break down when surface conduction is significant.

The reduction in conductivity in part comes from the reduction in volume of the conducting phase, i.e. the replacement of conducting water with insulating solid. The formation factor should therefore be related to the interconnected porosity (ϕ) of soils and rocks,

$$\phi = \frac{V_{v\_i}}{V_T}, \tag{2.18}$$

where V_{v_i} is the volume of the interconnected voids and V_T is the total volume of the porous medium. Sundberg (1932) first proposed that the ratio shown in Equation 2.17 should be related to the interconnected porosity. However, it was Archie (1942) who first presented experimental datasets describing the dependence of F on ϕ. He found compelling evidence for a simple power law,

$$F = \phi^{-m}, \tag{2.19}$$

where m was defined as the cementation exponent, in recognition of his primary interest in cemented sandstone reservoirs (Box 2.3). Equations 2.17 and 2.19 combined describes Archie's first law. Although empirically derived on sandstones, the theoretical basis of this critical relationship in electrical geophysics has since been well established (e.g., Accerboni, 1970; Sen et al., 1981). The cementation exponent accounts for the fact that the reduction in conductivity due to the presence of the insulating solids also depends on the connectivity and tortuosity of the electrolytic conduction paths through the pore space. Factors such as particle shape, sorting, orientation and packing influence connectivity and therefore determine the value of m (Jackson et al., 1978; Mendelson and Cohen, 1982). A value of $m = 1.5$ is the case for a soil composed of perfect spheres (Sen, Scala and Cohen, 1981), and m increases with ellipticity and preferential grain orientation. Sen (1984) expressed m in terms of particle shape,

$$m = \frac{5 - 3\widetilde{N}_i}{3\left(1 - \widetilde{N}_i^2\right)}, \tag{2.20}$$

where $\widetilde{N}$ is a depolarization factor. For non-spherical particles aligned parallel to the electric field,

$$\widetilde{N} = \frac{1}{1 + 1.6(a/b) + 0.4(a/b)^2}, \tag{2.21}$$

where a/b is the aspect ratio of the particles. A value of $m = 1$ would be the case where the void space is composed of a bundle of tubes crossing the sample in a straight line. Higher values of m are associated with increasingly poorer connectivity of the pore network, making the conductivity of the porous medium ($\sigma_{el[s]}$) more sensitive to changes in porosity.

Glover (2009) presents a reformulation of Archie's law where F^{-1} defines the 'connectedness' of the medium and m defines the rate of change of this parameter with both porosity and the arrangement of the pore space (i.e. its connectivity). This formulation provides a physical basis for linking m to the connectivity. Box 2.3 shows how F varies with the interconnected porosity and the cementation exponent. Figure 2.2 shows Archie's law fit to an experimental database made up of sandstone and mixed sandstone/dolomite rock cores. Archie's law describes the relationship between F and ϕ well with a best estimate of m equal to 1.86. The scatter in the relationship indicates that m varies substantially between individual samples.

An empirical extension of Archie's law involves the addition of an extra fitting parameter a that was originally introduced to improve the fitting of Archie's law to experimental datasets,

$$F = a\phi^{-m}. \tag{2.22}$$

This modification to Archie's law was originally proposed in reservoir petrophysics (Winsauer et al., 1952) and continues to appear in soil physics literature today (Shah

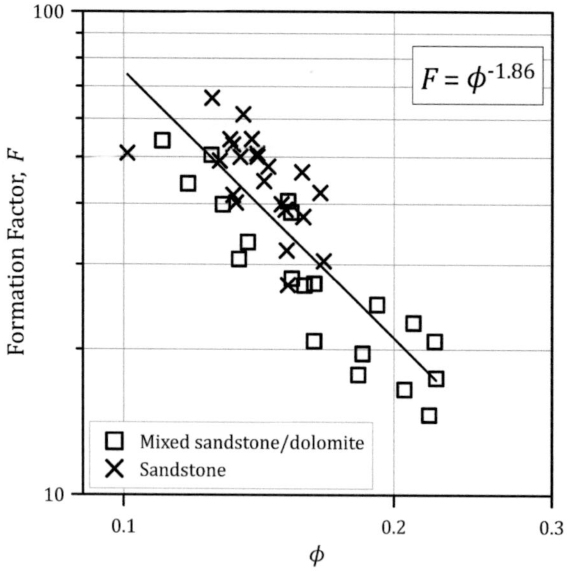

Figure 2.2 Archie's law fitted to an experimental dataset of rock cores with $m = 1.86$. The saturating fluid was an 80 S/m NaBr solution so the assumption that surface conduction is insignificant is reasonable. F and ϕ are unitless.

and Singh 2005). However, it is only under the condition $a = 1$ that $\sigma_{el} = \sigma_w$ at 100% porosity. This condition must hold as the bulk material is then only composed of the water phase. Consequently, Equation 2.22 is analytically incorrect and should be avoided (Revil, 2013; Glover, 2015), although Glover (2016) suggests that it may serve as a useful measure of data quality in petrophysical investigations applying Archie's law.

Electrolytic conduction can only occur through those fluid-filled pores that are interconnected. Porosity of a rock can be measured in many ways. Some methods (e.g. image analysis of thin sections) are likely to overestimate the porosity contributing to conduction. A strong analogy inevitably exists between the flow of electric current and the flow of fluids through the interconnected pore space. Therefore, porosities derived from methods that involve fluids invading the pore space (e.g. mercury intrusion) will better represent the porosity term in Archie's law.

Capillary bundle models used to describe the flow of fluids through porous media have been applied to describe the flow of electric current (Mualem and Friedman, 1991). These models highlight the importance of the tortuosity, which defines the deviation of the medium from a bundle of straight capillary tubes. The tortuosity is a dimensionless term ≥ 1. A modified Archie-type equation that includes the electrical tortuosity (T_e) from using a capillary bundle model takes the form (Wyllie and Rose, 1950),

$$\sigma_{el[s]} = \sigma_w \frac{\phi}{T_e}. \tag{2.23}$$

This leads to the following definition of the formation factor,

$$F = \frac{T_e}{\phi}. \tag{2.24}$$

The above relationships define F from a geometric point of view, i.e. ϕ and T_e are geometric terms. This electrical tortuosity is often assumed to equal the tortuosity of fluid flow as discussed later. Theoretical approaches based on solving for the distribution of normalized electric potentials throughout the porous medium under appropriate boundary conditions (Pride, 1994) yield an interesting alternative definition of the formation factor,

$$F = \phi_{eff}^{-1}, \tag{2.25}$$

where ϕ_{eff} is an effective porosity that is only a fraction of ϕ (Revil, 2013). The effective porosity represents the interconnected pore space that is probed by the current flow lines and does not include dead-end pores through which the current does not penetrate (Figure 2.3).

Based on Equations 2.17, 2.19 and 2.25, the conductivity of a saturated porous medium due to electrolytic conduction alone is

$$\sigma_{el[s]} = \frac{1}{F}\sigma_w = \phi^m \sigma_w = \phi_{eff}\sigma_w. \tag{2.26}$$

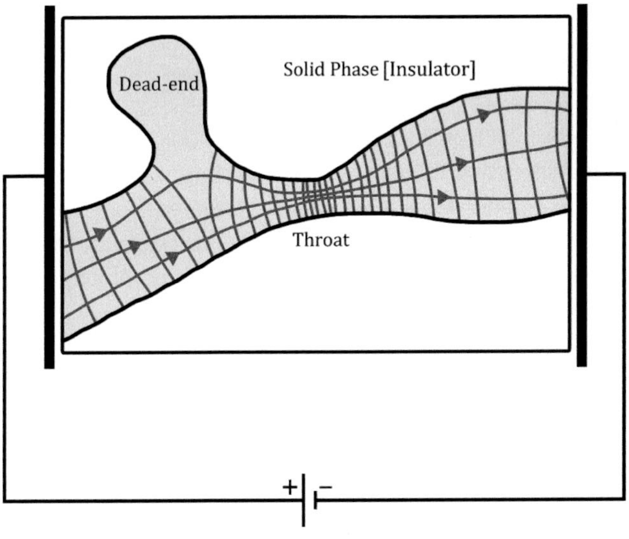

Figure 2.3 Concept of effective porosity sensed by electrical resistivity as part of the interconnected porosity. Modified from Revil (2013).

The conductivity of a porous medium will decrease when the pores are filled with an insulating fluid (e.g., a gas or a non-aqueous phase liquid) instead of a conducting liquid. This decrease in conductivity again results from both the decrease in volume of the conducting liquid and the additional tortuosity (reduction in connectivity) due to the insulating fluid filling the pore space. Archie (1942) showed that the electrolytic conductivity of a partially water-saturated rock ($\sigma_{el[ps]}$) is inversely proportional to the conductivity of the saturated rock for constant σ_w,

$$\frac{\sigma_{el[ps]}}{\sigma_{el[s]}} = \frac{1}{I_r}, \tag{2.27}$$

where I_r was defined as the resistivity saturation index describing the rate of increase in conductivity with increasing saturation (Archie's second law). Archie (1942) found that $1/I_r$ shows a power law dependence on fractional water saturation (S_w),

$$\frac{1}{I_r} = S_w^n, \tag{2.28}$$

where n is known as the saturation exponent. Similar to m, it is related to the tortuosity/reduction in connectivity of the pore space but specifically associated with the insulating fluid replacing conducting water. Figure 2.4 shows example datasets of resistivity as a function of saturation (water the conducting liquid, air the insulating fluid) for four samples of a Sherwood sandstone with fitted saturation exponents and showing uncertainty bounds.

Combining Equations 2.26–2.28, the conductivity of a partially saturated porous medium due to electrolytic conduction alone is

$$\sigma_{el[ps]} = \frac{1}{F}\sigma_w S_w^n = \phi^m \sigma_w S_w^n = \phi_{eff}\sigma_w S_w^n. \tag{2.29}$$

Values of n tend to be similar to m for sandstones and $n \approx 2$ is often assumed. Glover (2015) points out that, in the case where the water is displaced from a previously saturated rock in a uniform manner, the volume fraction of water decreases from ϕ to ϕS_w, leading to $\sigma_{el[ps]} = \sigma_w(\phi S_w)^m$. However, this assumes that the connectivity of the conducting phase ϕS_w is the same as the previous conducting phase ϕ, which is unlikely to be valid (Glover, 2015). Consequently, in general it must be assumed that $n \neq m$.

2.2.4.2 Surface Conduction and the Parallel Conduction Paths Model

We next consider the physical explanation of surface conduction that is generally assumed to add in parallel to electrolytic conduction (Waxman and Smits, 1968; Rink and Schopper, 1974) (Figure 2.5). We recall that Archie's law incorporates the conductivity of the conducting phase (σ_w) and a formation factor (F) that accounts for the volume and degree of connectivity (or tortuosity) of that conducting phase. The same is done for the surface conductivity. From Equation 2.16 and again considering the case of a saturated [s] porous material,

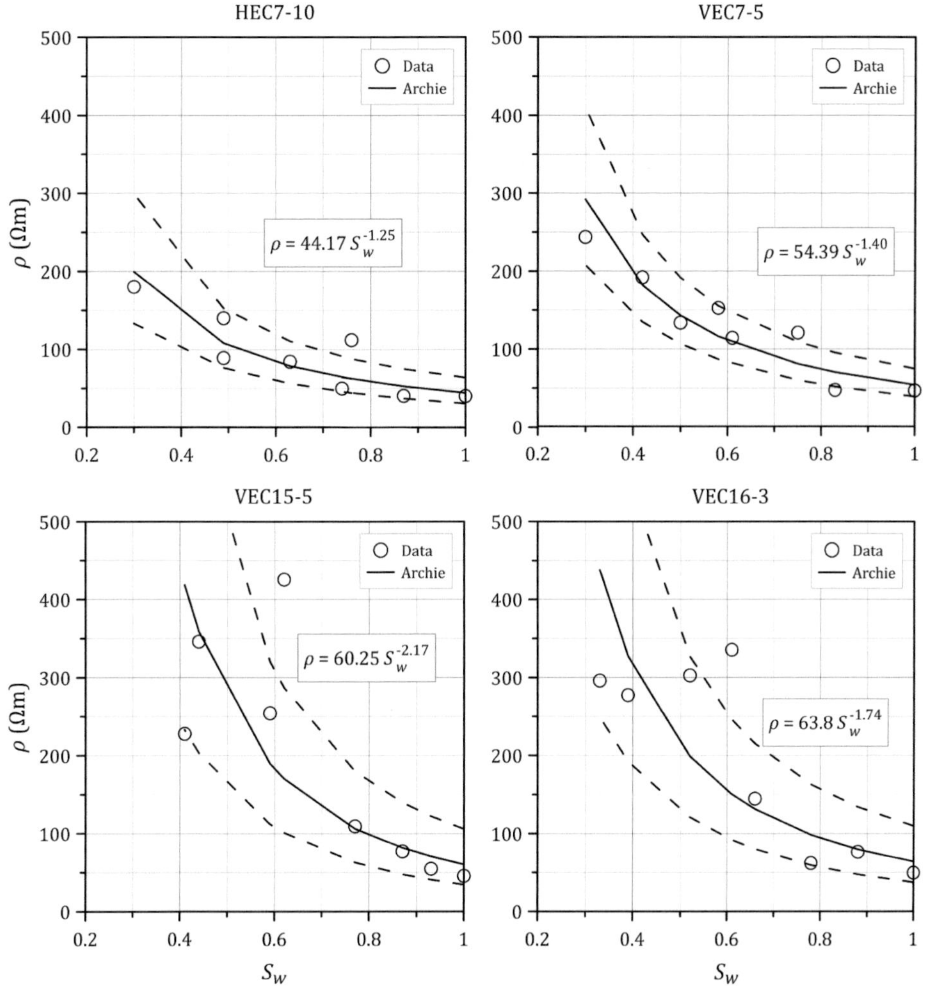

Figure 2.4 Experimental results of resistivity versus water saturation (S_w) for four sandstone cores showing uncertainty bounds. Data from Tso et al.(2019).

$$\sigma_{[s]} = \sigma_{el[s]} + \sigma_{surf[s]} = \frac{1}{F}\sigma_w + \frac{1}{F_s}\sigma_{EDL}, \tag{2.30}$$

where F_s is an equivalent formation factor for the conduction within the EDL of the interconnected pore space and σ_{EDL} represents the electrical conductivity of the EDL. It is often assumed that $F_s = F$ (note, this is no longer F as defined by Archie which assumes no surface conduction), which implies that the same interconnected pores support both electrolytic conduction and EDL conduction. This assumption is a generalization of an important parallel conduction model for shaly sands introduced by Waxman and Smits (1968), where electric current via counterions in the EDL of clays lining pores travels along the same tortuous path as the current traveling via the pore-filling electrolyte. However, the

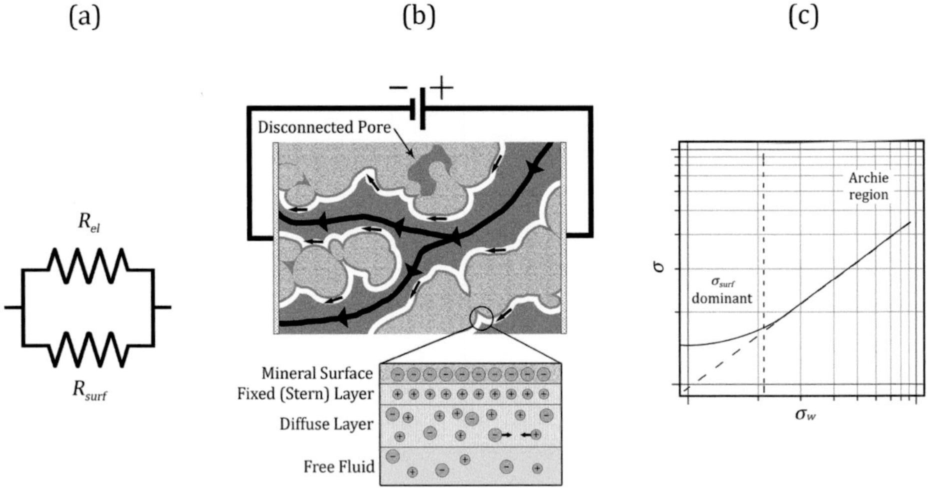

Figure 2.5 Parallel electrolytic (*el*) and surface (*surf*) conduction pathways in a porous medium: (a) equivalent circuit representation; (b) conceptual representation of parallel electrolytic and surface conduction pathways in a porous medium; (c) measured conductivity versus fluid conductivity showing regions where electrolytic conduction and surface conduction are dominant (note the double logarithmic scale).

condition $F_s \neq F$ might arise in the case where the EDL remains connected between closely spaced mineral grains that do not support electrolytic conduction.

Equation 2.30 emphasizes the inherent ambiguity of the measured electrical conductivity ($\sigma_{[s]}$) in that a single measurement is the sum of two additive terms, i.e. the electrolytic and surface conductivity. This creates a dilemma in the interpretation of field-scale electrical geophysical datasets, i.e. to what degree are measured variations in $\sigma_{[s]}$ due to changes in $\sigma_{[el]}$ as opposed to being due to changes in $\sigma_{surf[s]}$? We will shortly see that $\sigma_{surf[s]}$ is strongly controlled by the surface area of the interconnected pore space of a porous medium. Finer grained materials have more interconnected surface per unit pore volume and therefore more surface conductivity. This surface conductivity effect was historically first associated with clays in shaly sandstones (Waxman and Smits, 1968; Clavier et al., 1984), although it was subsequently recognized that all soils and rocks exhibit surface conductivity including clean sandstones (Rink and Schopper, 1974; Revil and Glover, 1998).

Traditionally, separation of the surface conductivity from the electrolytic conductivity is based on ignoring the dependence of σ_{EDL} on σ_w and fitting measurements of $\sigma_{[s]}$ as a linear function of σ_w to determine F and $\sigma_{surf[s]}$ (Equation 2.30). This requires careful, time-consuming laboratory measurements of $\sigma_{[s]}$ over a range of fluid salinities. The formation factor (F) is predicted from the inverse of the gradient of this linear relationship and the intercept gives a single salinity-independent estimate of $\sigma_{surf[s]}$. Under the assumption that $F_s = F$, the intercept and gradient can be used to estimate σ_{EDL}. A double logarithmic plot of $\sigma_{[s]}$ versus σ_w highlights how the relative importance of surface conductivity versus electrolytic conductivity depends on σ_w (Figure 2.5c). At high salinities, an 'Archie domain'

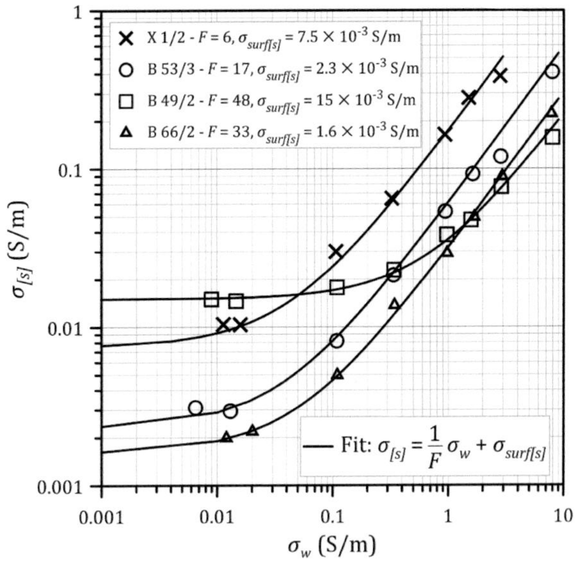

Figure 2.6 Electrical conductivity ($\sigma_{[s]}$) versus fluid conductivity (σ_w) for four saturated [s] sandstone samples. Formation factor (F) and salinity-independent surface conductivity ($\sigma_{surf[s]}$) are estimated from the linear relationship given by Equation 2.30. The double-logarithmic plot serves to highlight the relative importance of $\sigma_{surf[s]}$ at low salinities. Reproduced from Rink and Schopper (1974).

can be defined where surface conductivity is relatively insignificant and the plot of $\sigma_{[s]}$ versus σ_w approximately obeys Archie's law with a gradient $\approx 1/F$. At low salinities, the relative importance of surface conductivity increases and the double-logarithmic plot asymptotically reaches $\sigma_{surf[s]}$ as σ_w approaches zero. Figure 2.6 shows such plots for four sandstone samples measured by Rink and Schopper (1974). The variation in surface conductivity is immediately apparent between the shaly sandstone (sample B 49/2) versus the two clean sandstones (samples B 53/3 and B 66/2). Later, we will discuss how the addition of IP measurements can reduce the inherent ambiguity in the measurement of $\sigma_{[s]}$ and potentially provide an independent estimate of $\sigma_{surf[s]}$.

One common mistake in the interpretation of electrical resistivity measurements arises when surface conduction is ignored. A popular use of the method in environmental investigations is to determine variations in the fluid conductivity (e.g. mapping contaminant plumes, investigating saline intrusion). In the oil exploration industry, resistivity is used to estimate porosity from well logs. In both cases, the reliable estimation of these properties of interest from resistivity measurements alone will require that surface conduction is low and not dominating the measured resistivity signal. Whether surface conductivity can be ignored primarily depends on the trade-off between the magnitude of the surface conductivity and the magnitude of the electrolytic conductivity associated with the pore fluids (Figure 2.5). It is often assumed that surface conductivity in unconsolidated sediments can be ignored in coarser grained materials. In well logging, it is commonly assumed that

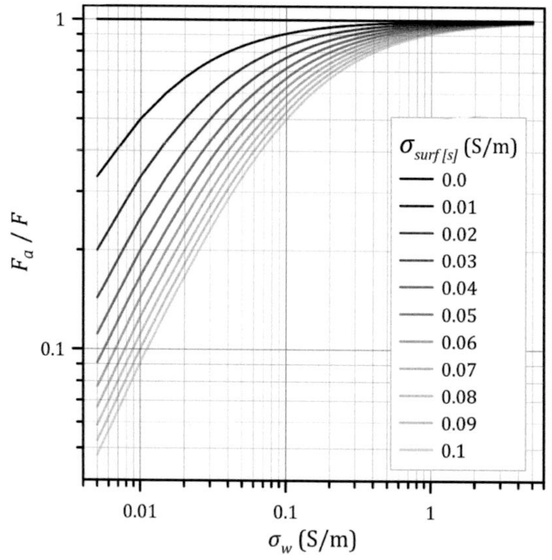

Figure 2.7 Ratio of measured formation factor to Archie formation factor as a function of fluid conductivity and surface conductivity from Equation 2.31.

surface conductivity can be neglected in clean (clay-free) sandstones saturated with brines. In fact, the effect of surface conduction is probably frequently underestimated in the interpretation of resistivity datasets. The significance of the problem can be assessed by examining how the measured formation factor (F_a) at a given salinity varies with salinity (Lesmes and Frye, 2001; Weller et al., 2013).

Figure 2.7 shows the ratio

$$\frac{F_a}{F} = \frac{\sigma_w \left(\sigma_{el[s]} + \sigma_{surf[s]}\right)^{-1}}{\sigma_{[w]} \sigma_{el[s]}^{-1}} = \frac{\sigma_{el[s]}}{\left(\sigma_{el[s]} + \sigma_{surf[s]}\right)} \tag{2.31}$$

plotted as a function of fluid conductivity for different values of the surface conductivity ranging from 0 to 0.1 S/m (F is the Archie formation factor in the absence of surface conduction). Neglecting the effect of surface conductivity results in an underestimate of the Archie formation factor. This underestimate is severe at low salinity and for high values of the surface conductivity. However, it is generally significant over a wide range of fluid conductivity and even for small values of surface conductivity. Similarly, ignoring surface conduction will result in erroneous values of the cementation exponent, in extreme cases causing it to become negative (Worthington, 1993), if Archie's law (Equation 2.19) is assumed to hold.

Further consideration of the electrical properties of the EDL highlights that the assumption of a salinity-independent $\sigma_{surf[s]}$ (Equation 2.30 and Figure 2.6) is only a first-order approximation. In fact, σ_{EDL} is related to the electrochemical properties of the charge carriers (dependent on the fluid chemistry and temperature) and the interfacial geometry.

This geometry can be quantified by the characteristic length-scale (Λ) (Johnson et al., 1986) being approximately equivalent to twice the pore volume divided by the pore surface area (being exactly equal to the radius of a cylindrical pore). Using this concept,

$$\sigma_{EDL} = \frac{2\Sigma}{\Lambda}, \tag{2.32}$$

where Σ is the surface conductance (in siemens). One simple way to describe Σ is using Equation 2.5, which states that Σ will depend on the charge density ($\hat{n}$), valence ($\hat{Z}$) and mobility (β) of the ions in the EDL (Glover, 2015). For a single surface ion i,

$$\Sigma_i = \hat{n}_{(s)i}\hat{Z}_{(s)i}e\beta_{(s)i}, \tag{2.33}$$

where subscript s denotes charges in the EDL at the mineral surface. More sophisticated models to describe the surface conductance consider the relative contributions from ions in the diffuse and Stern layer, the possible additional contributions from protons and the electrochemical reactions between the mineral surface and the fluid (e.g. Revil and Glover, 1998). These are utilized in mechanistic models for IP (e.g. Leroy et al., 2008) described later in Section 2.3.5.1. Equation 2.33 indicates that Σ should be a function of σ_w as ion exchange and sorption processes driven by disequilibrium between the ions in the pore fluid and the EDL will change Σ. It will also depend on temperature due to the influence of temperature on ion mobility as discussed in Section 2.2.3.1. Laboratory datasets on unconsolidated sediments and sandstones show that σ_{EDL} tends to increase with σ_w at low salinities but reaches a high-salinity asymptote above 1 S/m (Weller et al., 2013). However, increases in σ_w tend to foremost increase $\sigma_{el[s]}$ (a linear relation is predicted per Equation 2.17) with a secondary (often ignored) weaker increase in $\sigma_{surf[s]}$. We return to the salinity dependence of the surface conductivity later in this chapter when we discuss IP properties (see Section 2.3.3).

With the common assumption that $F_s = F$,

$$\sigma_{[s]} = \frac{1}{F}\left(\sigma_w + \frac{2\Sigma}{\Lambda}\right). \tag{2.34}$$

It is important to note that F in Equation 2.34 (and following equations in this chapter) will only equal F as defined by Archie (1942) when the surface conductivity (second term in brackets) is zero. Other geometrically based models for σ_{surf} invoke a dependence of σ_{EDL} on the pore volume normalized surface area (S_{por}) of the porous material. S_{por} is measurable using gas-adsorption methods and is an important property for permeability estimation (see Section 2.3.6). Rink and Schopper (1974) used a capillary bundle model that attributed σ_{EDL} to the diffuse layer, giving

$$\sigma_{EDL} = S_{por}\delta\sigma_{diff}, \tag{2.35}$$

where σ_{diff} is the intrinsic conductivity of the diffuse layer and δ is the thickness of the double layer. Recognizing that $\Lambda = 2/S_{por}$ for a capillary bundle (Weller and Slater, 2012),

$$\sigma_{EDL} = \frac{2\Sigma}{\Lambda} \cong S_{por}\Sigma. \tag{2.36}$$

Consequently, Equation 2.34 can be recast as

$$\sigma_{[s]} = \frac{1}{F}\left(\sigma_w + \frac{2\Sigma}{\Lambda}\right) \cong \frac{1}{F}\left(\sigma_w + S_{por}\Sigma\right). \tag{2.37}$$

An alternative expression for $\sigma_{[s]}$ comes from an extensively used model developed for oil exploration to explain the 'excess' conductivity (i.e. above what is predicted by Archie's law) of shaly sandstones (Waxman and Smits, 1968),

$$\sigma_{[s]} = \frac{1}{F}\left(\sigma_w + \sigma_{surf[s]}\right) = \frac{1}{F}\left(\sigma_w + \hat{B}Q_v\right), \tag{2.38}$$

where F is then a formation factor for a shaly sand, $\hat{B}$ is the equivalent conductance of sodium clay-exchange cations (S cm^2 meq^{-1}) and Q_v is a shaliness factor describing the excess charge involved in surface conduction. Similar to Equation 2.37, the electric current transported via counterions in the EDL of clay minerals is assumed to travel along the same tortuous path as the current transported via the pore-filling electrolyte. The term $\hat{B}Q_v/F$ represents the 'excess conductivity' over electrolytic conduction. $\hat{B}$ is a function of σ_w and the mobility of sodium ions (and therefore temperature), and thus represents the effect of the fluid chemistry on surface conductivity. This model emphasizes the importance of ion exchange at the clay mineral–electrolyte interface. Q_v (the cation-exchange capacity (CEC) per unit volume in meq cm^{-3}) can be estimated from CEC (in meq g^{-1} of dry clay), a measure of the ability of clay minerals to release cations when the density of the solid phase (ρ_s, in g cm^{-3}) and porosity (ϕ) are known,

$$Q_v = CEC\left(\frac{1-\phi}{\phi}\right)\rho_s. \tag{2.39}$$

Cation exchange is thus the physical basis for conductance at a clay–water interface in the Waxman and Smits model. A strong correlation between specific internal surface area and CEC is expected: CEC is a surface-based phenomenon as cations are primarily exchanged at broken bonds or on cleavage surfaces of the clay minerals (Schön, 2011). The term $\hat{B}$ acts to transform the CEC (via Q_v) into a conductivity term.

The dependence of σ_{surf} on saturation (S_w) is difficult to directly determine experimentally as it is challenging to separate out the influence of the saturation-dependent electrolytic conductivity on measured $\sigma_{[s]}$. Following Archie's second law, an equivalent saturation resistivity index for the surface conductivity (I_s) in the case of constant σ_w can be defined,

$$\frac{\sigma_{surf[ps]}}{\sigma_{surf[s]}} = \frac{1}{I_s}, \tag{2.40}$$

where $\sigma_{surf[ps]}$ is the surface conductivity of a partially saturated rock. As we discuss in Section 2.3.3, IP measurements provide a way to investigate the controls of surface conductivity on saturation. These measurements, along with theory linking IP measurements to surface conductivity, suggest that, similar to the electrolytic conductivity, surface conductivity should also follow a power law dependence on saturation, i.e.

$$\sigma_{[ps]} = \frac{1}{F}\left(\sigma_w S_w^n + \frac{2\Sigma}{\Lambda} S_w^p\right) \cong \frac{1}{F}\left(\sigma_w S_w^n + S_{por} \Sigma S_w^p\right), \tag{2.41}$$

where p is a surface conductivity saturation exponent. Vinegar and Waxman (1984) acquired IP measurements on shaly sandstones where

$$1/I_s = S_w^p \approx S_w^{n-1}. \tag{2.42}$$

The Waxman and Smits (1968) model for partially saturated (or hydrocarbon-containing) shaly sands is

$$\sigma_{[ps]} = \frac{1}{F}\left(\sigma_w S_w^2 + \hat{B} Q_v S_w\right), \tag{2.43}$$

being the specific case of $p = n - 1$ when the Archie exponent is assumed to equal 2.

2.2.4.3 Conduction in Frozen Soils

Understanding the electrical properties of frozen ground is of increasing importance given concerns over climate change. Permafrost soils cover approximately 20% of the Earth's land surface at high latitudes. Permafrost soils are in a sub-zero °C state for at least two consecutive years, although they may be much older. A shallow active layer that contains unfrozen soil forms, varying in thickness depending on the time of year. Geophysical research has focused on characterizing permafrost and active layer properties, particularly the relative concentrations of frozen versus unfrozen water in permafrost soils.

The decrease in resistivity in frozen soils is in part related to the temperature effect discussed in Section 2.2.3.1. However, the electrical conductivity of soils and sediments decreases dramatically with the transition of pore fluids from liquid water to ice (an insulator) across the freezing point (Figure 2.8b). Pure ice is an insulator and thus the replacement of pore water with ice is akin to the replacement of pore water by air, natural gas or oil. Permafrost soil retains a continuous unfrozen layer of absorbed water on the grain surfaces and so it continues to conduct current (Figure 2.8a).

The variation in conductivity that results from replacing unfrozen water with ice around a temperature of zero degrees can be represented by a modified version of Archie's law that accounts for a reduction in unfrozen water saturation. Upon freezing of soils, there is both a decrease in saturation (of the liquid water phase) and also an increase in fluid conductivity as ions are excluded from the frozen state. The electrical properties of a frozen soil (σ_F) containing both frozen and unfrozen water (Figure 2.8a) can be represented by the following Archie-type expression (King et al., 1988; Oldenborger and LeBlanc, 2018),

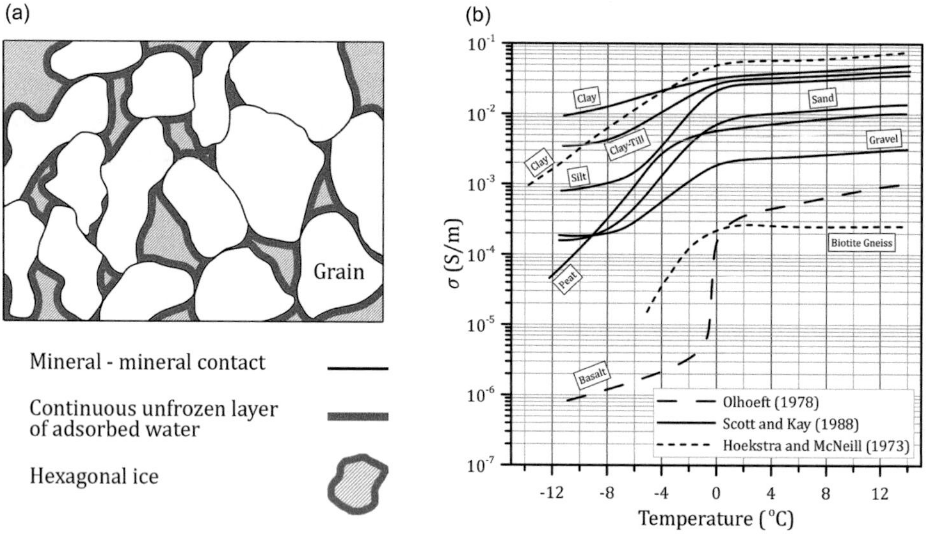

Figure 2.8 (a) Schematic representation of permafrost soil showing the unfrozen water layer that remains below the freezing point (after King et al., 1988); (b) changes in conductivity measured for a range of materials in previous studies (modified from Scott et al., 1990).

$$\sigma_F = \sigma_{w[F]}\phi^m S_F^n = \sigma_{w[0]}\phi^m S_F^{n-1} = \sigma_0 S_F^{n-1}, \tag{2.44}$$

where $\sigma_{w[F]}$ is the conductivity of the water remaining in the frozen state, $\sigma_{w[0]}$ is the conductivity of the water before freezing, σ_0 is the conductivity in the unfrozen state and S_F is the saturation degree for the frozen state. This equation assumes that the saturation of the unfrozen soil (S_0) is equal to 1 and that $\sigma_{w[0]}/\sigma_{w[F]} = S_F/S_0$, the latter assuming exclusion based on a linear electrolyte and no density effects. Figure 2.9 shows examples of the predictions of Equation 2.44 with expected soil types superimposed, along with measurements made on frozen, consolidated sandstones (King et al., 1988). The fact that the measurements on sandstones do not follow the theoretical curves likely results from the numerous simplifications, including ignoring surface conduction in this simple model.

2.2.4.4 Other Models for Predicting the Conductivity of Soils and Rocks

2.2.4.4.1 Empirical Models Developed in Soil Physics

Driven by the economic engine of the oil exploration industry, Archie's laws and the parallel surface/electrolytic conduction model form the most common framework for modelling and interpreting electrical conductivity data. However, alternative approaches have been pursued in other scientific disciplines. In the soil science community, the model of Rhoades et al. (1976) (and derivations thereafter) is popular for relating the partially saturated electrical conductivity (σ_{ps}) to soil volumetric water content ($\theta = S_w\phi$). The theoretical part of this model assumes parallel electrolytic and surface conduction paths defined by a simple geometrical description of the pore geometry. The surface conductivity is considered independent of both θ and σ_w for simplicity. An empirical function is used to define a

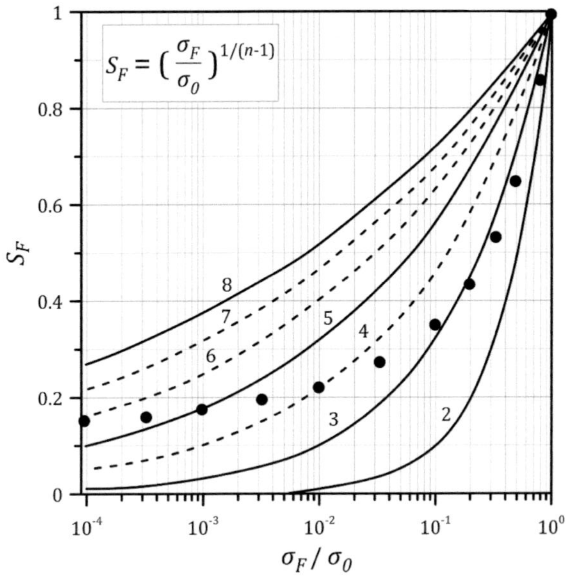

Figure 2.9 Fraction of pore space saturated with water (S_F) in frozen soils as a function of the ratio of the conductivity of the frozen (σ_F) to unfrozen (σ_0) states (for a constant temperature at freezing) predicted by Equation 2.44 showing different integer values of the saturation exponent n (adapted from King et al., 1988). Measurements on frozen sandstone samples (Pandit and King, 1979) are also shown as solid circles.

dimensionless transmission coefficient $T_c(\theta)$, representing a volumetric water content-dependent factor associated with the tortuous pore geometry of the soil, resulting in the following predictive equation:

$$\sigma_{[ps]} = \sigma_w \theta T_c(\theta) + \sigma_{surf} . \tag{2.45}$$

This semi-empirical model is attractive to soil scientists as the transmission coefficient can be determined for specific soils by calibration to moisture-content measurements (e.g. made with a time domain reflectometry (TDR) probe). The transmission coefficient is assumed to be linearly related to θ,

$$T_c(\theta) = a\theta + b , \tag{2.46}$$

where a and b are defined for different soil types. This empirical approach removes the need to know porosity (and cementation exponent) through the use of parameters a and b that can be defined based on soil classification (e.g. for clay soils $a = 2.1$ and $b = -2.5$). Rhoades et al. (1976) showed that this empirical model could describe soil data over a limited range of fluid conductivity from 0.25 to 5.6 S/m. Nadler and Frenkel (1980) proposed an additional salinity-dependent parameter to serve as a multiplier on surface conductivity to account for an increase in surface conductivity with salinity. Further refinements of the Rhoades et al. (1976) model led to some physical insights into the parameters in this model, e.g. that the transmission coefficient (T_c) is foremost related to the fraction of the water that is in well-connected (large) pores (Rhoades et al., 1989). Other developments included expansion to a three-path parallel conduction model that differentiated between relatively well-connected and poorly connected pores of the soil matrix (Rhoades et al., 1989). This model removes the empirical terms appearing in Equations 2.45 and 2.46, being replaced by physical parameters of soils that are more readily measurable.

2.2.4.4.2 Mixing Models

Theoretical mixing models for the electrical conductivity of soils and rocks are used across a wide range of disciplines. They involve prediction of the conductivity of a mixture (e.g. a soil) from the electrical properties of the individual constituents (e.g. soil particles, water and air) and assumptions about the geometrical arrangement of the constituents and how their electrical properties interact. Such theoretical models have provided valuable insights into how soil/rock structure affects soil/rock conductivity. However, application of these models at the field-scale tends to be restricted because the parameters entering into these models are often poorly constrained.

The simplest mixing models consider parallel, perpendicular or random geometric arrangements of conducting phases. In the parallel case, the effective conductivity of the mixture is

$$\sigma_{eff} = \sum_{i=1}^{n} (\phi_v)_i \sigma_i \tag{2.47}$$

where ϕ_v represents a volume fraction of the mixture, and subscript i represents the ith phase and n is the total number of phases. This parallel conduction model is the basis of the two-phase parallel surface and electrolytic conduction model described in Section 2.2.4.2 and Figure 2.5. The less commonly used perpendicular conduction model assumes that the conducting constituents of a soil/rock add in series,

$$\sigma_{eff}^{-1} = \sum_{i=1}^{n} (\phi_v)_i \sigma_i^{-1} . \tag{2.48}$$

This perpendicular arrangement has relevance in cases where the solid phase behaves as a conductor rather than an insulator as assumed in Archie's law. For example, Lévy et al. (2018) used a series addition of conductors to explain the electrical conductivity of clay-rich volcanic rocks where smectite was assumed to be a conductor due to the presence of interstitial water of high conductivity.

Other arrangements of the conducting phase and the presence of multiple conducting phases can be modelled with the Lichtenecker–Rother equation (Lichtenecker and Rother, 1931). This equation was originally defined for two dielectric phases but Glover (2015) presents a generalized form in terms of conductivity,

$$\sigma_{eff} = \left(\sum_{i=1}^{n} \sigma_i^{1/d} (\phi_v)_i \right)^d , \tag{2.49}$$

where d depends on the arrangement of the conducting constituents. When $d = 1$, Equation 2.49 is equivalent to the parallel arrangement expressed in Equation 2.47 and when $d = -1$, Equation 2.49 is equal to the perpendicular arrangement expressed in Equation 2.48. Glover (2015) points out that Equation 2.49 is equivalent to Archie's law (with $d = m$) for a two-constituent (pore fluid, rock matrix) medium when one constituent (the rock matrix) has zero conductivity. This again highlights how m relates to the geometric arrangement (i.e. the connectedness) of the conducting phases. Singha et al. (2007) took a different approach to describe a dual-porosity system composed of a relatively well-connected (mobile to fluids) and a relatively poorly connected (immobile to fluids) domain based on a parallel averaging of two conducting phases, each described individually by Archie's law. Assuming a single common cementation exponent (m),

$$\sigma_{eff} = (\phi_m + \phi_{lm})^{m-1} (\phi_m \sigma_m + \phi_{lm} \sigma_{lm}), \tag{2.50}$$

where subscript m represents the connected (mobile) domain and subscript im represents the less-connected (immobile) domain. Day-Lewis et al. (2017) improved the mixing formulation to relax the improbable constraint whereby both domains have the same value of m.

Another popular class of mixing models is derived from effective medium theory. These models provide theoretical approximations of the conductivity of a composite soil/rock based on averaging the multiple values of the conductivity of the individual constituents. The original theories were based on embedding spherical inclusions of a component in a homogeneous host medium. The self-consistent (SC) effective medium theory derived by Bruggeman (1935) and Landauer (1952) can be expressed as

$$\sum_{i=1}^{N} \phi_{vi} \frac{\sigma_i - \widetilde{\sigma}_{eff[sc]}}{\sigma_i + 2\widetilde{\sigma}_{eff[sc]}} = 0 , \tag{2.51}$$

where $\phi_{v1}, \ldots, \phi_{vN}$ are the volume fractions of the constituents with conductivity $i_1, \ldots, i_N$ and $\widetilde{\sigma}_{eff[sc]}$ is an approximation of the true effective conductivity σ_{eff}. Equation 2.51 must be solved iteratively.

The differential effective medium (DEM) approach is theoretically powerful, being based on iteratively computing the updated effective conductivity for incremental additions of an inclusion into a host medium, where the initial starting point is a homogenous host medium of a single constituent. For a two-constituent system where y is the volume fraction of the inclusion and starting with the condition $\widetilde{\sigma}_{eff}(y = 0) = \sigma_1$, the DEM approach is given as (Berryman, 1995)

$$\left(\frac{\sigma_2 - \widetilde{\sigma}_{eff}(y)}{\sigma_2 - \sigma_1}\right)\left(\frac{\sigma_1}{\widetilde{\sigma}_{eff}(y)}\right)^{1/3} = 1 - y. \tag{2.52}$$

The effective medium theories can be generalized to incorporate non-spherical inclusions through the use of depolarization factors that are typically defined along the three principal axes of an ellipsoidal inclusion. Sen et al. (1981) showed that, for the case of aligned ellipsoids representing an anisotropic porous medium, the exponent 1/3 in the DEM for spherical inclusions (Equation 2.52) is replaced by the depolarization factor for the axis of the ellipsoids aligned with the direction of the applied electric field.

The effective medium theories have proven particularly powerful for improving the understanding of the electrical properties of soils and rocks. For example, Sen et al. (1981) used the DEM approach to provide a theoretical justification for Archie's classical empirical law. An important result from the work of Sen et al. (1981) is that the cementation factor is predicted to have a value of 1.5 for spherical particles.

2.2.4.4.3 Circuit Models and Pore Network Models

Equivalent circuit models, easily combined with pore network models, have also been used to describe the resistivity of soils and rocks (Fatt, 1956). The approach is based on representing the geometry of the interconnected porous medium as a network of electrically conductive pipes/tubes of known dimensions. Ohm's law is used to describe the current flowing in individual pipes/tubes in response to an electric field across them. A potential difference across the entire network is assumed and Kirchhoff's laws are implemented to determine the potentials at the junctions of individual pipes/tubes through the network. The total current flowing through all pipes/tubes is calculated and used along with the total voltage drop across the sample to determine the resistance and resistivity.

Greenberg and Brace (1969) showed how 2D and 3D geometrical arrangements of resistors reproduced the response of rocks as predicted from Archie's law. They demonstrated the value of the approach for exploring how the resistivity of a rock changes as pore channels are blocked/disconnected (e.g. from compression). Suman and Knight (1997) used a 3D pore network model to better understand the electrical properties of a rock as a function of saturation. Different saturation states were simulated by removing progressively smaller pores from the electrically conducting network as saturation decreased. The approach led to insights into the role of wettability in controlling resistivity as well as the nature of hysteresis in resistivity-saturation curves.

Pore network models are also very popular for simulating the flow of fluids (rather than electric current) and determining the transport properties of porous materials (Bernabé, 1995). Bernabé et al. (2010) developed a pore network model for permeability (resistance to fluid flow) that related the permeability to a power law relationship involving a critical coordination number (z_c), defining the connectivity of nodes in the network at the percolation threshold (when the porous medium becomes connected over a large distance). Bernabé et al. (2011) applied this pore network model to electric current flow and argued for the existence of a 'universal' power law of the form

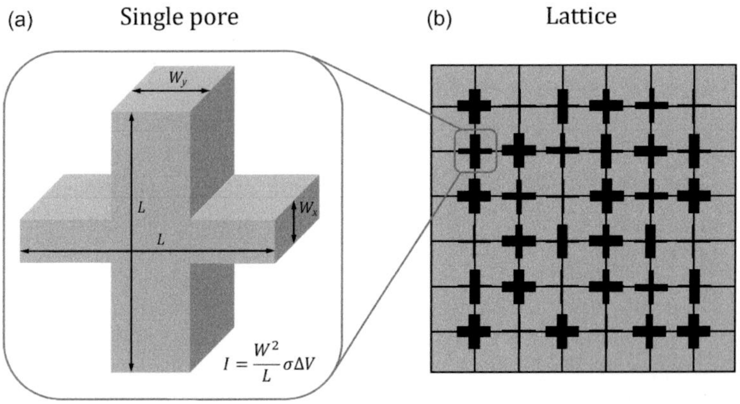

Figure 2.10 Pore network model approach to modelling the electrical conductivity of rocks (a) electric current (I) is calculated from the conductivity of the pore-filling fluid (σ), the pore width (W) and the electrical field strength ($\Delta V/L$) using Ohm's law; (b) lattice model based on pores of various width in two dimensions where Kirchhoff's laws are applied (with permission from Frederick Day-Lewis, USGS).

$$\frac{1}{F} \propto (z - z_c)^{\gamma}, \tag{2.53}$$

where z is the average coordination number (mean number of pore conduits/throats attached to a node/pore in the network) and γ is a function of the pore radius distribution. Such pore network models emphasize better the importance of connectivity (that is not immediately obvious in Archie's law) in controlling the electrical properties of porous media.

The pore network model is an attractive way to couple electrical properties to hydrogeological properties. Day-Lewis et al. (2017) recently demonstrated how the coupling of electric current and fluid flow through a porous medium can be used to better understand solute transport in dual-porosity systems where mass is transferred by diffusion between relatively mobile and immobile (for fluids) domains. This pore network model used a primary pipe lattice to represent the pore space of the mobile domain between particles and a secondary lattice to represent the intra-particle porosity associated with the particles themselves (Figure 2.10). The model showed that the assumption made by Singha et al. (2007) and the mixing model represented by Equation 2.50, being that the electrical connectedness of the mobile and less-mobile domains is identical and that a single formation factor represents both domains, is not rigorously correct. Day-Lewis et al. (2017) therefore developed a modified mixing model based on the DEM theory (Equation 2.52) that incorporated the different connectivities of the mobile and less-mobile domains. Given the increasing use of geophysics in hydrogeology and the growing need to better understand flow and transport using geophysical datasets, applications of pore network models in resistivity modelling (and also for IP, e.g. (Maineult et al., 2017a)) are likely to grow.

2.2.4.4.4 Models Based on Fractal Theory

The irregular nature of a porous medium displays fractal characteristics (statistical self-similarity). Fractal theory has therefore been extensively used to analyse pore geometry and pore-scale processes in porous media. Fractal models for electric current flow through Earth materials have yielded additional insights into the physical significance of the Archie model parameters, demonstrating that they depend on microstructural properties of the porous medium (Cai et al., 2017).

2.3 Induced Polarization

In order to incorporate IP, the electrical properties must be extended to account for the temporary, reversible storage of charge by the porous medium in addition to the electromigration of charge by conduction described earlier. In terms of an analogy to an electrical circuit, the medium must now be considered equivalent to a resistor (conduction)–capacitor (charge storage) network (Box 2.4). We will see that the analogy with a true capacitor is not technically correct and we will need to consider a 'leaky' capacitor, but the basic analogy holds for now. This can be efficiently done by representing measured electrical conductivity as an effective complex property (σ^*) where the real part (σ') represents electromigration

Box 2.4
Electrical Polarization

In the context of electric charge, 'polarization' refers to the displacement of bound, charged elements in response to an alternating external electric field. Unlike in the case of conduction, the charges are restricted from freely moving in the material. Instead, positive charges are 'displaced' in the direction of the electric field and negative charges are displaced opposite to the direction of the field, producing an electric dipole moment ($\Delta \boldsymbol{p}$). Polarization classically refers to the behaviour observed for a dielectric material (an insulator that can be polarized by an electric field), where the charged elements are bound to molecules (i.e. molecular-scale polarization). The concept equally applies to the restricted movement of charges occurring at other scales.

In IP, it is not a dielectric but a porous medium that is polarized. The scale of this polarization (distance over which charges are displaced) is much larger than the molecular scale of charge displacement observed in a dielectric. Furthermore, the polarization process is slower than observed for a dielectric, in part because the charges move over much larger distances. It is the ionic charges in the EDL that are primarily responsible for the low-frequency (below 1,000 Hz) polarization observed in porous media (the additional role of electrons is discussed in Section 2.3.7). In order for polarization to occur, the ions must be restricted in their mobility such that transport is limited over a length-scale associated with the geometry of the porous material. One popular conceptual representation of this length-scale that has been extensively used in mechanistic models of the IP response is the grain size. This model attributes the polarization to the fixed ionic charges in the very thin Stern layer surrounding each grain. The Stern layers of each individual grain are considered to be disconnected, thereby restricting the charge movement. Other models consider the polarization of a porous material to result from the formation of ion-selective zones. The analogy with a capacitor is only partially correct as it involves diffusive transport of charges rather than true dielectric polarization. We will represent this polarization of a porous medium by a 'leaky' capacitance.

Box 2.4 (cont.)

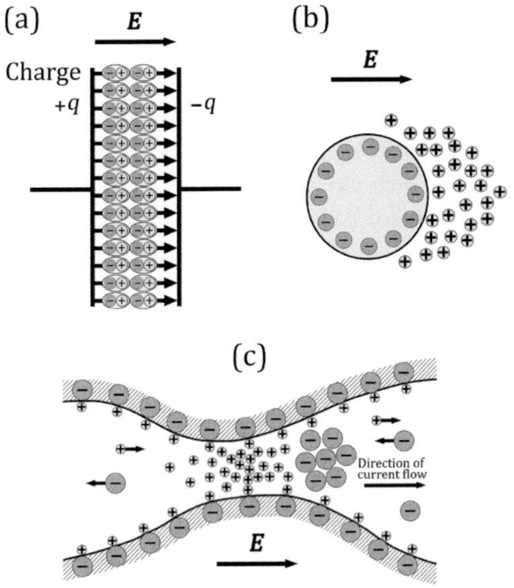

Polarization of a dielectric (a), a mineral grain (b) and a pore throat constriction (c).

The intensity of the polarization ($\boldsymbol{P}$) is given by

$$\boldsymbol{P} = \frac{\Delta \boldsymbol{p}}{\Delta V_a},$$

where $\Delta \boldsymbol{p}$ is the induced dipole moment in response to the electric field and ΔV_a is the volume of the material. The induced dipole moment is a measure of the separation of the positive and negative charges (SI units of coulomb m). It is formally defined for a dielectric but the same concept can be used to quantify the polarization strength for other mechanisms.

and the imaginary part (σ'') represents the temporary, reversible charge storage associated with a variety of mechanisms that dominate at different frequencies.

It is important to note that this convenient definition of an effective measured complex conductivity can include contributions from multiple fundamental mechanisms of charge storage depending on frequency (f in Hz, angular frequency $\omega = 2\pi f$ in radian/s). At the low frequencies ($f = < 10^3$ Hz) where resistivity and IP measurements are mostly made, the dominant mechanism is a diffusion-controlled polarization of the EDL of primary concern here. At higher frequencies (10^2 Hz$< f < 10^8$ Hz), a Maxwell–Wagner mechanism (an interfacial space-charge polarization that arises from discontinuities in electrical conductivity) is often invoked (Box 2.5). At very high frequencies ($f > 10^8$ Hz),

Box 2.5

Maxwell–Wagner polarization

The IP response of a porous medium is foremost assumed to arise from the polarization of the EDL at the mineral–fluid interface within the interconnected pore space, as discussed in detail in the main text. However, researchers in spectral IP (SIP) measurements sometimes refer to a 'Maxwell–Wagner' (MW) polarization mechanism to partly explain dispersion curves observed at high frequencies (over 100 Hz) for SIP measurements. The MW polarization exists due to the bulk electrical properties of soil and rock components and is not related to the complex surface conductivity of the mineral surface. Instead, this polarization mechanism results from the discontinuity in conductivity at interfaces between the different phases (solid, liquid, gas) of the porous medium. Free charge distributions form near the interface between the phases of the porous medium in response to the applied electric field. This effect, associated with the geometric arrangement of the phases, differs from the diffusion-limited EDL polarization that is foremost attributed to IP signals. It can be predicted from the bulk partial volumes and electrical properties of the individual components, along with their microstructure, using effective medium theories (Chelidze and Gueguen, 1999). Whether or not the MW polarization is truly observed in SIP datasets is hard to determine, particularly as the effect becomes dominant in the higher-frequency range where measurement errors due to electrode effects (discussed in Chapter 3) grow large. In fact, it is likely that the additional dispersion observed in many SIP datasets and attributed to MW effects has resulted from errors associated with the electrodes and instrumentation. At frequencies above 10^8 Hz, far from the range of SIP measurements, the dipolar polarization of water molecules dominates.

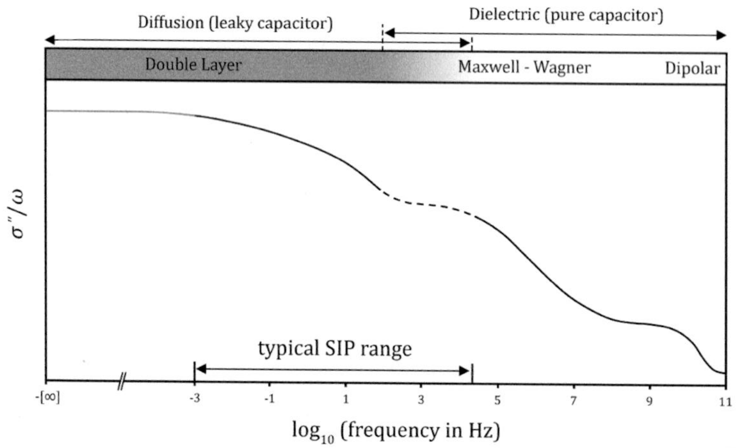

Polarization mechanisms across a broad range of frequencies from 10^{-3} to 10^{11} Hz. Multiple mechanisms may overlap in the intermediate range represented by the dashed line.

the dipolar polarization of water molecules dominates. The interpretation of IP data in general inherently attributes the measured polarization to the diffusive EDL mechanism described later.

2.3.1 Complex Resistivity/Conductivity Definitions

In terms of an effective (measured) complex conductivity, Equation 2.3 becomes

$$\boldsymbol{J} = \sigma^* \boldsymbol{E} = \frac{1}{\rho^*} \boldsymbol{E} = \omega \varepsilon^* \boldsymbol{E} = \omega \kappa^* \varepsilon_0 \boldsymbol{E}, \tag{2.54}$$

where ρ^* is the equivalent effective complex resistivity, ε^* is the equivalent effective complex dielectric permittivity, ε_0 is the dielectric permittivity of free space (8.854×10^{-12} F/m) and κ^* is the equivalent effective complex relative dielectric permittivity. The effective properties appearing in Equation 2.54 vary with frequency. They do not refer to specific classical mechanisms of polarization but instead represent the quantities actually measured on soils and rocks that may incorporate multiple mechanisms (Fuller and Ward, 1970; Vinegar and Waxman, 1984). Thus, we can choose to represent the measured effective properties in terms of a (complex) conductivity, resistivity or dielectric permittivity. In each case, the effective properties can be split into real and imaginary components (Box 2.6). In terms of the σ^* terminology most commonly used for IP,

$$\sigma^* = |\sigma| e^{i\varphi_c} = \sigma' + i\sigma'', \tag{2.55}$$

where $|\sigma|$ is the conductivity magnitude, φ_c is the phase (the subscript c is to indicate that phase is expressed in terms of conductivity space and therefore positive) and i is the imaginary number equal to $\sqrt{-1}$. The phase represents the lag of the induced alternating current (AC) electric field behind the injected AC current. The real part of σ^* is associated with electromigration and the imaginary part is associated with any charge storage mechanisms, i.e. the polarization effect. The fact that Equations 2.54 and 2.55 refer to effective properties and not specific mechanisms is sometimes misunderstood. For example, a common point of confusion relates to the very large values of effective dielectric permittivity that result from translating the measured effective complex conductivity into an equivalent effective dielectric permittivity. The real part of the effective dielectric permittivity ε' is related to σ'' according to

$$\kappa' = \frac{\varepsilon'}{\varepsilon_0} = \frac{\sigma''}{\omega \varepsilon_0}. \tag{2.56}$$

This results in very large values of κ' at low frequencies in the presence of both Maxwell–Wagner (Box 2.5) and EDL (what we measure with IP) polarization mechanisms (κ' up to 10^9) relative to the high-frequency relative dielectric permittivity associated with molecular polarization (κ_∞) (Chelidze and Gueguen, 1999), which varies from 1 in air to approximately 80 in water.

Box 2.6

Complex numbers and complex conductivity

The effective electrical properties of a porous material are conveniently expressed as complex numbers. They take the form $a + bi$ where a is the real part, b is the imaginary part and the imaginary number $i = \sqrt{-1}$. The term 'imaginary' comes from the fact that i is a solution to the equation $x^2 = -1$: no real numbers can satisfy this equation. Despite this somewhat confusing terminology, the imaginary numbers are just as important as the real numbers. Complex numbers provide a framework for representing a wide range of physics and electrical engineering concepts, including the analysis of time, varying voltages, and currents needed to explain resistors, capacitors and inductors. Electrical engineers represent the complex number by j instead of i (primarily to avoid any confusion with symbols that represent the electric current).

The easiest way to make sense of a complex number is geometrically using a 2D complex plane. Here, the complex number is represented by a pair of numbers (a, b) that describe a vector plotted on what is known as an Argand diagram. The value of the real number is plotted on the x-axis and the value of the imaginary number is plotted on the y-axis. Purely real numbers lie on the horizontal axis of the plane and purely imaginary numbers lie on the vertical axis of the plane. The complex number can then also be quantified using polar coordinates, where the magnitude of the complex number is given by the vector distance from the origin and the phase (φ) is given by the angle of the vector relative to the horizontal (real) axis. In IP measurements, the magnitude and phase are recorded by the instrument, and the real part (representing conduction or electromigration) and the imaginary part (representing polarization) are computed. In terms of a complex conductivity, φ is positive (as depicted in the figure) but most induced polarization instruments report a complex resistivity or impedance as the raw reading, in which case the phase is negative ($-\varphi$).

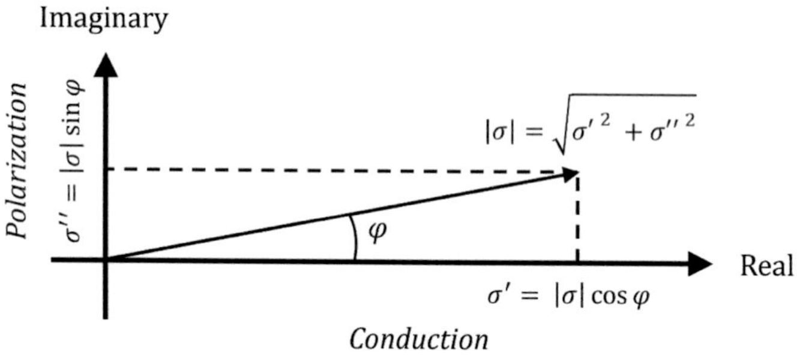

Graphical representation of complex conductivity (σ^*). The real part represents the conduction strength (what is essentially measured with the resistivity method because $|\sigma| \approx \sigma'$) and the imaginary part represents the polarization strength (a direct measure of the IP effect).

The magnitude and phase are related to the real and imaginary parts of the complex conductivity,

$$|\sigma| = \sqrt{(\sigma')^2 + (\sigma'')^2}\,, \tag{2.57}$$

$$\varphi = \tan^{-1}\left[\frac{\sigma''}{\sigma'}\right]. \tag{2.58}$$

The real and imaginary components of the complex conductivity can be expressed as (Box 2.6)

$$\sigma' = |\sigma| \cos\varphi \tag{2.59}$$

$$\sigma'' = |\sigma| \sin\varphi. \tag{2.60}$$

The real conductivity represents the conduction (electromigration) strength (what is essentially measured with the resistivity method because $|\sigma| \approx \sigma'$) and the imaginary conductivity is a fundamental measure of the polarization strength, i.e. a direct measure of the strength of the IP effect.

At low frequencies used in resistivity and IP, the charge storage is small relative to electromigration such that

$$\varphi = \tan^{-1}\left[\frac{\sigma''}{\sigma'}\right] \approx \left[\frac{\sigma''}{\sigma'}\right]. \tag{2.61}$$

The approximation in Equation 2.61 is valid for $\varphi < 0.1$ radians, which is usually the case for porous media. The major exception to this rule is when electron conductors (e.g. ores) are present, as polarization processes resulting from the presence of electron conductors (discussed in Section 2.3.7) can easily result in $\varphi > 0.1$ radians. Equation 2.61 states that the phase angle is proportional to the polarization strength of the medium divided by the electromigration strength of the medium. In Chapter 3 we will see that other measurements of the IP response recorded by field instrumentation are, like the phase, proportional to the polarization strength of the medium divided by the electromigration strength of the medium.

Due to historic developments in the method, the effective complex resistivity (ρ^*) is often used to describe IP in the mineral exploration literature,

$$\rho^* = \frac{1}{\sigma^*} = \rho' - i\rho''. \tag{2.62}$$

The magnitude ($|\rho|$) and phase (φ_r) of the complex resistivity are directly related to the magnitude ($|\sigma|$) and phase (φ_c) of the complex conductivity,

$$|\rho| = \frac{1}{|\sigma|} = \sqrt{(\rho')^2 + (\rho'')^2}, \tag{2.63}$$

$$\varphi_r = -\varphi_c = \tan^{-1}\left(\frac{-\rho''}{\rho'}\right) \approx \left(\frac{-\rho''}{\rho'}\right), \tag{2.64}$$

where the approximation in Equation 2.64 is again for $\varphi_c < 0.1$ radians. Note that $\rho' \neq 1/\sigma'$ and $\sigma'' \neq 1/\rho''$.

2.3.2 Polarization Mechanisms

A number of distinct polarization mechanisms are needed to explain the complex effective properties when measured over a broad frequency range (e.g. from 10^{-3} Hz to 10^{11} Hz). At frequencies above 10^4 Hz, the polarization mechanisms are primarily determined by the bulk electrical properties of the components, the relative volumes of these components and the configuration of the components. These polarization mechanisms can be computed from extensions of the effective medium approaches described in Section 2.2.4.4.2 for modelling electrical resistivity. For example, Hanai (1968) extends an effective medium model defined by Bruggeman (1935) to describe the complex effective dielectric permittivity of a two component system in terms of the complex dielectric permittivity of each component,

$$\frac{\varepsilon^* - \varepsilon_2^*}{\varepsilon_1^* - \varepsilon_2^*}\left(\frac{\varepsilon_1^*}{\varepsilon^*}\right)^{1/3} = 1 - \phi_{v2}, \tag{2.65}$$

where ϕ_{v2} is the volume fraction occupied by the second component. At frequencies above 10^8 Hz, the MW polarization mechanisms become insignificant and the dipolar polarization of water molecules dominates instead.

Although the MW mechanism may be captured in some high-frequency spectral IP (SIP) measurements, completely different polarization effects associated with diffusion-driven decays of ionic concentration gradients established by an applied electric field are responsible for the low-frequency polarization sensed with IP. These diffusive decays result from the redistribution of ions from an excited state back to an equilibrium position. Two main mechanisms, both associated with the EDL, have been proposed to drive such ionic concentration gradients in rocks devoid of electron conducting minerals (Vinegar and Waxman, 1984). These mechanisms are (1) tangential displacement of counterions in the Stern layer of the EDL forming at mineral surfaces (Stern layer polarization); (2) blockage of ions in the diffuse layer at sites within the interconnected pore space where localized concentration excesses and deficiencies occur (membrane polarization). In the early IP model of Marshall and Madden (1959), both effects were conceptualized to occur in connection with alternating clay-rich and clay-free zones of a rock, where high concentrations of fixed negative charge in clay-rich zones enhance cation transport relative to anion transport. However, the two concepts have since been generalized to explain the IP signatures for a wide range of rock types. Both mechanisms have been invoked to explain frequency-independent IP measurements as well as frequency-dependent (spectral) IP measurements. The Stern layer polarization (SLP) has been argued to dominate IP measurements except at very high salinities where the membrane polarization may dominate (Vinegar and Waxman, 1984; Revil, 2012). In the case of electron conducting minerals, the tendency of the mineral to transport electrons requires more complex electrochemical models to describe IP signals. These models for polarization of electron conductors (often referred to as electrode polarization) are discussed separately in Section 2.3.7.

The equivalent electrical circuit description for current flow in a porous medium shown in Figure 2.5 can be extended to provide a conceptual representation of any of the proposed

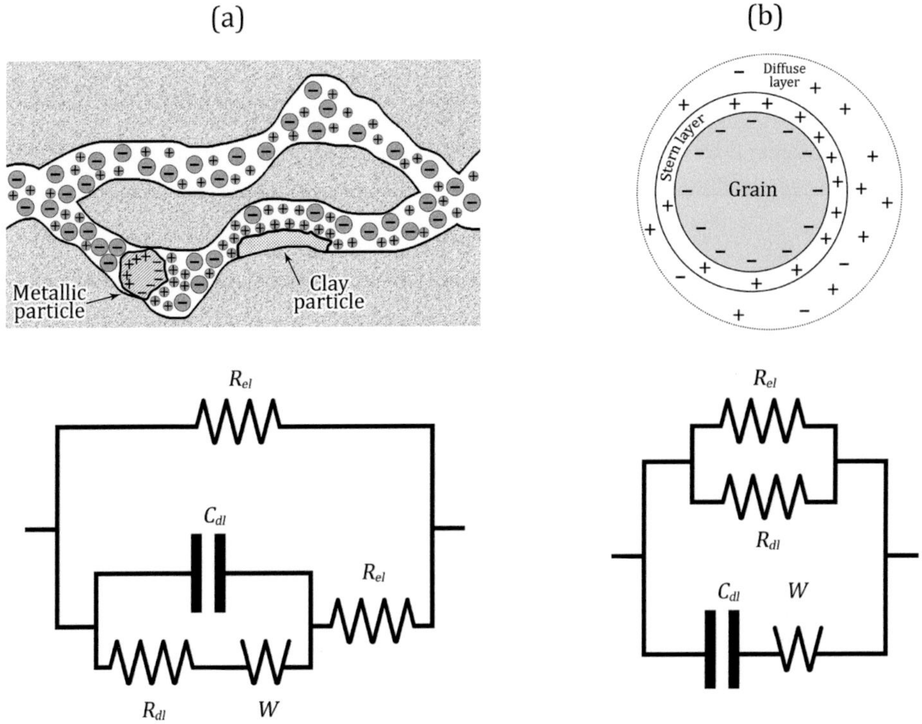

Figure 2.11 Equivalent circuit model representations of polarizable elements proposed in soils and rocks (a) electrode- or membrane-type polarization associated with elements blocking pores (Dias, 1972, 2000); (b) Stern layer polarization with elements parallel to pores (modified from Revil et al., 2017a). Both invoke the Warburg impedance (*W*) description for a leaky capacitance where the impedance varies inversely with the square root of frequency. R_{el} – pure resistor due to electrolyte; C_{dl} – normal capacitance of double layer; R_{dl} – pure resistor due to Ohmic resistance across double layer.

IP mechanisms (SLP, membrane and electrode). Whereas high-frequency dielectric polarization (MW and dipolar) mechanisms can be incorporated into such circuits as a capacitor and resistor combination, the diffusion-driven EDL mechanisms measured with IP do not behave as a perfect capacitance but must instead be incorporated into such circuit models as a 'leaky' capacitance. Geophysicists have adopted the use of a Warburg impedance (Grahame, 1952) to describe pure diffusion of ions across/along the EDL. This impedance varies inversely with the square root of frequency (whereas a perfect capacitor varies inversely with frequency). Although originally adopted to describe the polarization associated with electron conducting minerals (Marshall and Madden, 1959), the Warburg impedance circuit component was subsequently adopted to also represent membrane polarization mechanisms (Dias, 1972, 2000). More recently, the Warburg impedance has been invoked to generate equivalent circuit representations of the Stern layer polarization model. Figure 2.11 contrasts equivalent circuit models for (a) electrode- or membrane-type polarization, where the polarizable element essentially blocks pore throats (Dias, 1972,

2000), and (b) Stern layer polarization, where the polarizable elements are essentially in parallel with the pore throats (Revil et al., 2017a). As we shall shortly see, the frequency dependence of the IP response of a porous medium composed of such components will depend on the relative abundance of such components of different sizes within a soil or rock. We next consider both frequency-independent (typically an approximation) and frequency-dependent polarization models.

2.3.3 Frequency-Independent IP Model in the Absence of Electron Conducting Particles

The diffusion-dominated, reversible EDL polarization mechanisms recorded with IP occur over some length-scale that characterizes the distance over which ions diffuse in response to an applied electric field (and decay back to an equilibrium upon removal of the inducing signal). When a narrow distribution of length-scales exists in a soil or rock, the polarization effect may often show a peak at a frequency characterizing the time required for this diffusion of charge to occur over the dominant length-scales. In this case, it will be necessary to incorporate this dominant relaxation time (represented by a time constant (τ)) and length-scale in IP models. Such models relate the length-scale to either grain size (convenient for unconsolidated sediments and soils) or pore size (more appropriate for consolidated rocks) as shown in Figure 2.12. These models are described further in Section 2.3.5. However, it will sometimes be sufficient to model the polarization process occurring in soils and rocks as frequency independent. This is the case when the porous medium is characterized by a broad distribution of length-scales (grain size, pore size) as the individual

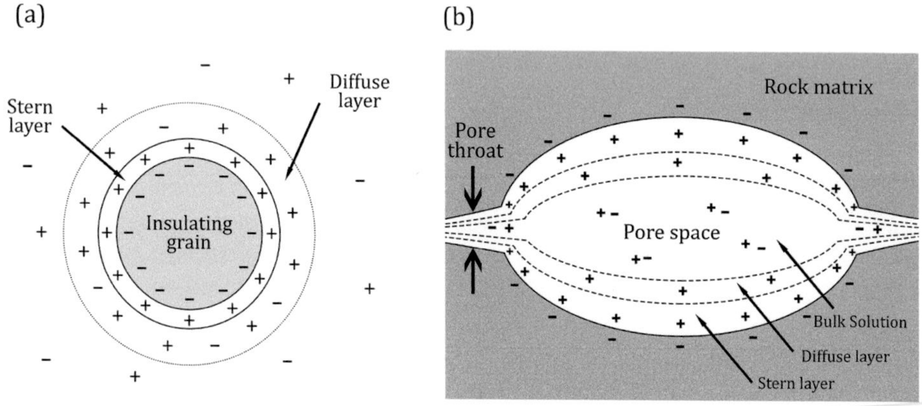

Figure 2.12 Conceptual model for the polarizable electrical double layer in a porous medium for two possible polarizable elements: (a) concept of a polarizable mineral grain where the Stern layer is considered discontinuous between grains; (b) concept of a polarizable pore connected to pore throats where pore constrictions restrict free movement of charge due to merging of the double layer. The Stern and diffuse layers are very thin and not drawn to scale.

polarization peaks merge to form a flat (or at least weakly frequency-dependent) spectral response (Vinegar and Waxman, 1984). These frequency-independent models for the IP response are discussed here.

Irrespective of the exact mechanism, it is the EDL that is ultimately responsible for IP signals. Given that the EDL is the cause of the surface conductivity described in the interpretation of resistivity measurements in Section 2.2.4.2, it is valuable to reformulate this basic model for electric charge transport to include a complex surface conductivity (σ^*_{surf}) that represents both conduction and polarization occurring in the EDL.

When the matrix is devoid of electron conducting particles, the basic model for a parallel addition of electrolytic and surface conduction presented in Equation 2.30 can be extended to account for surface polarization (Vinegar and Waxman, 1984; Lesmes and Frye, 2001),

$$\sigma^* = \sigma_{el} + \sigma^*{}_{surf} = \sigma_{el} + \sigma'_{surf} + i\sigma''_{surf}, \tag{2.66}$$

where the real part of the surface conductivity (σ'_{surf}) represents electromigration in the EDL and the imaginary part (σ''_{surf}) represents the reversible temporary charge storage. The electrolytic (Archie) conductivity associated with the fluid-filled pores of the interconnected pore network is assumed to be non-polarizable. This is a reasonable assumption at frequencies less than 1,000 Hz where the dipolar polarization of the water molecules is very weak (Box 2.5).

Using this basic model, the real and imaginary parts of the effective (measured) complex conductivity are related to the electrolytic and surface conductivity mechanisms as follows:

$$\sigma' = \sigma_{el} + \sigma'_{surf}, \tag{2.67}$$

$$\sigma'' = \sigma''_{surf}. \tag{2.68}$$

Equation 2.67 is consistent with Equation 2.30 described in the DC resistivity section, where the measured conductivity is the sum of the conductivities resulting from the parallel electrolytic and surface conduction pathways within the porous medium. Equation 2.68 makes an important statement that highlights the value of IP measurements, i.e. that the imaginary conductivity only senses the surface pathway and is thus independent of the electrolytic (Archie) conduction pathway. Equations 2.67 and 2.68 indicate that IP measurements might resolve the inherent ambiguity of DC resistivity measurements, whereby the relative contribution of surface conduction to the total measured conductivity is unknown if a strong relationship between σ''_{surf} and σ'_{surf} exists. Evidence for this important relationship will be presented shortly. This suggests a potentially powerful use of IP, i.e. as an extension of the DC resistivity method to help improve subsurface interpretation of resistivity datasets (see example applications in Chapter 6).

Vinegar and Waxman (1984) were one of the first to utilize this potential of IP measurements to resolve such ambiguity in their efforts to model the IP response of shaly sands. They were motivated by the need to determine shaliness as quantified from their parameter Q_v independent from the other parameters appearing in the Waxman and Smits model

(Equation 2.38). Vinegar and Waxman (1984) used the Waxman and Smits model to represent the real part of the complex conductivity,

$$\sigma' = \frac{1}{F}\left(\sigma_w + \hat{B}Q_v\right). \tag{2.69}$$

They modelled the imaginary part as

$$\sigma'' = \frac{1}{F_q}\hat{\lambda}Q_v, \tag{2.70}$$

where $F_q = \phi F$ provided a better fit to the data than using a single formation factor to describe both the real and imaginary conductivity dependence on pore geometry. In Equation 2.70, $\hat{\lambda}$ represents the equivalent imaginary conductance of the ions in the EDL of clay minerals in the same way as $\hat{B}$ represents the conductance of the ions in the EDL of clay minerals in the Waxman and Smits model. Equation 2.70 highlights the direct dependence of σ'' on Q_v and the opportunity to reduce the inherent ambiguity of interpretation when applying the Waxman–Smits equation to resistivity measurements alone.

It is intuitive to expect that the real and imaginary parts of the surface conductivity will be related. The relationship is not easy to test as it requires accurate estimates of the formation factor to determine a reliable estimate of σ'_{surf}. Börner (1992) was the first to confirm the existence of a relationship of the form

$$\sigma''_{surf} = l\sigma'_{surf}, \tag{2.71}$$

where l is a proportionality factor that Börner et al. (1996) found to vary from 0.01 and 0.15 for their limited database of sandstone samples. Weller et al. (2013) examined this relationship for a comprehensive database of samples from different datasets where $l = 0.042$ with a coefficient of determination (R^2) of 0.91 (Figure 2.13). Other experimental datasets confirm the linear proportionality given by Equation 2.71 as shown by the addition of new data (not used in the calibration) in Figure 2.13, along with measurements on a wide range of material types from soil to volcanic rock with some variation in l (Revil et al., 2017b).

This relationship has important implications for the use of IP datasets to improve the interpretation of conventional resistivity measurements as it holds the key to overcoming the inherent ambiguity in a single measurement that depends on two conduction pathways. Substituting Equation 2.71 into Equation 2.67,

$$\sigma' = \sigma_{el} + \sigma'_{surf} \approx \sigma_{el} + \frac{\sigma''_{surf}}{l}. \tag{2.72}$$

As σ''_{surf} is known from the measurement of σ'', σ_{el} can be isolated when both resistivity and IP are measured (Börner et al, 1996),

$$\sigma_{el} = \frac{1}{F}\sigma_w \approx \sigma' - \frac{\sigma''_{surf}}{l} = \sigma' - \frac{\sigma''}{l}. \tag{2.73}$$

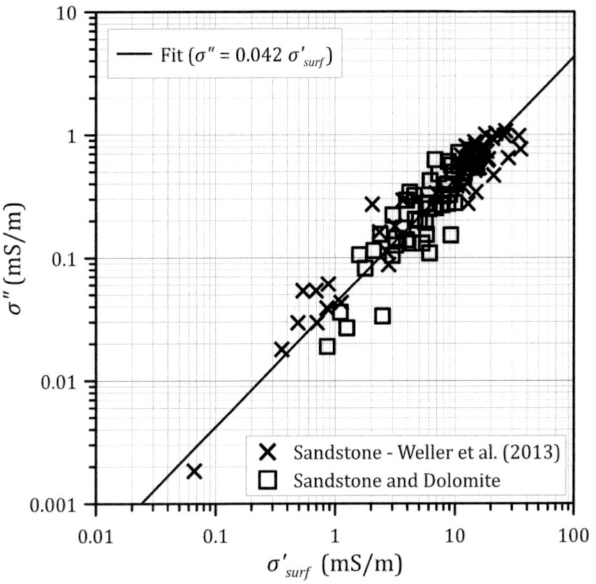

Figure 2.13 The dependence of the imaginary conductivity at 1 Hz on the surface conductivity for a wide database of 63 samples consisting of sandstones and unconsolidated sediments as originally reported by Weller et al. (2013) along with 58 additional sandstone and sandstone/dolomite samples. All measurements are for a pore fluid conductivity of 0.1 S/m. A single value of $l = 0.042$ was fit ($R^2 = 0.911$) to the 63 samples from Weller et al. (2013). This calibrated value fits the additional samples very well.

Equation 2.73 indicates that (1) F can be directly measured with resistivity and IP in the case of a known groundwater conductivity σ_w, or (2) σ_w can be determined if an estimate of the formation factor is known (Weller et al., 2013).

According to Equation 2.71, σ''_{surf} is a scaled version of σ'_{surf}. Consequently, σ''_{surf} can be directly related to the same physicochemical properties describing surface conductivity presented in Equations 2.34–2.38. Therefore, we can expect

$$\sigma''_{[s]} = \sigma''_{surf[s]} \approx l \frac{1}{F} \frac{2\Sigma}{\Lambda} \approx l \frac{1}{F} S_{por} \Sigma. \tag{2.74}$$

Mechanistic models offer further validation of the close association between the surface conductivity and the measured imaginary conductivity. Revil (2012) introduced his POLARIS model for the complex conductivity of shaly sands, which extends the model of Vinegar and Waxman (1984). The model is a based on a solution for the frequency-independent complex conductivity calculated from effective medium theory for grains coated with an EDL and in solution. In this model, the real part of the surface conductivity is foremost associated with the diffuse layer because the mobility of counterions in the Stern layer of clay particles is assumed to be only 1/350th of the mobility of counterions in the diffuse layer and also ions in the bulk solution. The polarization (and hence the imaginary conductivity signal) is attributed to the Stern layer. The model yields the following expressions for the real and imaginary parts of the surface conductivity,

$$\sigma'_{surf} = \frac{2}{3}\frac{\phi}{1-\phi}\beta_P Q_v, \tag{2.75}$$

$$\sigma''_{surf} = \frac{2}{3}\frac{\phi}{1-\phi}\lambda_P Q_v, \tag{2.76}$$

where β_P and λ_P are apparent ionic mobilities that act similar to the ionic conductance terms appearing in the Vinegar and Waxman (1984) equations (Equations 2.69–2.70). Equations 2.75 and 2.76 satisfy Equation 2.71 with $l = \lambda_P/\beta_P$.

Revil (2012) also reformulated the POLARIS model in terms of a linear dependence on S_{por},

$$\sigma''_{surf} = \frac{2}{3}\frac{\phi}{1-\phi}\lambda_P Q_s S_{por}, \tag{2.77}$$

where Q_s is the surface charge density. This predicted linear dependence of $\sigma''_{[s]}$ on S_{por} from Equations 2.74 and 2.77 can be tested using measurements of S_{por} from the gas adsorption technique (Brunauer et al., 1938), although more laborious methodologies based on the adsorption of aqueous dyes provide better resolution of the internal surface area between clay minerals (Yukselen and Kaya, 2008). Börner and Schön (1991) were the first to experimentally confirm a relationship between σ'' and surface area, although they established the relationship with the surface area normalized by the total sample volume (S_{tot}). Weller et al. (2010a) compiled an extensive database of sandstone and unconsolidated sediments (114 samples from nine independent datasets) to confirm a strong linear relationship between $\sigma''_{[s]}$ and S_{por} (Figure 2.14) at a near-constant conductivity of the pore fluid (0.1 S/m). Numerous subsequent studies support this relationship, including measurements on volcanic rocks (Revil et al., 2017a). Weller et al. (2010a) introduced the concept of the 'specific polarizability', being the polarization strength per unit S_{por},

$$c_p = \frac{\sigma''}{S_{por}} \tag{2.78}$$

to represent the effect of the EDL chemistry on the polarization independent of the pore geometry. However, Equation 2.74 suggests that c_p should also be a function of the connectivity of the pore space associated with F (Börner et al., 1996). Measurements on consolidated rocks with a wide range of F confirm that the correct pore geometric parameter defining the polarization magnitude is S_{por}/F (Niu et al., 2016a), resulting in the modified definition of the specific polarizability (Weller and Slater, 2019),

$$c_p = \frac{F\sigma''}{S_{por}}. \tag{2.79}$$

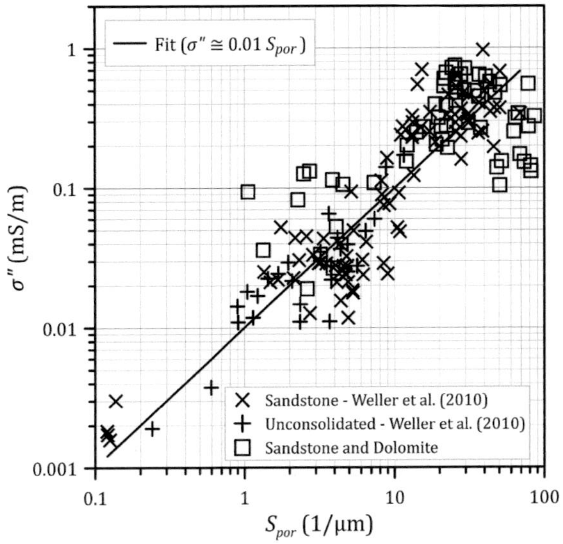

Figure 2.14 Imaginary conductivity (σ'') versus pore volume normalized surface area (S_{por}) for a broad database of samples. Sandstone (×) and unconsolidated (+) samples are from a database compiled by Weller et al. (2010a). Sandstone and dolomite (open squares) are additional samples not used in the calibration and illustrate the predictive capability of Equation 2.78.

The specific polarizability (and thus σ'' for a single soil/rock of constant pore geometry) should foremost depend on Σ_s and therefore reflect the dependence of σ'' on the EDL chemistry. Previous arguments that σ'' is essentially a scaled estimate of σ_{surf} suggest that the dependence on EDL chemistry observed for σ'' should also represent the dependence of σ_{surf} on chemistry.

Lesmes and Frye (2001) investigated the effect of pore fluid conductivity and pH on σ'' of clean sandstones, finding that σ'' increased at low salinity. However, this increase in σ'' lessened to reach a plateau at higher salinity, and even started to decrease as salinity increased further. Lesmes and Frye (2001) attributed the increase in σ'' with increasing σ_w at low salinity to the effect of the increasing charge density as sorption/ion exchange between the free fluid and EDL occurs. They attributed the reduction in the rate of increase of σ'' with σ_w, sometimes reversing to a decrease of σ'' with σ_w, to a reduction in the mobility of the charges with increasing σ_w at high salinities. The rate of increase in σ'' with σ_w at low salinities is experimentally found to approximate a power law with an exponent roughly equal to 0.5 (Weller et al., 2011) as shown in Figure 2.15a.

At higher salinities, there is a trade-off between the effects of increases in surface charge density and decreases in ionic mobility as σ_w increases. At low salinities, the effect of charge density is assumed to dominate, with the effect of ionic mobility only becoming dominant at very high salinities. Capturing both effects requires high-salinity measurements over a wide range of salinities, as illustrated by measurements on sandstone samples shown in Figure

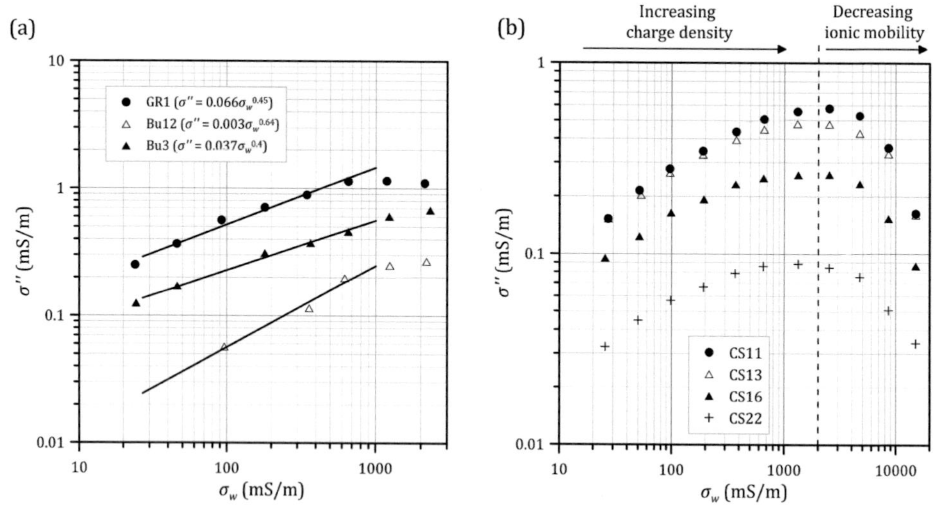

Figure 2.15 Examples of the dependence of the imaginary conductivity (σ'') on σ_w of sandstone samples: (a) low-salinity dependence of σ'' where an approximate dependence on the square root of σ_w is observed (data from Weller et al. (2011)); (b) broad-salinity dependence of σ'' on σ_w where a decrease in σ'' is observed at high salinities (data from Weller et al. (2015a)).

2.15b (Weller et al., 2015a). As σ'' is essentially a scaled version of σ_{surf}, it follows that σ_{surf} exhibits the same dependence on σ_w as σ'' (Weller et al., 2013; Niu et al., 2016b). Slight changes to the proportionality between σ'' and σ_{surf} as a function of salinity have been modelled in terms of ion exchange processes between Stern and diffuse layers (Niu et al., 2016b). The shape of the σ'' dependence on σ_w, shown in Figure 2.15b, has been reproduced in membrane polarization models (Bücker et al., 2016).

In the case of a partially saturated medium and referring back to Equation 2.41,

$$\sigma''_{[ps]} = \sigma''_{surf[ps]} \approx l\frac{1}{F}\left(\frac{2\Sigma}{\Lambda}S_w^{n-1}\right) \approx l\frac{1}{F}\left(S_{por}\Sigma S_w^{n-1}\right), \tag{2.80}$$

for the specific case $p = n - 1$.This dependence of $\sigma''_{[ps]}$ on S_w^{n-1} has been observed for measurements made on oil-bearing sandstones and found consistent with theoretical models (Vinegar and Waxman, 1984; Schmutz et al., 2010). Other studies on unconsolidated sediments have shown that saturation exponents for the imaginary conductivity are less than those for measured real conductivity, although not necessarily demonstrating the exact n–1 dependence and depending on whether samples are dried evaporatively or by pressure drainage (Ulrich and Slater, 2004). At the time of writing, more research is needed to better constrain the dependence of the surface (and hence imaginary) conductivity on saturation.

2.3.4 Frequency Dependence of the Complex Conductivity

Until this point, the dependence of the electrical properties on frequency has been ignored. The electrical properties needed to describe the resistivity method are based on DC resistivity theory, with both σ_{el} and σ_{surf} treated as DC values. Section 2.3.3 similarly considered frequency-independent IP measurements. However, the complex conductivity describing the IP phenomenon is actually frequency dependent. Extracting additional information on physicochemical properties of the porous medium from this frequency dependence is the goal of spectral induced polarization (SIP). The mechanistic description of the pore-scale polarization mechanisms causing such dispersion in the effective complex conductivity remains an active area of research. Numerous models have been proposed, and we return to them shortly, paying attention to those that integrate the concept of surface conductivity. However, we first consider simple empirical fitting models to describe the $\sigma^*(\omega)$ behaviour recorded in SIP datasets.

Box 2.7

A note on characteristic relaxation times and time constants

The acquisition of frequency-dependent IP measurements provides additional information beyond a single measure of the polarization strength based on the shape of the frequency response. The most valuable information is a measure of the characteristic relaxation time (τ), which is related (via a diffusion coefficient) to a characteristic length-scale over which charges are temporarily displaced. A characteristic time can be defined from the spectrum of frequency-dependent measurements in a number of ways. When the measurements contain a clear peak in the phase spectrum, a time $\tau_p = 1/2\pi f_p$ can be defined where f_p is the frequency where the peak occurs.

Phenomenological relaxation models (discussed in Box 2.8) are often fit to frequency-dependent IP datasets. Common models are the Debye, Cole–Cole and Davidson–Cole models, although many variants of these models have been proposed (Dias, 2000). These models contain a time constant τ_0 that depends on the form of the selected model. The Cole–Cole model (Equation 2.83) τ_0 is the inverse of the angular frequency at the peak of imaginary conductivity $f_{\sigma''}$ ($\tau_0 = 1/2\pi f_{\sigma''}$) (Tarasov and Titov, 2013). The frequency peak in the phase spectrum is given as

$$f_{p[CC]} = \frac{1}{2\pi\tau_0}(1 - \widetilde{m})^{1/2c},$$

where $\widetilde{m}$ and c are model parameters described in the main text. The popular model of Pelton et al. (1978), extensively used to fit IP data, results in a distinctly different relationship between f_p and the model parameters (Tarasov and Titov, 2013),

$$f_{p[P]} = \frac{1}{2\pi\tau_0}\frac{1}{(1 - \widetilde{m})^{1/2c}}.$$

An alternative time parameter results when more flexible curve fitting routines, such as the Debye decomposition approach (Nordsiek and Weller, 2008), are used to describe frequency-dependent datasets. In this case, a spectrum of individual relaxation times is computed by

Box 2.7 (cont.)

assigning each measurement to a Debye relaxation associated with the corresponding frequency. The weighted (by the polarization strength) average of these relaxation times is computed as an integral parameter defining the mean relaxation time τ_{mean}. Although τ_p, τ_0 and τ_{mean} should be closely related, it is important to recognize that they will not be numerically equal. Therefore, care must be taken in comparing results from different publications depending on how the time constant or mean relaxation time is defined.

The simplest description of the frequency dependence of the complex conductivity is the constant phase model CPA (Dissado and Hill, 1984; Börner et al. 1993; Börner et al., 1996),

$$\sigma^*(\omega) = \sigma_n(i\omega)^{1-q}, \tag{2.81}$$

where,

$$\varphi = \frac{\pi}{2}(1-q). \tag{2.82}$$

The frequency exponent $1-q$ describes a proportional power law increase in the real and imaginary parts of the complex conductivity. Most commonly, ω is normalized to its value at 1 s^{-1} such that σ_n' and σ_n'' are frequency-independent variables equal to σ' and σ'' measured at 1 Hz. The constant phase model is often a good approximation when soils and rocks are characterized by a wide distribution of grain or pore sizes. In such situations the frequency-independent polarization model described in Section 2.3.3 is valid. The fit of the CPA model to a dataset conforming to a broad range of relaxation times is shown in Figure 2.16.

However, in many cases the frequency dependence of σ^* does not conform to CPA behaviour. Example datasets showing such dependence (with fits to two types of models discussed later) are shown in Figure 2.16. Often the phase response contains a dispersion peak resulting from a dominant length-scale associated with the polarization mechanism (Figure 2.16, Cole–Cole entry). Such IP spectra must be described not just by the strength of the polarization but also by a characteristic relaxation time (τ) that defines the time-scale over which the polarization mechanism is strongest. In mechanistic models discussed shortly, such time-scales are related via a diffusion coefficient to the length of the mineral–fluid interface over which charges are temporarily redistributed. Polarization processes occurring over short distances have small relaxation times (observed at high frequencies), whereas polarization processes that occur over larger distances have large relaxation times (observed at low frequencies). These relaxation times have been correlated with grain size (Pelton et al., 1978; Klein and Sill, 1982) and pore size (Scott and Barker, 2003; Binley et al., 2005; Niu and Revil, 2016) of saturated porous media. In partially saturated media, the relaxation times are also a function of degree of saturation (Binley et al., 2005) and may exhibit hysteresis (i.e. different values of τ at the same saturation degree depending on drainage or imbibition) (Maineult et al., 2017b). The diffusion coefficient

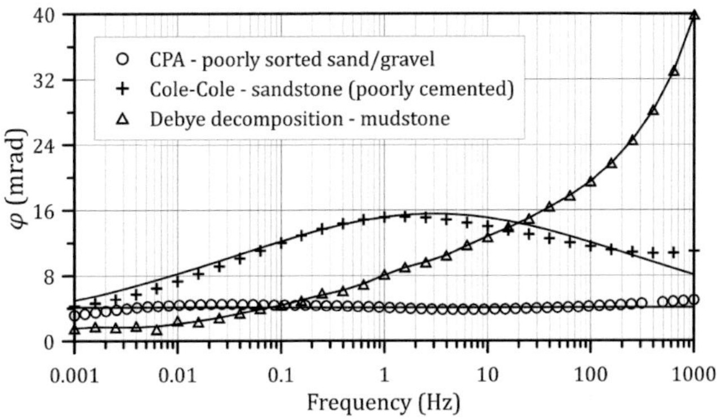

Figure 2.16 Examples of the fit of different phenomenological models for describing the frequency dependence of the phase (in conductivity space, $+\varphi$): constant phase angle (CPA) fit to a poorly sorted sand/gravel (Slater et al., 2014); Cole–Cole fit for a sandstone sample with distinct characteristic pore size (Robinson et al., 2018); Debye decomposition fit to a low-permeability mudstone sample (unpublished dataset).

adds further complexity, resulting in some dependence of relaxation time on salinity that is weak in the absence of electron conducting minerals but pronounced in the presence of such minerals (e.g. Slater et al., 2005). The diffusion coefficient also imparts a temperature dependence on the relaxation time for both non-electron conducting (Olhoeft, 1974; Zisser et al., 2010a; Bairlein et al., 2016) and electron conducting mineral (Revil et al., 2018a) polarization mechanisms. In both cases, the characteristic relaxation time decreases with temperature. The characteristic relaxation times derived from analysing frequency-dependent complex conductivity data can get confusing, as they can be derived in different ways and take different values. Box 2.7 discusses this issue.

Relaxation models originally developed to describe the broad-frequency range electrical response of dielectrics (Cole and Cole, 1941) have been extensively used to fit the observed variation in the complex electrical properties of porous materials containing a dominant relaxation peak (e.g. Pelton et al., 1978). They are useful for defining an effective (observed) time constant ($\tau_0 = 1/2\pi f_0$, where f_0 is the critical frequency) that can characterize the dominant length-scale of the polarization process occurring in the porous medium. These models are commonly represented in terms of an effective complex permittivity $\varepsilon^*(\omega)$, effective complex resistivity $\rho^*(\omega)$ or an effective complex conductivity $\sigma^*(\omega)$.

The commonly used Cole–Cole expression (Cole and Cole, 1941) is written as complex conductivity (Box 2.8):

$$\sigma^*(\omega) = \sigma_\infty - \frac{\sigma_\infty - \sigma_0}{1 + (i\omega\tau_0)^c}, \tag{2.83}$$

where σ_0 and σ_∞ are the low- and high-frequency values of the conductivity and c is the Cole–Cole exponent describing the steepness of the dispersion. The term $\sigma_\infty - \sigma_0$ quantifies the strength of the polarization. It is also known as the normalized chargeability, a term that

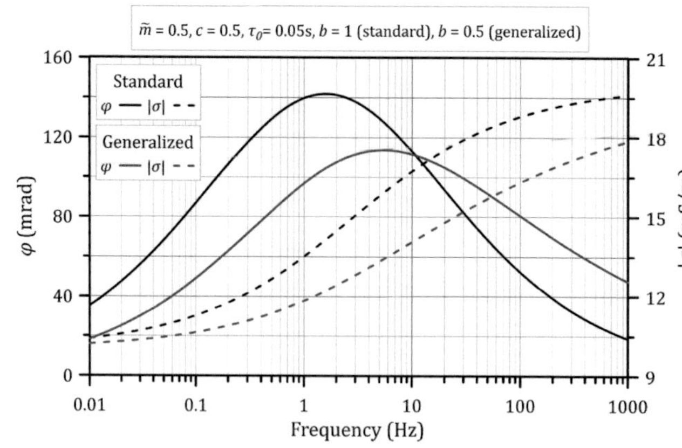

Figure 2.17 Plot of the conductivity magnitude and phase (in conductivity space, $+\varphi$) for the Cole–Cole model parameterized in terms of chargeability $\widetilde{m}$ (Box 2.8) showing the standard (symmetric) form where $b = 1$ and the generalized asymmetric form (in this case for $b = 0.5$).

is used later. Figure 2.17 shows the dependence of $|\sigma|$ and φ predicted by the Cole–Cole model. The phase peak occurs at the inflection point of the curve of $|\sigma|$ increase. SIP data are limited in terms of measured frequency range, so only estimates of $\sigma_\infty - \sigma_0$ are typically obtained from fitting the low- and high-frequency end of the range of observations. However, many porous materials display a characteristic time constant (τ_0) such that Equation 2.83 can be reliably fit to experimental data. An extension of the Cole–Cole model includes an additional fitting parameter to account for asymmetry in the dispersion around the characteristic time constant,

$$\sigma^*(\omega) = \sigma_\infty - \frac{\sigma_\infty - \sigma_0}{\left(1 + (i\omega\tau_0)^c\right)^b}. \tag{2.84}$$

The original Cole–Cole model is obtained when $b = 1$, giving a symmetric phase curve. Setting $c = 1$ in Equation 2.84 gives the Davidson and Cole (1951) model.

Although these models have no theoretical basis and primarily represent a curve-fitting procedure, the model parameters have been successfully correlated with pore geometric properties in soils and rocks. The term $\sigma_\infty - \sigma_0$, also known as the normalized chargeability m_n, is a measure of the overall polarization strength and so, similar to σ'' (the polarization strength at a specific frequency), is related to surface conductivity and the factors such as S_{por} that control it. In contrast, τ_0 has been well correlated with grain or pore size for a wide range of materials (Pelton et al., 1978). As described in Box 2.8, these models are frequently formulated in terms of the intrinsic chargeability $\widetilde{m} = (\sigma_\infty - \sigma_0)/\sigma_\infty$ (Seigel, 1959), which is well correlated with the concentration of electron conductors in a rock (Pelton et al., 1978) as discussed in Section 2.3.7.

Box 2.8
Phenomenological relaxation models

Phenomenological models for describing frequency-dependent IP data provide a convenient way to fit measured spectra to a curve that is defined by a few parameters needed to represent the observed dispersion in the complex electrical properties. These models were originally developed in materials science to describe the electrical properties of dielectrics, but have since been adopted by geophysicists as a convenient way to fit frequency-dependent IP datasets. The dielectric model of Cole and Cole (1941) forms the foundation of such phenomenological models,

$$\varepsilon^*(\omega) = \varepsilon^* + \frac{\Delta\varepsilon}{1 + (i\omega\tau_0)^{1-\alpha}},$$

where $\Delta\varepsilon = \varepsilon_0 - \varepsilon_\infty$ is known as the dielectric increment determining the strength of the polarization and τ_0 is a characteristic time constant. The fitting parameter α describes the shape of the relaxation around the characteristic frequency, specifically the steepness of the phase curve around τ_0. The Cole–Cole model has been extensively used to describe frequency-dependent IP datasets, and rewritten as a complex conductivity,

$$\sigma^*(\omega) = \sigma_\infty - \frac{\sigma_\infty - \sigma_0}{1 + (i\omega\tau_0)^c},$$

where $c = 1 - \alpha$. τ_0 is directly related to the critical frequency defined as the peak in imaginary conductivity.

In IP studies, it is popular to write the Cole–Cole model in the form

$$\sigma^*(\omega) = \sigma_0\left(1 + \frac{\widetilde{m}}{1-\widetilde{m}}\left(1 - \frac{1}{1 + (i\omega\tau_0)^c}\right)\right),$$

where $\widetilde{m}$ is the chargeability,

$$\widetilde{m} = \frac{\sigma_\infty - \sigma_0}{\sigma_\infty}.$$

The frequency of the maximum in the phase peak ($f_{p[CC]}$) is related to the time constant by (Tarasov and Titov, 2013)

$$f_{p[CC]} = \frac{1}{2\pi\tau_0}(1 - \widetilde{m})^{1/2c}.$$

Pelton et al. (1978) introduced a slight modification of the conventional Cole–Cole model formulated in terms of complex resistivity that has been popular in IP studies. Tarasov and Titov (2013) rewrote this model in terms of complex conductivity,

$$\sigma^*(\omega) = \sigma_0\left(1 + \frac{\widetilde{m}}{1-\widetilde{m}}\left(1 - \frac{1}{1 + (i\omega\tau_0)^c(1-\widetilde{m})}\right)\right).$$

In this expression, the frequency of the maximum in the phase peak ($f_{p[P]}$) is now related to the time constant by (Tarasov and Titov, 2013)

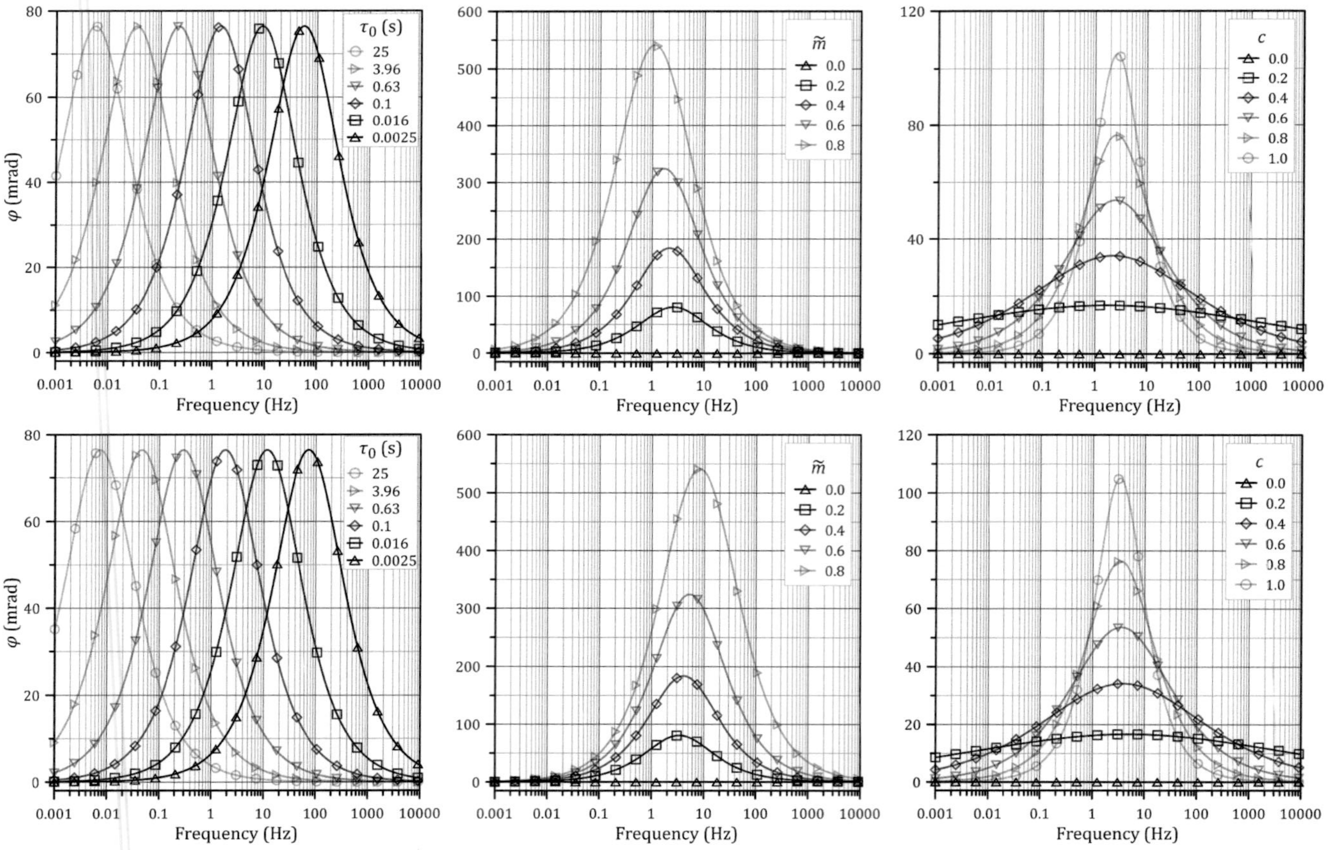

Examples of the Cole–Cole model (top row) and Pelton (bottom row) model behaviour for different values of the shape parameters (phase plotted in conductivity space, $+\varphi$). Constant parameter values are $\sigma_0 = 0.00166$ S m^{-1}, $\widetilde{m} = 0.19$, $\tau_0 = 0.05386$ s and $c = 0.8$.

Box 2.8 (cont.)

$$f_{p[P]} = \frac{1}{2\pi\tau_0} \frac{1}{(1-\widetilde{m})^{1/2c}}.$$

The figure shows how variations in the relaxation model parameters shift the phase spectra.

The top images show the Cole–Cole model, whereas the bottom images show the reformulation of the Pelton et al. (1978) model in terms of complex conductivity. The comparison highlights the fact that relaxation models can provide different parameter estimates depending on the model formulation. This will have implications for the prediction of physical and hydrogeological properties from such parameters. In the case of the Cole–Cole model, the phase peak shifts to slightly lower frequencies with increasing $\widetilde{m}$, whereas the opposite is observed for the Tarasov and Titov (2013) expression, although for small $\widetilde{m}$ the differences will be negligible. Recently, Fiandaca et al. (2018a) reformulated the Cole–Cole model with the maximum imaginary conductivity replacing $\widetilde{m}$ based on weaker equivalence between the model parameters.

Other scenarios frequently arise where $|\sigma|$ and φ measurements do not exhibit Cole–Cole model or CPA model behaviour. In some cases, it may be possible to superimpose two Cole–Cole type models to describe measurements where there are two distinct peaks in the φ spectra. However, in order to fit any arbitrarily shaped SIP datasets, Nordsiek and Weller (2008) developed an approach they named 'Debye decomposition'. This approach involves fitting the spectra to a superposition of Debye models, which represent a specific form of the Cole–Cole model for the case when $c = 1$, in terms of a continuous distribution $g(\tau)$ of relaxation times τ (Fuoss and Kirkwood, 1941; Weigand and Kemna, 2016a),

$$\sigma^*(\omega) = \sigma_\infty + (\sigma_0 - \sigma_\infty) \int_0^\infty \frac{g(\tau)}{1 + i\omega\tau} d\tau, \tag{2.85}$$

where,

$$\int_0^\infty g(\tau) = 1 . \tag{2.86}$$

The discrete form of the distribution function $g(\tau)$ for a finite set of relaxation times N is given by a combination of Dirac δ functions,

$$g(\tau) = \sum_{k=1}^{N} g_k \delta(\tau - \tau_k) \Delta\tau_k \tag{2.87}$$

where $\sum_k^M p_k = 1$ and $p_k = g_k \Delta\tau_k$ (Ustra et al., 2015). The discrete form of Equation 2.87 is then

$$\sigma^*(\omega) = \sigma_\infty + \Delta\sigma \sum_{k=1}^{N} p_k \left(\frac{1}{1 + i\omega\tau_k} \right), \tag{2.88}$$

where $\Delta\sigma = \sigma_\infty - \sigma_0$.

Integrating parameters can be used to determine a measure of the total polarization strength and the mean representative relaxation time of the polarization mechanisms. In the formulation presented in Equations 2.85–2.86, the overall polarization strength is given by $\Delta\sigma = \sigma_\infty - \sigma_0$, i.e. the normalized chargeability (m_n). Nordsiek and Weller (2008) use the weighted logarithmic values ($\ln(\tau_k)$) of the k relaxation times to define a mean relaxation time in the case of Equation 2.8 being,

$$\tau_{mean} = \exp\left(\frac{\sum_{k=1}^{N_\tau} \Delta\sigma_k \ln\tau_k}{\sum_{k=1}^{N_\tau} \Delta\sigma_k} \right). \tag{2.89}$$

Ustra et al. (2015) describe an alternative approach where a small number of dominant relaxation times are associated with specific mechanisms to adequately represent the entire spectrum. In all the models presented in this section, it is important to emphasize that they do not provide a mechanistic understanding of the frequency dependence of the IP response.

2.3.5 Mechanistic Models for the Frequency-Dependent Complex Conductivity in the Absence of Electron Conducting Particles

The extension of theoretical explanations for single-frequency IP measurements to multi-frequency IP measurements requires a physical explanation for the frequency dependence of $\sigma^*(\omega)$. Section 2.3.3 explained how the magnitude of the frequency-independent IP response is related to the total surface area of the interconnected pore space that can be polarized. Mechanistic descriptions of the frequency-dependent IP response are based on explaining how σ^* varies with frequency. Critical to developing this mechanistic understanding is the role of electromigration current densities ($\boldsymbol{J}_{mig}$) in generating ionic concentration gradients and the resulting back diffusion current densities ($\boldsymbol{J}_{diff}$) that develop in response to these concentration gradients and are responsible for the observed polarization. As introduced in Section 2.3.4, the frequency dependence is associated with a length-scale in the porous medium that defines a dominant relaxation time attributed to the polarization. The concept of spatial discontinuities that support polarization (i.e. support diffusion current densities) is inherently important in developing mechanistic understanding of the phenomenon.

Whereas measures of the polarization strength (σ'' for a single frequency, m_n for spectral measurements) are foremost related to the total interfacial surface area of the pore space that contributes to conduction and polarization, the time-dependence relates to how the total polarization response is distributed across a range of length-scales. Mechanistic models associate the distribution of these geophysical length-scales to geometric length-scales that describe the porous medium. These can be categorized into two types of length-scale models: (1) grain-size-based or pore-size-based descriptions of the complex surface conductivity; (2) pore-throat-based models. These models differ not only in how they define the length-scale of the porous medium controlling polarization but also in how they attribute the surface conduction and polarization mechanisms to the diffuse versus the fixed (Stern) layers of the EDL forming at the mineral–fluid interface. It is common to attribute the polarization foremost to the Stern layer (Schwarz, 1962; Leroy et al., 2008), as the diffuse layer is assumed to be a continuous phase in the same way as the interconnected pore space and thus only able to support electromigration currents (Figure 2.18). Other models consider polarization to occur in the diffuse layer (Dukhin and Shilov, 1974). Some authors have considered both Stern and diffuse layer contributions (Lima and Sharma, 1992; Lesmes and Morgan, 2001; Bücker et al., 2019). Others advocate for large polarization enhancements due to fluxes of charge between the EDL and the bulk electrolyte, normal to the surface on either side of a particle, resulting in diffuse charge clouds in the electrolyte (Chelidze and Gueguen, 1999).

2.3.5.1 Grain- and Pore-Size-Based Models of the Surface Conductivity

The grain-size-based models define the $\sigma^*(\omega)$ response of a single mineral grain and then compute the integrated $\sigma^*(\omega)$ response of the material through a convolution with the particle-size distribution (Lesmes and Morgan, 2001; Leroy et al., 2008). They represent an extension of the surface conductivity theory described in Section 2.3.3 whereby the frequency dependence of the complex surface conductivity is associated with a length-scale related to the grain size. Lesmes and Morgan (2001) argued that the polarization of the Stern layer around a mineral grain according to the model of Schwarz (1962) (Box 2.9) is much stronger than the polarization of the diffuse layer according to the model of Fixman (1980). Grain-based polarization models are therefore based upon the model of Schwarz (1962) and its extension by Schurr (1964) to account for a diffuse layer that contributes a surface conductivity but does not polarize as the charges in the diffuse layer exchange readily with the electrolyte (de Lima and Sharma, 1992). The polarization is exclusively attributed to diffusion current densities ($\boldsymbol{J}_{diff}$) in the Stern layer, whereas ions in both the Stern and the diffuse layers contribute to electromigration current densities ($\boldsymbol{J}_{mig}$) (Figure 2.18). Such models are generalized to account for a distribution of grain size through a convolution operation that relates the grain-volume distribution to the relaxation-time distribution (Lesmes and Morgan, 2001). These models generally assume a smooth mineral grain. Grain-surface roughness may result in an additional, smaller length-scale that manifests itself as a polarization enhancement at high frequencies (Lesmes and Morgan, 2001; Leroy et al., 2008). The models also assume a fully saturated medium.

De Lima and Sharma (1992) provide an expression for the complex surface conductivity of a mineral grain with a single characteristic time constant (τ_o), being associated with a single grain size (diameter d_0) coated with a conductive, fixed layer that can be written as (Leroy et al., 2008)

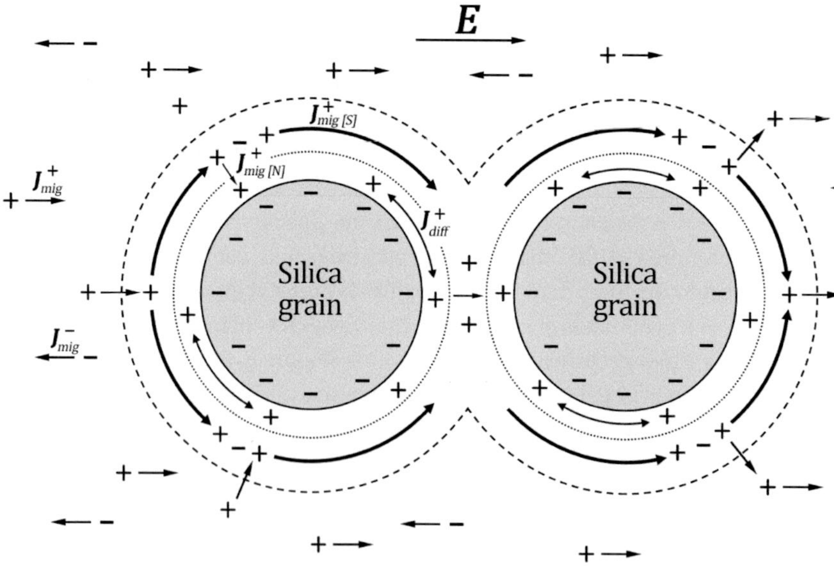

Figure 2.18 Schematic of the polarization model of Leroy et al. (2008) based on silica particles in a 1:1 electrolyte. The applied electric field $\boldsymbol{E}$ drives migration and polarization current densities. Electromigration of ions generates the migration current densities in the free electrolyte ($\boldsymbol{J}^{+}_{mig}$), and in the diffuse part of the double layer ($\boldsymbol{J}^{+}_{mig[S]}$, $\boldsymbol{J}^{+}_{mig[N]}$). Back diffusion of ions in the Stern layer generates diffusion current densities $\boldsymbol{J}^{+}_{diff}$ and a resulting polarization. Ions in the diffuse layer are assumed not to polarize because the diffuse layer is continuous between grains, whereas the Stern layer is discontinuous. Figure reproduced from Okay et al. (2014).

$$\sigma^{*}_{surf[d_0]} = \frac{4}{d_0}\left(\Sigma^{d} + \Sigma^{s}\right) - \frac{4}{d_0}\frac{\Sigma^{s}}{1 + i\omega\tau_0}\,, \tag{2.90}$$

where Σ^{d} is the surface conductance of the diffuse layer and Σ^{s} is the surface conductance of the Stern layer. Equation 2.90 has the form of a Debye relaxation with a characteristic τ_0 that can be expressed as

$$\tau_0 = \frac{d_0^2}{8D_{+}}\,, \tag{2.91}$$

where D_{+} is the diffusion coefficient (in m^2/s) of ions in the Stern layer (Leroy et al., 2008; Revil and Florsch, 2010). The diffusion coefficient is related to the mobility of ions in the Stern layer β_{+} ($m^2\ s^{-1}\ V^{-1}$) via the Nernst–Einstein relationship,

$$D_{+} = k_b T \beta_{+} / |q_{(+)}|\,, \tag{2.92}$$

where q_{+} is the charge of the counterions (in C).

The low-frequency ($\sigma^0_{surf[d_0]}$) and high-frequency ($\sigma^\infty_{surf[d_0]}$) asymptotic values of the surface conductivity are (de Lima and Sharma, 1992; Revil and Florsch, 2010)

$$\sigma^0_{surf[d_0]} = \frac{4}{d_0}\Sigma^d, \tag{2.93}$$

$$\sigma^\infty_{surf[d_0]} = \frac{4}{d_0}\left(\Sigma^d + \Sigma^s\right). \tag{2.94}$$

Equation 2.90 then becomes (Revil and Florsch, 2010)

$$\sigma^*_{surf[d_0]} = \sigma^\infty_{surf} + \frac{\sigma^0_{surf[d_0]} - \sigma^\infty_{surf[d_0]}}{1 + i\omega\tau_0}. \tag{2.95}$$

A significant and potentially powerful development in this mechanistic approach to the description of frequency-dependent IP measurements is the coupling of this expression of the complex surface conductivity of a mineral grain to an electrochemical model for the electrical double (or triple) layer (Leroy et al., 2008; Leroy and Revil, 2009). This approach provides a rigorous electrochemical description of Σ^d and Σ^s, yielding insights into how complex conductivity spectra depend on salinity, pH and valence. Leroy et al. (2008) used an electrical triple layer (ETL) model description for silica in contact with a binary symmetric electrolyte (e.g. NaCl) to model the complex conductivity response of glass beads. Leroy and Revil (2009) modified the approach to consider the complex conductivity of clay minerals through integration of an ETL model for kaolinite developed by Leroy and Revil (2004). Incorporation of the ETL model describes the distribution of the counterions at the mineral/fluid surface, specifically through a partitioning coefficient representing the fraction of the counterions located in the Stern layer versus all counterions (being a critical parameter in this model). Revil and Skold (2011) used this approach to model the salinity dependence of imaginary conductivity recorded for sandstones with an analytical solution for the ETL model parameterized in terms of total site density of surface charges, pH and the sorption coefficient for cations in the Stern layer.

The coupling of expressions such as Equation 2.90 with an electrochemical model of the double or triple layer also allows the effect of sorption processes on σ^* measurements to be understood (Figure 2.19). Vaudelet et al. (2011) include a complexation model for the sorption of copper and sodium ions onto silica in their model for the complex surface conductivity of a mineral grain, applying the model to σ^* measurements acquired during sorption of copper and sodium ions in sands. Further implementation of such a coupled Stern layer polarization ETL model is likely to yield new insights into how IP measurements can be used to probe electrochemical processes in soils and rocks.

Once an expression for the complex surface conductivity of a single grain is defined (e.g. Equation 2.95), the total complex surface conductivity of a granular porous medium is subsequently determined from a convolution of this expression for a single grain size with the grain-size distribution of the sample (Lesmes and Morgan, 2001; Leroy et al., 2008),

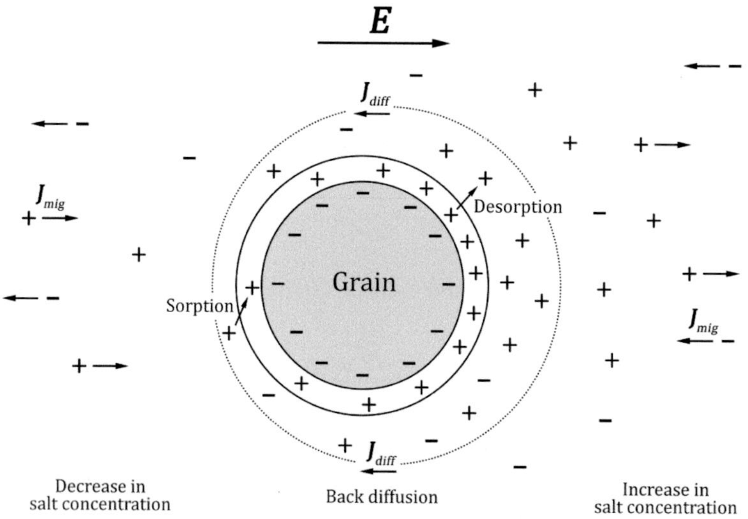

Figure 2.19 Schematic polarization of the electrical double layer of a mineral grain in response to an electric field ($\boldsymbol{E}$). Diffusion currents ($\boldsymbol{J}_{diff}$) (here shown to occur in both the Stern and diffuse layers) result from the developed charge concentration gradients, with sorption/desorption occurring on long time-scales (modified from Revil and Florsch, 2010).

$$\sigma^*_{surf} = \sigma^{\infty}_{surf[d]} + \left(\sigma^0_{surf[d]} - \sigma^{\infty}_{surf[d]}\right) \int_0^{\infty} \frac{g(\tau)}{1 + i\omega\tau} d\tau, \tag{2.96}$$

$$\int_0^{\infty} g(\tau) d\tau = 1. \tag{2.97}$$

The full expression for the complex conductivity of the porous medium can then be obtained by assuming parallel surface and electrolytic conduction paths and adding the purely real electrolytic conductivity contribution as per Equation 2.66. Alternatively, mixing models discussed in Section 2.2.4.4.2 can be used to embed the complex surface conductivity associated with the distribution of grain sizes into an effective medium for the entire rock or soil sample. For example, using the Bruggeman–Hanei–Sen (BHS) effective medium model (Bruggeman, 1935; Hanai, 1960; Sen et al. 1981) gives (Leroy et al., 2008),

$$\sigma^* = \sigma_w \phi^m \left(\frac{\left(1 - \sigma^*_{surf}/\sigma_w\right)}{1 - \sigma^*_{surf}/\sigma^*} \right)^m. \tag{2.98}$$

Modifications of this polarization model have been based on the realization that the grain size may not be the optimal geometrical length-scale of a porous material to link to the

Box 2.9

The Schwarz (1962) model for the polarization of colloidal suspensions

Schwarz (1962) developed a theory for the polarization of the counterions in the EDL that balances the fixed layer charges on the surface of a charged spherical particle in a relatively conductive electrolytic solution (i.e. a colloidal suspension). The theory assumes that, upon application of an electric field, the counterions (charges) will only migrate tangential to the surface of the spherical particle. The model has the form of a Debye relaxation, which in terms of complex conductivity is

$$\sigma^*(\omega) = \frac{\sigma_\infty - \sigma_0}{1 + (i\omega\tau_0)}$$

and

$$\tau_0 = \frac{R^2}{2\beta k_B T} = \frac{R^2}{2D},$$

where R is the radius of the sphere, D is the diffusion coefficient of the counterions and β is the mobility of the counterions. Schurr (1964) extended the model to include the contribution of a DC surface conductivity.

Although developed for a colloidal suspension, the expression linking the time constant τ_0 to R and D has been extensively adopted to explain the frequency-dependent IP response of porous media. Lesmes and Morgan (2001) embedded the Schwarz theory for a single particle within an effective medium theory to develop a model for predicting the complex conductivity of sedimentary rocks based on a distribution of grain sizes. They assumed that the counterions in the Stern layer dominate the EDL polarization. Leroy et al. (2008) adopted the same approach to model the IP response of glass beads, extending it to incorporate a triple-layer electrochemical model and to account for the Maxwell–Wagner polarization (Box 2.5). However, experimental datasets have indicated that it may actually be the pore size, rather than the grain size, that controls the relaxation time in porous media (Scott and Barker, 2003). After all, a rock is quite different from a colloidal suspension. Researchers have since assumed that the expression for the time constant appearing in the Schwarz model can be transferred to represent the polarization of a pore rather than a grain. For example, Revil et al. (2014) related a time constant τ_0 (in this case found from fitting a Cole–Cole model) to a dynamic pore throat radius (Λ),

$$\tau_0 = \frac{\Lambda^2}{2D_+},$$

with the diffusion coefficient (D_+) specifically related to the ions in the Stern layer.

In general, we assume that the time constant of a porous medium is related to the square of the dominant interfacial length-scale (l_e) over which the counterions migrate in response to an applied field and a diffusion coefficient for the counterions at this interface,

$$\tau_0 \propto \frac{l_e^2}{D}.$$

geophysical length-scale controlling τ_0. Measurements on consolidated rocks indicate that the distribution of relaxation times might be better linked to the pore-size distribution, and that the dominant relaxation time might be associated with a dominant pore size (Scott and Barker, 2003). In such models, Equation 2.91 is expressed in terms of a characteristic pore radius or length (a_0) instead of a characteristic grain diameter,

$$\tau_0 = \frac{a_0^2}{2D_+}. \tag{2.99}$$

In the same manner as described for the grain-size distribution by Equations 2.96–2.97, σ^*_{surf} of the porous medium can be determined by a convolution of the expression for the complex surface conductivity of a single pore with the pore-size distribution (Niu and Revil, 2016).

Whether expressed as a grain size or a pore size, these mechanistic descriptions of the complex surface conductivity provide a direct link between the relaxation time (or time constant) and the geometry of the porous medium. However, Equations 2.91–2.92 highlight the fact that the relaxation times depend not just on the pore geometry but also on the electrochemical properties of the EDL through the diffusion coefficient (D_+). The diffusion coefficient is directly dependent on the mobility of ions in the Stern layer, β_+ (Equation 2.92). Revil (2012) argues that, for clay minerals, β_+ is 350 times lower than the mobility of ions in the bulk pore water, whereas, for silica, β_+ is approximately equal to the ion mobility in the bulk pore water. Specific values of D_+ for clay-rich ($D_+ = 3.8 \times 10^{-12}$ m^2/s) versus clay-free ($D_+ = 1.3 \times 10^{-9}$ m^2/s) materials have hence been proposed in order to constrain the interpretation of relaxation times in terms of geometric properties of the porous medium (Revil, 2012, 2013). However, experimental observations indicate that diffusion coefficients may in fact vary over many orders of magnitude due to lithological variation, complicating efforts to determine grain diameters or pore sizes from relaxation times (Kruschwitz et al., 2010; Weller et al., 2016). Furthermore, the diffusion coefficients are temperature-dependent, resulting in a decrease in relaxation time with increasing temperature (Zisser et al., 2010a).

2.3.5.2 Pore-Throat-Based Models

This second type of mechanistic model is based on an extension of the classical 'membrane polarization effect' discussed in Section 2.3.2, where the polarization results from the presence of ion-selective zones along the current pathway and the build-up of salt concentration gradients when an electric field is applied (Marshall and Madden, 1959). In this generalization from the original model where clay-rich zones result in ion selectivity, constrictions in the interconnected pore space resulting from small pore throats, narrow passages and/or clay minerals are deemed responsible for resulting in ion-selective zones with unequal mobilities of anions and cations (Box 2.10). The modelling approach was largely developed by the Russian IP community, with a time domain solution for the IP response in a capillary medium developed from the Marshall and Madden model by Kormiltsev (1963). Titov et al. (2002) used these concepts to develop a short narrow pores (SNP) model to explain time domain IP data. Hallbauer-Zadorozhnaya et al. (2015)

developed a mathematical model based on membrane polarization to estimate the pore-size distribution of rocks. Bücker and Hördt (2013b) developed an analytical model in cylindrical coordinates to describe the SIP response in terms of a model of short and narrow pores. Here the ion-selective properties are specifically associated with the EDL lining pores. When the EDL thickness is large relative to the pore radius, the double layer behaves as an ion (usually cation)-selective zone. These ion-selective zones have different transport numbers for cations and anions, resulting in local concentration gradients when an electric field is applied. A comparison of the classic membrane polarization model and this modified model is shown in Figure 2.20.

In the Bücker and Hördt (2013b) model, the frequency response is related to two time constants associated with two ($i = 2$) pore lengths (L_1 and L_2)

$$\tau_i = \frac{L_i^2}{8D_{pi}t_{ni}}, \tag{2.100}$$

where t_{ni} represents two transference numbers defined by differences in the mobility of anions and cations (a measure of the anion selectivity) in each pore. D_{pi} is the diffusion coefficient for each pore,

$$D_{p(i)} = \frac{\beta_p \bar{c}_{(p)i} k_b T}{e}, \tag{2.101}$$

where $\bar{c}_{(p)i}$ are normalized integrated cation concentrations in each pore and β_p is the mobility of the cations. Thus, similar to the surface conductivity-based model, this model

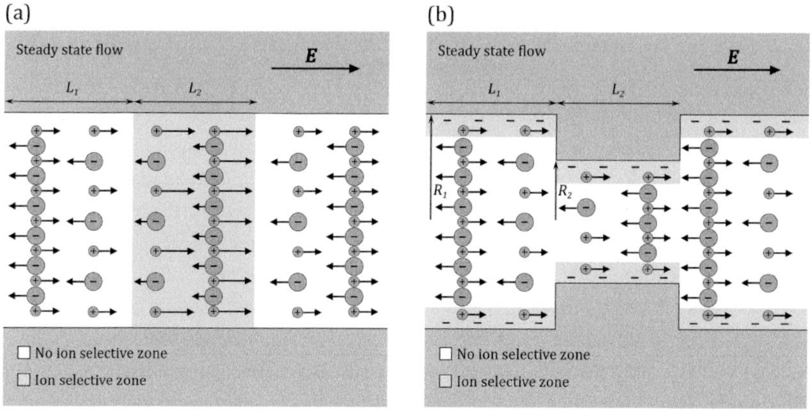

Figure 2.20 Comparison of the membrane polarization models of (a) Marshall and Madden (1959), where ion selectivity is associated with varying ion mobility in the free electrolyte of the two pore types; and (b) Bücker and Hördt (2013a), where the selectivity is associated with the double layer properties and becomes significant when the EDL thickness is large relative to the pore throat radius. Arrow lengths depict relative ion mobility. Modified from Leroy et al. (2019).

also represents the frequency dependence of the complex conductivity in terms of a geometric length-scale (L_i) and a diffusion coefficient (D_p) describing the electrochemistry of the polarization depending on both the fluid chemistry and temperature.

2.3.6 Estimation of Hydraulic Properties from Electrical Properties in the Absence of Electron Conducting Particles

By now, it should be clear that the electrical properties of porous media are strongly dependent on the geometric properties of the pore space. These are the same geometric properties that control the flow of fluids through porous media. Therefore, it comes as no surprise that a large body of research has been dedicated to the effort of determining hydraulic properties from electrical measurements (e.g. Katz and Thompson, 1986; Bernabe, 1995). The equations presented in this chapter have highlighted the possibility of estimating porosity, grain size and surface area from electrical measurements. A logical extension of this work is the electrical estimation of the permeability (k), or hydraulic conductivity (K_h), of soils and rocks.

In a homogenous, isotropic porous material, K_h (in m/s) defines the relationship between the fluid flux $\boldsymbol{q}$ (in m/s) and the gradient of the hydraulic head (h),

$$\boldsymbol{q} = -K_\mathrm{h}\nabla h, \tag{2.102}$$

where the negative sign indicates that the direction of fluid flow is in the direction of decreasing head. The hydraulic conductivity is defined by both the geometric properties of the porous medium and the fluid properties,

Box 2.10

Membrane polarization

'Membrane polarization' traditionally refers to an alternative IP mechanism to the tangential migration of counterions in the EDL around a particle. This alternative mechanism is attributed to the formation of ion-selective zones (associated with variations in ion concentration) in the interconnected pore space. Marshall and Madden (1959) introduced the theory based on a sequence of zones where variations in ionic mobilities between these pores drive ion concentration gradients. They envisaged the formation of cation-selective zones, where the electric current is primarily transported by cations. These zones of different transport numbers could form at pore constrictions or where clay particles locally increase ionic concentration. Diffusion currents associated with the ionic concentration gradients around these zones oppose the currents driven by an applied electric field, resulting in a frequency-dependent conductivity. The concept has been extended to polarization of pore throats by similar ion selectivity effects (e.g. Titov et al., 2002). Revil et al. (2014) propose a more generalized 'membrane polarization effect' arising from the differences in ion concentration in the free electrolyte that are caused by tangential charge movement in the EDL and any sorption/desorption processes occurring between ions in the EDL and in the free electrolyte. As noted in the main text, the membrane polarization models, similar to the EDL models, result in a frequency dependence of the IP effect that can be described by a characteristic length-scale and a diffusion coefficient.

Box 2.10 (cont.)

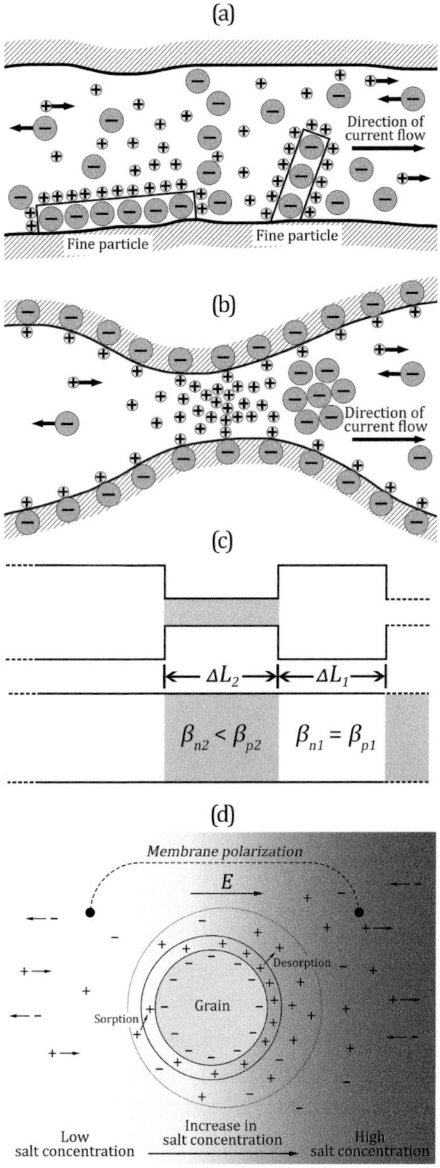

Schematic illustrations of (a) clays lining pore throats; and (b) pore constrictions, both resulting in locally increased cation concentrations and reductions in mobility; (c) conceptual models for sequences of pore throats of varying mobility with constrictions where anion mobility (β_n) is less than cation mobility (β_p); (d) postulated membrane polarization mechanism due to salt concentration gradients in the electrolyte resulting from EDL polarization of a mineral grain (a modified from Ward and Fraser, 1967; d modified from Revil et al., 2014).

$$K_h = \frac{k\rho_g g}{\eta}, \tag{2.103}$$

where ρ_g is the fluid density (kg m^{-3}), g is the gravitational acceleration (m s^{-2}) and η is the dynamic viscosity (N s/m^2).

Geometric models to determine k were initially based on representing the porous material as a bundle of capillary tubes via the Kozeny–Carman relation (Carman, 1939; Bear, 1972). Based on geometric considerations and applying the Hagen–Poiseuille equation for flow through a long cylindrical pipe, k can be related to a measure of the effective pore radius r (Pape et al., 1999),

$$k = \frac{\phi r^2}{8T_h}, \tag{2.104}$$

where T_h is the hydraulic tortuosity of the capillary tubes. Assuming that the electrical tortuosity (T_e) is the same as the hydraulic tortuosity, ϕ/T_h can be replaced with F (Equation 2.24). Recognizing that $2/S_{por}$ equals the hydraulic radius for a capillary bundle, Equation 2.104 can be reformulated:

$$k = \frac{1}{2FS_{por}^2}. \tag{2.105}$$

This model assumes cylindrical capillaries with smooth surfaces. Pape et al. (1987) accounted for the fractal nature of the internal surface area of sedimentary rocks, resulting in a modified Kozeny–Carman-based model for permeability prediction known as the PaRiS equation,

$$k_{PaRiS} = \frac{a_{PaRiS}}{FS_{por}^{3.1}}, \tag{2.106}$$

where a_{PaRiS}= 475, when permeability is in units of 10^{-15} m^2 and S_{por} is in units of 1/μm. The larger exponent on S_{por} than predicted by theoretical considerations for a capillary bundle is consistent with experimental observations (Börner et al., 1996).

Another model for k prediction results from application of percolation theory to flow in porous media (Katz and Thompson, 1986). Percolation models describe a random system with a broad distribution of conductances where transport is dominated by those conductances that exceed a threshold value (Ambegaokar et al., 1971). This characteristic threshold conductance defines the largest conductance such that the set of conductances forms an infinite connected cluster. Application of this percolation scheme for the prediction of permeability of a porous medium is based on defining a characteristic length-scale l_c that defines the permeability (Katz and Thompson, 1986),

$$k = \frac{l_c^2}{cF}, \tag{2.107}$$

where $c = 226$. The characteristic length-scale l_c is related via a scaling constant to the dynamical length-scale (Λ) introduced in Section 2.2.4.2 to describe the surface conductivity and that can be estimated from mercury injection porosimetry (Katz and Thompson, 1987). This results in the following permeability equation,

$$k = \frac{\Lambda^2}{8F}. \tag{2.108}$$

As Λ defines a pore radius (Section 2.2.4.2), Equation 2.108 is equivalent to Equation 2.104.

Extensive efforts have been made to establish empirical relationships between resistivity and permeability (e.g. Huntley, 1986). However, such relationships fail because the form of the electrical resistivity/permeability relationship varies depending on whether surface conduction or electrolytic conduction dominates (Purvance and Andricevic, 2000). When surface conduction dominates, electrical conductivity tends to be inversely related to the permeability as finer grained materials decrease the hydraulic conductivity but increase the electrical conductivity. When electrolytic conduction dominates, electrical conductivity tends to be directly related to k as both electrical and hydraulic conductivity increase with increasing porosity.

Equations 2.104–2.108 show that classical estimation of k requires a measure of a geometric length-scale along with the tortuosity (related to F). The Kozeny–Carman-based model uses $1/S_{por}$ as a measurable length-scale, whereas the percolation threshold-based model uses a characteristic pore radius from mercury porosimetry. These geometric length-scales are properties that cannot be directly measured in situ but require destructive laboratory analysis of samples. Estimation of k from resistivity and IP measurements is based on an equivalent geophysical length-scale replacing the geometrical length-scale and the simultaneous estimation of a proxy of F (Börner et al., 1996; Slater and Lesmes, 2002b; Robinson et al., 2018).

Electrical equivalents of the Kozeny–Carman model are based on substitution of c_p/σ'' (a geophysical length-scale) for $1/S_{por}$ (a geometrical length-scale) that is justified given the strong empirical relationship observed between σ'' and S_{por} (Weller et al., 2010a) as shown in Figure 2.14. Imaginary conductivity-based empirical k prediction models then take a generalized form

$$k = \frac{a}{F^b(\sigma'')^d}, \tag{2.109}$$

where a, b and d are fitting constants (Börner et al., 1996; Weller et al., 2015b). The model is dependent on the fluid salinity as this exerts a control on the specific polarizability. Weller et al. (2015b) fit such a model to an extensive database of consolidated sedimentary rock samples saturated with a fluid of constant fluid conductivity (constant c_p), finding $b = 5.35$

and $d = 0.66$ with σ'' measured at 1 Hz. Weller et al. (2015b) also fit Equation 2.109 to a database of unconsolidated sediments, finding, $b = 1.12$ and $d = 2.27$ (again constant c_p and σ'' measured at 1 Hz). The large difference between these power law exponents for the sandstone database versus the unconsolidated sample database highlights that such empirical models will only provide reliable predictive estimates of k when applied to materials similar to the calibration dataset.

The use of Equation 2.109 is based on a single frequency measurement of σ''. The frequency dependence of σ'' can therefore affect the prediction. The model can be modified to utilize 'global' estimates of the total polarization strength derived from fitting SIP data to relaxation type models described in Section 2.3.4. For example, Weller et al. (2015b) gave the following equivalent k prediction model using the normalized chargeability ($m_n = \sigma_\infty - \sigma_0$, Equation 2.83 and supporting text) derived from Debye decomposition fitting of the spectra (Section 2.3.4) in place of a single frequency σ'',

$$k = \frac{a}{F^b (m_n)^d}. \tag{2.110}$$

For the integrated database of consolidated rock and unconsolidated samples, $b = 3.68$ and $d = 1.19$. A comparison of the true k values to those predicted with Equation 2.110 is shown in Figure 2.21. Although the form of the model describes the data relatively well, some outliers fall beyond the lines representing 1 order of magnitude deviation from the true k.

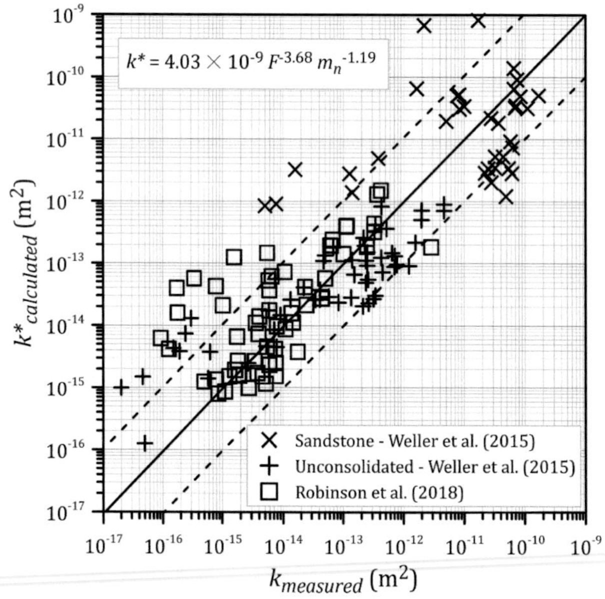

Figure 2.21 Comparison of predicted versus measured permeability for an empirical model calibrated on an extensive database described in Weller et al. (2015b). The data shown as open squares were not used in the calibration and indicate the predictive capability of the model.

These empirical predictive equations result in much larger exponents on F (e.g. 3.68 in Equation 2.110) than predicted by the geometric models (i.e. 1 in Equations 2.105–2.108) (Weller et al., 2015b). Weller and Slater (2019) present theoretical evidence for a larger value of the coefficient b than predicted by Equation 2.110 in the case of porous materials where the tortuosity (hence F) exerts a strong influence on the imaginary conductivity that cannot be neglected (Equation 2.74). Indeed, F may exert the overriding control on k in sedimentary rocks where secondary processes (e.g. cementation) result in high-tortuosity current flow paths. In contrast, σ'' appears to exert a much stronger control on k in unconsolidated sediments and clean sandstones where formation factors are typically low and only vary over a relatively limited range. Recent field-scale studies support the good predictive capability of such models when applied to unconsolidated sediments.

Electrical equivalents of the Katz and Thompson (1986) model (Equation 2.108) are based on the substitution of l_c with a geophysical length-scale defined from the product of the time constant and the diffusion coefficient. The approach is motivated by strong reported correlations between the relaxation time and permeability (Binley et al., 2005; Tong et al., 2006). A mechanistic formulation to this approach follows from the Schwartz model (Box 2.9) where a geophysically defined pore length (or radius) Λ_{IP} is given by,

$$\Lambda_{IP} = \sqrt{2D_+\tau_0}. \tag{2.111}$$

Substitution of Equation 2.111 into Equation 2.108 results in the following time-constant-based mechanistic model for permeability estimation (Revil et al., 2015b),

$$k = \frac{\Lambda_{IP}^2}{8F} = \frac{\tau_0 D_+}{4F} . \tag{2.112}$$

This approach is elegant and at first glance does not involve empirical calibration of fitting parameters. However, one limitation of applying this model regards uncertainty in the values for the diffusion coefficients. End values for D_+ have been proposed: 1.3×10^{-9} m^2/s for clean sands and 3.8×10^{-12} m^2/s for clays (Revil, 2012, 2013). However, estimated diffusion coefficients show a wide range of values, particularly in the presence of low-permeability materials (Kruschwitz et al., 2010). Weller et al. (2016) reported estimated diffusion coefficients spanning 6 orders of magnitude, with values for high surface area materials orders of magnitude lower than the value of 3.8×10^{-12} m^2/s proposed for clays. This uncertainty will restrict the accuracy of the permeability estimate when a wide range of lithological variability exists in the subsurface.

Both geophysical length-scales provide a possible path forward for non-invasive estimation of the permeability. Laboratory studies show that the geophysical length-scales may provide similar predictive accuracy to the geometric length-scales (Osterman et al., 2016; Robinson et al., 2018; Weller and Slater, 2019). Laboratory calibrations of k prediction equations have highlighted the uncertainty that will result from variations in the electrochemical parameters that appear in the geophysical length-scales. However, reported variations in c_p approximately span only about an order of magnitude, whereas variations in D_+ appear to span 5–6 orders of magnitude (Weller et al., 2016). Another uncertainty results from the differing dependence of the complex conductivity and k on the organization of clays (e.g. dispersed versus aggregates) (Osterman et al., 2019), something that is not considered in the k prediction models discussed here.

Field-scale permeability estimates that are accurate to within an order of magnitude might be achievable using the relationships described in this section. The σ''-based approach is more readily transferable to the field-scale as it uses a single-frequency measure of the polarization strength that is relatively straightforward to obtain from field IP datasets. Fiandaca et al. (2018b) used Equation 2.109 with the fitting parameters for unconsolidated sediments reported in Weller et al. (2015b) to predict permeability from IP logging-while-drilling datasets. Maurya et al. (2018) used the same equation to predict permeability from 2D IP imaging measurements in unconsolidated sediments. In contrast, the τ-based model requires that IP measurements are made over a sufficiently wide range needed to capture the relaxation peak in the spectra. As discussed in Chapter 3, this is a challenging task using currently available field instrumentation.

2.3.7 Polarization of Soils and Rocks Containing Electron Conducting Particles

The prior treatment of the IP effect neglected the role of electron conducting particles. In this case, a unifying framework for the joint interpretation of resistivity and IP datasets is provided through the concept of a complex surface conductivity that is described by the EDL at the mineral–fluid interface, along with the Archie-type electrolytic conduction via the pore space. As discussed in the introduction to this chapter, the origins of the IP effect in fact lie in mineral exploration, where the strong induced dipole moment associated with electron conducting particles in an electrolyte results in a very large polarization magnitude relative to insulating minerals. The effect is commonly described as 'electrode polarization' as the electron conducting grains have historically been considered to act like electrodes throughout a rock. To illustrate this effect, Figure 2.22 shows IP measurements (as phase) on the same sand material mixed with 4% (by volume) of clay mineral (kaolinite) versus the same volume concentration of magnetite, an electron conducting mineral. The response for a pure sand sample is also shown. In each case the porosity is 38 ± 2%. The polarization enhancement for the clay mineral (relative to the sand) results from the stronger EDL polarization associated with the larger surface area of the clay. The much stronger polarization enhancement for the magnetite mineral results from the presence of an electron conductor and how this electron conductor affects the distribution of ionic charges (both in the EDL and in the bulk electrolyte) around it under the application of an electric field.

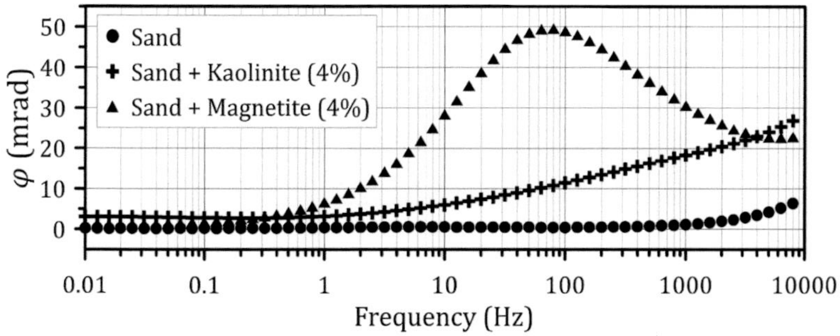

Figure 2.22 Comparison of the phase (in conductivity space, $+\varphi$) spectrum of an identical sand mixed with 4% (by volume) magnetite and 4% (by volume) kaolinite clay. Response of the same pure sand also shown.

Two different types of conduction occur in a porous medium containing electron conducting particles: (1) electron conduction inside the mineral, where electrons are the charge carriers in the case of conductors and both electrons and holes are the charge carriers for semiconductors, e.g. pyrite (Revil et al., 2015a, c); (2) electrolytic (or ionic) conduction, where ions are the charge carriers in the electrolyte. In the presence of an external electric field, polarization mechanisms develop at the interface between the electrolyte and the electron conducting particle; these act as a barrier for the ions in the electrolyte and for electrons in the particle.

Electrochemistry has played an important role in understanding the polarization response at the interface of an electron conductor and the electrolyte. Inspired by electrochemistry studies of metals in electrolytes, Seigel (1959) attributed the IP response to an 'overvoltage' effect. Equivalent circuit models were popular for providing a simple representation of the electrical properties of an interface representing an iron mineral (Angoran and Madden, 1977). One example is Randles' circuit (Randles, 1947) describing a single electrode in contact with a fluid (Figure 2.23). Each component of the circuit represents a part of the double layer at the electrode. R_s is the resistance due to the fluid in the pores, whereas the other three terms relate to the double layer forming at the metal interface. These are the dielectric capacitance of the EDL (C_{dl}), the Warburg impedance (W) representing the leaky capacitance due to diffusion-controlled processes (discussed in Section 2.3.2) and the charge transfer resistance (R_{ct}) attributed to the finite rate of electron transfer at the electrode surface. However, unlike in the case of electrochemistry-based measurements of a single electrode, charge is not necessarily transferred across the interface of electron conducting particles distributed throughout a rock. Some IP models described next inherently assume that this charge-transfer process occurs. Others consider it possible

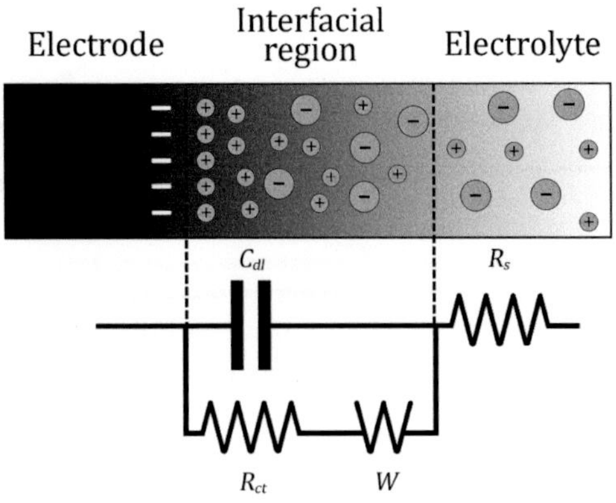

Figure 2.23 Randles' circuit used to represent the impedance of an electrode–electrolyte interface across which charge is transferred. R_s – Resistance due to fluid in pores; C_{dl} – double-layer capacitance; R_{ct} – charge transfer resistance; W – Warburg impedance.

in the presence of ions that promote redox reactions at the particle surface, but not a requirement to generate an IP effect.

Because of the complexity of the polarization mechanisms in the presence of electron conducting particles, IP research in support of mineral exploration during the 1970s and 1980s largely focused on either fitting phenomenological relaxation models to the observations or using equivalent electrical circuits to represent electrical current transport through an ore body. Figure 2.24 is an early well-cited example of this approach from Pelton et al. (1978) where electron conducting particles are envisaged to block electrolytic current conduction paths. The equivalent circuit model includes a complex impedance ($i\omega X^{-c}$) to simulate the lossy capacitance ($c < 1$) of the electron conductor-ionic interface. The case of $c = 0.5$ gives the Warburg impedance. Given that the electron conducting particles were assumed to block pore passages, and that the host matrix is an insulator, this conceptual model implies that charge is transported across the electron conductor–fluid interface via favourable redox reactions. The IP effect was attributed to the very large polarization impedance associated with the charge transfer from electrolytic to electron conduction.

Pelton et al. (1978) expressed their conceptual model (Box 2.8) in terms of a modified Cole–Cole expression for the complex impedance of an equivalent circuit

$$Z(\omega) = R_0\left[1 - \widetilde{m}\left(1 - \frac{1}{1 + (i\omega\tau_0)^c}\right)\right]. \tag{2.113}$$

Box 2.8 gives the complex conductivity formulation equivalent of Equation 2.113. As noted by Macnae (2015), the parameters of this model influence the spectra in a subtly different way when formulated as a complex conductivity versus a complex resistivity (or impedance). The model parameters of Equation 2.113 are related to the equivalent circuit,

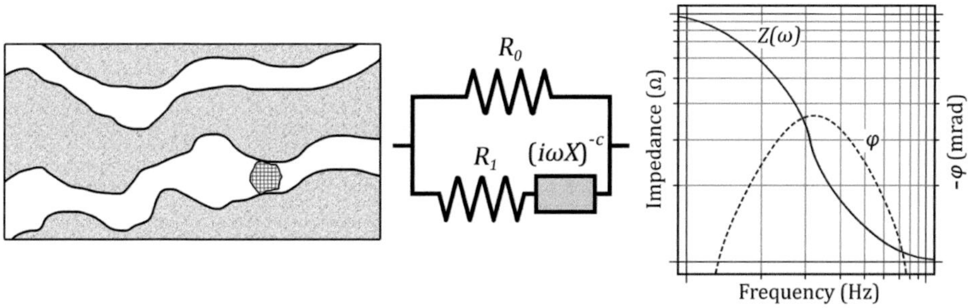

Figure 2.24 Conceptual and equivalent circuit model for polarization in a rock containing electron conducting minerals modified from Pelton et al. (1978). Frequency-dependent IP data are described by a modified Cole–Cole formulation (here expressed as complex impedance).

$$\tilde{m} = \frac{1}{1 + \frac{R_1}{R_0}}, \tag{2.114}$$

and,

$$\tau_0 = X\left(\frac{R_0}{\tilde{m}}\right)^{1/c}. \tag{2.115}$$

Mechanistic models to explain IP effects associated with electron conducting minerals were initially developed continuing the concept of current transport across the fluid–mineral interface. De Witt (1979) introduced a mechanistic model that attributed the polarization to charge separation occurring within a thin diffuse layer in between the electrolyte and the mineral grains. This model was the first to predict some important characteristics of polarization due to electron conducting minerals. These include: (1) the chargeability is foremost controlled by the volumetric concentration of the electron conducting particles; (2) the time constant is proportional to both the square of the particle diameter and the resistivity of the host medium (and thus the fluid chemistry); and (3) the time constant also depends on a third parameter representing the electrochemistry involved in the polarization, in this case being quantified by a Warburg impedance.

Wong (1979) introduced an electrochemical model to interpret IP responses from disseminated conductor and semiconductor mineral deposits. The model describes the polarization of a dilute suspension of infinitely conducting particles in a non-polarizable matrix. Although it assumes infinitely conducting particles, Wong (1979) argued that the model could be applied to particles with a conductivity 100 times or more larger than the background conductivity. This was a significant advancement in the theoretical treatment of IP that was extended by Wong and Strangway (1981) for the more complicated case of elongated electron conducting particles. The Wong model solves the Poisson–Nernst–Planck system of differential equations for the polarization of a single infinitely electron conducting particle embedded inside an electrolyte both in terms of diffusion and migration currents, along with reaction currents across the particle–fluid interface in the presence of electroactive cations. Wong used a mixing model (Section 2.2.4.4.2) to determine the macroscopic response of a medium characterized by dispersed particles at low concentrations (less than 10% by volume).

The Wong model considers both the size of the grains and the electrolyte composition. Unlike the previously discussed models, the Wong model does not require transport of charge across the electron conductor–fluid interface to occur, although it is assumed to occur in the presence of redox active ions in solution. Bücker et al. (2018) point out that Wong (1979) did not provide a comprehensive description of the contributions of different polarization mechanisms involved in his theory and he also did not fully describe the relationships between the mechanisms and key model parameters. Bücker et al. (2018) re-evaluated and extended the original Wong (1979) model, providing a more complete conceptual understanding in terms of two simultaneously acting polarization mechanisms: (1) charging of the diffuse layer induced over the poles of the electron conductor and associated diffusion current densities ($\boldsymbol{J}_{diff}$); (2) volume diffusion generated by reaction current densities ($\boldsymbol{J}_{reac}$) crossing the electron conductor–fluid interface in the presence of redox active cations (Figure 2.25). Bücker et al. (2018) show that the diffuse layer mechanism dominates the macroscopic response in the absence of reactive cations, but the volume diffusion effect may be significant for large particle sizes in the presence of reactive cations.

The electrochemical part of the Wong model uses a total of ten parameters to describe the effective conductivity of disseminated electron conducting spheres in an electrolyte. The geometrical parameters are the diameter and the volume concentration of the spheres. The complex conductivity of the non-polarizing background in which the particles are embedded is also needed. The remaining parameters describe the electrolyte chemistry as well as the electrochemistry of the mineral–fluid interface. They include the background concentration of cations in the electrolyte, the mobility of the ions in the electrolyte, the diffusion coefficient for active and passive ions, and three coefficients that describe the electrochemical reactions occurring at the interface of the electron conducting grain and the background medium. The challenge in using the Wong model is that many of the electrochemical model parameters are poorly constrained. However, the model does produce one simple important prediction, being that the intrinsic chargeability ($\widetilde{m}$) depends only on the chargeability of the background matrix ($\widetilde{m}_b$) and the volume fraction of the electron conducting particles ($\hat{v}$) (Gurin et al., 2015),

$$\widetilde{m} = 1 - (1 - \widetilde{m}_b)\frac{2(1-\hat{v})^2}{(2+\hat{v})(1+2\hat{v})}. \tag{2.116}$$

In the case of a non-polarizable background matrix, Equation 2.116 simplifies to

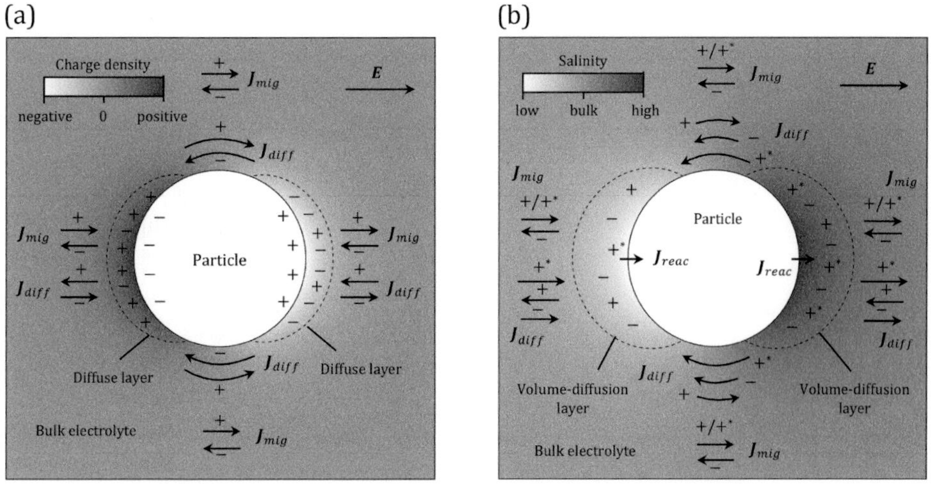

Figure 2.25 Reinterpretation of the two polarization mechanisms associated with electron conducting particles based on the electrochemical model of Wong (1979): (a) diffuse layer charging mechanism whereby migration currents $\boldsymbol{J}_{mig}$ locally deplete ions in the diffuse layer, charging the diffuse layer and inducing charges on the particles. Diffusion currents $\boldsymbol{J}_{diff}$ are driven by resulting concentration gradients; (b) volume diffusion mechanism whereby reaction currents $\boldsymbol{J}_{reac}$ and migration currents change the concentration of active cations ($+^*$) to the point that concentration gradients in the electrolyte drive diffusion currents. Reproduced from Bücker et al. (2018).

$$\widetilde{m} = \frac{9v}{2 + 5\hat{v} + 2\hat{v}^2}\,, \tag{2.117}$$

i.e. the chargeability is only a function of the volume concentration of the electron conductors and is independent of the fluid chemistry. Revil et al. (2015a) noted that this expression can be simplified to

$$\widetilde{m} = \frac{9}{2}\hat{v}\,, \tag{2.118}$$

being valid when volume concentrations of electron conducting particles are below 10%. The fit of the prediction of the Wong model for a non-polarizing background (Equation 2.117) to experimental datasets is shown in Figure 2.26a. This prediction of the Wong model suggests a potentially very powerful use of IP measurements for determining volumetric concentrations of electron conducting minerals in the subsurface.

Wong (1979) did not derive a direct predictive relationship between the relaxation time and the size of the electron conducting particles. However, he did suggest that the frequency of the phase peak is inversely proportional to the square of particle

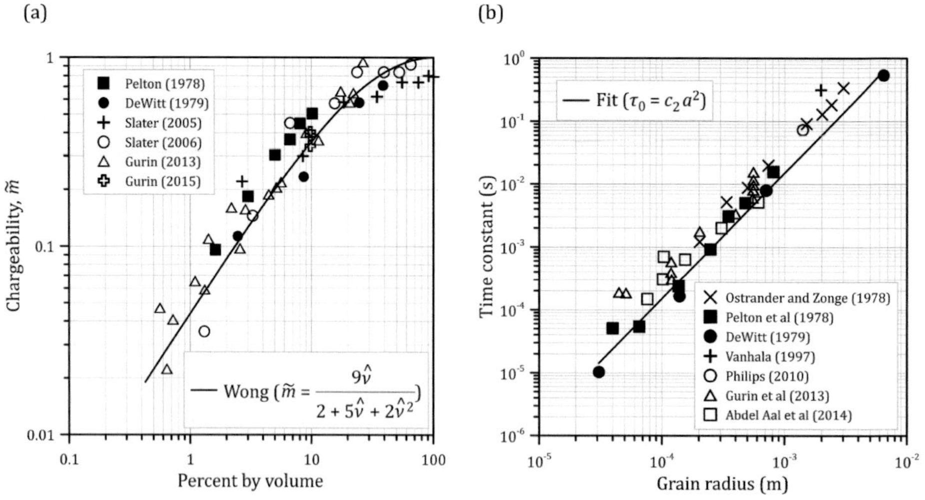

Figure 2.26 (a) Chargeability dependence on concentration of electron conducting minerals and fit to Wong (1979) prediction for a non-polarizing matrix; (b) time-constant dependence on radius of electron conducting particles. Pore fluid conductivity varies between data sources and likely explains some of the scatter in the data.

radius for large particles and inversely proportional to the particle radius for small particles. In both cases, the frequency of the relaxation peak was also proportional to the ionic diffusion coefficient. In their reassessment of the Wong model, Bücker et al. (2018) derived expressions for the relaxation time of the dominant diffuse layer mechanism and the volume diffusion mechanism that may be important for larger particles in the presence of reactive cations. For the diffuse layer mechanism, the relaxation time is proportional to the particle radius, and in the volume diffusion mechanism it is proportional to the square of the particle radius, being consistent with statements in Wong (1979).

Revil et al. (2015a) and Misra et al. (2016) introduce a model for semiconductive particles in the absence of redox reactive ions that removes the inherent limitation of an infinite particle conductivity as assumed by Wong (1979). They develop an approximate solution of the Poisson–Nernst–Plank equations to model the induced dipole moment for a single electron conducting particle and also for a non-conducting particle, both in contact with an electrolyte. Like Wong (1979), effective medium theory for dilute solutions is used to upscale the response for a porous medium containing semiconductive particles. Whereas the Wong model considers only the polarization of the charge carriers outside the electron conductor, Revil et al. (2015a) also consider the polarization of the charge carriers (p and n) within the semiconductors (Figure 2.27). The model results in very similar predictions to the Wong (1979) model for the dependence of chargeability on the volume concentration

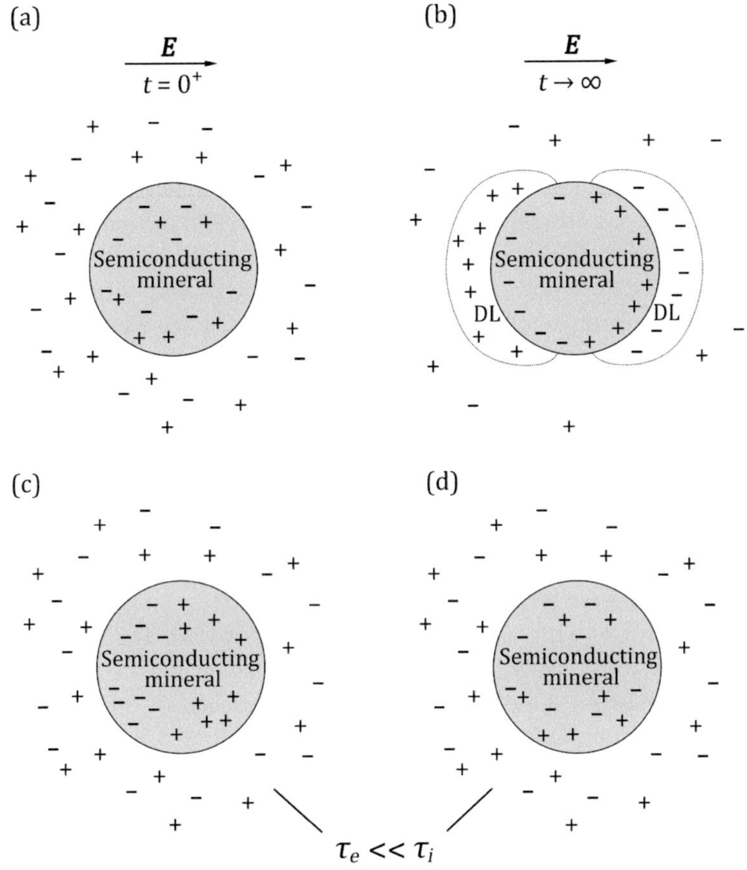

Figure 2.27 Conceptual representation of polarization mechanisms involved in the model for large semiconducting particles proposed by Revil et al. (2015a). (a) Conducting particle shortly after application of electrical field $\boldsymbol{E}$; (b) insulating entirely polarized particle at large time after application of $\boldsymbol{E}$ (DL = double layer generated by surface charges); (c) relaxation of the electrolyte with a time τ_e; (d) relaxation of the charges in the semiconductor with a time τ_i for small particle $\tau_e >> \tau_i$. Modified from Revil et al. (2018a).

of the electron conductors. The model of Revil et al. (2015a) also predicts a dependence of the relaxation time on the square of particle radius. Based on the large effective diffusion coefficients (on the order of 10^{-6} m^2/s) determined from the relaxation times, Revil et al. (2015a) hypothesized that the relaxation of the p and n charge carriers (Figure 2.27d), rather than ions external to the particle (Figure 2.27c) (with diffusion coefficients typically around 10^{-11} m^2/s), may dominate the measured relaxation time. Revil et al. (2018a) argued that the internal relaxation of the charge carriers will dominate for large particles, whereas the relaxation of the external ions will dominate for small particles. The model also assumes that the particle behaves as a conductor immediately after application of an electric field (or equivalently at high frequencies) (Figure 2.27a) but as an insulator when entirely polarized after long application of the field (or equivalently at low frequencies) (Figure 2.27b) per the model of Wong (1979).

Experimental datasets arc needed to validate these IP models for electron conducting particles. Petrophysical measurements made on both artificial materials representing analogues of ore bodies and ores themselves date back to the 1970s, showing the effects of the volume concentration and the size of electron conducting particles on SIP data (Pelton et al., 1978; De Witt, 1979; Revil et al., 2015c). The chargeability shows a strong dependence on the volume of the electron conducting particles as predicted by the Wong model (Equation 2.117). Chargeability is relatively independent of the fluid conductivity and temperature (Revil et al., 2018a) although it will depend somewhat on particle shape (Wong and Strangway, 1981; Gurin et al., 2018).

The time constant is shown to be proportional to the square of the radius (r) of the particles (Figure 2.26b), but also inversely proportional to the conductivity of the pore-filling fluid (σ_w), illustrating the strong role of the electrochemistry in controlling the polarization response of electron conducting minerals (Slater et al., 2005; Gurin et al., 2015). Similar to the case for the polarization of non-electron conducting minerals, the time constant decreases with temperature (Revil et al., 2018a). The scatter in the data shown in Figure 2.26b is likely partly attributed to variations in pore fluid conductivity and temperature between the different data sources. In contrast, the time constant is almost independent of the fluid conductivity for the polarization of non-electron conducting minerals.

To illustrate this significant difference between polarization of electron conducting and insulating minerals, Figure 2.28 shows the phase spectra for a uniform sand mixed with 4 per cent (by volume) magnetite versus the same sand packed with 4 per cent (by volume) kaolinite clay. In each case, measurements were made with the sample saturated with an electrolyte with three distinct fluid conductivities: 50 μS/cm, 500 μS/cm and 5,000 μS/cm. Figure 2.28a shows the typical polarization response for electron conducting minerals, where the phase peak shifts to higher frequencies as the fluid conductivity increases. It also shows that the magnitude of the phase is almost independent of the fluid conductivity. Figure 2.28b shows that, in contrast, the shape of the phase spectrum is independent of the salinity for polarization of insulating minerals. However, the magnitude of the phase is strongly dependent on the fluid conductivity in this case.

The contrasting models put forth to explain the IP response in electron conducting minerals have motivated numerous experiments in electrochemistry to better understand the nature of the microscopic relaxation mechanisms at the mineral–fluid interface. Some experiments focused on the relative importance of active versus inactive ions in controlling the polarization, as well as whether diffusion of charges in the electrolyte versus surface diffusion/adsorption phenomena contribute to the IP signal. Techniques that are standard in electrochemistry investigations of metal–fluid interfaces were employed in some of this research (Angoran and Madden, 1977; Klein et al., 1984). Much of this work was driven by the possibility of identifying the mineralogical composition of rocks although this remains a challenge (Seigel et al., 2007; Hupfer et al., 2016).

Gurin et al. (2015) took a semi-empirical approach to describe how relaxation time depends on both grain size and pore fluid conductivity for the polarization of electron conducting minerals, yielding

$$\tau = a_s \frac{r^2}{\sigma_w}, \tag{2.119}$$

where a_s (units of Fm^{-3}) was termed the specific volumetric capacitance describing the role of particle mineralogy and surface chemistry on the IP response. This inverse linear dependence on fluid conductivity is apparent in Figure 2.28a, where an order of magnitude increase in the pore fluid conductivity results in an equivalent order of magnitude decrease

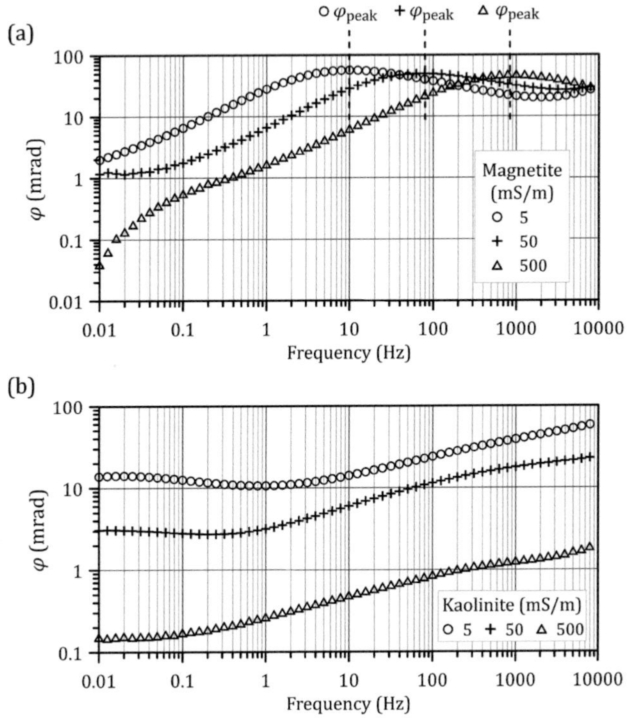

Figure 2.28 Comparison of the effect of the pore fluid conductivity on the relaxation-time distribution (i.e. the shape) of IP spectra for (a) a sand–magnetite mixture (4% by volume magnetite); (b) a sand–clay mixture (4% by volume kaolinite). The strong effect of the pore fluid conductivity on the location of the φ peak (Equation 2.119) for the polarization of the electron conducting mineral (magnetite) is evident from the shift in the phase peak (φ_{peak}) (phase in conductivity space, $+\varphi$).

in the position of the phase peak. However, the electrochemical properties of the interfacial polarization controlling a_s are uncertain and more work is still needed to improve understanding of the control of the electrochemistry on the IP effect in the presence of electron conducting minerals. Certainly, the role of the mineralogy of the electron conducting minerals on spectral IP measurements deserves more attention. Abdulsamad et al. (2017) describe a numerical study of a single semiconducting particle in a single electrolyte in the absence of any redox active species where the relaxation time depends on mineralogy. Indeed, interest in using spectral IP measurements to infer mineralogy has captivated the mineral exploration community for decades (e.g. Pelton et al., 1978), with more recent studies returning to address this possibility (Bérubé et al., 2018).

2.3.8 Electrical Properties of Contaminated Soils and Rocks

Contaminants in the pore space may influence the electrical properties of soils and rocks. Environmental geophysics research increased in the 1980s in recognition of the need to detect and map contaminant plumes. The extent to which contaminants modify electrical properties is inherently complex, depending on many factors including the contaminant type (e.g. aqueous phase or non-aqueous phase), the concentration and length of time that the contaminant has been in the ground and subject to degradation by biogeochemical processes (Atekwana and Atekwana, 2009). Some contaminants modify electrical properties in a predictable, well-understood way. For example, inorganic contaminants (e.g. salt plumes) increase the pore fluid conductivity (σ_w) and Archie's law (Equation 2.17) can be used to estimate the resulting change in the conductivity of the porous medium. Organic contaminants, such as hydrocarbon spills, cause a time-varying change in the electrical properties (Sauck, 2000). A fresh hydrocarbon is highly resistive and fresh hydrocarbon spills will decrease conductivity of soils due to the replacement of ion-rich groundwater with hydrocarbon. Electrical resistivity and IP measurements are sensitive to the presence of fresh non-aqueous phase liquid (NAPL) in the pore space of soils (Olhoeft, 1985; Börner et al., 1993). IP measurements have been successfully used to characterize NAPL contaminated field sites (Chambers et al., 2005). However, natural attenuation of the hydrocarbon over time transforms the electrical signature from low to high conductivity (relative to the native soil). The complex biogeochemical processes involved in transforming the hydrocarbon into inorganic compounds release ions and produce organic acids that tend to increase electrical conductivity over time (Atekwana et al., 2000; Sauck, 2000). Heenan et al. (2014) argue that the electrical properties of soils in the presence of hydrocarbon contamination will follow a distinct curve representing the progress of degradation (Figure 2.29). This curve also indicates that contaminants at a certain stage of aging may exert a weak (or even no) influence on the electrical properties, representing the midway point between a relatively young (resistive) spill to a mature (conductive) spill (vertical dashed line in Figure 2.29). IP measurements are sensitive to the sorption/desorption of cations (Section 2.3.5.1) hinting at opportunities to monitor reactive transport processes associated with contaminants in the subsurface (Hao et al., 2015).

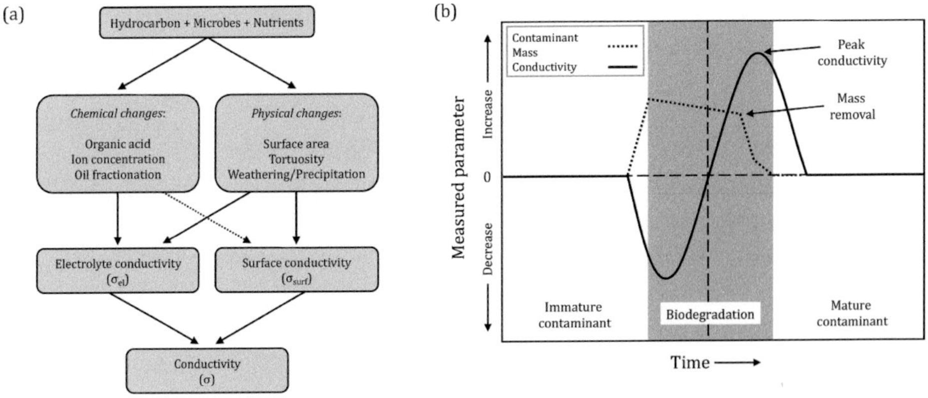

Figure 2.29 (a) Summary of hydrocarbon contaminant transformations causing changes in electrical conductivity with time. (b) Idealized evolution of the electrical conductivity of a hydrocarbon spill. Vertical dashed line indicates point in time when hydrocarbon ageing results in no anomaly (modified from Heenan et al. (2014)).

2.3.9 Non-linear IP Effects

During advances of the IP method in the 1970s–1980s, substantial interest in the possible presence of non-linear IP effects was generated because of the possibility of linking measures of non-linearity to electrochemical processes (charge transfer reactions accompanying oxidation-reduction reactions and ion exchange) involving electron conducting minerals, clays and organic contaminants (Olhoeft, 1985). Non-linearity means that the system output does not scale with the input which, in the case of electrical measurements, means that the linear relationship between strength of the IP response and applied voltage does not hold. Signal-processing techniques to assess evidence for non-linearity in spectral IP datasets have been developed, foremost, based on assessing harmonic distortion of the voltages recorded on samples (Olhoeft, 1979). Non-linear IP effects have been reported for measurements on mineralized rocks (Anderson, 1981; Olhoeft, 1985) and also proposed to occur as a result of interactions between organic contaminants and clays (Olhoeft, 1985). However, others have found no evidence for non-linear IP effects in the presence of clay minerals (Klein and Sill, 1982). Interest in non-linear IP effects waned in the early 2000s, mostly as a result of inconclusive results. However, Hallbauer-Zadorozhnaya et al. (2015) recently put forward both observations and a modelling framework based on membrane polarization to support the non-linear IP effect. They went even further to suggest that the linearity between current density and applied voltage represented by Ohm's law (Equation 2.1) also may not always hold, although independent measurements on the same samples showed no evidence for non-linearity (A. Weller, personal communication). Although our discussion of electrical properties has been limited to assumed linearity, non-linear IP responses may yet be proven and utilized in electrical surveys. However, it will be important to ensure that the observations are in fact from the sample rather than artefacts that result from the poor design of sample holders for the accurate measurement of electrical properties or even from defective instrumentation. These issues are addressed in Chapter 3.

2.4 Closing Remarks

The electrical properties of porous materials with a non-conducting matrix (i.e. absence of electron conducting minerals) are related to the pore network geometry (porosity, connectivity, surface area) and the pore-filling fluids (relative concentrations, distributions and temperature). An electron conducting matrix provides an additional control on the electrical properties. This makes the electrical resistivity and IP methods powerful tools for investigating a wide range of subsurface properties and processes. The contribution of these different factors to the electrical conductivity recorded with a resistivity measurement alone is hard to decipher, resulting in substantial uncertainty and potential for misinterpretation of recorded variations in resistivity. This uncertainty is substantially reduced when a measurement of IP is included. The polarization of the EDL of the pore network is exclusively measured by the imaginary part of the complex electrical conductivity. With this additional information, the complex electrical conductivity for a porous material with a non-conducting matrix can be broken into individual components that relate to (1) conduction by the fluid-filled interconnected pore space, and (2) conduction and polarization of the EDL at the pore fluid–mineral grain interface. This allows what is known as Archie-type conduction to be separated from surface conduction, thereby reducing the ambiguity of interpretation, e.g. allowing for more confidence in the interpretation of changes in fluid conductivity versus changes in soil structure (such as clay content). The electrical properties of partially saturated soils are less well understood than saturated soils, particularly the dependence of surface conduction and IP on degree of saturation and how the saturating fluid is distributed through the pore space.

IP is particularly powerful for investigating subsurface properties and processes involving electron conducting minerals due to the strong polarization enhancement that results from electron transport in the minerals. Both the volume concentration of the minerals and information on the particle size are potentially extractable. Consequently, IP has long been a valuable technology for mineral exploration. Interest in this aspect of IP for near-surface studies has grown over the last twenty-five years in response to recognition of the value of the methodology for investigating environmental questions involving the transformation of metals. Innovative recent applications include monitoring of pore-clogging by nano-sized iron particles used in environmental remediation (Flores Orozco et al., 2019a) and mapping of hotspots of biogeochemical activity associated with iron-mineral precipitates in reduced zones of floodplain sediments (Wainwright et al., 2016). However, understanding the role of mineralogy on the IP response remains incomplete. The role of mineralogy of electron conducting particles has long intrigued mineral exploration researchers, but it is also increasingly apparent that mineralogy may exert a strong control on IP mechanisms in non-electron conducting rocks (Chuprinko and Titov, 2017). Large mineralogical variations may cause some of the most important petrophysical relations, e.g. the linear proportionality between polarization and surface conduction discussed in this chapter, to break down (Revil et al., 2018b).

The estimation of permeability represents an intriguing potential application of resistivity and IP measurements, being a challenging physical property to estimate in situ. A geophysical approach to estimating permeability would provide hydrogeologists with a powerful new technology for field-scale understanding of flow and transport processes. The additional information provided by the IP measurement improves permeability estimation from electrical measurements and allows for mechanistic and empirical formulations describing the link between the electrical properties and permeability. The formation factor appearing in permeability prediction equations remains challenging to estimate in situ, although constraining the surface conductivity using IP measurements can improve formation factor estimation (Weller et al., 2013). However, the equations developed to link IP measurements to permeability are not general and may perform poorly on new datasets (Razavirad et al., 2018).

In Chapter 3, we describe resistivity and induced polarization instruments used to acquire the measurements needed to implement the petrophysical relationships covered in this chapter. We consider laboratory measurements, where most fundamental petrophysical relationships have been developed or validated, and field instruments, where these relationships can be used to link the field observations to variations in the physical and chemical properties of the subsurface.

3

Instrumentation and Laboratory Measurements

3.1 Introduction

Geophysical instruments have been developed to measure the electrical properties of the subsurface over multiple scales. Laboratory instrumentation has also been developed to provide researchers with the tools needed to characterize the electrical properties of soil samples and rock cores. Such laboratory instruments have been used to acquire the datasets needed to develop the petrophysical relationships introduced in Chapter 2. Although electrical conductivity (and complex electrical conductivity) is directly related to the charge transport (conduction and polarization) mechanisms in the subsurface, geophysicists measure resistance (or impedance) and historically report resistivity. We follow this historic convention by using resistivity terminology in describing instrumentation and measurements reported in this chapter. The measured resistivity is simply the reciprocal of the measured conductivity ($\rho_m = 1/\sigma_m$). Similarly, the measured complex resistivity is the reciprocal of the measured complex conductivity ($\rho_m^* = 1/\sigma_m^*$).

The underlying 'nuts and bolts' of an electrical resistivity measurement are relatively simple (Figure 3.1). A power source is used to drive a current into the ground and the electric field strength needed to drive this known current is recorded. Originally, much of the resistivity and induced polarization (IP) instrumentation development was foremost driven by the mineral exploration industry. High-power transmitters were developed to drive sufficient current into the subsurface to resolve deep (+100 m) structures over relatively large scales. First single, and then multi-channel (typically up to 10), receivers were developed to record the resulting electric potentials at the surface from a small number of electrodes. The data produced by these instruments were typically interpreted as 1D soundings, 1D profiles or 2D pseudosections of apparent resistivity (see Section 4.2.2.5). The major technological advances in the field-scale instruments over the last few decades were driven by the development of 2D (and then 3D) imaging algorithms that require multi-dimensional datasets to be acquired from a large number of electrodes placed on the Earth's surface and/or in boreholes. The development of these imaging algorithms coincided with the rapid growth of the environmental characterization/remediation industry, along with growing interest in shallow subsurface hydrological processes and water resources. This resulted in the evolution from high-power instruments allowing only a limited number of electrodes to be addressed via a small number (often just one) of measurement channels to portable, lower-power resistivity imaging systems that address a large number of electrodes. These imaging instruments use some combination of multiple channels and multiplexing (the ability to switch between electrodes) to automatically address large grids (typically 100s) of electrodes connected to the instrument via multicore electrode cables. This development has made 2D and even 3D resistivity/IP imaging efficient and economically viable, at least for shallow (<100 m) surveys. A survey that may have taken days to complete 20 years ago might now be completed in a few hours.

Other technological advances have included the development of instrumentation for continuous electrical imaging from an array of towed electrodes. Continuous electrical imaging is easiest to implement when towing the array of electrodes in water. This has led to increasing use of electrical imaging for shallow marine surveys, with popular applications being the investigation of saline intrusion and groundwater–surface water exchange (Day-Lewis et al., 2006; Mansoor et al., 2006). Towed arrays have also been developed for land use, although this is challenging when relying on traditional galvanic contact provided by standard electrodes. The need to perform continuous surveys over areas of resistive ground (examples being asphalt/concrete and frozen ground) has led to the development of instrumentation that overcomes the need for galvanic contact via capacitively coupled electrodes (Geometrics, 2001; Kuras et al., 2007). Another advancement has been the development of resistivity monitoring systems, whereby automatic data-acquisition systems have been deployed to continually collect data and infer subsurface processes occurring within the investigation volume of the monitoring arrays (Bevc and Morrison, 1991; Van et al., 1991).

The fact that measuring the induced polarization (IP) response is much more challenging than measuring resistivity cannot be overemphasized. Some practitioners may argue that IP measurements should be acquired routinely as most resistivity instruments record an IP

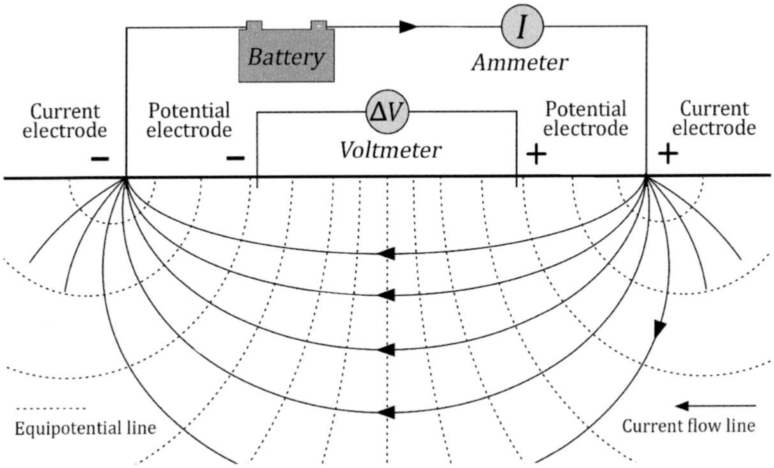

Figure 3.1 Basic elements of a resistivity measurement.

measurement, making this 'free' additional information. In fact, considerable additional effort and attention to detail is required to acquire meaningful IP data. Given that recording IP data always lengthens the data-acquisition time (relative to resistivity measurements alone), it is well worth this additional attention to detail so that quality IP data are recorded. Otherwise, it is quite easy to waste survey time recording a lot of essentially useless information. The additional challenges in IP data acquisition in large part result from the signal-to-noise ratio (SNR) often being 100–1,000 times smaller relative to the resistivity signal. Another challenge is that the field-scale measurement of IP is easily corrupted by coupling effects that are associated with the wiring and supporting hardware used to connect the instrument to the ground. Recommendations for field-scale data acquisition to avoid such problems are described in recent review papers on the IP method (Kemna et al., 2012; Zarif et al., 2017).

In the laboratory, instrumentation exists to acquire the spectral induced polarization (SIP) response over a range of frequencies. However, acquisition of precise laboratory SIP measurements requires rigorous attention to the placement of electrodes, the geometry of the sample holder and other intricate factors that the expert must consider (Vanhala and Soininen, 1995; Kemna et al., 2012). Field-scale SIP systems were developed for mineral exploration, where very large phase signals could be observed above the significant sources of error associated with the instrumentation. More recently, instruments have been developed to make field-scale near-surface measurements of the smaller SIP signals that are recorded in the absence of electron conducting minerals (Schlumberger's background effect) (Radic et al., 1998; Radic, 2004), although this remains the domain of the specialist.

The informed measurement of IP also requires a solid understanding of how the measurement provided by the instrument relates to the physical properties (primarily the magnitude of the polarization response) that we are interested in (Section 2.3). Whereas it is straightforward to translate the information recorded with a resistivity instrument to an estimate of the resistivity, this is not always the case for IP measurements. The IP measurement recorded with time domain IP instruments will depend not just on the physical properties of the subsurface but also on how the instrument is configured to measure the IP effect. This problem is alleviated when a frequency domain IP measurement is made, although important steps remain to translate the measurement to the physical properties of interest.

In this chapter, we describe the main principles used to acquire electrical resistivity and IP datasets. We consider the entire system, which includes the power transmitter, the receiver, the electrodes used to drive current into the ground and measure resulting voltages in the subsurface, and the electrode arrays used to connect the instrument to the Earth. As with other chapters in this book, we start by considering electrical resistivity and dedicate the latter part of the chapter to the extension of the approach to IP. This considers not just the instruments but also how the measured IP parameters relate to the properties of interest. Special attention is given to the additional challenges associated with IP data acquisition.

3.2 Resistivity Measurements

3.2.1 Resistance, Resistivity and the Geometric Factor

We recall from Chapter 2 that the resistivity is an intrinsic property of the material that describes the resistance it exerts to the conduction of electrical current. This intrinsic property is determined from measurements of the transfer resistance ($\Delta V/I$) and a geometric factor (K),

$$\rho = R \cdot K = \frac{\Delta V}{I} K. \tag{3.1}$$

The transfer resistance is determined by measuring the voltage difference (ΔV) between two points that results from the injection of an electric current (I) into the material. The geometric factor is determined by knowing (1) the relative locations of the current injection and voltage measurement positions, and (2) the geometry of the current flow lines in the medium. In certain cases, a simple analytical formula for K can be defined (see Chapter 4). These include 1D current flow in a homogenous material and radial current flow to a point (again in a homogenous material). In more complex cases, the geometric factor might be determined experimentally (in the laboratory) or numerically (in the laboratory or field). We will visit such cases later, but here we first focus on the case of 1D current flow that is most commonly used to determine the resistivity of Earth materials in the laboratory. Such measurements are typically used to assess the intrinsic resistivity of a sample, which can then be correlated with the physical and chemical properties of the sample. These measurements form the foundation of the petrophysical relations introduced in Chapter 2.

3.2.2 Laboratory Measurements

3.2.2.1 Measurement Cells and the Four-Electrode Measurement

Laboratory resistivity measurements are most usually performed with a cell that generates a 1D current flow and associated electric field. McCollum and Logan (1915) provide an early example of the measurement of cylindrical soil samples using this strategy. The situation is analogous to current flowing in a wire, where the resistivity is related to the resistance by

$$\rho = R \cdot K = R \frac{A}{L} = \frac{\Delta V}{I} \frac{A}{L}, \tag{3.2}$$

where A is the cross-sectional area of the wire and L is the length of the wire between the two points where the voltage drop (ΔV) is measured. In this case, there is a simple analytical expression for the geometric factor, $K = A/L$. The same equation is commonly used to determine the resistivity of soils or rocks (Figure 3.2).

In common practice, the voltage drop (ΔV) is measured with a separate pair of electrodes from those used to drive the current into the sample. This is necessary as, at the low frequencies used for current injection, the electrode contact resistance associated with the metal electrode–electrolyte contact can be large relative to the sample resistance. This electrode–electrolyte contact is characterized by an impedance ($Z^* = R + iX$), a

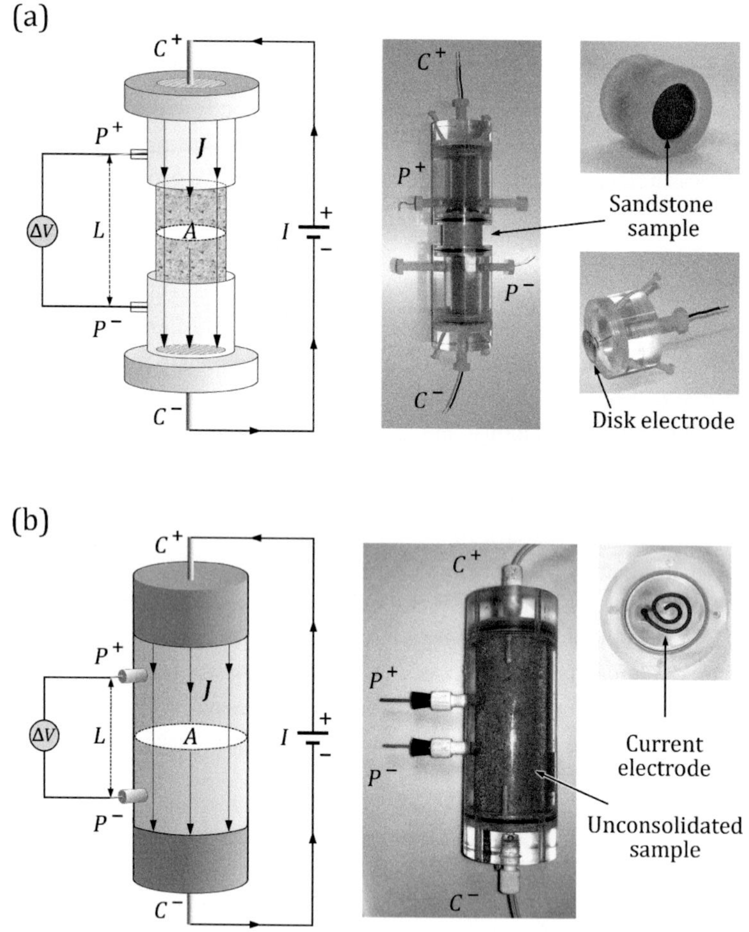

Figure 3.2 Laboratory measurement of electrical resistivity using a 1D flow cell: (a) example rock core sample holder where both current and potential electrodes are housed in end caps; (b) example unconsolidated sample holder where potential electrodes are within the unconsolidated sample.

complex-valued generalization of the resistance, where the real part is the ohmic resistance (R) and the imaginary part is the reactance (X), which describes any non-ohmic resistance to charge transport (due to capacitive or inductive effects). For simplicity, we refer to just the ohmic resistance here. When the voltage drop is measured across the current electrodes, the sum of the sample resistance and the two electrode contact resistances is recorded. Driving current across the electrode–electrolyte interface polarizes the electrode and makes these contact resistances very large, such that they may dominate the total resistance recorded. This problem is overcome by using a four-electrode configuration, where ΔV is measured between a pair of potential electrodes (Box 3.1). Electrode contact resistances also exist at each of these potential electrodes (and will be important in the discussion of IP later), but they do not dominate the total impedance as negligible current flows through these electrodes due to the high-input impedance on the receiver channels of resistivity instrumentation. Consequently, the recorded voltage drop accurately reflects the difference in voltage between the two points along the length of the soil sample and gives the resistance of the material through which the current is flowing. The four-electrode measurement is the standard practice for laboratory and field-scale resistivity measurements.

In the case of soils and unconsolidated sediments, the four electrodes can be embedded into the sample at locations along it (Figure 3.2b). However, embedding electrodes into the sample is not always practical, e.g. in the case of rock cores. The alternative is to place the sample between two-electrode end caps that will contain both the current electrodes and the potential electrodes as discussed later (Figure 3.2a). Equation 3.2 produces a single value of resistivity, representing the 'true' intrinsic resistivity of a soil/rock at this measurement scale. Any heterogeneity within the soil/rock is not considered and is effectively integrated within the single estimate of resistivity determined from this procedure.

Making such laboratory resistivity measurements is relatively straightforward, although attention must be given to the design and placement of electrodes. In the sample holders shown in Figure 3.2, the objective is to generate a 1D electric field (i.e. current flow in one direction along the long axis of the sample holder). The best way to do this is have current electrodes that span the entire cross-section of the sample holder. This can be achieved with a wire mesh, although this can sometimes generate problems due to the trapping of gas bubbles in the mesh. Another option is to use a spiral of wire (Figure 3.2b) to result in an equivalent even distribution of the source voltage throughout the cross-section of the sample holder. Point current electrodes must be used with caution as the electric field may not meet the 1D assumption close to the electrodes.

The potential electrodes can either be point electrodes, ring electrodes (that follow the circumference of the sample holder) or mesh/spiral electrodes that cover the cross-section of the sample holder (the latter is not appropriate for IP measurements as discussed in Section 3.3.1). The standard laboratory measurement, shown in Figure 3.2, gives the equivalent resistivity of a homogenous sample. In the case of a sample that exhibits significant heterogeneity, the current flow lines may be distorted and result in a biased potential difference between point electrodes. The argument for the use of potential electrodes that cover the entire cross-section of the sample holder is that the recorded potential difference will average out such bias. The ring electrodes bring similar benefits over point electrodes, but to a lesser extent, as the potentials are sampled just around the circumference of the sample rather than the entire cross-section. However, the point electrodes offer some important advantages with respect to making IP measurements as discussed later in Section 3.3.1.

The impedance at the current injection electrodes must be low enough that the total resistance (R_{tot}) between the current injection pair does not exceed the limitations of the instrumentation. Laboratory instruments work by providing a constant source of current or a constant voltage between the injection electrodes. In the case of a constant voltage source, the increase in R_{tot} decreases the current flowing in the circuit following Ohm's law ($I = \Delta V/R$). In the case of a constant current source, the increase in R_{tot} increases ΔV to a point that may exceed the maximum voltage provided by the instrumentation.

Box 3.1

The four-electrode measurement

In two-electrode measurements the measured resistance is the sum of the contact resistances (R_c) at the current electrodes and the resistance of the sample between electrodes (R_{sample}). As current is driven through the current injection electrodes, the electrodes polarize and the impedance to current flow across the electrode–ground interface builds up. To overcome this problem, a four-electrode measurement is used, where R_{sample} is directly recorded. Although there is still a contact resistance associated with the two potential electrodes, no significant current flows across this electrode–sample interface (due to the very high input impedance of the recording channel) and the resistance of the sample is reliably recorded. The electrode–sample contact in fact represents an impedance ($Z^* = R + iX$), a complex-valued generalization of the resistance, where the real part is the ohmic resistance (R) and the imaginary part is the reactance (X), which describes any non-ohmic resistance to charge transport. At the low frequencies used for resistivity measurements the ohmic resistance is the dominant term.

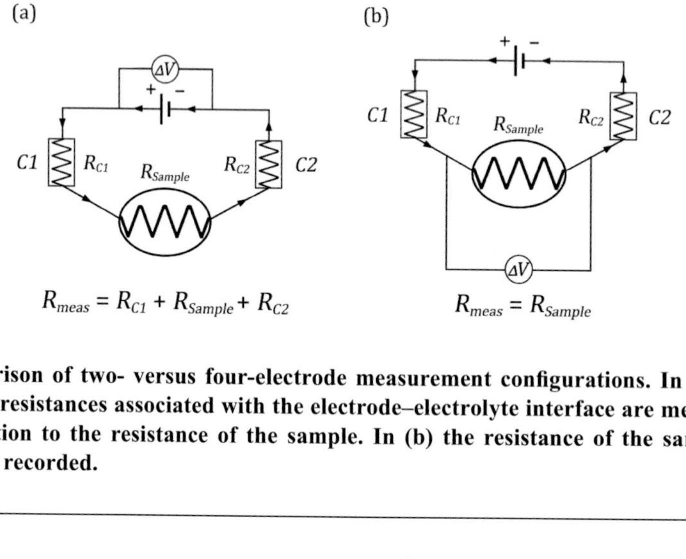

Comparison of two- versus four-electrode measurement configurations. In (a) the contact resistances associated with the electrode–electrolyte interface are measured in addition to the resistance of the sample. In (b) the resistance of the sample is directly recorded.

3.2.2.2 Types of Sample Holders

As shown in Figure 3.2, there are two basic types of sample holders used for resistivity measurements: holders designed for repacked unconsolidated material and holders designed for rock cores and/or undisturbed unconsolidated sediments. The holders designed for repacked sediments can have the current electrodes embedded in the sediments (Figure 3.2b). The potential electrodes are either also embedded in the sediments or, for fully saturated sediments, placed in fluid-filled chambers on the edge of the sample in a way that electrolytic contact is ensured. The geometric factor for the sample holder is analytically determined assuming 1D current flow, i.e. $K = A/L$ (Figure 3.2a). Measurements on unsaturated samples can be obtained with electrodes directly embedded in the sample, but this approach fails at relatively low levels of saturation due to high contact resistances at the current electrodes, particularly in coarse-grained material. One way around this is to employ porous ceramics in conjunction with electrodes placed in fluid-filled chambers on the edge of the sample (Ulrich and Slater, 2004). The ceramic maintains an electrolytic connection between the electrode and the pore fluid in the unsaturated pore space of the sample.

The holders designed for rock cores and/or undisturbed unconsolidated material have the current electrodes and potential electrodes embedded in end caps that attach to either side of the sample (Figure 3.2a). The end caps are usually filled with an electrolyte (liquid or a gel) to establish electrical contact with the sample. Ideally, the internal diameter of the end cap should be equal to the diameter of the core to maintain 1D current flow. Unsaturated samples are measured using an electrolytic gel in the end caps (Taylor and Barker, 2002; Binley et al., 2005). End caps are the most practical way to acquire measurements on cores and also prevent disturbance of 'undisturbed' unconsolidated materials acquired during drilling. The potential electrodes should be as close as physically possible to the edge of the sample, otherwise a correction for the additional resistance between the potential electrode pair due to the electrolyte in the end cap between potential electrode and edge of core is needed (as discussed in Section 3.2.2.3).

When using end caps, saturated samples can be simply wrapped in Parafilm® (or some other water-retaining film) and placed between the caps. Alternatively, samples can be cast in resin or placed in a high-strength rubber known as a Hassler sleeve. This sleeve is used to ensure that water flows through a rock core (rather than along the sides) in permeability measurements on rock cores. It is also valuable for making sure that electric current is transmitted through the core. Lateral pressure applied to the outside of the sleeve presses the sleeve firmly against the core. Casting samples in resin is useful for poorly cemented/fragile rock cores (e.g. Binley et al., 2005). One challenge with end caps is accurately determining the geometric factor. Unless the potential electrodes can be exactly located against the edge of the core, it is generally not possible to determine a precise estimate of the geometric factor from the analytical solution for 1D current flow.

3.2.2.3 Determining the Geometric Factor

Accurate determination of the geometric factor is critical to reliable resistivity estimation. Errors of a few per cent or more in resistivity can translate into unacceptable errors in the estimation of petrophysical parameters, e.g. the Archie cementation exponent. As described earlier, there is a simple analytical expression for the geometric factor when dealing with 1D current flow through a sample. However, this analytical expression may not accurately reflect the true geometric factor of the sample. The most significant error relates to the uncertainty in the potential electrode locations: the finite size of the electrode means that it is challenging to accurately determine the correct distance (L) between the potential electrodes required for the analytical geometric factor to be precise. It is more problematic when sample holders that are used do not create a 1D current flow along the column, e.g. when point electrodes are used for the current injection and result in a 3D current distribution close to the electrode.

A more robust approach is to experimentally determine the geometric factor. This can be done by filling the sample holder (and end caps when used) with a number of different fluids of precisely known electrical conductivity. The simplest fluid to use is a binary salt (e.g.

NaCl) solution of varying concentrations, typically spanning a few orders of magnitude of conductivity change. The electrical conductivity of the fluid (σ_w) is recorded for each solution. The resistance between the potential electrode pair is then recorded with a resistivity meter. This measurement should ideally be made in an environmental chamber to avoid errors associated with temperature differences and/or the need to apply a temperature correction (see Section 2.2.3.1). Once enough pairs of measurements are made, the best estimate of K is determined from the reciprocal of the slope of the best fitting linear relation,

$$R = \left(\frac{1}{K}\right)\frac{1}{\sigma_w}. \tag{3.3}$$

An identical procedure is used to calibrate what is commonly called the 'cell constant' of specific conductance probes.

Another option for computing the geometric factor, especially when 1D flow is not supported by the sample holder, is to numerically determine it using a 3D solution to the Poisson equation. In this way, 3D current flow pathways within the sample are modelled and K is determined from the numerical solution of R between the potential electrodes for a given sample resistivity (Figure 3.3).

End caps create specific challenges for accurate determination of K when the potential electrodes cannot be placed right up against the core/sample. In this case, the end caps will add an extra resistance in series with the sample resistance (Figure 3.4). An experimentally determined K, using fluids of known electrical conductivity, will now not be accurate as, during a measurement on a sample of unknown resistivity, there will be a resistivity contrast

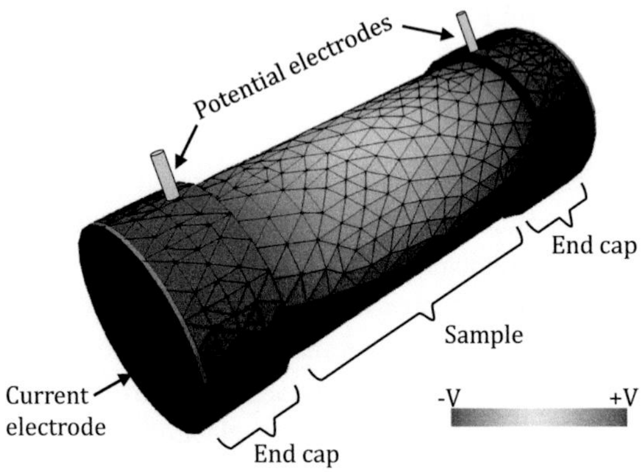

Figure 3.3 Example of a numerical model for the potential field in a sample holder and its end caps. The sample (of homogeneous resistivity) is 2 cm long and 1 cm in diameter. The end caps are 5 mm long and 11 mm in diameter. The end caps have a resistivity of 20% of that of the sample, which results in lower potential gradients within the end caps. (A black and white version of this figure will appear in some formats. For the colour version, please refer to the plate section.)

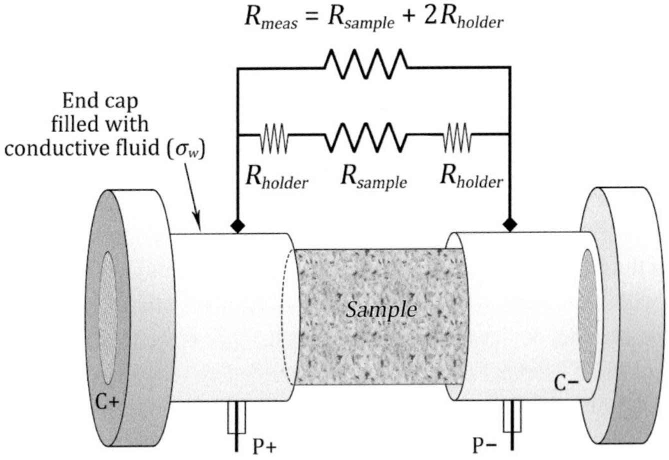

Figure 3.4 The potential electrodes for core sample holders should be placed in the end caps so that they are as close as possible to the edge of the core. The additional resistance (R_{holder}) associated with any finite distance from the sample edge to the potential electrode will usually be small but should be calibrated to obtain the most accurate resistivity measurements possible.

at the end of the core that will cause current bending. In this case, the resistance associated with the end caps must be determined and subtracted from the total resistance recorded to give the corrected resistance across the sample holder. The geometric factors of the end caps can be determined experimentally via a calibration procedure carried out with the end caps filled with fluids of known resistivity and with the sample resistivity also known. This way, the resistance correction can be computed by measuring the conductivity of the fluid (or gel) used to fill the end caps.

3.2.2.4 Laboratory Instrumentation

The laboratory measurement of resistivity is relatively straightforward and it is possible to purchase resistivity meters specifically designed to measure the electrical properties of soils/rocks. These may be packaged with a sample holder and/or accessories for mounting cores/soils. With a little electrical engineering experience/confidence, it is also possible to custom build a resistivity measurement system from inexpensive hardware components (Florsch and Muhlach, 2017). This custom-build approach is increasingly attractive, as inexpensive hardware components continue to grow in availability, supported by a user community that develops the necessary software to run the instruments. The basic requirements for a laboratory resistivity instrument are a controlled source of electric current and a precise recording of a differential voltage (ΔV), both across the sample and across a resistor (R_{ref}) in series with the sample. The voltage difference (ΔV_{ref}) recorded across the resistor is used to measure the electrical current flowing in the circuit from Ohm's law,

$$I = \frac{\Delta V_{ref}}{R_{ref}}. \quad (3.4)$$

Small power sources (e.g. a few watts) usually provide accurate laboratory resistivity measurements on soils and rocks. A few milliamperes of current produced by a constant voltage (e.g. 12 V or less) power supply is sufficient. Low current densities are less likely to alter the biogeochemical characteristics of samples, a consideration in some environmental applications of the method. Laboratory resistivity measurements can therefore be made with a data logger with a constant voltage output. Laboratory resistivity instruments may be as simple as single-channel devices, where just one resistance is recorded between a single pair of electrodes, or multi-channel devices for simultaneously recording any number of resistances along a column or in an experimental tank. Field instruments described later can be used to acquire laboratory measurements, although caution is necessary to make sure that the current density is limited by using minimal power-output settings and/or using a shunt resistor to limit the current. Excessive current densities may drive unwanted electrochemical reactions at the current electrodes and even heat up the sample. Current density is a consideration for IP measurements as discussed later.

3.2.2.5 Current Sources

The output voltage providing the source of the current is either a low-frequency (<100 Hz) alternating square or sine wave source (a frequency domain measurement) or a DC source that is repeatedly switched on and off but with the direction of the source repeatedly switched between each off-period (a time domain measurement) (Figure 3.5). The voltage due to the current injection (V_p) is recorded during the current on-period. The off-period is important for measuring any residual (secondary) voltage differences (V_{sp}) between the electrodes that are not caused by the impressed current. These result from (1) open-circuit potential differences due to electrochemical disequilibrium between the potential electrodes and (2) natural sources of electric current in the sample known as 'self-potentials' which are discussed in Section 3.2.2.7. Figure 3.5b shows an example of the modified switched square wave used in the time domain. The reversal of current provided by both methods is important to avoid excessive polarization of the current electrodes that will occur when driving the current continuously in a single direction for extended periods of time. Reversing the current direction reverses the coupled anodic and cathodic reactions associated with current transfer across the metal–fluid interface and reduces the build-up of a large contact-resistance barrier at the current electrodes.

3.2.2.6 Potential Recordings

The channels used to record voltage differences between electrodes must have a high input impedance to avoid current leakage through the electronic circuitry, particularly when measuring highly resistive samples. The precision and resolution of the voltage-recording channel should be sufficient to record the voltage with 0.1 per cent accuracy or better. As resistivity varies over many orders of magnitude and resistance is proportional to resistivity,

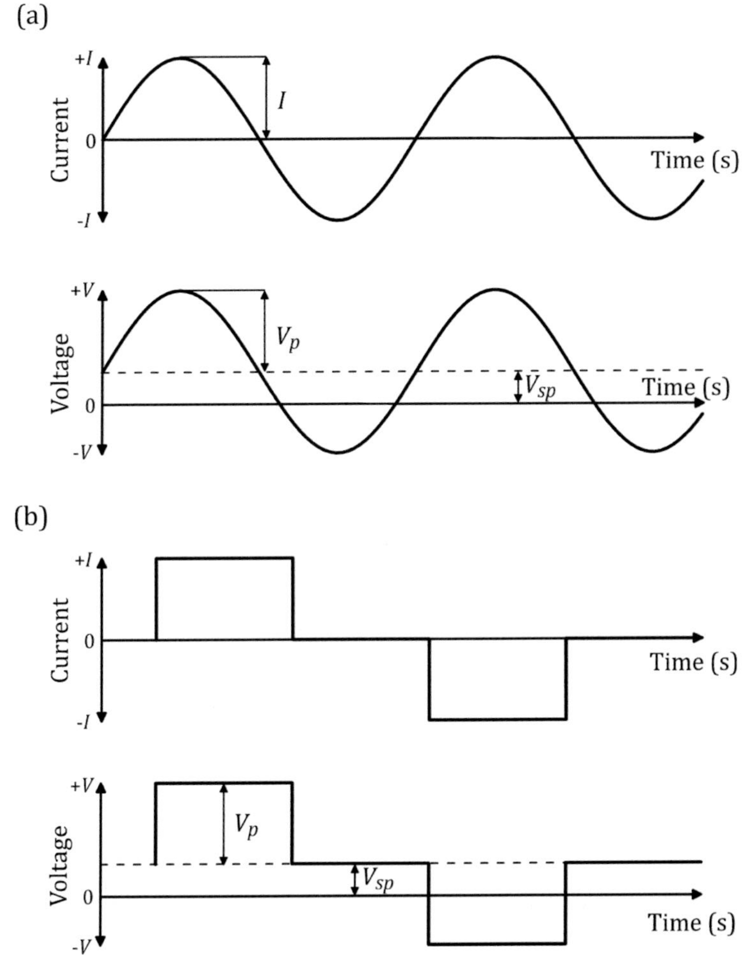

Figure 3.5 Standard output waveforms and voltage signals: (a) sine wave source used for a frequency domain resistivity measurement; (b) square wave source for a time domain resistivity measurement; primary (V_p) and secondary (V_{sp}) signals are shown in each case.

the instrumentation needs to accurately measure over a wide range (i.e. many orders of magnitude) in voltage differences. As the measurement resolution is a function of the full-scale range of the receiver, variable gains (multiplications of the signal amplitude on the receiver) are used to more accurately measure this wide range in voltages. Higher gains are applied to smaller voltages in order to normalize all voltages over a narrower range with higher resolution.

3.2.2.7 Laboratory Electrodes

The composition of the electrodes is rarely a consideration for resistivity measurements. Some studies have identified significant differences in data quality depending on electrode composition (LaBrecque and Daily, 2008), but there is no general consensus on this issue. As noted in Section 3.3, electrode composition is important when considering IP measurements. Stainless steel electrodes are commonly used as a relatively inert, inexpensive metal electrode. Copper is another common choice. Metal chemistry may become important when performing experiments that involve chemical conditions that cause redox reactions on the surface of the metals. Degradation of the electrodes may occur, although the effect on the resistivity measurements may not be obvious. Graphite, a crystalline form of elemental carbon, is an excellent electrode in corrosive conditions.

Irrespective of the electrode material used, the potential electrodes will inevitably have an open-circuit voltage difference that arises from variable redox conditions in the fluid local to electrode surfaces. This open-circuit potential arises in the absence of any source of current applied to the column. It can be recorded with a voltmeter connected across the potential electrodes when the resistivity meter is off (Slater et al., 2008). Superimposed on the electrodic potential is an additional source of voltage difference that arises if there are natural sources of current present in the column material that are unrelated to the current injected by the instrument. Such voltage differences are commonly referred to as self-potentials, representing the concept that the potential differences are associated with the electric fields generated by current sources in the material itself. The self-potential geophysical technique (see Revil and Jardani (2013) for review) relies on measuring such voltages caused by these natural current sources. In resistivity measurements, the sum of the electrodic and self-potentials represents noise (V_{sp}), which must be removed from the total voltage difference recorded during current injection to accurately know the voltage difference across the potential pair just from the injection of external electrical current (Figure 3.5). When using a continuous square or sine wave function with a high enough frequency, these unwanted voltages represent the DC offset of the alternating voltage (Figure 3.5a). When using time domain measurements that rely on turning the current on and shutting it off for an extended period of time, these unwanted voltages are measured during the off-period (Figure 3.5b).

3.2.3 Field Instruments

Field resistivity instruments employ similar concepts to those described for laboratory instruments in Section 3.2.2. The main differences are (1) the current flow is now 3D rather than 1D, (2) much larger currents must be injected into the ground, and (3) additional hardware for connecting a large number of electrodes to the instrument is needed. The simplest field instrument is a transmitter with a single-channel receiver that connects directly to the four electrodes (the two current electrodes and the two potential electrodes). Up until the late 1980s, this was the most common field resistivity instrument and was used extensively (Figure 1.1). Each new measurement required that all four electrodes were moved. In the 1990s, single-channel imaging instruments that can switch the current injection and potential reading locations between a limited number (e.g. less than 100) of electrodes became commercially available. Today, more sophisticated instruments address large (100 or more) numbers of electrodes using multiple channels (Figure 3.6). Field resistivity instruments most commonly use the time domain measurement approach described earlier (Figure 3.5b).

Two key elements of modern resistivity instruments are (1) capacity to address multiple receiver channels simultaneously, and (2) multiplexing capability. Multiple channels and multiplexing capability both support fast acquisition of a large set of measurements suitable for performing 2D or 3D resistivity imaging (Figure 3.6). Some instruments just rely on a large number of channels (+100) to perform a single set of transfer resistances during the injection of current between a pair of electrodes. However, it is more common for a transmitter/receiver to incorporate a multiplexer, which is a mechanical switching unit. The mechanical switching process involves turning on or off a bank of switches, where each switch is associated with a single electrode. The switching unit diverts the current to a specified pair of injection electrodes and identifies which pairs of electrodes are being used to measure voltage differences on the receiver channel(s). The switching takes a finite amount of time, and the speed of the multiplexer is important in determining the data acquisition rate.

3.2.3.1 Field Transmitters

The power output by a field resistivity instrument must be high enough to ensure that enough current is injected to produce measurable (above the noise) voltages at the receiving electrodes. The amount of power output by a transmitter in a field resistivity instrument is typically a compromise made by the manufacturer. On the one hand, the greater the power, the higher the voltage signal (and thus the SNR) that can be recorded with the receiver. The compromise is with respect to wanting to maintain instrument portability, along with

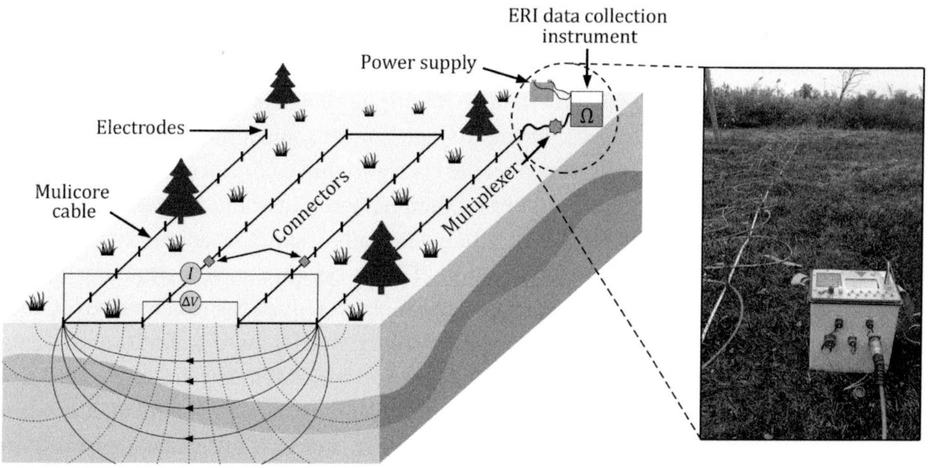

Figure 3.6 The main components of field-scale resistivity (and IP) instrumentation.

minimizing safety risks and reducing costs of components. Most modern portable resistivity meters are designed to be carried (preferably not too far) by an individual. These meters usually have the power source (i.e. a battery) and transmitter electronics included along with the receiver electronics in a single instrument. The power output of such instruments is typically 200–250 watts, although for shallow applications (e.g. archaeology) lower-power units are used. These instruments inject up to a few amperes of current into the ground using voltages less than 1,000 V. They are designed for investigating the top ~50 m of the subsurface when electrodes are placed on the surface. Greater investigation depths, and higher SNR, are achieved by using higher-power external transmitters that are synchronized with the resistivity receiver. Such transmitters are capable of providing +10 kW of power and designed to provide +20 amperes of current. These external transmitters are heavy and drastically reduce field portability. Use of high-power transmitters also significantly increases the danger of injury from electric shock. Robust safety procedures should be developed when working with external transmitters.

3.2.3.2 Field Receivers

Similar to laboratory instruments, a key requirement of the receiver is high input impedance (typically +100 MΩ). Field receivers are also characterized by their resolution, precision and measurable voltage range. The resolution of the recorded reference voltage is usually better than 1 μV with a peak-to-peak maximum voltage around 10–15 V. This means that reliable resistivity measurements can be determined from voltage differences on the order of a few mV, assuming that noise levels are low enough to ensure an adequate SNR. As noted earlier for laboratory instruments, the measurement resolution is a function of the full-scale range of the receiver, and variable gains are used to measure this wide range in voltages more accurately. This is even more of an issue for the field receivers as the wide range in electrode geometries used for acquiring field data results in a very wide range in measured voltage differences. Some manufacturers argue that higher (e.g. nV) resolution on the receiver and associated signal processing can compensate for a lower-power transmitter to get comparable results to those obtained from standard transmitters.

Resistivity receivers are configured with multiple channels to allow simultaneous measurement of multiple voltages during a single current injection. Depending on the manufacturer, there may be restrictions on the configuration of the electrodes that can be addressed with the multiple receivers. For example, some instruments can only utilize multiple channels when adjacent channels (e.g. channels 1 and 2, channels 2 and 3, etc.) share a common electrode (Figure 3.7a and b). This is because such instruments do not have multiple true differential voltage input channels to reduce instrument costs. Other instruments that have been developed are even more restrictive, e.g. limiting the data acquisition to pole–pole-type measurements, where all measurements are referenced to a single potential electrode as shown in Figure 3.7c (see Chapter 4). However, some instruments with full matrix multi-channel capability, i.e. true differential capabilities (e.g. the five channels selected in Figure 3.7d can be simultaneously acquired), are available.

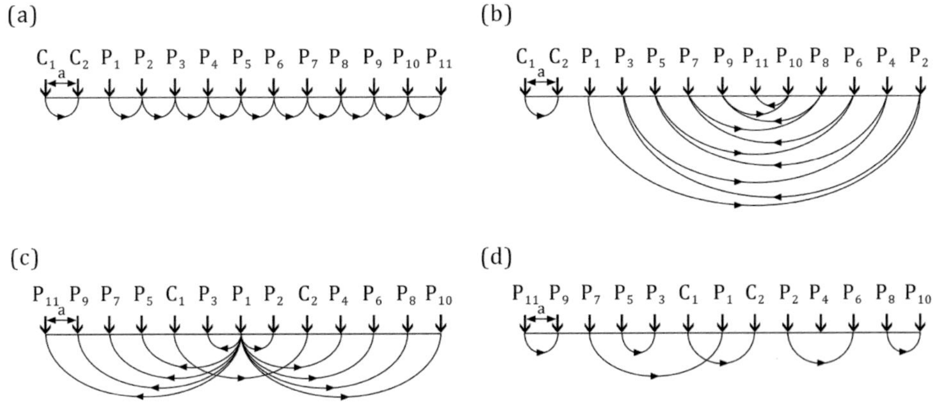

Figure 3.7 Examples of some possible measurement configurations for a ten-channel resistivity system. Some receivers require a common electrode to be shared across adjacent channels (as is the case in (a)–(c)). Configuration (d) would require a receiver with fully independent differential channels.

3.2.3.3 Multiple Transmitter Instruments

A recent instrumentation development in field-scale electrical imaging involves simultaneous injection of currents across multiple channels (e.g. Yamashita and Lebert, 2015a). In fact, this concept has been extensively explored by the biomedical tomography community for some time (Gisser et al., 1987). Such biomedical applications have focused on developing an optimal set of multiple current injections to enhance the resolution of a target (see also the pioneering concepts of Lytle and Dines (1978)). In contrast, the motivation for recent field instrument developments is to reduce the data-acquisition time by removing the standard limitation on resistivity instruments that only a single pair of electrodes can be used for current injection at one specific time. Yamashita and Lebert (2015) also argue that SNR can be improved. This novel, and to date little-used, approach to field measurements requires very specific waveforms and code-division multiple-access (CDMA) coding of the injected currents, thereby allowing the contribution of each current (transmitted at a specific pair of injection electrodes) to be determined in the total voltage recorded at the potential electrode pairs (Yamashita et al., 2014, 2017). The Syscal Multi-Tx instrument (Iris Instruments, France) incorporates three channels to simultaneously inject between three electrode pairs, with six receiver channels allowing a total of eighteen measurements to be acquired simultaneously.

3.2.4 Monitoring Systems

Automated resistivity monitoring systems (Van et al., 1991) can be left in place for extended periods of time to capture the evolution of changes in the resistivity of the subsurface associated with a wide range of hydrogeological and biogeochemical processes (LaBrecque et al., 1996a). Automated resistivity monitoring involves establishing a monitoring system that is programmed to acquire data periodically and to be left in place for some period of time. Some systems (e.g. MPT-Iris, USA) are specifically designed with long-term monitoring as the primary application. Manufacturers of some conventional resistivity meters provide functionality to utilize the instrument in a monitoring mode, e.g. the Iris Instruments (France) and GuidelineGeo/ABEM (Sweden) systems.

There is an increasing interest in the development of systems that have low power requirements to facilitate autonomous data acquisition in remote places 'off the grid'. Figure 3.8 shows a 'permanent' resistivity monitoring system that was established to monitor solute transport from an agricultural field into a drainage ditch. This system ran for over two years, with the temporal sampling interval modified so that repeat datasets were rapidly acquired during storm events, with much less frequent sampling performed during dry conditions (Robinson et al., 2019). The system addressed 192 electrodes on a 3D grid and acquired almost 16,000 measurements in less than half an hour.

Figure 3.9 shows the application of a resistivity monitoring system deployed for an ~18-month period to record the transformation of hydrocarbons that contaminated beach sediments following the Deepwater Horizon oil spill that occurred in the Gulf of Mexico (Heenan et al., 2014). This monitoring system provided information on the biodegradation of the oil, which resulted in a progressive decrease in resistivity over time (Figure 3.9c). This monitoring system recorded data twice daily at a site on an uninhabited island, with power supplied by a bank of 300 W solar panels. The monitoring system required two maintenance trips during the 18-month period to address hardware malfunctions. However, the transmitter/receiver unit of this system was a conventional geophysical instrument that

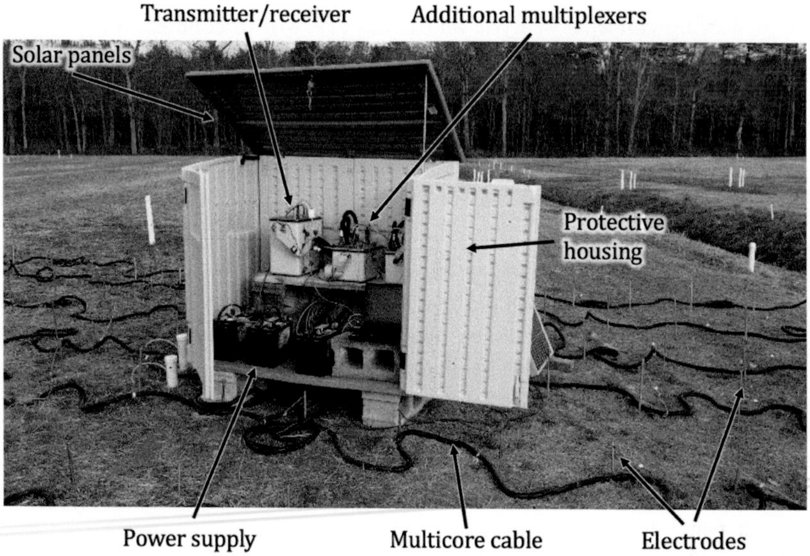

Figure 3.8 Basic components of an electrical resistivity monitoring system. In this case, the system addresses a total of 192 electrodes on a rectangular grid, with power to the batteries supplied either from the mains or solar power.

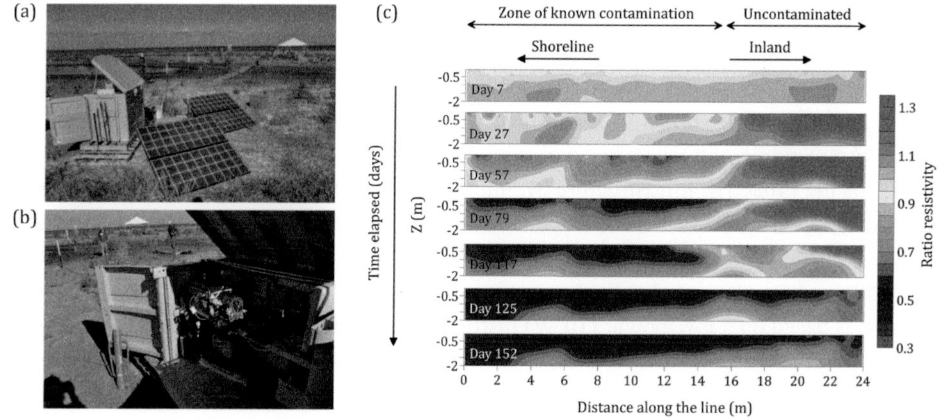

Figure 3.9 Long-term electrical resistivity monitoring system set up at a site on the Grand Terre (GT1) barrier islands off the southeastern coast of Louisiana, USA, to monitor the degradation of oil-contaminated beach sediments: (a) site photo showing instrument storage, solar panels and electrode array; (b) close-up of resistivity meter being used for monitoring; (c) resulting time-lapse sequence of ratio changes in resistivity (unitless) recorded with this system. Data from Heenan et al. (2014). The system addressed a total of 96 electrodes (48 on the surface and 48 in boreholes) twice daily for an 18-month period. (A black and white version of this figure will appear in some formats. For the colour version, please refer to the plate section.)

could have been replaced with a lower-cost data acquisition system that would have lessened the power requirements.

Although this was a successful monitoring experiment, conventional resistivity/IP instruments are often not well configured for a dedicated monitoring role, resulting in developments of dedicated monitoring hardware and software better suited for the job. In cases where the application is relatively small-scale and shallow, it is possible to construct a resistivity monitoring system around a general data-logging instrument, as long as it includes an analogue channel output and multiplexer functionality. Many relatively small-scale applications of resistivity do not require the +200 W transmitters used in conventional resistivity instruments. Numerous hydrological and biogeochemical processes of interest may be occurring in the upper ~5 metres of the subsurface that can be investigated with low-power (e.g. ~20 W) instruments. Sherrod et al. (2012) describe one such system that was constructed around a Campbell Scientific data logger to address vertical borehole arrays containing 96 electrodes installed within the unsaturated zone. The development of open-source hardware and software provides further opportunities to construct inexpensive resistivity logging instruments for shallow studies. These loggers can be powered from a modest solar panel assembly, making them suitable for long-term monitoring in remote environments, with routine wireless communication of system diagnostics and acquired data to a remote server. These systems are much less expensive than the conventional resistivity and IP monitoring systems on the market, especially when open-source hardware is used to develop the monitoring interface.

Example applications of resistivity monitoring systems include investigations of (1) surface water–groundwater exchange along river channels (Johnson et al., 2012a), (2) soil moisture dynamics in the unsaturated zone (Winship et al., 2006, discussed in Section 6.1.4), and (3) effectiveness of environmental remediation technologies (Ramirez et al., 1993). An example of a recently developed, commercially available resistivity monitoring system is the Proactive Infrastructure Monitoring and Evaluation (PRIME) system (Figure 3.10) developed by the British Geological Survey (Chambers et al., 2015). This system was originally developed to detect early evidence of problems with railway embankments in the United Kingdom associated with changes in slope stability, e.g. due to moisture dynamics. PRIME is based on a low-power (10 W) instrument developed around a modular design. It is configured to simultaneously record information on environmental sensors (e.g. rain gauges, moisture probes) that can trigger resistivity data acquisition events during times of interest (e.g. during rainfall events). The commercial package supports autonomous data acquisition with pre-configured software for remote transfer of command files and datasets.

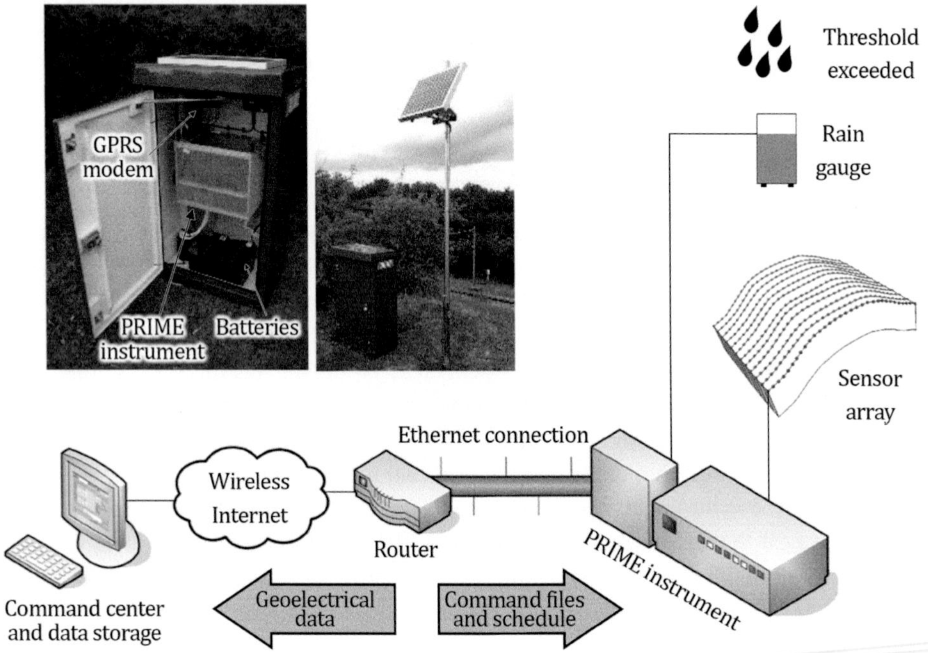

Figure 3.10 The low-power PRIME monitoring system developed by the British Geological Survey (BGS). Photos show close-up of system with off-the-grid deployment on a railway embankment. Flowchart shows data acquisition workflow, including measurements triggered by a threshold recorded with an environmental sensor (in this example, a rain gauge). Figure based on concept provided by Jon Chambers (British Geological Survey).

3.2.5 Surface Electrode Equipment

3.2.5.1 Surface Cables

Electrodes (Figure 3.11a) are usually connected to the resistivity meter via multicore cables (Figure 3.11b). Electrode cables are typically constructed considering a trade-off between functionality and weight/durability. It is desirable to use a multicore cable constructed from wire of the smallest acceptable diameter (gauge) that will support the maximum current output by the instrument after an appropriate safety margin is applied. The multicore wire is embedded in a rugged, waterproof casing to increase durability in field deployment. Longer cables allow greater separation distances between electrodes and facilitate imaging to greater depths. However, longer cables are bulkier and more challenging to transport/deploy in the field. Each wire in the multicore cable connects to an electrode 'take-out'. The electrode take-outs are spaced at regular intervals (defining the electrode spacing) along the cable (Figure 3.11a). Electrodes either normally attach directly to the take-out on the multicore cable or via a jumper lead spanning between electrode and cable (Figure 3.11b). The cables connect to a centralized multiplexer that controls the switching of electrode functioning.

Multicore electrode cables are manageable by one field operator, although for health and safety reasons at least two operators are recommended. Cables are usually dual-ended and the instrument is typically placed in the centre of the survey line. A single multicore cable capable of addressing 24 electrodes with 5 m spaced electrode take-outs will typically weigh 15 to 20 kg (including cable drum), making a 48-electrode survey easy to manage on relatively flat terrain. Multicore cables capable of being connected to larger arrays will often require more cable reels to ensure the cables are manageable. For example, a 96-electrode system with 5 m take-outs may comprise 6 cable reels, each connected to 16 electrodes: 3 reels running to the left, 3 reels running to the right, with connector boxes to join each cable triplet. A 96-electrode system with 10 m take-outs would then require 12 cables in total (around 200 kg in total for the cables and cable drums), thus requiring significant labour.

3.2.5.2 Smart Electrode Take-Outs

An alternative to the multicore cable is to use smart electrode take-outs where there are switches and data controllers at each location of a cable where the electrode is connected (Figure 3.11c). In this way, the switching is done by these distributed data controllers instead of at a central multiplexer. One advantage of this approach is that the number of cores in the surface cable is substantially reduced relative to a passive multicore cable. Four wires are needed to provide the connection between the instrument and the current electrodes and potential electrodes. A few additional wires are needed to communicate with the controllers at each electrode (to turn them on and off as an active potential or current electrode as needed to populate the survey). This can result in lighter cables with obvious benefits for field data acquisition. Another advantage of this technology is that two different electrodes can be used at each location, with one assigned only for current injection and the other assigned only for potential readings. Such an approach can improve resistivity (and IP in particular) data quality by avoiding voltage sensing on electrodes that have been polarized as a result of being used for current transmission. This is not possible with a conventional multicore cable and central multiplexer. The disadvantage of this approach is the increased cost, since each electrode requires a controller unit.

3.2.5.3 Surface Electrodes

Similar to the laboratory, the composition of the electrode is often not a major concern in resistivity measurements, although LaBrecque and Daily (2008) identified differences in data quality with metal type. It is standard practice today to use stainless steel electrodes, although most metals (or graphite) can be used. The electrodes are most often metal rods machined to a point to approximate the point source of current injection assumed in modelling resistivity data obtained during a field survey (see Section 4.2.1). However, resistivity electrodes can never be a true point, as more surface area of the metal is needed to reduce the contact resistance between the electrode and the ground. Consequently, electrodes are hammered into the ground, resulting in some length of metal electrode being in electrical contact with the soil. All other things equal, the contact resistance of the electrode decreases with the length of electrode inside the ground and the cross-sectional area of the electrode. In dry ground, contact resistance can be very high. In these situations, it is common to infiltrate water with high ionic content (e.g. a salt solution) into the soil around the electrode, which can dramatically reduce electrode contact resistance. A similar improvement can be obtained by packing wet clay around the electrode, having the advantage of retaining moisture close to the electrode for longer than achievable by infiltrating water into the soil. However, in some instances the percolation of water into the subsurface or the placement of a clay pack can disrupt the imaging study, e.g. in hydrogeophysical work focused on examining shallow flow and transport processes. Other approaches to reducing contact resistance include large plate electrodes in larger-scale surveys. This can present a problem in the modelling of resistivity datasets, which normally assumes a point source of current injection and a point measurement of the electric field strength at the potential electrodes. Ochs and Klitzsch (2020) highlight how 3D modelling of non-point source electrodes is needed when conducting shallow (upper few metres) subsurface investigations where electrode length in contact with the ground is a significant fraction of the electrode separation. As a general rule, the electrode length or width should be less than 5 per cent of the electrode separation to approximate the point source assumption. Advanced modelling approaches discussed in Section 4.2.2.8 accommodate non-point source electrodes (see also Johnson and Wellman, 2015), but this is not routinely implemented in commercially available software packages.

Hammering rods into the ground becomes impractical in the presence of hard surfaces (e.g. exposed bedrock, concrete, ice). Such surfaces also tend to be electrically resistive, resulting in high contact resistances even if rods are hammered in. A number of approaches can be taken to address this problem. One involves the use of metal pads, where the large surface area of the pad placed onto the ground reduces the contact resistance relative to a rod electrode. Pads can be mounted on a saturated ion-rich medium (e.g. a clay pad or even a sponge) to further reduce the contact resistance. Another option is to use porous pots, where

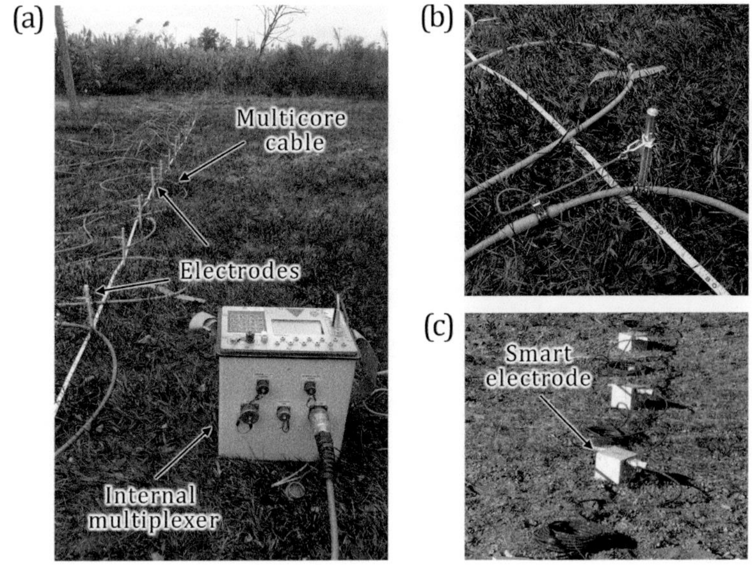

Figure 3.11 Examples of electrodes and cables for resistivity measurements: (a) standard setup of a resistivity system with internal central multiplexer connected to an array of electrodes by multicore cable; (b) close-up of take-out on multicore cable connecting to electrode; (c) smart electrode setup where distributed switching occurs at each electrode.

Box 3.2

Contact resistance

The contact resistance refers to the resistance to current flow (strictly an impedance) exerted by the electrode–ground interface. The contact resistance associated with the current injection electrodes limits the amount of current injected into the ground by a resistivity instrument. It therefore also limits the SNR recorded between potential electrode pairs. Good contact resistances are on the order of a few kΩ or less, although contact resistances in the 10s of kΩ range are still usually acceptable for resistivity measurements. However, such contact resistances may severely limit the measurement of the much smaller IP signals discussed in Section 3.3. The contact resistance is a function of the ground conditions (particularly the water content) and the electrode. Different electrodes will result in different contact resistances at the same location. In the figure, the porous pot electrode has a significantly higher contact resistance than the stainless steel and graphite electrodes. Contact resistance decreases with increasing electrode surface area in contact with the soil. The figure shows that the contact resistance along a single line can vary substantially even when field conditions appear uniform. Resistivity instruments record the total resistance between successive pairs of electrodes to be used for current injection and alert the operator to problematic electrodes. The total resistance between the pair of electrodes recorded by the instrument is the sum of the two contact resistances (one at each electrode) and the resistance of the subsurface material between the electrodes per the two-electrode measurement shown in Box 3.1.

Box 3.2 (cont.)

Contact resistances associated with three types of electrode materials along a 2D resistivity line where site conditions appear uniform. The Cu-$CuSO_4$ pot has a notably higher contact resistance than the metal stakes driven into the ground.

the metal electrode sits in a fluid-filled chamber that makes electrical contact with the ground through a porous membrane (e.g. a ceramic or wood). Such electrodes are often used in self-potential measurements, where they are referred to as non-polarizing electrodes when both metals are identical and submerged in a saturated solution of a salt of that metal (Petiau, 2000). Common non-polarizing electrodes are Cu-$CuSO_4$ and Pb-$PbCl_2$ pots. In the latter case, the $PbCl_2$ is in the form of a gel rather than a liquid. These porous pots can be used to inject current as well as measure the resulting voltage differences. For resistivity measurements, it is not necessary for the pots to be identical in composition or chemistry, as the electrodes are polarized when they are used for current injection.

Surface water is usually an excellent ionic conductor and facilitates implementation of resistivity surveys by reducing the contact resistance at electrodes. In fact, electrode cables can be placed in surface water with the cable take-out (if appropriately designed) serving as the electrode. Shallow water (e.g. wetlands, edges of stream/river channels and lakes) measurements can be done this way. Cables can be configured to float on the surface or sink to the bottom (assuming the cable insulation stays watertight under pressure). Floating arrays of electrodes have been designed to be pulled behind a boat for waterborne continuous resistivity surveying (Figure 3.12a), the use of which is discussed in Chapter 4. In the case of marine resistivity surveys in saline water, graphite (a non-metal element that is a good conductor) electrodes are an excellent choice as they are resistant to corrosion. Pulled arrays that rely on galvanic contact can also be used on land when the ground surface is relatively conductive, e.g. in ploughed agricultural fields after recent rainfall (Figure 3.12b). For agricultural applications, pulled array systems have been developed based on rotating discs as electrodes: these discs cut into the soil and result in excellent electrical contact in agricultural fields (Figure 3.12c). With the technology pulled behind a tractor, large areas can be rapidly covered (see Section 4.2.2.2).

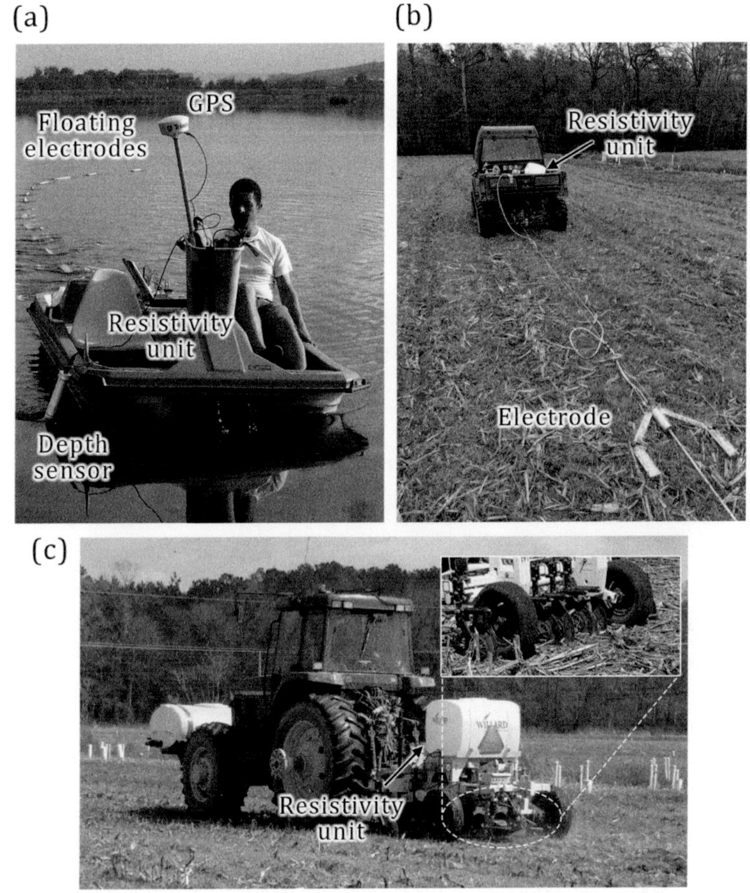

Figure 3.12 (a) Pulled 2D array over water in a shallow water wetland (Mansoor and Slater, 2007); (b) pulled 2D array over an agricultural field; (c) VERIS 3100 (Veris Technologies, USA) proximal soil sensing system where four rotating blades serve as the electrodes for a continuous apparent resistivity measurement.

A relatively recent technological development is capacitively coupled electrodes designed to permit continuous resistivity profiling across resistive ground where it is challenging (often impractical or impossible) to make adequate direct galvanic contact (Walker and Houser, 2002). Capacitively coupled electrodes facilitate continuous surveying with pulled arrays over concrete/asphalt, bedrock, frozen ground (Hauck and Kneisel, 2006) and other high-resistivity surfaces (Figure 3.13). Capacitively coupled electrodes usually result in lower-quality (noisier) data than would be obtained with standard galvanic measurements, and reliable IP measurements are currently not feasible with capacitively coupled electrodes. However, these novel electrodes have resulted in new applications of resistivity surveys that would otherwise not be practical.

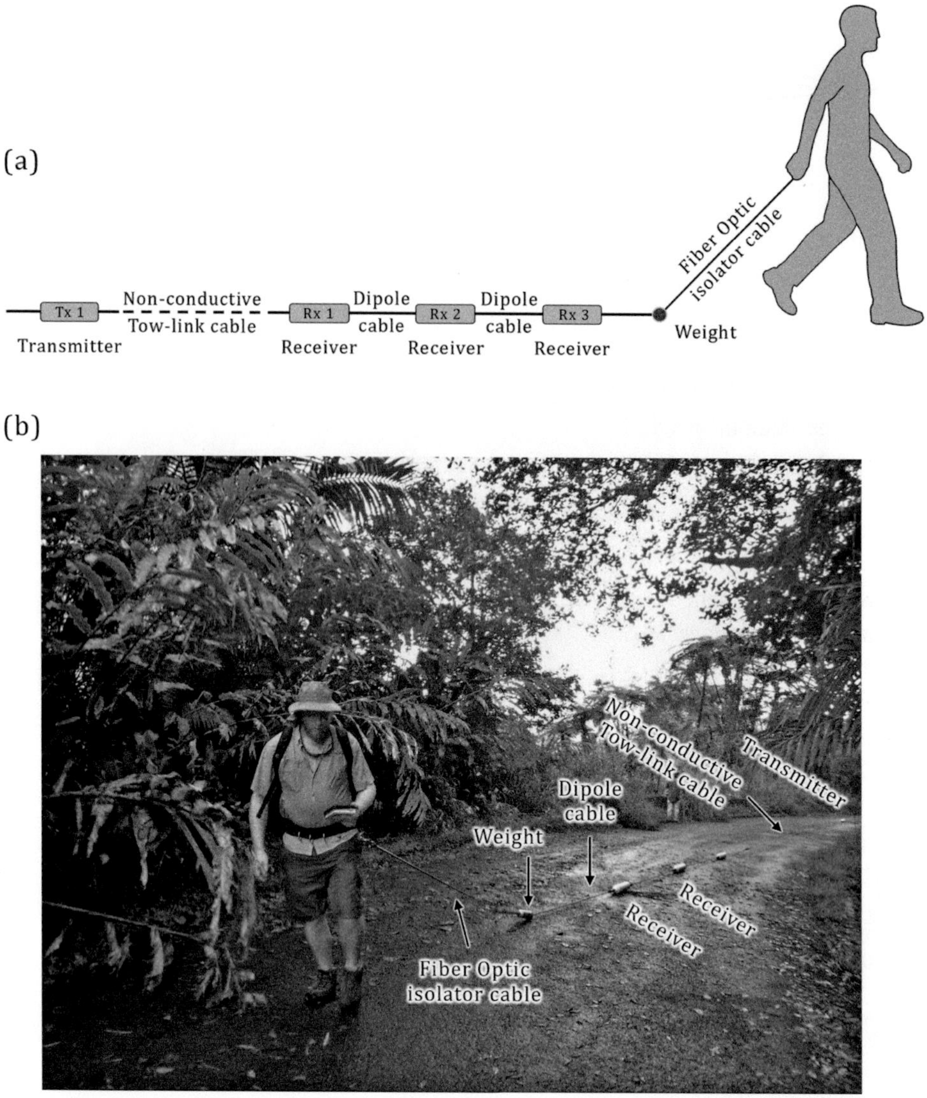

Figure 3.13 Continuous resistivity profiling over asphalt using the OhmMapper (GeoMetrics Inc., USA), a capacitively coupled resistivity system; (a) schematic of setup and (b) field implementation.

3.2.6 Borehole Electrode Arrays

Borehole electrode arrays can be constructed in a variety of ways, depending on whether the arrays are to be used below the water table or above it. Borehole electrode installations ideally require open (uncased) boreholes. Pre-existing boreholes are most likely to be open in bedrock locations, although in unconsolidated sediments above bedrock they are usually cased to prevent infilling. PVC-cased boreholes can be used, but electrodes can only be placed in areas where the casing is slotted, thereby making electrical contact with the formation. Groundwater monitoring wells may be accessible for deploying electrode arrays over the screened interval. Borehole arrays should generally not be deployed in pre-existing wells that are cased with metal, as casing will dominate the electrical current flow and prevent any useful information on the resistivity structure of the formation from being obtained. One exception is in specialized imaging surveys, where the casing is energized as a long electrode (Ramirez et al., 1996; Rucker et al. 2011) (see also Section 4.2.2.7.2). This unconventional approach requires the numerical modelling of a line source of current flow instead of the conventional point source assumption.

Deployment of electrodes below the water table is more straightforward than above the water table, as direct galvanic contact is made between the electrode and the water filling the hole. Some equipment manufacturers supply electrode cables specifically designed for implementation in water-filled boreholes. However, it is relatively straightforward to construct low-cost borehole electrode arrays from a combination of PVC tubing, wire, stainless steel mesh (or other electrode material) and an ample supply of cable ties/duct tape (Figure 3.14). One concern with borehole arrays is that the injected electrical current will be preferentially channelled along the relatively electrically conductive borehole rather than flowing into the more resistive sediments/rock (Osiensky et al., 2004; Nimmer et al., 2008). This effect can be reduced by using resistive packers to help electrically isolate individual electrodes from each other (Binley et al., 2016). In exceptional circumstances, it may be worth the effort to use inflatable packers (Figure 3.14c) to fully isolate electrodes from one

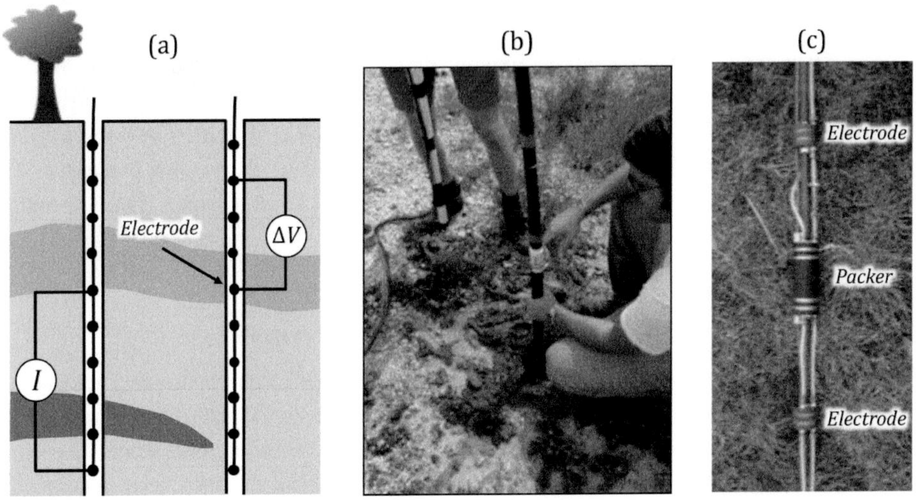

Figure 3.14 (a) Concept of cross-borehole electrical imaging; (b) example of inexpensive electrode arrays constructed from PVC pipe being inserted into a borehole; (c) advanced cross-borehole array complete with packers for isolating sections of the borehole to reduce current channelling through the conductive borehole fluid.

another along the hole (Robinson et al., 2016). This effort should be saved for highly resistive environments, such as imaging in bedrock where the resistivity contrast between the water-filled borehole and the surrounding rock is high.

Borehole electrode arrays are much more challenging to use above the water table. As electrodes suspended in air will be disconnected, the borehole must be backfilled with material that will provide adequate electrical contact (i.e. a sufficiently low contact resistance) at the electrodes. The challenge is to backfill the open borehole in a way that all electrodes are in contact with the backfill. In unconsolidated sediments, it may be sufficient to backfill the borehole with the native material left over from drilling the borehole. Alternatively, an electrically conductive grout (e.g. a clay slurry) can be poured (or pumped) down the borehole. Contact resistance checks (Box 3.2) between pairs of electrodes should be used to assess those electrodes that have been successfully connected to the ground versus those where contact has not been established. In most cases, backfilling the holes means sacrificing these electrode arrays for a one-time use (in loose, unconsolidated materials, it may be possible to retrieve electrode arrays by pulling them out of the backfill). Another solution above the water table is to install the electrodes on the outside of a FLUTe® flexible liner (Keller, 2012). These liners are filled with water so that the outside of the liner makes a secure connection to the borehole wall.

Although resistivity arrays are generally not installed in metal-cased holes as the casing severely alters the electrical current flow, attempts have been made to perform four-electrode measurements in cased wells by accounting for the highly conductive casing in the modelling (e.g. Schenkel (1994); see also Section 4.2.2.7.2). As noted earlier, it is possible to use the metal borehole casing as long electrodes in specialized imaging surveys. It may also be possible to install electrode arrays into unconsolidated sediments via a direct push technology. Direct push relies on pneumatically pushing instruments and sensors into the ground without drilling a borehole. Existing direct push tools include sensors for making a single four-electrode resistivity measurement, as the head of the device is advanced through the subsurface (Schulmeister et al., 2003). Pidlisecky et al. (2013) describe a direct push strategy for installing small electrode arrays for studying vadose zone processes. We discuss borehole imaging approaches further in Section 4.2.2.7.2. Chapter 6 also provides several case studies illustrating their use.

3.3 Induced Polarization Measurements

Most commercially available resistivity instruments are designed to simultaneously measure the IP effect in addition to the resistivity of the subsurface. However, the acquisition of reliable IP measurements is much more challenging than acquisition of resistivity data alone. The extra challenge is related to the much smaller signal associated with the interfacial polarization relative to the signal associated with electromigration of the charge carriers. The measured imaginary conductivity (σ'') is on the order of 0.1 to 0.001 times σ' for small phase (φ) angles less than 100 mrad (typical when in the absence of electron conducting minerals) and assuming a minimum measurable φ of 1 mrad (Equation 2.61). Consequently, the SNR of the IP measurement is orders of magnitude smaller than the resistivity measurement. Furthermore, IP measurements are susceptible to systematic errors associated with the instrumentation (e.g. composition of electrodes, layout of cabling) that do not exert a significant effect on the resistivity. These problems are compounded when SIP measurements are made to higher frequencies, where most instrumentation errors become more significant.

IP measurements can either be obtained in the frequency domain or the time domain. Pullen (1929) describes some of the earliest laboratory observations of the IP effect. He observed the resistivity variation with time that is indicative of IP and also observed that resistivity varies as a function of frequency. Box 3.3 discusses the differences between time and frequency domain measurements and explains why frequency domain measurements are generally preferred for laboratory instrumentation and time domain measurements are more often made in field applications. Field IP measurements can be acquired in the frequency domain, where φ is directly recorded in addition to $|\rho|$. However, it is more common for field IP instruments to measure in the time domain, where in the case of the simplest measurements, the recorded parameters are indirectly related to φ and depend on the instrument configuration. Another challenge with IP measurements is a meaningful interpretation of the acquired measurements. Such challenges are discussed in this section.

3.3.1 Laboratory Measurements

The objective of laboratory IP measurements is to obtain the best possible information on the complex electrical properties of the sample. As noted in Box 3.3, the most accurate characterization of the sample is achieved by directly measuring the frequency-dependent phase spectrum over the widest possible frequency range in the frequency domain. A swept-sine signal is used to measure the complex impedance as a function of frequency, usually reported as a magnitude ($|Z|$), and phase angle (φ). The phase angle of a complex impedance measurement is negative for a polarizable medium (although see discussion on negative IP effects in Section 4.3.1), consistent with the impedance of a capacitor in an electrical circuit. The magnitude and phase angle are combined with the geometric factor to express the measurement as a complex resistivity (ρ^*) or a complex conductivity (σ^*) (Box 2.6). $|Z|$ and φ are measured over a range of discrete frequencies. The larger the frequency range, the more information can potentially be extracted from the measurements by analysing the frequency dependence. Time domain laboratory IP instruments have been developed (Hallbauer-Zadorozhnaya et al., 2015), although time domain measurements are most commonly made in field studies as described later.

3.3.1.1 Sample Holders

Acquisition of reliable IP data requires more attention to the characteristics of the sample holder if measurements need to be made with 0.1 mrad or better resolution. Most importantly, the electron conductive parts (usually a metal but can be carbon in the case of graphite) of the potential electrodes must be 'outside' of the direct path of the current flow lines through the sample (Vanhala and Soininen, 1995). If they are not, it is very easy for these electrodes to be polarized and cause an anomalous low-frequency phase response unrelated to the physical properties of the soil/rock under investigation. The electrode is polarized when the electronically conductive part of the electrode extends far enough into the sample to be transected by current flow lines. If a voltage gradient exists across the metal electrode (almost inevitable when electrodes fall inside the current pathway) then a current is induced in the metal. Ions diffuse around the metal and generate a measurable electrode polarization (Section 2.3.7) effect resulting in φ errors that can reach 10 mrad when the electrical potential difference across the electrode is large. Consequently, the electron conducting part of the electrodes must always be placed outside of the direct flow of the current through the sample holder. In this case, there is no voltage difference across the

Box 3.3

Ways of measuring induced polarization

The objective of IP measurements is to determine the reversible charge storage properties of soils and rocks. In laboratory investigations, the objective is increasingly to collect the full magnitude and phase spectrum over a wide frequency range. The phase spectrum is measured in the frequency domain, whereby a sinusoidal waveform is swept across a range of frequencies, and the phase lag (φ) of the recorded voltage waveform relative to the current waveform is accurately measured. The problem of the frequency domain swept-sine measurements is that the time taken to collect the data gets long as the low frequencies are approached. For example, a swept-sine measurement from a few kHz down to 0.001 Hz might take up to 8 hours, whereas the time might be reduced to approximately 40 minutes when terminating the lower frequencies at 0.01 Hz. This time constraint is generally not a concern when the objective is to obtain gold standard spectral measurements on laboratory samples.

When acquiring field datasets, survey time (proportional to survey cost) is often a concern and 'time domain' IP measurements are generally preferred, although more specialized equipment for making frequency domain field measurements is available. In the time domain, the charge storage effect is captured in the transient voltage decay following current shut-off. Depending on the length of the duty cycle used for the IP measurements, a single time domain current injection/measurement will only take 10s of seconds to a minute to acquire. Most often, the objective of the time domain survey is to measure just the magnitude of the time domain chargeability effect, which quantifies the strength of the secondary voltage decay relative to the primary voltage. Another option is to model the decay curve in terms of a relaxation (e.g. Cole–Cole) model as described in Section 2.3.4. However, with the development of time domain IP instruments that record the full waveform, there is increasing interest in determining the equivalent phase spectrum from the time domain curve. Digital signal processing techniques can be used to transform the time domain response to an equivalent phase spectrum, e.g. through a Fourier transformation. In theory, this means that there is equivalent information in the time domain decay curve and the frequency domain spectrum. In practice, it is hard to sample the decay immediately after current shut-off with sufficient time resolution to capture the higher frequency part of the phase curve, especially given the presence of noise in the data. However, the bigger limitation of attempting to acquire full spectral data in field time domain measurements is the phase errors that result from capacitive and EM coupling effects associated with the wiring and electrodes.

Box 3.3 (cont.)

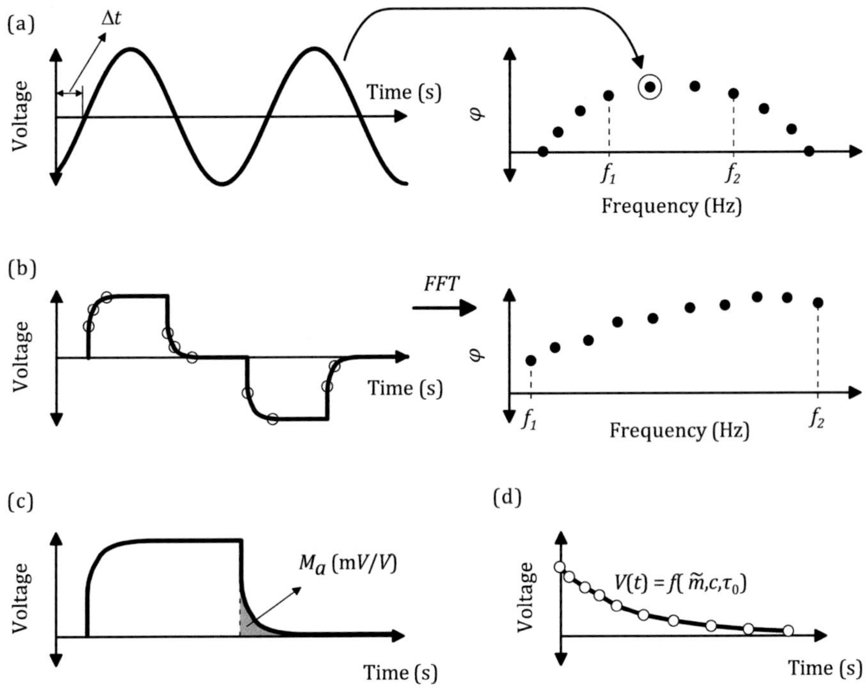

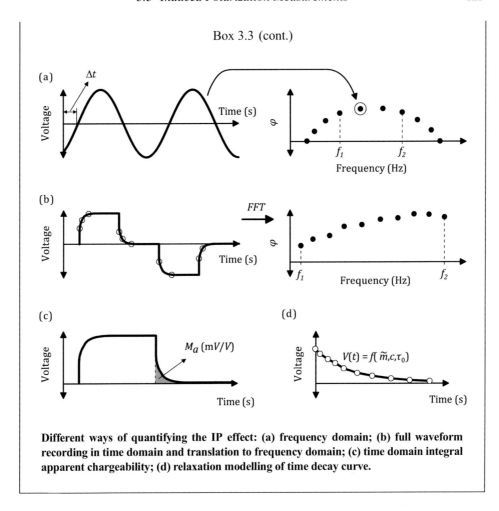

Different ways of quantifying the IP effect: (a) frequency domain; (b) full waveform recording in time domain and translation to frequency domain; (c) time domain integral apparent chargeability; (d) relaxation modelling of time decay curve.

electron conducting part of the electrode. As the electron conducting part of the electrode is pushed further into the sample, the low-frequency errors grow progressively larger. As a result of this effect, potential electrodes that cover the cross-section of the sample holder (permissible for resistivity measurements alone with some advantages as discussed earlier) should never be used in IP measurements. Point electrodes placed in chambers on the edge of the sample holder are a common solution, with the electron conducting part of the electrode in contact with the sample via the fluid filling the chamber (Vinegar and Waxman, 1984) and far enough from the current flow path. Some researchers prefer the use of ring electrodes placed in grooves around the circumference of the sample holder (Zisser et al., 2010a). This helps to average out the recorded potential difference between electrode positions in the case of significant sample heterogeneity. However, a limitation of the ring electrode for IP measurements is that any voltage difference along the potential electrode may result in a spurious phase shift due to polarization of the electrode.

Figure 3.15 illustrates the importance of removing the potential electrodes from the direct current path by showing measurements on a water sample when the metal part of the electrode is fully contained in a fluid-filled chamber maintaining electrolytic contact with the sample versus when the electron conducting part extends into the sample. The polarization of the potential electrodes is more pronounced when the potential electrodes are not perfectly orthogonal to the 1D current flow as the resulting voltage difference across the electrode driving the polarization is larger.

Some researchers also argue that the distance between the current injection electrodes and the potential electrodes needs to exceed a certain minimum distance below which phase artefacts are generated (Kemna et al., 2012). Zimmermann et al. (2008a) argue that the distance between current and potential electrodes on each side of the sample should be at least twice the sample width to avoid errors associated with the polarization of the current electrodes. Such anomalous phase errors can be removed by increasing the distance between current and potential electrodes.

As various aspects of the sample holder have the potential to cause spurious phase errors, it is critical to calibrate the sample holder and ensure adequate performance prior to acquiring research-grade IP datasets. This can be done experimentally by filling the sample

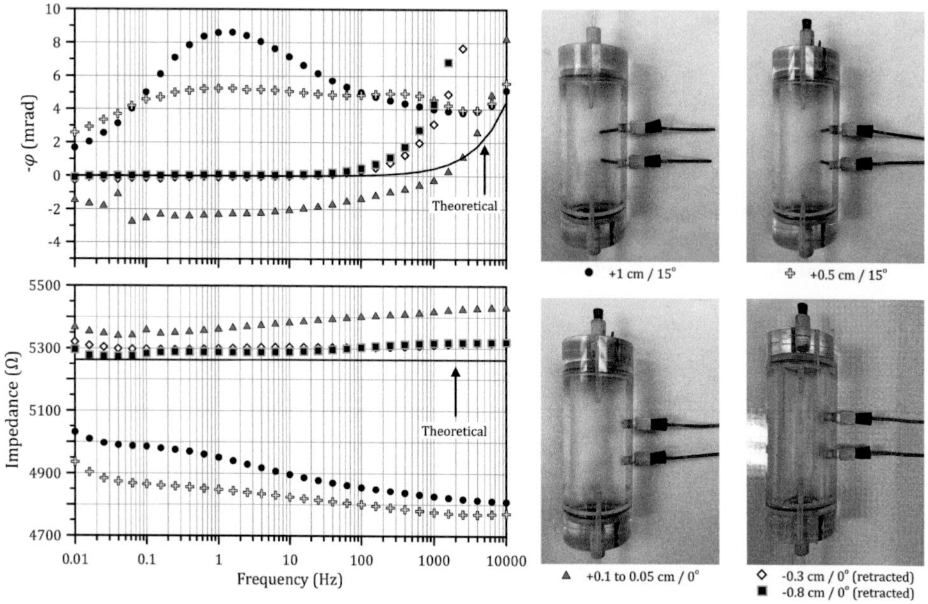

Figure 3.15 Example of the low-frequency errors in IP data that result from polarization of the electron conducting part of the potential electrode when it is inserted into the sample holder and cut by current path lines. The data are shown for a water sample, where the theoretical impedance and phase (calculated from Equation 3.5) are shown ($-\varphi$ plotted in impedance (Z) space). Photos show potential electrode configurations with the metal varying from 1 cm inserted into current flow path (top left), all the way to 0.8 cm retracted away from the current flow path (bottom right) (data credit Chen Wang, Rutgers University Newark).

holder with water of a precisely known conductivity and dielectric permittivity (Vanhala and Soininen, 1995). Water is a conductor but does not have any interfaces, so no IP phenomenon will exist. However, the dipolar polarization of water molecules does exist and can result in measurable phase signals above 1,000 Hz when the water is of sufficiently low conductivity. The theoretical response of a water sample can be modelled as

$$\sigma^* = \sigma_0 + i\omega\kappa\varepsilon_0, \tag{3.5}$$

where σ_0 is the DC conductivity, ε_0 is the permittivity of free space (equal to 8.854×10^{-12} F/m) and κ is the unitless relative dielectric permittivity associated with the dipolar polarization mechanism. The value for σ_0 can be determined from a measurement made with a specific conductance probe. The value of κ for water is temperature dependent and equal to 80.1 at 20° C. Figure 3.16 shows magnitude and phase measurements over the frequency range 1 Hz to 20 kHz for a water sample where σ_0= 0.01 S/m along with the theoretical response based on Equation 3.5 and corrected values based on a procedure described in Wang and Slater (2019). The absolute error in φ increases with frequency due to the impedances associated with the potential electrodes and also the impedance between the negative potential electrode and instrument ground. However, phase errors below 1,000 Hz are very low and indicate that the sample holder is well designed and devoid of electrode polarization effects.

3.3.1.2 *Laboratory Instruments*

Laboratory IP instruments use the same principles as resistivity instruments but must reliably record the polarizability of the sample in addition to the conductivity. As previously noted, laboratory IP instruments usually operate in the frequency domain, where the phase lag of the voltage waveform across the sample relative to the injected current waveform

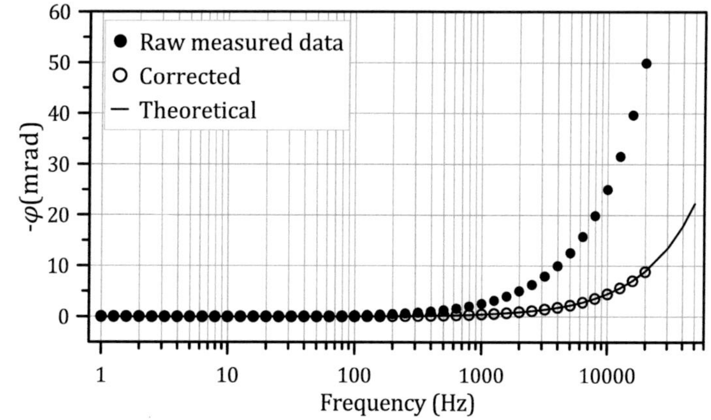

Figure 3.16 Testing the performance of a sample holder by making measurements on a NaCl water sample with a known electrical conductivity of 0.01 S/m with a theoretical phase response according to Equation 3.5 (solid line). Filled circles are raw measurements and open circles are corrected values after applying a correction procedure described in Wang and Slater (2019). (measured phase in impedance space, $-\varphi$).

must be accurately recorded (Figure 3.17). The current waveform is recorded on a precision resistor and provides the waveform against which the recorded voltage waveform on the sample is referenced. Single or multi-channel instruments are available, with some instruments providing multiple current source channels in addition to multiple receiver channels (e.g. the PSIP by Ontash & Ermac, USA). The instruments typically employ a swept-sine wave function similar to that used in impedance spectroscopy testing of electronic components. A sine wave is generated for a number of discrete frequencies across the measured frequency range, and the magnitude and phase is recorded for each frequency. One consideration in such measurements is the time it takes to reliably record the phase difference (and magnitude) at the low frequencies. Whereas this measurement is obtained in a few seconds or less at frequencies above 1 Hz, the measurement time increases by a multiple of 10 with each successive order of magnitude decrease in frequency. A common lowest measurement frequency is 10^{-3} Hz, where it may take more than an hour to acquire an accurate phase measurement.

Laboratory IP instruments must record the sample impedance only and not be contaminated by the impedances associated with the electrodes and wires making connections to the sample or the instrument electronics itself. A key requirement of IP instruments is a very large input impedance on the receiver channels (typically 10^9 ohms or greater) so that no

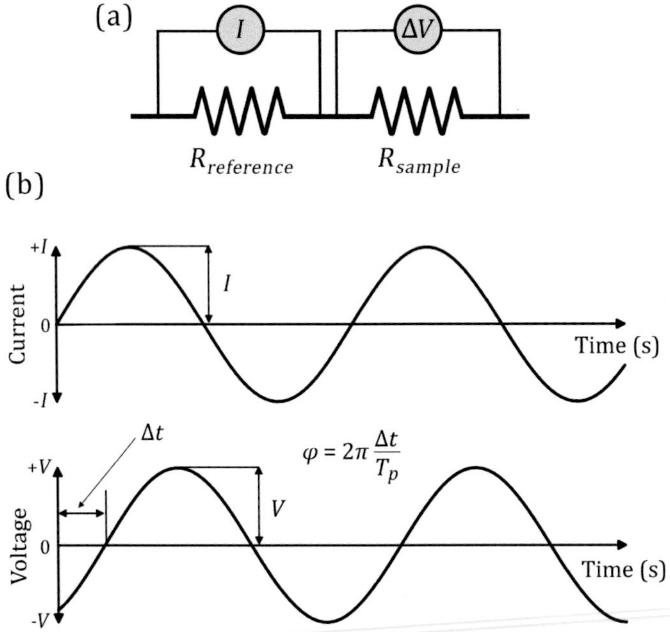

Figure 3.17 Concept of frequency domain SIP measurement where a sinusoidal current waveform is output and the resulting voltage waveform is recorded across the sample with the current recorded on a precision reference resistor (a). The magnitude (V) of the voltage waveform and the phase lag (φ) of the voltage relative to the current waveform of period T_p are recorded (b).

leakage of current occurs through them. This also prevents polarization of the potential electrodes.

Current density was considered under laboratory resistivity measurements where the trade-off between larger current densities improving SNR versus driving excessive polarization of the current electrodes was discussed. Current density is an important consideration in IP acquisition as the SNR of IP measurements is typically 2–3 orders of magnitude smaller than for the resistivity measurements. Excessively high current densities can result in non-linear impedances at the current electrodes that are manifested as phase errors in the low-frequency part of the phase spectrum. These non-linear impedances are expressed as voltages not scaling proportionally to applied current (following Ohm's law) and the creation of excitation currents with harmonics for constant voltage supplies (Zimmermann et al., 2008a). These errors decrease with increasing frequency as the more rapid reversals of the current direction limit the charge-up at the electrode. An extreme example of the problem is shown in Figure 3.18, where SIP measurements are made as a function of current density for a brine ($\sigma_w = 18$ S/m) with a theoretical phase response of ≈ 0 mrad. The erroneous, large, low-frequency phase angles recorded at high current densities decrease as the current density is reduced.

The community that developed the IP method for mineral exploration in the 1970s avoided high current densities due to this reason. Sumner (1976) recommended avoiding the use of current densities greater than 1 mA/m^2, which seems a very conservative estimate. However, the IP research community of the last few decades has not expressed great concern about non-linear effects at high current densities, in part because they are

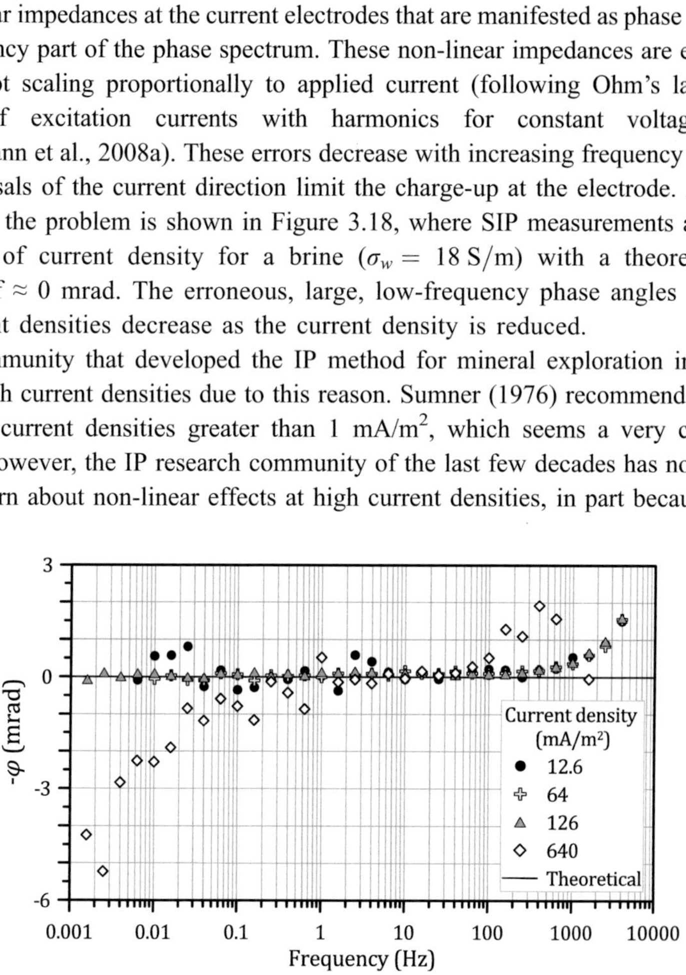

Figure 3.18 Influence of current density on phase spectra for measurements made on a highly conductive (18 S/m) brine. Low current density (12.6 mA/m^2) data are noisy because the recorded signal is very low. High current density (640 mA/m^2) data contain large low-frequency phase errors due to polarization of the current injection electrodes. The intermediate 64 and 126 mA/m^2 data are devoid of errors other than those at high frequencies resulting from the electrode impedance discussed in Section 3.3.1.3. (phase in impedance (Z) space, $-\varphi$).

rarely observed with modern instrumentation. Vanhala and Soininen (1995) found that the IP effect of their samples did not change for current densities varying from 0.01 mA/m^2 to 200 mA/m^2.

The quality of the IP measurements is foremost dependent on the sample holder design rather than the instrument. The performance of the instrument itself can be easily evaluated by making measurements on an electrical circuit composed of precisely measured components where the theoretical magnitude and phase response can be calculated.

3.3.1.3 Electrodes for Laboratory Measurements

The design of electrodes for IP measurements requires more consideration than for resistivity measurements alone. As previously noted, the placement of the electrodes in the sample holder is a critical factor; potential electrodes in a chamber outside of the current path are an absolute necessity. Beyond this, the main consideration is the impedance of the potential electrodes, as these impedances generate an additional phase unrelated to the sample response that limits the reliability of high-frequency measurements. Ag-AgCl electrodes are popular potential electrodes as this junction can reduce the electrode impedance relative to a native wire electrode (all other factors being equal). These can be made by immersing silver wire in household bleach for a few hours. Commercially manufactured Ag-AgCl electrodes developed for biomedical sensing are also available. Tables describing the impedance properties of different metals can be found in the electrochemistry literature. However, the composition of the metal electrode is less important than the surface area of the metal that is in contact with fluid. The greater this surface area, the lower the electrode impedance (all other factors being equal).

The phase errors associated with the electrode impedances, as well as any additional impedance between the negative potential electrode and instrument ground, increase with frequency and become a major limitation on the acquisition of high-quality IP data beyond a few hundred Hz, particularly for more resistive samples. These impedance errors can be quantified following procedures described in Zimmermann et al. (2008a) and Wang and Slater (2019). The phase associated with the additional impedances is calculated from an approach that involves additional measurements to estimate these impedances along with an assumption on, or calculation of, the input capacitance of the measurement device. The phase errors can include negative IP effects (i.e. where the polarity of the phase is opposite of that associated with a charge storage effect and consistent with an induction effect). In laboratory measurements, these apparent negative IP effects result when there is a large difference in the impedance of the two potential electrodes (Wang and Slater, 2019). Negative IP effects also originate at the field-scale, as will be discussed in Section 4.3.1. Using such correction procedures, accurate four-electrode IP measurements can be acquired up to the 10s of kHz range. The importance of applying such correction procedures will in large part depend on the electrical properties of the sample and the desired upper limit of the measured frequency range. In the case of highly conductive samples, these unwanted impedances are low and corrections may not even be necessary. However, these correction procedures are critical, even at frequencies of a few hundred Hz, when measurements on relatively resistive samples (e.g. low porosity rocks, unsaturated soils/rocks) are made (Figure 3.19).

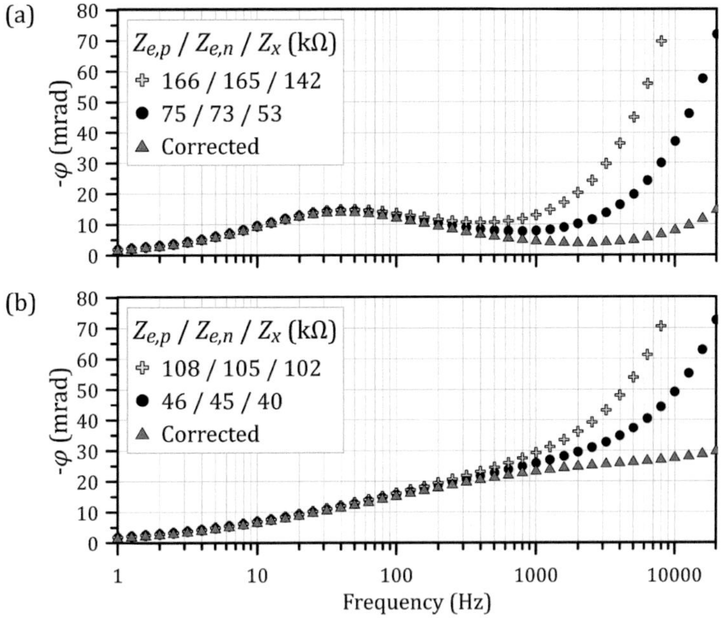

Figure 3.19 Examples of effects of sample holder potential electrode impedances on high-frequency phase measurements. The figures show measured phase spectra for different potential electrode impedances ($Z_{e,p}$ is positive electrode and $Z_{e,n}$ is negative electrode). The corrected phase after removal of the effects of these electrode impedances (Wang and Slater, 2019) is shown: (a) pyrite-sand sample with 5% pyrite by volume (b) kaolinite-sand sample with 10% kaolinite by volume. In both cases, the sample is saturated with a 0.01 S/m NaCl solution and Z_x denotes the sample impedance (phase in impedance (Z) space, $-\varphi$).

3.3.1.4 Two-Electrode Dielectric Spectroscopy Measurements

The complex electrical response of colloidal suspensions is traditionally measured with a two-electrode (Box 3.1) technique commonly referred to as dielectric spectroscopy (see Asami, 2002, for review). This technique can also be used to study polarization processes in porous media, most typically from about 10 kHz up to the MHz range (Knight and Nur, 1987; Chelidze et al., 1999). Polarization of the current electrodes is minimized at high frequencies, such that a reliable measurement of the electrical properties of a sample is obtained. The two-electrode measurement is challenging at low frequencies due to the large contribution of electrode polarization to the measured response (Box 3.1). The main advantage of the two-electrode technique is its simple implementation, especially on smaller samples. Dielectric spectroscopy measurements are normally presented as a complex dielectric permittivity ($\varepsilon^* = \sigma^*/\omega$). They capture Maxwell–Wagner polarization mechanisms (Box 2.5) across a wide-frequency range and may also record electrical double-layer polarization (the IP effect) towards the lower-frequency range of the measurements. At the high-frequency end, the molecular polarization of the constituents making up the sample is captured (Box 2.5). Electrode polarization removal techniques have been developed to extend the two-electrode approach to frequencies below 100 Hz (Prodan and Bot, 2009). Although the approach is typically not used in IP measurements, when combined with four-electrode measurements, it provides a strategy for broadband measurements across the widest possible frequency range (Lesmes, 1993).

3.3.2 Field Instruments

3.3.2.1 Time Domain Systems

Field IP transmitters and receivers utilize the principles previously described for resistivity measurements except that a reliable measurement of the polarizability of the subsurface must be captured. Although frequency domain field IP instruments do exist and utilize the same principles as described for laboratory IP instruments, field IP measurements are commonly made in the time domain. This requires a modification to the recording of the time domain resistivity waveform such that the IP response is captured. Most commonly, the transient voltage decay following charge-up and subsequent shut-off due to the applied current waveform is recorded. In the case of a completely non-polarizing medium, the voltage recorded across the sample would immediately drop to zero on termination of the applied current. When the subsurface is polarizable, a transient voltage decay is instead recorded in response to the discharging of localized charge perturbations in the electrical double layer (EDL). An identical, except inverted, charge-up curve is observed when the current is turned on, representing the analogous charge-up of the EDL (Figure 3.20).

The most common time domain IP waveform is a 50 per cent duty cycle (the fraction of a period that the signal/instrument is active) square wave (Figure 3.20). As with the standard waveform used in resistivity, the polarity of the waveform is reversed to minimize current electrode polarization effects. At least two pulses with opposite signs are injected, although the pulse train is usually repeated (and stacked) to improve the SNR. Ideally, the on-period (charge-up) and shut-off period (discharge) would be long enough to fully capture the polarization effect, causing some researchers (particularly from the Russian IP community) to advocate for long pulse durations (Sumner, 1976). However, this is usually impractical and relatively short duration pulses (typically 1–8 seconds) are used. Shorter duration pulses may not allow sufficient off-period for full charge/discharge to occur during a cycle (Figure 3.21). An additional problem is that polarization in the first cycle may manifest in the second cycle, and even subsequent cycles (see e.g. Fiandaca et al., 2012).

The selection of the pulse duration and number of stacks also has a big impact on the measurement time. For 2-second pulses involving two stacks the measurement would take 16 seconds. Recently, use of a 100 per cent duty cycle has been proposed, whereby the IP measurements are taken during the current on-period, avoiding a need for the off-period (Olsson et al., 2015). This effectively results in a superposition of the charge-up and discharge IP response, and has the added benefit of reducing data acquisition time (theoretically halving it as there is no need for the off-period). The charge-up curve can also be used to quantify the IP effect, although common practice remains to use the decay curve.

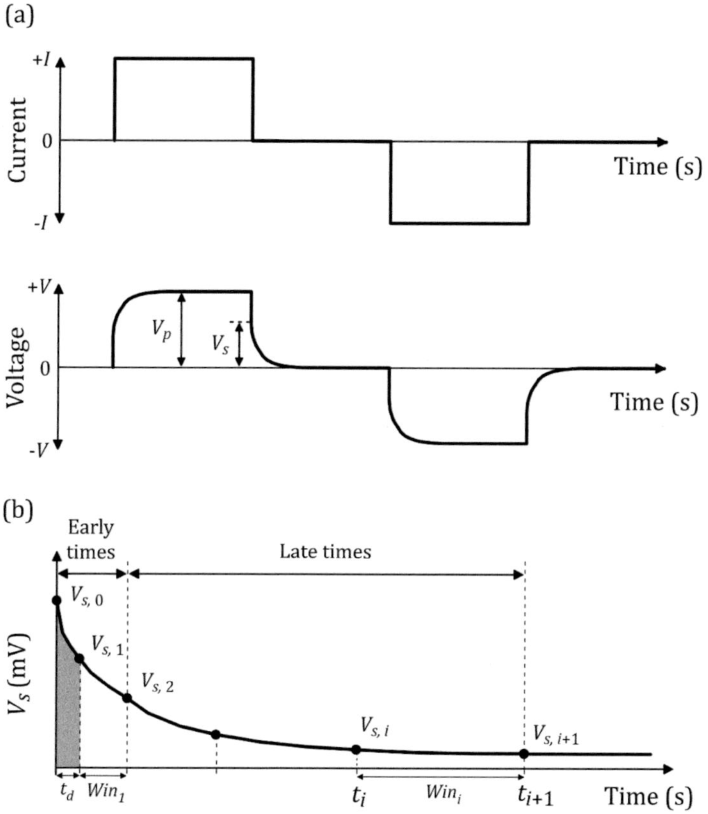

Figure 3.20 (a) Standard 50% duty cycle time domain IP waveform with (b) enlargement of decay curve showing the sampling of the voltage (V_s) decay via IP windows (*Win*) following current shut-off used to estimate the time domain apparent chargeability (t_d is a delay time).

Schlumberger (1939) first proposed quantifying the strength of the polarization from the ratio of the voltage (V_s) recorded after current shut-off relative to the primary voltage recorded during the transmitter on time (V_p). It was A. S. Polyakov, a Russian geophysicist, who first referred to this ratio as 'chargeability' (Seigel et al., 2007). IP instruments employ a delay time (t_d) prior to start of sampling to minimize the impact of higher-frequency inductive and capacitive coupling effects (Section 3.3.2.4) on the polarization measurement. We refer to these time domain measures of the IP effect as apparent chargeability (M_a) to distinguish them from the intrinsic chargeability of the medium ($\hat{m}$) (Box 2.8).

The unitless measure of the instantaneous time domain IP apparent chargeability is

$$M_a = \frac{V_s}{V_p}, \tag{3.6}$$

where V_s is the instantaneous secondary voltage recorded immediately at current shut-off. Although unitless, it is common practice to report M_a with the units of mV/V as V_s is typically 2–3 orders of magnitude smaller than V_p (except in the case of metals where V_s can reach 100s of mV/V). In practice, it is not easy to reliably record the instantaneous value of V_s at current shut-off, and it is most common to quantify the IP effect from the integral of the decay curve broken into time windows (fractions of a signal defined between two times). For a single time window defined between two times t_2 and t_1 after shut-off,

$$M_a = \frac{1}{(t_2 - t_1)} \frac{\int_{t_1}^{t_2} V_s dt}{V_p}, \tag{3.7}$$

where M_a is once again unitless but commonly expressed in mV/V. In some cases, the time domain IP parameter is expressed as

$$M_a = \frac{\int_{t_1}^{t_2} V_s dt}{V_p}, \tag{3.8}$$

which has units of time and is conventionally expressed in milliseconds.

Time domain IP measurements are often a source of confusion because M_a will differ to some degree between instruments, depending on how the manufacturer and/or operator configures the instrumentation to quantify the decay curve. For historical reasons, the units of mV/V (Equation 3.7) are more frequently used in the environmental and engineering community, whereas the mining community prefers the use of the units of msec (Equation 3.8). Different values of M_a will be returned depending on the choice of the integral times t_1 and t_2. Shorter charge-up periods will result in smaller apparent chargeabilities than longer charge-up periods (Figure 3.21). This largely explains attempts by the mining community to adopt standards (e.g. the Newmont Standard, Box 3.4) to standardize acquisition of time domain IP datasets (in terms of waveform and the time windows used to sample the decay curve) so that it

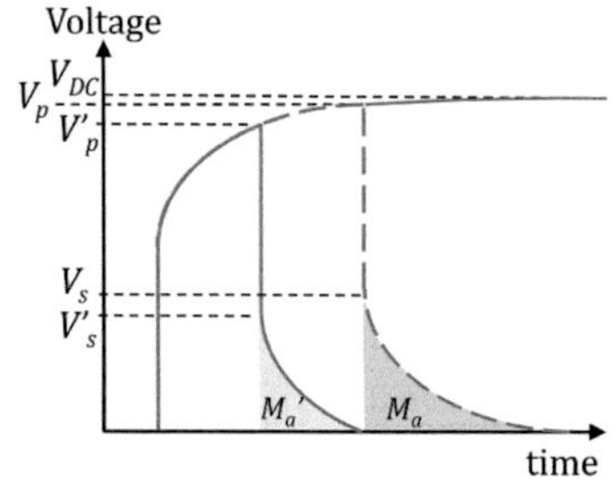

Figure 3.21 Effect of time on period on an IP measurement. The solid grey line shows the measured voltage over a short time on period, the dashed grey line shows the measured voltage over a longer time on period. V_{DC} is the DC voltage (obtained for an infinitely long current step). V_p and V_s are primary and secondary voltages for a relatively long charge period. V'_p are V'_s are equivalent voltages for a relatively short charge period. The apparent chargeability for the short period (M_a') is less than for the long period (M_a).

is possible to directly compare apparent chargeability measurements from different surveys. However, these limitations of time domain IP measurements made with different instruments and different settings are not so well acknowledged by geophysicists in the environmental/engineering sector, making it sometimes challenging to quantitatively compare results between different environmental/engineering surveys.

The quality of time domain IP measurements foremost increases with the IP signal amplitude, although other factors that promote data quality include use of longer duration current pulses and avoidance of electrode configurations that result in large geometric factors (Gazoty et al., 2013).

3.3.2.2 Estimating Relaxation Model Parameters from Time Domain Measurements

Time domain decay curves can, in principle, be modelled directly based on relaxation model (e.g. Cole–Cole, see Box 2.8, Chapter 2) behaviour. Swift (1973), Tombs (1981) and Johnson (1984) outlined such a method (see also Duckworth and Calvert, 1995), which has received recent attention (e.g. Fiandaca et al., 2012). Tombs (1981) explored the time domain response of the Cole–Cole model of Pelton et al. (1978) (Box 2.8, Chapter 2) written in terms of a measured impedance ($Z(\omega)$) in ohms,

$$Z(\omega) = R_0\left[1 - \widetilde{m}\left(1 - \frac{1}{1 + (i\omega\tau_0)^c}\right)\right], \tag{3.9}$$

where R_0 is the DC resistance. The time domain voltage response ($V(t)$) for a finite current pulse (I_0) of duration (t_p) is given by

$$V(t) = \widetilde{m}I_0R_0\sum_{n=0}^{\infty}\frac{(-1)^n}{\Gamma(nc+1)}\left[(t/\tau)^{nc} - \left(\left(t + t_p\right)/\tau_0\right)^{nc}\right], \tag{3.10}$$

where Γ is the gamma function. Tombs (1981) showed that the relaxation model parameters in Equation 3.10 are poorly resolved when fitting time domain curves resulting from finite current pulse durations typical of field instruments (e.g. a few seconds). Furthermore, the series in Equation 3.10 converges very slowly (i.e. the summation must be made over a large number of values of n), making it difficult to use in practice. Guptasarma (1982) developed a digital linear filter to represent Equation 3.10 based on 21 filter coefficients, offering a more practical alternative, particularly if multiple relaxations are to be accounted for in the model. The situation improves for an infinite current pulse, but measurements approximating this condition are impractical in field applications. Tombs (1981) reached a somewhat negative conclusion regarding estimation of relaxation model parameters directly from time domain measurements, stating that the approach is '... *unlikely to be able to perform any useful discriminatory function except for recognition of electromagnetic coupling*'.

Such limitations are partly overcome by the approach of Komarov (1980) based on the differential polarizability,

$$\eta_d(t) = \frac{d\eta(t)}{d(\log t)}, \tag{3.11}$$

where $\eta(t) = (V_t/V_0)$ is the polarization-induced time variation in the voltage in response to a current step of infinite duration and V_0 is the voltage at the end of the current on-period (Figure 3.22). Long pulse duration (10s of seconds of more) measurements can be used to approximate an infinite time step. Alternatively, Titov et al. (2002) showed that $\eta_d(t)$ curves calculated from pulses of different durations can be reliably superimposed to represent a broad range of time-scales. Unlike the monotonous decay of $\eta(t)$, $\eta_d(t)$ contains a maximum at a time that is close to the inverse of the critical frequency of the relaxation observed in the frequency domain. In fact, the shape of $\eta_d(t)$ becomes similar to the shape of $\varphi(1/\omega)$ recorded with a frequency domain measurement. The differential polarizability has been successfully used to determine relaxation model parameters from time domain IP datasets (Titov et al., 2002, 2010a). More recently, Gurin et al. (2013) inverted IP time domain decays for relaxation time parameters of a Debye decomposition model (Section 2.4). Tarasov and Titov (2007) present an approach to capture the full relaxation time distribution from time domain IP measurements.

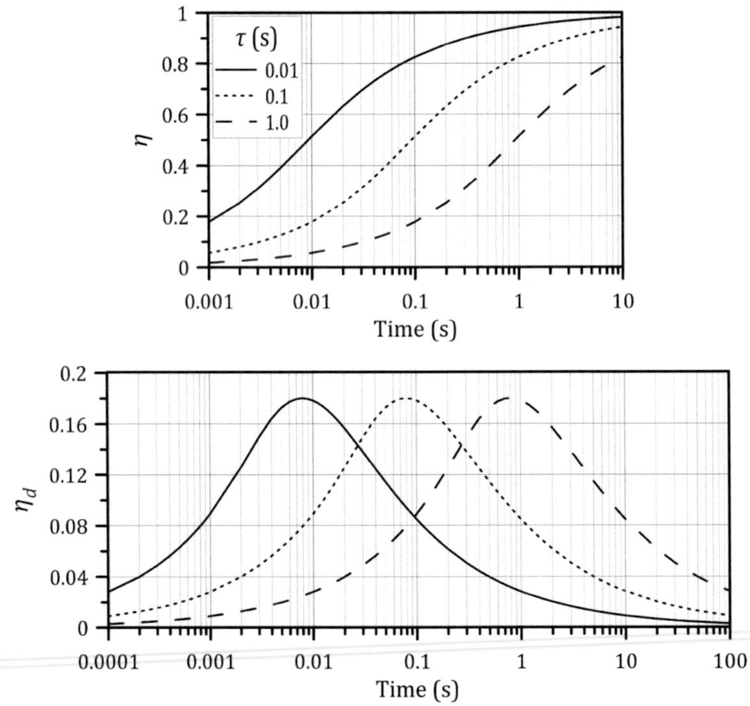

Figure 3.22 Direct estimation of relaxation times from IP decay curves using the differential polarizability concept introduced by Komarov (1980) for different dominant relaxation times (τ). Modified from Titov et al. (2002).

3.3.2.3 Equivalent Frequency Domain Information from Full Time Domain Waveforms

Signal processing techniques can improve the information extractable from time domain measurements. Instead of recording just a portion of the decay curve and estimating IP parameters from Equations 3.6–3.8, the full time domain waveform is recorded with a high temporal sampling density. Full waveform processing also allows flexibility in the definition of the time windows used to quantify the IP effect, e.g. the use of tapered and/or overlapping IP time windows. Kemna (2000) illustrates the approach where high-frequency sampling of the time domain waveform of injected current and measured voltage is converted to an equivalent complex impedance measurement through Fourier analysis (Figure 3.23). Time series analysis of the full waveforms also allows for a reduction in noise levels (Olsson et al., 2016). Digital filters can be applied to remove data spikes and reduce both harmonic noise and background drift. The frequency content of the recovered spectrum from the time domain waveform will depend on the sampling rate (high-frequency limit) and length of the current on/off period (low-frequency limit). High sampling rates are required around the time of current shut-off to obtain high-frequency content. Although the spectral content of this approach may be uncertain, the phase angle at the primary frequency of the time domain waveform is recoverable using this method. Maurya et al. (2017) present results in support of obtaining comparable spectral information from advanced processing of full time domain waveforms relative to that obtained from field-based frequency domain measurements.

3.3.2.4 Frequency Domain Systems

Frequency domain SIP field instruments are more specialized than time domain systems. Ideally, these field instruments would provide the same broadband spectral information as obtained using laboratory SIP instruments described in Section 3.3.1.2. In practice, the high-frequency range of reliable field SIP measurements is limited by phase errors associated with the cables and supporting hardware needed to connect the electrodes to the receiver over much larger distances than used in the laboratory. Capacitive and inductive coupling between the wiring connecting the potential electrodes, the ground and the wiring connecting the current injection electrodes can generate spurious phase errors that increase with increasing frequency. Inductive coupling is worse in high-conductivity ground but is generally only a major concern for large electrode spacings, e.g. as used in 1D soundings (Section 4.2.1.3). In contrast, capacitive coupling is a major problem for typical 2D and 3D instrumentation designed for near-surface applications (Radic et al., 1998). In addition, the impedances at the potential electrodes generate high-frequency errors similar to those described for the laboratory systems (Section 3.3.1.3). The combined phase errors associated with such effects can limit the range of useful information in field SIP measurements to less than 100 Hz, and frequently to less than 10 Hz (depending on the instrumentation and ground conditions) without very careful consideration to data acquisition.

Capacitive coupling arises from current leakage from high- to low-potential surfaces/conductors (Dahlin and Leroux, 2012). Multicore cables worsen the capacitive coupling between the current and potential wires, between each of the two current wires, and between

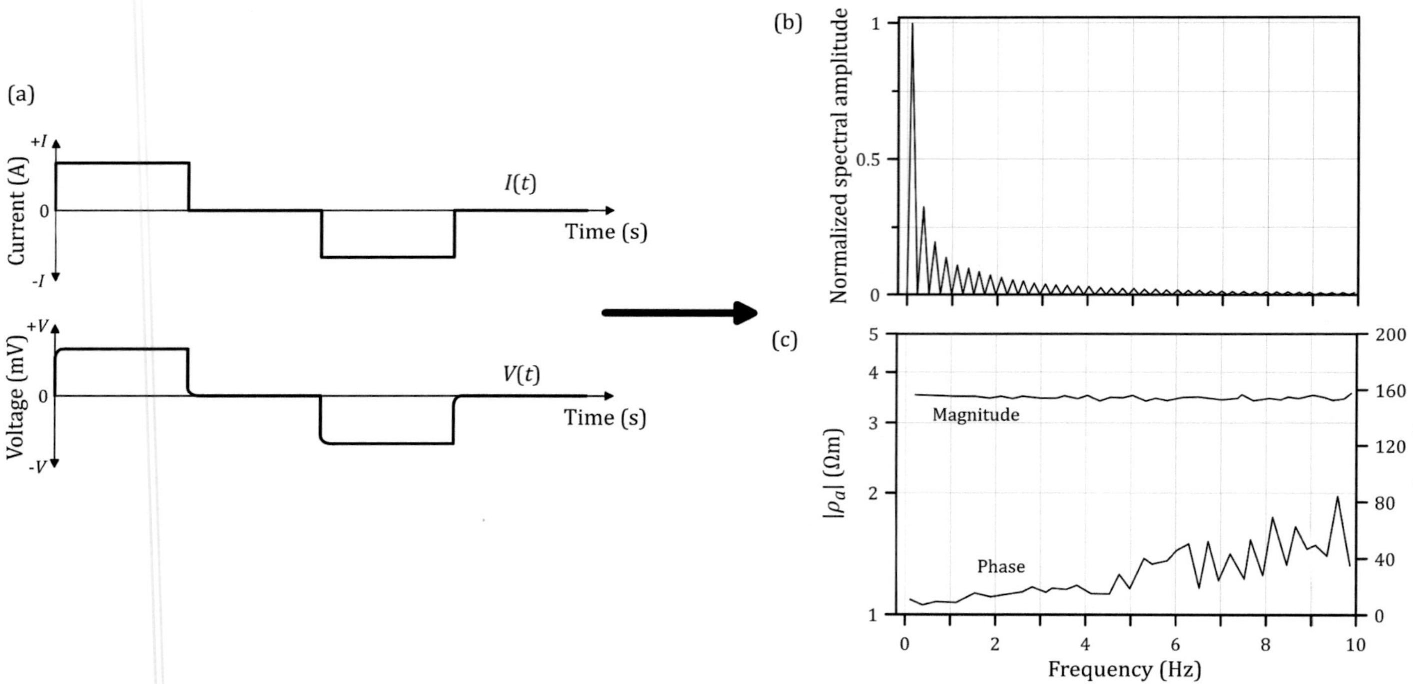

Figure 3.23 Concept of time to frequency domain conversion of a fully digitized waveform (after Kemna (2000). ($-\varphi$ indicates IP effect as plotted in resistivity space).

Box 3.4

The ambiguity of conventional time domain IP measurements

Quantification of the IP effect from the apparent chargeability (M_a) defined from the integration of the decay curve following current shut-off (Figure 3.20) depends on how the measurement is made. Longer current periods (i.e. longer on-period and off-period) will result in larger measured chargeabilities (all other factors being equal) (Figure 3.21). How the time windows are selected to compute M_a will also slightly change the computed value. Therefore, it is important to keep the time domain IP (TDIP) instrument settings constant throughout a survey. A common convention used to be to integrate over one log cycle (Sumner, 1976). Direct comparisons of M_a measurements made using different instruments/operators will only be meaningful if common TDIP instrument settings are used. The mining community identified this problem during the rapid development of TDIP receivers for mineral exploration. They recognized the need for a standard TDIP acquisition configuration, leading to the widespread adoption of the configuration used by the Newmont receiver (Newmont Mining Company, CO, USA), a popular TDIP receiver during the 1970s boom in mineral exploration. This configuration, commonly known as 'the Newmont Standard', is based on a 2-second waveform with a 50% duty cycle. The apparent chargeability is calculated from the integral of the decay curve sampled between 0.45 and 1.1 seconds after current shut-off as shown below. The output of the Newmont receiver was sometimes normalized to another standard known as M_{331} (Sumner, 1976), being the equivalent M_a for a 3-second waveform with a 50% duty cycle and a 1-second integration time after current shut-off. The mining community also developed early measures of the shape of the decay curve, recognizing that the steepness of the curve is controlled by a distribution of relaxation times associated with the polarizing components of the material. The figure shows the Newmont standard, where the ratio of L to M_a (the latter an apparent chargeability per Equation 3.7) provides a measure of steepness. These simple metrics have been eclipsed by more rigorous methods to transform the decay into an equivalent distribution of relaxation times described in Sections 3.2.2.2–3.2.2.3.

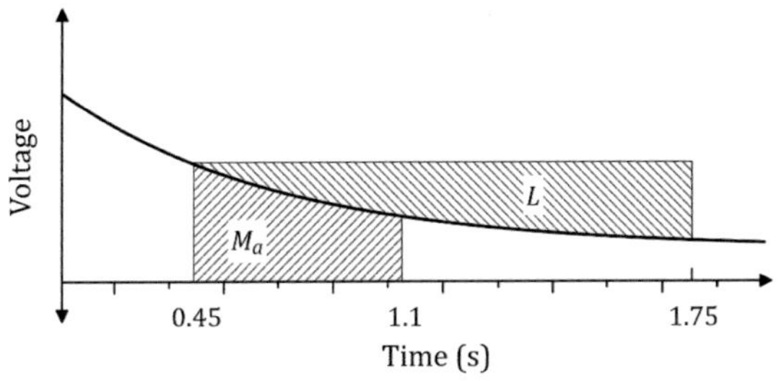

The Newmont standard involves measuring the decay time between 0.45 s and 1.1 s after current shut-off. The ratio of the areas L/M_a provides a simple measure of the shape of the decay curve. Modified from Sumner (1976).

the current wire and the subsurface. The biggest coupling effect is between the current and potential wires (Radic, 2004). Increasing the distance between the current- and voltage-carrying wires substantially reduces this capacitive coupling effect. Separating the wires used for current injection from those used for voltage measurements, e.g. by using two separate multicore cables, can improve IP data acquisition (Dahlin and Leroux, 2012).

When using large electrode spacings, inductive coupling occurs via magnetic fields and results from the mutual impedance between two lengths of wires placed on the ground. In the case of field IP surveys, the mutual impedance between the current-carrying wires and the receiver wires is the major problem. Depending on the electrical properties of the ground, either positive coupling effects (anomalous phase increases with increasing frequency) or negative coupling effects (anomalous phase decreases with increasing frequency) may be observed. The degree of coupling will depend on (1) the electrical properties of the ground and (2) the geometry of the electrodes and cables laid out on the surface and/or in boreholes. Field procedures can reduce the severity of the coupling (e.g. orienting the current-carrying wires 90° to the potential recording wires). However, coupling effects will always exist at higher frequencies and it may be necessary to model them so that they can be removed (e.g. Hohmann, 1973). A popular approach has been to describe the coupling response (inductive and capacitive combined) as a special form of the Cole–Cole relaxation model (e.g. Pelton et al., 1978). Such models approximate the coupling effect so some residual coupling effects will likely be present.

Instrumentation manufacturers attempt to minimize high-frequency errors due to coupling (and electrode impedances) in different ways. Time domain IP receivers include a delay time prior to integration of the decay curve used to estimate apparent chargeability (Figure 3.20). This delay time minimizes the effect of coupling on standard time domain IP measurements, which are then inherently restricted to a limited low-frequency range. A more sophisticated approach developed for broadband frequency domain measurements where high-frequency information is desired involves recording the current and voltage signals at the electrodes and using fibre optic cables for data transmission (Radic, 2004) (Figure 3.24). This method minimizes the direct (wire-to-wire) coupling between current and potential wires by making them as small as physically possible. Capacitive coupling between the current-carrying wires and the ground can also be reduced by use of active shielding of the current-carrying wire (Radic and Klitzsch, 2012). A limitation of such methods is the cost of dedicated electronics at each electrode, along with the impracticality of using such boxes under adverse field conditions. Another approach is based on modelling of the coupling effects based on the system geometry to minimize the instrumentation errors.

3.3.2.5 Electrodes for Field Measurements

During the development of IP for mineral exploration, field IP measurements were conventionally acquired with metal stakes as current electrodes and non-polarizing porous pot potential electrodes, as discussed in Section 3.2.5.3. Porous pots are critical for the measurement of self-potential, where the (usually small) voltages induced by natural current sources in the subsurface must be accurately recorded (Petiau, 2000). Any open circuit potential will add to the voltages from natural current sources and therefore represent noise in the self-potential measurement. The use of porous pots in IP data acquisition was to prevent development of open circuit potentials and to minimize electrode polarization

Figure 3.24 Electrode array for SIP measurements using analogue to digital signal conversion at each electrode and subsequent fibre optic transmission of recorded voltages to reduce coupling errors (SIP256, Radic Research, Germany). Each recording position has two electrodes, one for current injection and one for recording voltages.

between the electron conducting metal component of the electrode and the ground (Sumner, 1976).

Polarization of the potential electrodes is insignificant in modern instrumentation due to the very high input impedance (negligible current is drawn) of the receiver channels. Furthermore, the sum of the open circuit and the self-potential (V_{sp}) is recorded during the late portion of the current off-period in time domain measurements and is removed from both the primary voltage (V_p) and the secondary voltages (V_s) used to determine the apparent chargeability (Figure 3.5). These potentials are a DC offset in frequency domain (AC) measurements and do not adversely impact the phase measurement. However, electrodes recently used for current transmission should not be used as potential electrodes whenever possible. The injection of current into the ground polarizes (charges up) the electrode. This transient electrode polarization will interfere with the polarization signal coming from mechanisms in the ground (i.e. the IP effect) when potential recordings are made on electrodes too soon after current injection. This effect generally does not result in significant errors in the primary voltages (and hence resistivity measurement) but is a source of error for the much smaller transient secondary voltages. Some instrument manufacturers have configured IP instruments so that separate electrodes can be used at each measurement location, with one electrode being exclusively dedicated to current transmission at that point and the second being exclusively dedicated to voltage recording (Figure 3.24).

Practitioners still sometimes adopt the use of porous pots for the potential electrodes based on the historical practice. In formulating recommendations for IP research, Ward et al. (1995) described their use as an '*article of faith that deserves questioning*' and advocated for the use of a universal electrode (for both current and potential electrodes) for more efficient operation and true reciprocal measurements. Indeed, there is no loss in IP data quality when using standard metal electrodes instead of porous pots as potential electrodes (all other elements of the data acquisition being kept identical) (Dahlin et al., 2002b; Zarif et al., 2017). In fact, these studies indicate that standard metal electrodes (e.g. stainless steel) or graphite electrodes may result in slightly higher data quality relative to porous pots. However, very few studies of the small differences in IP data quality as a function of electrode material exist. Morris et al. (2004) found that lead, stainless steel and graphite made good IP electrodes based on data quality checks including reciprocity (Section 4.2.2.1).

3.3.2.6 Electrode Cables

IP surveys are commonly performed with the same multicore cables used for resistivity imaging. In most instances, this is adequate when the objective is to obtain an integral apparent chargeability or a single measure of phase at low frequency, e.g. 1 Hz or less. Dahlin and Leroux (2012) showed that high-quality IP measurements can be acquired with multicore cables under favourable conditions (good signal strength, low electrode contact resistance). Under less favourable conditions, separating out current and potential cables can be advantageous. This significantly reduces the coupling effects between the current injection wires and the potential measurement wires relative to when a single multicore cable is used to carry both the current and receive the voltage signals. The improvement comes from the physical separation of the two sets of wires. This places additional demands and expense on data acquisition but is worth the effort if the objective is to obtain spectral information from the survey.

Contact resistances require extra consideration for IP surveys. Lower voltages at the receiving electrodes due to limited current injection caused by high contact resistances adversely impact IP measurements more than resistivity measurements. This is because V_s is typically 100–1,000 times smaller than V_p, so V_s quickly descends into the instrument noise (signals below the minimum measurable voltage resulting from the current injection) as contact resistance increases. Therefore, IP data acquisition may benefit from larger electrodes than used for resistivity measurements alone, even if they violate the point source assumption that is used in the modelling of resistivity and IP datasets.

3.3.2.7 Distributed Transmitter and Receiver Systems

A recent and currently underutilized development in resistivity and IP is the distributed system that consists of a standard transmitter, a full waveform current recorder and a set of full waveform voltage receivers such as the FullWaver system from Iris Instruments

(Truffert et al., 2019). These receivers record the full waveform at a specific location, which is synchronized with the full waveform current recorder via a GPS clock signal (Figure 3.25). The full waveform recorder removes the need to run long wires from a centralized IP receiver to the voltage recording electrodes and is thus an attractive method for reducing the complexity of surveying in rough terrain. The full waveform receiver is moved to different positions recorded by the internal GPS receiver, removing the need for surveying on a regular grid. Each receiver uses three electrodes to measure the electric field in two orthogonal directions. In order to be efficient with a 3D survey over complex terrain, a field crew will move ten or more full waveform receivers, along with the transmitter and current injection electrodes, across the terrain. The data acquisition system provides proposed coordinate locations to guide receiver and electrode placement during a survey, but necessary alterations due to field conditions are recorded and immediately incorporated into the data processing. The data for all receivers (and the current monitor) are stored in memory and subsequently transferred to a memory drive or directly to a server via an internet connection. Such an instrumentation setup will be expensive compared to a resistivity/IP survey using a standard multicore cable system. However, this instrumentation is likely to be cost-effective for large-scale surveys over complex, 3D terrain. The technology has been used for 3D imaging of mineral deposits, characterization of landslides

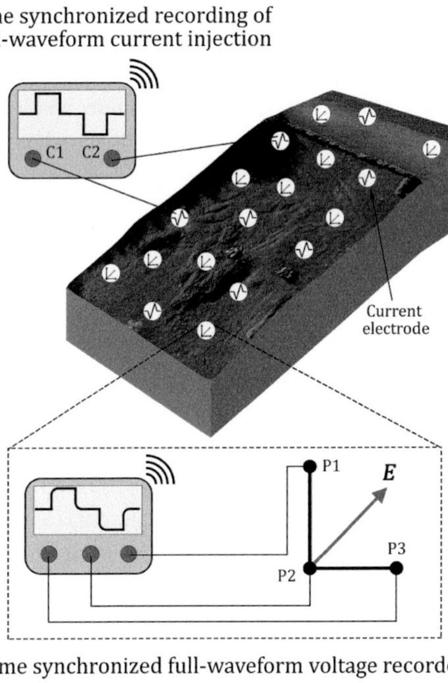

Figure 3.25 Concept of 3D resistivity and IP imaging using a fully distributed system. Figure based on the FullWaver system developed by Iris Instruments (France).

in mountainous regions and for mapping structures supporting intra-basin water flow (Ahmed et al., 2019; Truffert et al., 2019).

3.3.3 Relationships between Instrument Measurements

Frequency domain and time domain IP instruments both record a measurement of the ratio of the polarization strength to the conduction strength in a soil and rock. This is most obvious from a consideration of the phase angle,

$$\varphi = \tan^{-1}\frac{\sigma''}{\sigma'} \approx \frac{\sigma''}{\sigma'}, \tag{3.12}$$

where the approximation is reasonable for $\varphi \leq 100$ mrad. As discussed in Chapter 2, σ'' quantifies reversible charge storage (polarization) and σ' quantifies electromigration (conduction). It is clear from Equation 3.12 that φ can vary in response to changes in conduction independent of any changes in polarization. It is therefore important to distinguish relative polarization terms (e.g. φ) from absolute polarization terms (e.g. σ''). The absolute polarization is related to φ according to,

$$\sigma'' = \sigma' \tan\varphi \approx |\sigma| \tan\varphi \approx |\sigma|\varphi\ , \tag{3.13}$$

where the two approximations are again valid for $\varphi \leq 100$ mrad. Thus, the absolute polarization strength is obtained by multiplying the relative term (φ) by the conductivity ($|\sigma|$).

Time domain IP measurements are also relative measures of the polarization strength that, when multiplied by a measure of the conductivity magnitude, result in an absolute measure of the polarization strength (Lesmes and Frye, 2001; Slater and Lesmes, 2002a). The time domain absolute measure of the polarization is the apparent normalized chargeability,

$$M_{n(a)} = M_a|\sigma|. \tag{3.14}$$

It follows that (1) $M_{n(a)} \propto \sigma''$ and (2) $M_a \propto \varphi$. Given that the frequency domain measurements provide a direct quantification of the true electrical properties of a material, the time domain measurements only provide scaled measures of these properties. In order to facilitate direct interpretation of field time domain IP measurements in terms of complex conductivity (and to reduce potential ambiguity of the apparent chargeability measure), some researchers calibrate the proportionality constant between M_a and φ through laboratory tests on a range of samples (this requires access to a laboratory IP instrument) (Mwakanyamale et al., 2012). The calibration will change if the measurement settings for recording the integral of the decay curve are modified. It is also possible to mathematically compute an equivalent phase from a measured apparent chargeability when a constant phase model is assumed (Kemna et al., 1997). As a general guide, $M_a(\text{mV/V}) \approx -\varphi(\text{mrad})$.

One additional, now seldom-used, IP measurement is known as the 'percentage frequency effect' (*PFE*). Once popular in mineral exploration measurements, the *PFE* has been largely superseded by phase angle or apparent chargeability measurements. However, the *PFE* is a simple measurement to make as it quantifies the change in resistivity as a function of frequency. From Section 2.3.4, a polarizable material results in a decrease in resistivity with increasing frequency. The resistivity would be frequency independent in the case of a non-polarizing subsurface. The higher the polarization, the greater the decrease in resistivity with increasing frequency. The *PFE* quantifies this change in resistivity in the measurements,

$$PFE = 100\frac{\left(\rho_{f1} - \rho_{f2}\right)}{\rho_{f1}}, \tag{3.15}$$

where ρ_{f1} and ρ_{f2} are resistivities at two frequencies, $f_2 > f_1$. The metal factor (*MF*) was introduced by Marshall and Madden (1959) to 'magnify' the IP response from conductive ore bodies,

$$MF = a_{MF}\frac{1}{\rho_0}PFE, \tag{3.16}$$

where a_{MF} is a unitless constant (taken as $2\pi \times 10^5$ by Marshall and Madden (1959)).

Zonge et al. (1972) and Zonge and Wynn (1975) illustrate the relationship between different measures of polarizability and report that $PFE = -0.2\varphi$ (φ in mrad), using a frequency effect measurement over one frequency decade (i.e. one order of magnitude change in frequency). In general, we can expect an approximate proportionality between φ, M_a and *PFE*. Similarly, we can expect an approximate proportionality between σ'', $M_{n(a)}$ and *MF* (Lesmes and Frye, 2001).

3.3.4 Instrumentation for Imaging Tanks, Cores and Other Vessels

Dedicated measurement systems have been developed for imaging experimental tanks, soil columns and other vessels. These developments are reported in the fields of biomedical and industrial process tomography, the latter used for examining the distribution and mixing of fluids inside vessels, e.g. stirred tank reactors. In both communities the term 'electrical impedance tomography' (EIT) is commonly used. We refer the reader back to Chapter 1 for a discussion of the parallel development of electrical resistivity imaging and electrical impedance tomography.

Data acquisition systems designed for EIT are often well suited for small-scale imaging of porous media processes in tanks, cores and other vessels. Surprisingly, the opportunity to use such systems for small-scale imaging of subsurface materials has rarely been exploited (Binley et al., 1996a, 1996b; see also the case study in Section 6.1.5). These systems typically operate at much higher frequencies (10–100 kHz) than resistivity and IP instruments that have been developed for near-surface Earth applications. They utilize small currents (typically a few milliamperes) compared to what conventional resistivity imaging systems are capable of injecting, but this is sufficient to permit imaging of vessels over a

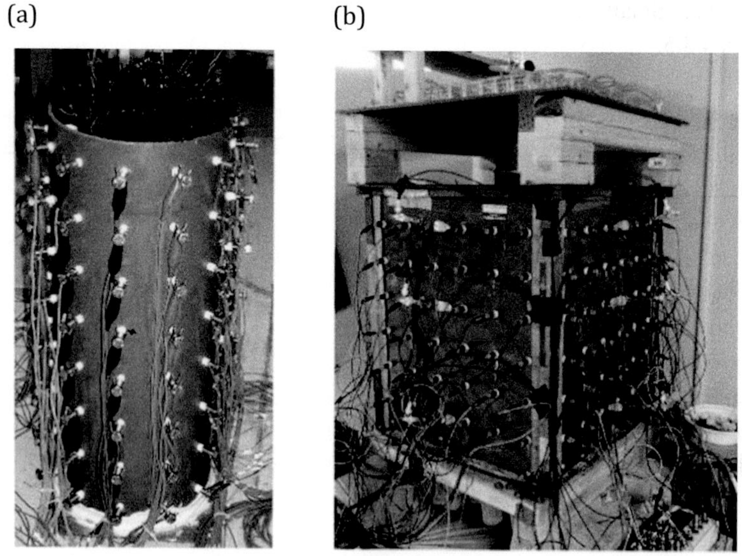

Figure 3.26 Examples of laboratory setups for imaging the internal structure of vessels: (a) cylindrical column of peat instrumented with 96 electrodes; (b) soil lysimeter instrumented with 144 electrodes for monitoring moisture content dynamics during infiltration events.

range of scales that are relevant to investigating processes occurring in soils and rocks. The high operating frequencies allow for much more rapid data acquisition than typically obtainable with conventional resistivity and IP systems. The systems vary considerably but typically address a large number of channels to further increase data acquisition time to better capture processes occurring on short time-scales. Modern EIT systems can support a temporal resolution that is on the order of milliseconds, whereas geophysical systems require at least a few minutes (and often longer) to collect the data needed for repeated imaging. In many respects, EIT systems are more advanced than the instrumentation that has been developed for geophysical applications. For example, some support current focusing whereby additional electrodes beyond the two conventional current injection electrodes are simultaneously energized to improve current density (and hence measurement sensitivity) in the central portions of vessels. Other systems have implemented CDMA coding (Section 3.2.3.3) to further speed up data acquisition (Yamashita et al., 2015a, 2015b). Figure 3.26 illustrates some applications of electrical imaging for laboratory-scale monitoring of processes occurring in vessels.

Developments in EIT systems parallel development in resistivity imaging systems. Multi-frequency EIT systems have been developed to acquire broadband information (again at much higher frequencies than used in resistivity imaging), where in the medical field the interest has been in improving discrimination between normal and abnormal cell tissues. As in the geosciences, the opportunity exists to better understand the composition of the imaged materials from broadband measurements (Kelter et al., 2015; Weigand and

Kemna, 2017). Small-scale resistivity and IP imaging is discussed further in Sections 4.2.2.8 and 4.3.5.

3.4 Closing Remarks

Resistivity and IP instruments permit measurements of the electrical properties introduced in Chapter 2 across a wide range of spatial and temporal scales. Laboratory instrumentation has advanced from devices designed to make a single resistivity measurement on a 1D cell to multi-channel instruments for making SIP measurements (typically from mHz to kHz) of dynamic processes occurring in a wide range of vessels. Whereas the acquisition of a laboratory resistivity measurement is relatively straightforward, high-accuracy laboratory IP measurements are critically dependent upon careful sample holder design, particularly the placement of electrodes. Poor sample holder design can lead to IP errors of the same magnitude as the measurement signal in the case of mechanisms involving non-conducting minerals. Erroneous IP measurements have resulted in some irreproducible results and incorrect interpretations of IP signals (Brown et al., 2003). In addition to the material discussed in this chapter, Kemna et al. (2012) provide some recommendations for laboratory strategies that encourage acquisition of IP signals coming exclusively from the sample under study rather than being corrupted by artefacts resulting from poor sample design. Modern laboratory IP systems make measurements in the frequency domain, where a swept-sine wave function accurately records the frequency response needed to characterize relaxation times.

The basic construction and operating principles of field resistivity and IP systems for studying the near-surface Earth advanced rapidly in the 1990s and early 2000s. Prior to this, major development focused on powerful but cumbersome (and dangerous) systems developed for deep mineral exploration. The advent of portable multi-channel, multi-electrode imaging systems at the time represented a paradigm shift in the methodology. This avoided the prior need for teams of operators to spend many hours relocating sets of electrodes between each measurement. A number of multi-electrode imaging systems have been developed and are well established in the market place. Most field-scale IP measurements are made in the time domain because of the lower cost of the electronics and the historic use of the voltage decay curve after current shut-off as a graphical representation of the IP effect. However, field-scale frequency domain systems working on the same principles as laboratory SIP systems have been developed. Early frequency domain field systems were developed for mineral exploration, but more recent developments have specifically targeted acquisition of high-accuracy SIP measurements for shallow applications. IP instruments that record the full waveform at a high sampling rate provide some spectral information from a time domain measurement (Kemna, 2000).

Technological advances in the last decade have been relatively limited, foremost focusing on faster data acquisition using larger numbers of addressable electrodes, along with increased flexibility in the configuration of electrodes (e.g. borehole installations). One limitation of resistivity and IP measurements is the need for adequate galvanic contact with the ground to ensure sufficient current injection. Resistivity and IP surveys are often impractical over very resistive ground. Commercially available capacitively coupled resistivity imaging systems have opened up applications of resistivity over highly resistive ground surfaces. Recent experiments indicate that SIP measurements may even be possible using capacitively coupled electrodes (Mudler et al., 2019). In the future, it may be possible to sense IP effects using airborne time domain electromagnetic (EM) systems. The presence of IP effects in land-based time domain EM systems, identified as a characteristic negative signal at late time, has been known for decades (Nabighian and Macnae, 1991). Such negative signals have been recorded in airborne EM datasets (Smith and Klein, 1996; Walker, 2008), mostly over large mineral deposits, inspiring modelling studies to better assess the information on IP potentially extractable from late-time airborne EM datasets (Macnae, 2016). However, the sensitivity of time domain EM measurements to small IP effects of relevance to near-surface investigations remains uncertain.

One promising trend is the development of remotely operated, automated subsurface monitoring systems (Chambers et al., 2015). There is an almost endless list of subsurface processes operating over multiple time-scales where permanently installed resistivity and IP monitoring could provide valuable information on system changes and return data to support informed decision-making. For example, resistivity and IP monitoring could guide (1) when/where to perform direct invasive sampling to confirm contaminant transport, (2) modifications to an active environmental remediation technology (e.g. addition of new amendment), (3) adjustments to groundwater-extraction strategies at locations experiencing saline intrusion and (4) emergency response decisions in locations prone to landslides. Commercial instruments specifically designed for autonomous, long-term monitoring are currently unavailable, although some field instruments do come with add-ons that support stand-alone monitoring. However, these systems have high power requirements and would be expensive to deploy long term at a single site. With growth in the internet of things, there is an opportunity to develop a new generation of relatively low-cost resistivity and IP monitoring systems for long-term deployment off the grid.

Field-scale acquisition of broadband SIP data similar to that recorded in the laboratory remains a challenging goal. Although instruments have been developed in pursuit of this ambitious objective, the errors in the phase data due to coupling effects that increase dramatically towards the higher frequencies may still thwart accurate data acquisition above a few 10s of Hz at unfavourable sites. Reliable field-scale acquisition of SIP data up to ~100 Hz may be possible under ideal field survey conditions. One must also consider the merit of acquiring broadband field SIP data in the low-frequency range, where it may take close to an hour to complete a single scan for a single current electrode pair if trying to reach frequencies of 10^{-2} Hz or lower. Although novel developments such as direct recording of currents and voltages at the electrodes coupled with fibre optic data transmission can help reduce the coupling problems, the long data-acquisition times needed to scan at low frequency remain a fundamental constraint. Given such limitations, along with the costs of these instruments, the potential user is encouraged to consider the worth of the additional information content obtained from a broadband SIP measurement relative to the single, intermediate frequency information more readily obtained from a time domain

IP measurement. It may often be the case that +90 per cent of the useful information is provided from a time domain IP resistivity system especially when full waveforms are recorded and processed.

In Chapter 4, we turn to the problem of using the instrumentation discussed in this chapter to acquire field-scale resistivity and IP measurements that can be meaningfully interpreted in terms of variations in subsurface electrical properties described in Chapter 2. Just as in the laboratory, where acquisition of quality IP data requires careful attention to sample design, the acquisition of meaningful field data depends more on the implementation of the measurement than on the instrumentation. We also discuss methods to assess the quality of field datasets and the sensitivity of measurements, and introduce some analytical models for interpreting the measurements in terms of subsurface structure.

4
Field-Scale Data Acquisition

4.1 Introduction

In Chapter 3, we showed how resistivity and induced polarization (IP) parameters (chargeability, complex resistivity) can be measured directly in a laboratory sample. Chapter 3 also showed that in the field, or a more general setting, we are not able to create a uniform pathway of current and thus need an alternative way of determining the resistivity or IP parameters. We have already discussed the concept of a four-electrode electrical measurement. In this chapter, we show how we can build on this and use additional concepts to allow the measurement of resistivity and IP in the field. This then permits us to measure variation of electrical properties in a generalized manner. We introduce the concept of apparent resistivity and chargeability and illustrate, for relatively simple cases, how the apparent resistivity is affected by non-uniformity of resistivity (e.g. a layered subsurface). We introduce the graphical presentation of apparent resistivity and IP measurements in a pseudosection for 2D problems. The general properties of field resistivity and IP instruments and associated components (cables, electrodes) for field surveys were covered in Chapter 3. In this chapter, we discuss some of the practical aspects of field measurements, including choice of electrode configuration and assessment of measurement errors. Although we provide extensive coverage of the more standard ground-based electrical methods which account for a vast proportion of electrical surveying, we illustrate how measurements can be made in 'non-standard' settings, such as between boreholes or for imaging laboratory-scale tanks and columns, and also discuss time-lapse measurement approaches. We also illustrate how potential fields using the same four-electrode configuration allows the mapping of electrical current, which has applications in the detection of fluid leaks, e.g. in landfills. As throughout the book, we have split the chapter into two sections: DC (direct current) resistivity and IP. The section on IP builds on some of the concepts covered in the DC resistivity section.

4.2 DC Resistivity

4.2.1 The Resistivity Quadrupole and Apparent Resistivity of Specific Resistivity Structures

In order to develop equations relating four-electrode measurements to the resistivity of the subsurface, we must first understand the spatial pattern of electrical potential due to current injected from an electrode. For a 3D, isotropic, electrical resistivity distribution, $\rho(x, y, z)$, the electric potential (voltage), $V(x, y, z)$, due to a single (point) current electrode, with strength I, located at coordinates x_c, y_c, z_c, is defined by a form of the Poisson equation:

$$\nabla.\left(\frac{1}{\rho}\nabla V\right) = -I\delta(x_c, y_c, z_c), \tag{4.1}$$

where $\nabla = \frac{\partial}{\partial x} + \frac{\partial}{\partial y} + \frac{\partial}{\partial z}$ and $\delta(x, y, z)$ is the Dirac delta function (which takes on a value of 1 at the position x, y, z and is 0 elsewhere).

Equation 4.1 is normally considered to be subject to boundary conditions:

$$\left(\frac{1}{\rho}\right)\frac{\partial V}{\partial n} = 0, \tag{4.2}$$

where n is the outward normal. Such conditions are called Neumann (or second type) and impose the condition of no flux into or out of the ground (except of course at the current electrode location).

For the case of homogenous resistivity, ρ, with current injection at a depth beyond the influence of the ground surface (which is flat and at $z = 0$), the solution of Equation 4.1 gives the voltage at coordinates x_p, y_p, z_p as

$$V(x_p, y_p, z_p) = \frac{I\rho}{4\pi r}, \tag{4.3}$$

where r is the distance between current source and potential measurement, i.e. $r = \sqrt{(x_p - x_c)^2 + (y_p - y_c)^2 + (z_p - z_c)^2}$.

The general case where the current is injected at a shallow depth (i.e. the potential field is affected by the non-conducting air at the ground surface) is easily derived by the method of images. For this case, the solution is the superposition of two solutions using Equation 4.3: one based on the real electrode at x_c, y_c, z_c, and the other based on an imaginary electrode at $x_c, y_c, -z_c$. The solution is

$$V(x_p, y_p, z_p) = \frac{I\rho}{4\pi r} + \frac{I\rho}{4\pi r_i}, \tag{4.4}$$

where r is defined as before and $r_i = \sqrt{(x_p - x_c)^2 + (y_p - y_c)^2 + (z_p + z_c)^2}$.

Equation 4.4 is needed for the general case, e.g. where electrodes deployed in boreholes are used, but for the common arrangement of surface electrodes ($z_c = 0, z_p = 0$) $r_i = r$, and so

$$V(x_p, y_p, 0) = \frac{I\rho}{4\pi r} + \frac{I\rho}{4\pi r} = \frac{I\rho}{2\pi r}. \tag{4.5}$$

Figure 4.1 illustrates the resultant potential field due to current injected at an electrode buried 2 m deep and at the ground surface.

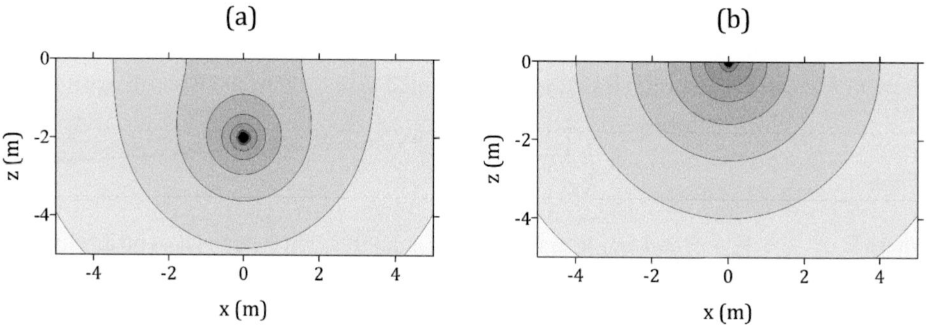

Figure 4.1 Example potential fields for (a) electrode buried at 2 m; (b) electrode at the surface.

As discussed in Chapter 3, resistivity (and IP) measurements are made using a four-electrode configuration: a quadrupole. Two electrodes serve to create the potential field (source and sink current electrodes), and two receiving electrodes are used to measure a potential difference. We can use the expressions in Equation 4.5 (or Equation 4.4 for the general case) to determine the relationship between the measured potential difference, the injected current and the resistivity of the subsurface.

The labels A, B, M and N are commonly used for the electrodes in a quadrupole: A is the current source, B is the current sink and voltage is measured between electrodes M and N. Examples of such quadrupoles are shown in Figure 4.2. We will discuss choice of quadrupole geometry in the next section, but first we will develop an expression for apparent resistivity, which is the resistivity of a homogenous subsurface that the measured voltage and current are equivalent to. That is, the apparent resistivity is only the true resistivity if the subsurface is homogenous and provided the assumptions used to compute the apparent resistivity are valid. Below we explain how an apparent resistivity is computed. For most field-based applications, the calculations are made with the assumption of a flat ground surface and infinite boundaries (a so-called, infinite half space).

Building on Equation 4.5, we can state, through superposition, that the difference in voltage between electrodes M and N due to current injected between electrodes A and B (with all electrodes on the ground surface, as in Figure 4.2a) is given by

$$\Delta V = V_M - V_N = \frac{I\rho}{2\pi}\left(\frac{1}{AM} - \frac{1}{BM} - \frac{1}{AN} + \frac{1}{BN}\right), \tag{4.6}$$

where AM is the distance between electrodes A and M, BM is the distance between electrodes B and M, etc., and the subsurface is homogenous with resistivity ρ.

Using Equation 4.6 an expression for the apparent resistivity, ρ_a, can be rewritten as

$$\rho_a = K\frac{\Delta V}{I}, \tag{4.7}$$

where K is the geometric factor, with dimensions of length,

$$K = \frac{2\pi}{\left(\frac{1}{AM} - \frac{1}{BM} - \frac{1}{AN} + \frac{1}{BN}\right)}. \tag{4.8}$$

The term $\frac{\Delta V}{I}$ is often referred to as a *transfer resistance* since it has units of ohms.

For the general case where electrodes are not on the ground surface (e.g. Figure 4.2b), we can follow the same principles using Equation 4.4 to derive

$$K = \frac{4\pi}{\left(\frac{1}{AM} + \frac{1}{A_iM} - \frac{1}{BM} - \frac{1}{B_iM} - \frac{1}{AN} - \frac{1}{A_iN} + \frac{1}{BN} + \frac{1}{B_iN}\right)}, \tag{4.9}$$

where A_iM is the distance between imaginary electrode A_i and electrode M, B_iM is the distance between imaginary electrode B_i and electrode M, etc., as shown in Figure 4.2b.

Apparent resistivity is a convenient value for field measurements as it has the same units as resistivity (Ωm), can allow the field operator to gauge variability of measurements directly in the field, and allow some immediate assessment of data quality (the apparent resistivity, for example, should be positive, even for a highly heterogeneous resistivity structure, provided the geometry designated is correct). Most field instruments report apparent resistivity during measurements, provided the user has supplied details of electrode geometry. However, it is important to note that the computation of apparent resistivity, as described above, assumes a flat ground surface and an infinite half space. If topographic variation exists at the site, or if the region of investigation is bounded in some way, then the reported apparent resistivity does not reflect the true resistivity of the subsurface even if it has a homogeneous resistivity. And in some such configurations the computed apparent resistivity may even be negative, which is clearly non-physically correct. An example of a bounded system is a vessel (see e.g. Figure 3.26) – in such a case, Equation 4.3 is not a correct solution to Poisson's equation and analytical solutions for the particular geometry

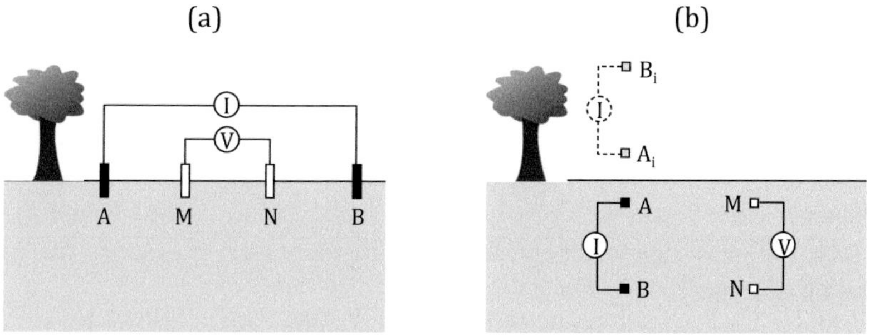

Figure 4.2 Example quadrupole configuration. (a) Surface electrodes. (b) Buried electrodes. A_i and B_i are imaginary electrodes in (b).

may not be possible. This may be overcome by deriving numerical approximations to Poisson's equation for the specific problem geometry.

As stated earlier, apparent resistivity is a convenient value for field measurements but the true measurement (that should be recorded for all surveys) is the transfer resistance, $R = \Delta V/I$: the ratio of measured potential difference to applied current. The transfer resistance (measured in Ω) can be positive or negative (a negative value implies a negative geometric factor) and can vary by orders of magnitude even if the resistivity is uniform. Recognizing that the transfer resistance may be negative is important for non-standard configurations of electrodes as some commercial instruments do not report the polarity of the transfer resistance (or voltage) under default settings.

For most applications, as discussed later, resistivity surveys consist of measuring apparent resistivity (or transfer resistance) for multiple positions (and arrangements) of the quadrupole in order to survey variations in resistivity vertically and horizontally. A collection of such measurements are then typically analysed using the modelling approaches discussed in Chapter 5. In a few cases, as shown later, the apparent resistivity (or transfer resistance) is used directly to assess resistivity variation at a site.

4.2.1.1 Electrode Array Geometries

A large number of quadrupole configurations are possible. The selection often depends on the type of survey being carried out, the nature of the target of interest and, in some cases, the flexibility of the instrument. Szalai and Szarka (2008) offer a classification of 92 different configurations. We focus here on surface electrode configurations and discuss quadrupoles for non-standard applications later in the chapter. Many commonly used arrays originate from several decades ago, when electrodes were moved manually, since older instruments could only connect to four electrodes at one time. The advent of multi-electrode devices, as discussed in Chapter 3, has led to a wider range of quadrupole geometries, although these are not necessarily fully exploited in most current applications.

Figure 4.3 shows schematics of the most commonly used surface array configurations, and Table 4.1 shows the resulting geometric factors from application of Equation 4.8. Each array can be moved horizontally to assess lateral variation in resistivity, or expanded to sense greater depths.

The Wenner array (attributed to Wenner, 1915) is one of the most common quadrupole geometries. In this configuration, the electrodes are equally spaced, distance a apart, and the current electrodes are located outside the potential dipole, ensuring good measurement signal strength. Sometimes the Wenner configuration A-M-N-B is referred to as Wenner α, with Wenner β and γ (again with equal spacing) configured as A-B-M-N and A-M-B-N, respectively. The term 'dipole–dipole' is more commonly used to describe the Wenner β configuration (see later).

The Schlumberger array is similar to the Wenner configuration except that the spacing between the current electrodes is much larger than the potential dipole spacing (distance $AB > 5 \times MN$, i.e. $n > 2$ in Figure 4.3). This array is commonly adopted for vertical soundings, as discussed later, because of its relative insensitivity to lateral variation in resistivity. It has the additional practical advantage that only one pair of electrodes is moved at a time.

The dipole–dipole array (attributed by Seigel et al. (2007) to the work of Madden in 1954, but note that the identical 'Eltran' array was discussed by West (1940)) is somewhat misleadingly named since all quadrupoles are some form of dipole–dipole. In this array, the current and potential dipoles are separated, which results in a weaker signal strength, in comparison to Wenner and Schlumberger configurations. The pole–dipole and pole–pole arrays utilize a remote electrode or pair of remote electrodes, allowing quicker manual movement of the mobile electrodes. The pole–pole configuration is widely used in archaeological studies (using the label 'twin array' or 'twin probe') – here a pair of fixed electrodes are mobilized over a site to map variation in resistivity at a shallow depth. In fact, for most applications of the twin array, the transfer resistance, rather than apparent resistivity, is reported.

The gradient array (see Dahlin & Zhou, 2006) is included in Figure 4.3 as an example of a quadrupole configured for multi-electrode measurement devices, specifically for 2D (horizontal-vertical) imaging of resistivity. The quadrupoles discussed so far are collinear arrays; other configurations exist, however. The square array (Figure 4.4) and extensions in a trapezoidal configuration have proved popular in archaeological surveys due to the greater mobility in comparison to the twin array (see e.g. Panissod et al., 1998; Gaffney et al.,

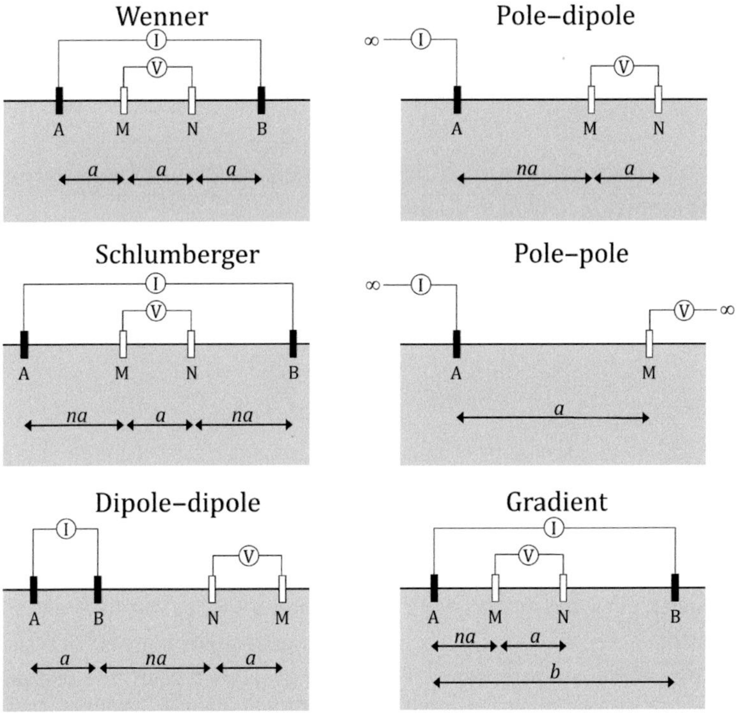

Figure 4.3 Common quadrupole configurations for surface electrode arrays.

Table 4.1 Geometric factors for quadrupoles shown in Figures 4.3 and 4.4

Array	Geometric factor, K
Wenner	$2\pi a$
Schlumberger	$\pi an(n+1)$ or πan^2 if $n \geq 10$
Dipole–dipole	$\pi an(n+1)(n+2)$
Pole–dipole	$2\pi an(n+1)$
Pole–pole	$2\pi a$
Gradient	$2\pi / \left(\left(\frac{1}{na}\right) + \left(\frac{1}{b-na}\right) + \left(\frac{1}{(n+1)a}\right) + \left(\frac{1}{b-(n+1)a}\right)\right)$
Square array	$2\pi a / (2+\sqrt{2})$

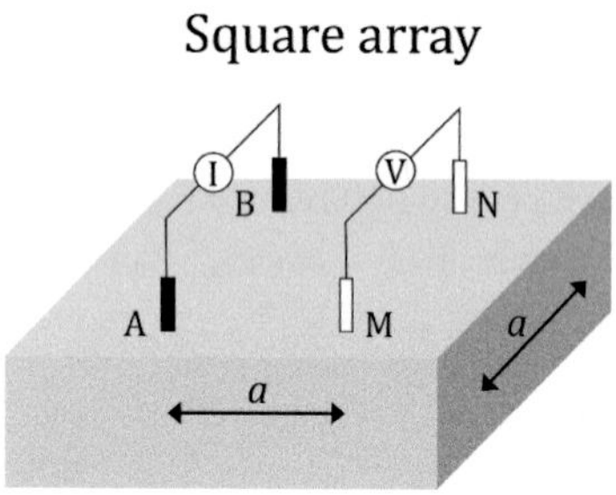

Figure 4.4 Square array configuration.

2015); they also allows assessment of directional (anisotropic) variation in resistivity by rotation of the array (e.g. Tsokas et al., 1997).

Each array has different sensitivity patterns, i.e. if the subsurface has spatial variation in resistivity, then the measurement of apparent resistivity will be affected differently for each array. If we define the sensitivity as

$$\text{Sensitivity} = \frac{\partial \log(\rho_a)}{\partial \log(\rho)}, \tag{4.10}$$

we can assess how different regions of the subsurface affect the measured apparent resistivity. In order to compute Equation 4.10 for the general case, we can use the numerical modelling techniques covered in Chapter 5. Figure 4.5 shows sensitivity patterns for three arrays for a homogenous resistivity, from which it is evident that some regions contribute in a positive manner and some in a negative manner, and some regions have no influence on the measurement. A negative sensitivity means that a localized increase in resistivity will reveal itself as a reduction in apparent resistivity. For this reason, the interpretation of observed apparent resistivity can be challenging without appreciation of sensitivity behaviour. In fact, one of the factors that have contributed to the popularity of the twin array (pole–pole) in archaeological studies is the more direct link between local resistivity and measurement (e.g. Clark, 1990; see also Section 4.2.2).

The different sensitivity to horizontal variation in resistivity can influence the choice of quadrupole. It is evident from Figure 4.5, for example, that the dipole–dipole configuration has greater sensitivity to lateral variation, in comparison to the Wenner arrangement. The Schlumberger array has even weaker sensitivity to horizontal variation in resistivity, making it a popular choice for investigating vertical (1D) profiles of resistivity, as discussed later.

The patterns in Figure 4.5 also reveal that the depth of sensitivity ('depth of investigation') of a measurement differs for each quadrupole. Gish and Rooney (1925) first proposed (incorrectly) that the depth of investigation can be considered to be the spacing between the electrodes. The study of Evjen (1938) provides estimates of a depth of investigation for surface arrays. Roy and Apparao (1971), Edwards (1977), Barker (1979) and Gómez-Treviño and Esparza (2014), amongst others, also discuss sensitivities of different arrays. As noted by Roy and Apparao (1971), the depth of penetration of current density is sometimes mistakenly used to assess depth of investigation; however, it is essential that the voltage response is also accounted for. Roy and Apparao (1971) developed analytical expressions for the voltage response of small perturbations in the resistivity of a small volume within a uniform half space (similar to that shown in Figure 4.5d). They then went on to compute depths of investigation for various four-electrode arrays and quote depths for Wenner, Schlumberger, dipole–dipole and pole–pole arrays as $0.11L$, $0.125L$, $0.195L$ and $0.35L$, respectively, where L is the longest distance between electrodes (for the pole–pole

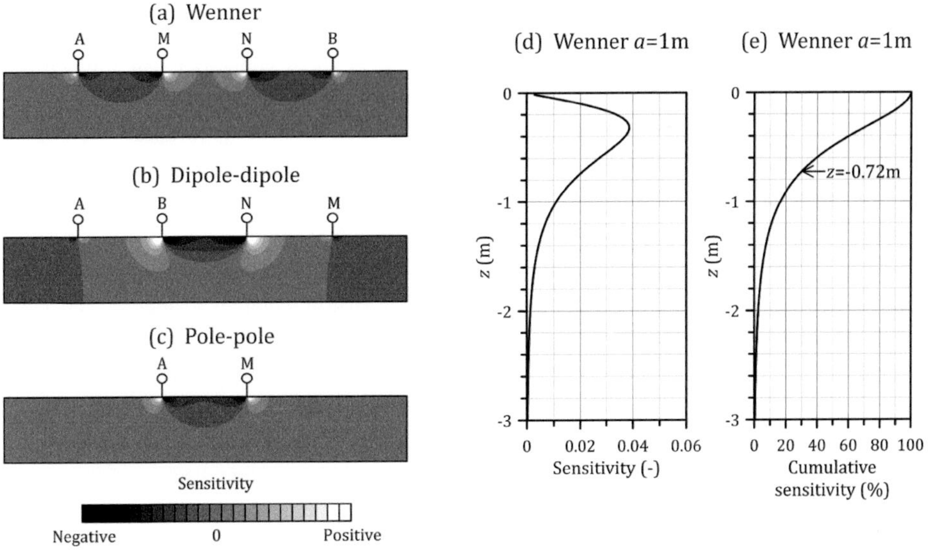

Figure 4.5 Quadrupole sensitivity patterns, assuming homogenous resistivity. (a) Wenner. (b) Dipole–dipole. (c) Pole–pole. (d) Sensitivity-depth profile for 1 m spaced Wenner configuration. (e) Cumulative sensitivity for profile in (d). Marked in (e) is the depth over which 70% of the sensitivity exists.

array, Roy and Apparao (1971) use the distance between the current and potential electrode for L; for the dipole–dipole, L is the distance between the centre of the two dipoles).

An alternative approach is to use the sensitivity profile (e.g. in Figure 4.5d), from which we can compute a depth of investigation based on the cumulative sensitivity. If we adopt 30% (a measure, e.g. used in other fields) as a threshold (i.e. the depth over which 70% of the sensitivity exists), then we find that the depth of investigation for a Wenner measurement is $0.72a$ (see Figure 4.5e). Similar analysis for the dipole–dipole and pole–pole arrays leads to a depth of investigation of $0.57a$ and $1.18a$, respectively. For a dipole–dipole array with $n = 2$ and $n = 3$, we estimate the depth of investigation to be $0.98a$ and $1.29a$, respectively.

As Roy and Apparao (1971) point out, depth of investigation does not equate to resolution; they go on to quote a vertical resolution (higher number equates to greater resolution) for the Wenner, Schlumberger, dipole–dipole and pole–pole arrays as 0.444, 0.408, 0.290 and 0.119, respectively, highlighting the relative weakness of the pole–pole configuration. However, again, such a characteristic of the pole–pole array is advantageous for shallow mapping of horizontal variations in resistivity in archaeological studies since the operator, in many cases, will not want signals impacted by vertical variability in resistivity.

The choice of configuration will also be strongly influenced by presence of noise at a site and the quality (sensitivity) of instrumentation. Because of the location of potential electrodes within the current dipole in the Wenner array, relatively high transfer resistances are assured, in contrast to the dipole–dipole array, where large separation of dipoles is often limited due to the relatively poor signal to noise ratio. As discussed in Chapter 3, many modern resistivity instruments are equipped with multi-channel capability allowing synchronous measurements of multiple potential dipoles (typically constrained by the need to have a common electrode between successive synchronous dipole measurements). Such capability can often mean greater efficiency (i.e. speed of survey) for particular types of quadrupoles.

Other logistical constraints may also influence the choice of quadrupole. Pole–pole arrays rely on remote electrode installations, which may not be achievable at a site, or may be constrained by safety factors – the operator should be aware of any risk to health (humans, livestock, etc.) within the controlled area of operation.

The selection of quadrupole must then be decided based on the type of investigation (e.g. to assess vertical variability), the site conditions and the instrumentation available. Table 4.2 summarizes some of the characteristics of the four most commonly used arrays; the depths of investigation in Table 4.2 are based on the cumulative sensitivity analysis discussed earlier. It should also be noted that traditionally a single quadrupole configuration was normally adopted but, for modern computer-controlled multi-electrode instruments, the option to combine configurations is straightforward and should be considered when designing a survey. Later in the chapter we discuss the concept of optimal measurements in survey design.

4.2.1.2 Apparent Resistivity of Laterally Variable Media

So far we have only considered uniform (homogenous) conditions. We now examine how the observed apparent resistivity will vary due to lateral variability in resistivity. Keller and

Table 4.2 Comparison of common electrode arrays for electrical resistivity measurements. Characteristics are ranked as H (high), M (medium) and L (low). Note that some multi-channel instruments do not have full flexibility in choice of concurrent potential dipoles, as discussed in Chapter 3. Where full flexibility is possible, the efficiency of collecting Schlumberger measurements may be greater than reported in the table.

	Wenner	Schlumberger	Dipole–dipole	Pole–pole
Depth of investigation	M	M	L	H
Vertical resolution	H	H	M	L
Signal strength	H	M	L	H
Suitability for vertical sounding	M	H	L	L
Suitability for lateral profiling	M	L	H	H
Measurement efficiency for multi-channel instruments	L	L	H	H

Frischknecht (1966) use analytical solutions of the Poisson equation for specific quadrupoles and geometrical settings to illustrate the lateral sensitivity to changes in resistivity. Here we use a more generalized and flexible approach, adopting modelling tools from Chapter 5. Figure 4.6 illustrates the response function for a 10 m spaced Wenner, dipole–dipole and pole–pole array moving across a sharp lateral contrast in resistivity. Several features of the response are worthy of note. First, as expected from the sensitivity patterns in Figure 4.5, a complex transition in apparent resistivity is seen for the Wenner and dipole–dipole arrays: three main stages of transition occur as the four electrodes pass the contrast in resistivity. In these cases, a local maximum in apparent resistivity is visible (+5 m and +15 m for Wenner and dipole–dipole, respectively). In the case of the dipole–dipole array, this represents an overshoot; note also the slight undershoot at −15 m for the dipole–dipole profile, a consequence of the sensitivity pattern shown in Figure 4.5. In Figure 4.7, the lateral response to a thin resistive dike-like structure is shown. For this case, the Wenner array reveals a double peak. Note also the damped response of the pole–pole array in comparison to the pole–pole response. Similar examples are illustrated in Keller and Frischknecht (1966). Such behaviour makes the direct use of four electrode measurements in a quantitative manner somewhat limited. The simpler profile for the pole–pole array adds further evidence of the value of this configuration for archaeological twin array surveys.

4.2.1.3 Apparent Resistivity of Layered Media

The effect of horizontal layering of resistivity has been extensively studied given the broad similarity of many geological environments. As electrode separation increases, measurements of apparent resistivity will reflect the impact of deeper layers in the profile, which forms the basis of vertical electrical soundings (VES), discussed later. Analytical solutions of the Poisson equation for horizontally layered systems can be derived following an extension of the concept of the method of images, referred to earlier. For a two-layered

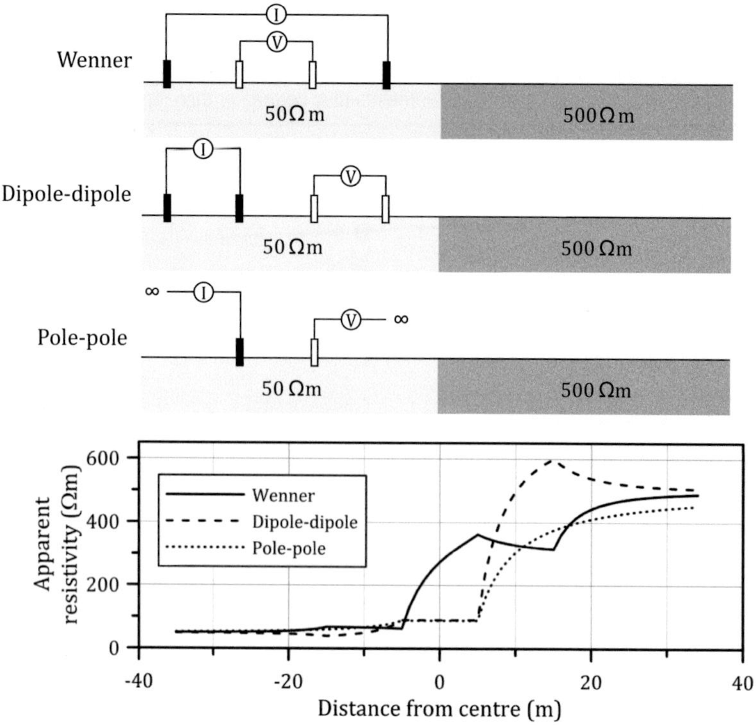

Figure 4.6 Impact of lateral variation in resistivity on apparent resistivity measured by Wenner, dipole–dipole and pole–pole arrays. The location of each measurement is plotted by the central position of the array.

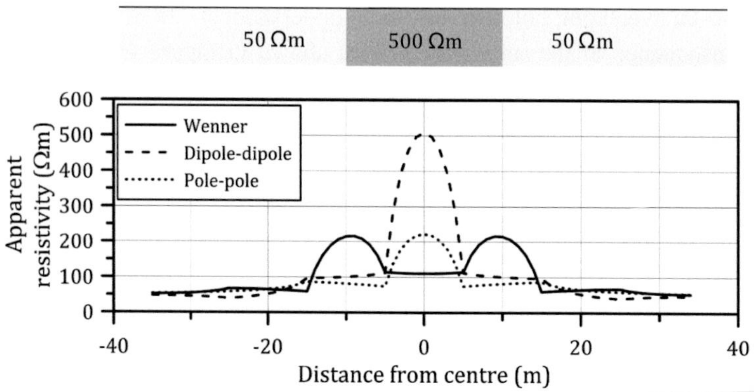

Figure 4.7 Lateral variation in apparent resistivity due to thin resistive dike structure. All arrays have 10 m spaced electrodes.

system (commonly referred to as the *overburden* case) with a layer of resistivity ρ_1 of thickness d overlying resistivity ρ_2, the potential, V at distance r from a current electrode, both of which are on the ground surface, can be written as the infinite series (Keller and Frischknecht, 1966):

$$V = \frac{I\rho_1}{2\pi r}\left[1 + 2\sum_{n=1}^{\infty} \frac{k_{1,2}^n}{\left(1 + \left(\frac{2nd}{r}\right)^2\right)^{1/2}}\right], \tag{4.11}$$

where $k_{1,2}$ is a *reflection coefficient* given by

$$k_{1,2} = \frac{\rho_2 - \rho_1}{\rho_2 + \rho_1}. \tag{4.12}$$

The value of $k_{1,2}$ reflects the distortion of the potential field due to the interface between layers 1 and 2; the infinite series in Equation 4.11 results from the infinite image electrodes above the ground layer and below the interface (see Keller and Frischknecht, 1966). Clearly, for the homogenous case, $k_{1,2} = 0$, resulting in Equation 4.5.

Armed with Equation 4.11, we can now derive apparent resistivities for a given quadrupole configuration, as before. Figure 4.8 shows apparent resistivities from a Schlumberger array with increasing current electrode spacing for two overburden cases. Note how the overburden

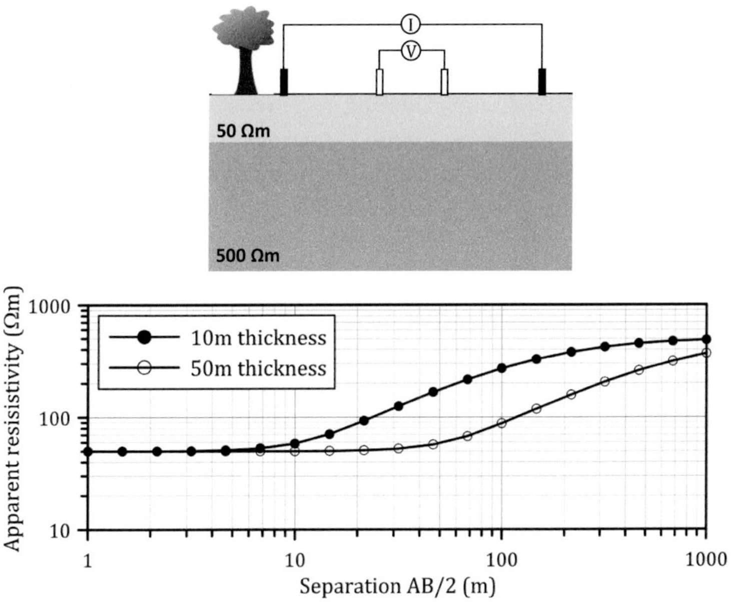

Figure 4.8 Apparent resistivity for Schlumberger array applied to a two-layer model with different upper-layer thickness.

resistivity impacts the apparent resistivity even at large electrode spacing. Box 4.1 shows a more specialized case where the electrodes are buried at the interface of two layers.

For the general, multiple horizontal layer case, Equation 4.11 can be written in an integral form, following Stefanesco et al. (1930), as (Telford et al., 1990):

$$V = \frac{I\rho_1}{2\pi r}\left[1 + 2r\int_0^\infty K_s(\lambda, k, d)J_0(\lambda r)d\lambda\right], \tag{4.13}$$

where $J_0(x)$ is the zero order Bessel function (an oscillatory function that decays from 1 to 0 with increasing x), λ is an integration variable and $K_s(\lambda, k, d)$ is referred to as the *Stefanesco kernel function*, which is governed by the reflection coefficients k and thicknesses d. Following Flathe (1955), for the two-layer case

$$K_s(\lambda, k, d) = \frac{k_{1,2}e^{-2\lambda d_1}}{1 - k_{1,2}e^{-2\lambda d_1}}, \tag{4.14}$$

and for the three-layer case

$$K_s(\lambda, k, d) = \frac{k_{1,2}e^{-2\lambda d_1} + k_{2,3}e^{-2\lambda d_2}}{1 + k_{1,2}k_{2,3}e^{-2\lambda(d_2 - d_1)} - k_{1,2}e^{-2\lambda d_1} + k_{2,3}e^{-2\lambda d_2}}, \tag{4.15}$$

where the upper layer has thickness d_1 and resistivity ρ_1, the second layer has thickness d_2 and resistivity ρ_2 and the lower unit has resistivity ρ_3, and

$$k_{i,j} = \frac{\rho_j - \rho_i}{\rho_j + \rho_i}. \tag{4.16}$$

Figure 4.9 illustrates the variation of apparent resistivity, via Equation 4.13, in a Schlumberger array with increasing electrode spacing for two three-layer cases. For the $\rho_3 = 500\ \Omega$m case, the impact of the second layer is seen; however, for the $\rho_3 = 50\ \Omega$m case, the apparent resistivity is insensitive to the second layer because of the effect of the deeper conductive unit.

4.2.1.4 Apparent Resistivity of Some Other Resistivity Structures

Expressions for the apparent resistivity can be derived for a range of specific resistivity inhomogeneity (Keller and Frischknecht, 1966). Historically, such expressions were used for the interpretation of resistivity data; however, with the availability of numerical tools (as covered in Chapter 5) such models are relatively redundant. A particular case worthy of note is the impact of inhomogeneity adjacent to the quadrupole (Figure 4.10). Such conditions may be encountered when measurements are made parallel to a water course or scarp. Knowledge of the impact of the inhomogeneity may then assist in survey design.

Keller and Frischknecht (1966) show that the method of images utilized earlier can also be applied for such a configuration and state, e.g. the apparent resistivity for a Wenner configuration with electrode spacing a measured d from a vertical feature (as in Figure 4.10) as

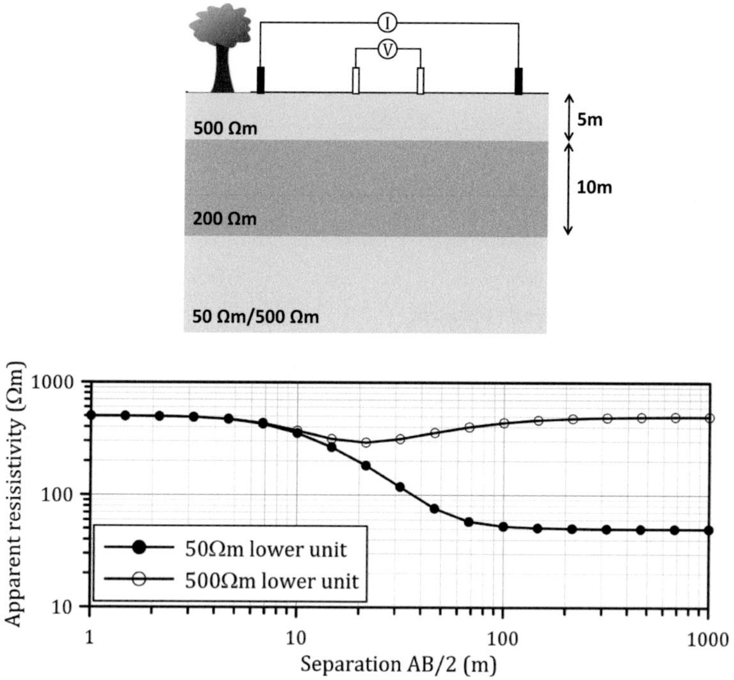

Figure 4.9 Effect of layering on apparent resistivity using the Schlumberger array.

Box 4.1

Apparent resistivity of a buried quadrupole

Keller and Frischknecht (1966) provide expressions for the potential at an electrode buried in a horizontally stratified system. The following figure shows the geometry of a two-layered body with current (A) and potential (M) electrodes placed at the interface, separated by distance r. According to Keller and Frischknecht (1966), the voltage at electrode M is given by

$$V = \frac{I\rho_1}{4\pi r}\left[1 + \frac{1}{\left[1+(2d/r)^2\right]^{1/2}} + k_{1,2} + 2\sum_{n=1}^{\infty}\frac{k_{1,2}{}^n}{\left[1+(2nd/r)^2\right]^{1/2}}\right.$$
$$\left. + \sum_{n=1}^{\infty}\frac{k_{1,2}{}^{n+1}}{\left[1+(2nd/r)^2\right]^{1/2}} + \sum_{n=1}^{\infty}\frac{k_{1,2}{}^n}{\left[1+(2(n+1)d/r)^2\right]^{1/2}}\right], \quad (4.17)$$

where $k_{1,2}$ is defined as before.

Equation 4.17 can be extended to multiple layers for the general kernel function. A specific application of this is the measurement of resistivity of the subsurface at the base of a water column (e.g. lakebed sediments). Electrodes placed at the base of the water column (upper layer in the figure) will be more sensitive to the resistivity of the lower layer than if floated on the water surface.

Box 4.1 (cont.)

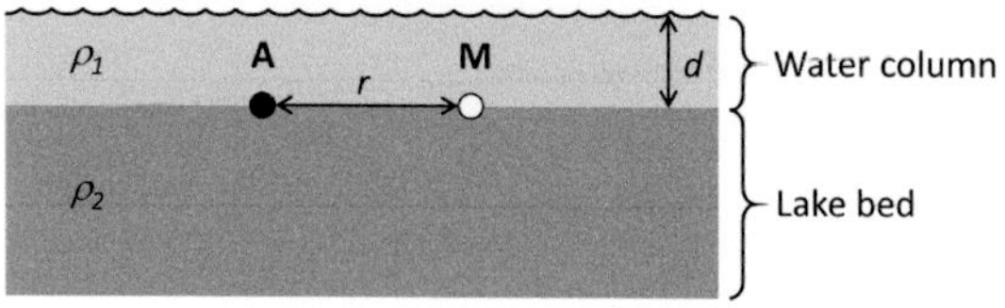

Geometry of buried electrode problem.

The figure below shows two examples of how a conductive layer 1 impacts on measured apparent resistivity for a Wenner array with spacing $a = 1$ m. Equation 4.17, and equivalents, could then be used to compute the value of ρ_2 from an apparent resistivity and given knowledge of water column resistivity ρ_1, water depth d. Alternatively, the impact of the water column on (somewhat easier) floating electrode deployment can be assessed (Lagabrielle, 1983).

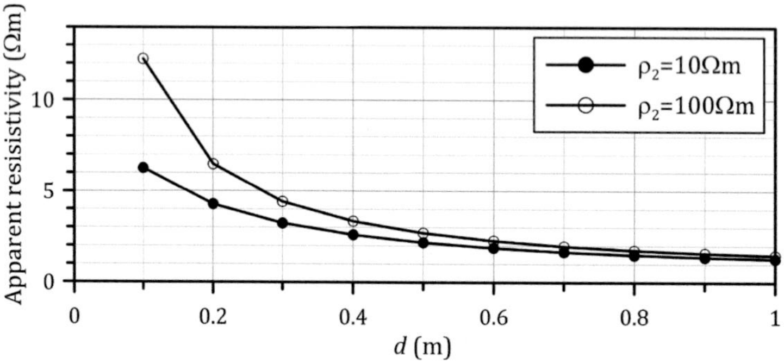

Apparent resistivity of a Wenner quadrupole installed at depth d at the base of a layer with resistivity ρ_1 = 1 Ωm.

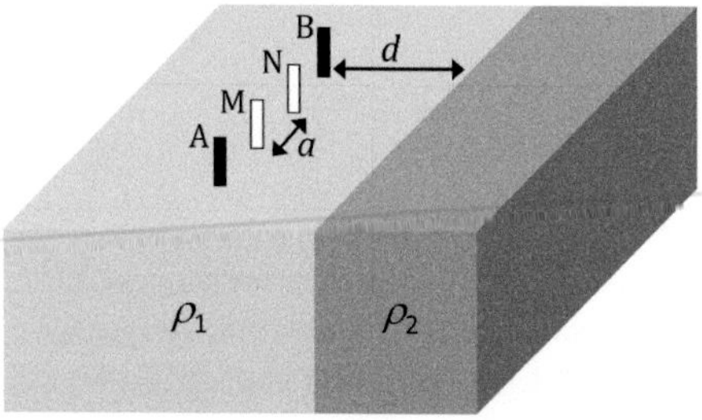

Figure 4.10 Electrode array adjacent to a vertical fault.

$$\rho_a = \left[1 + \frac{2k_{1,2}}{\left(1+\left(\frac{2d}{a}\right)^2\right)^{1/2}} - \frac{k_{1,2}}{\left(1+\left(\frac{d}{a}\right)^2\right)^{1/2}}\right], \tag{4.18}$$

where the reflection coefficient $k_{1,2}$ is defined as in Equation 4.12.

Figure 4.11 illustrates the use of Equation 4.18 and shows how minimal impact of the fault is seen for $d > a$. For the case of a relatively conductive fault, the impact of the fault can be significant for $d < a$. For the resistive fault case, the apparent resistivity curve plateaus to the upper limit of $2\rho_1$.

4.2.2 Measurements in the Field

We have outlined the basic principle of a four-electrode DC resistivity measurement and illustrated how the observed apparent resistivity is affected by inhomogeneity of resistivity in the subsurface. We now focus on the various approaches for combining such measurements for specific survey objectives. All field measurements are subject to errors and so it is first worthwhile outlining the nature of such errors and how they can be estimated.

4.2.2.1 Measurement Errors

DC resistivity instruments, like any measurement system, will be subject to errors (e.g. due to the tolerance of internal components or resolution of any digitization of the signal (current and voltage)). Error checks are easily carried out on the instrument using test resistors, although these are rarely done routinely. Like many geophysical instruments, DC resistivity instruments have a working lifetime of decades; when using old instruments, it is worthwhile checking consistency against test resistors. In most cases, errors due to instrumentation will be minor, and most causes of error are due to the field environment. However, it is important to appreciate the resolution and accuracy of an instrument under ideal conditions. Low-cost instruments may have a significantly reduced current source and

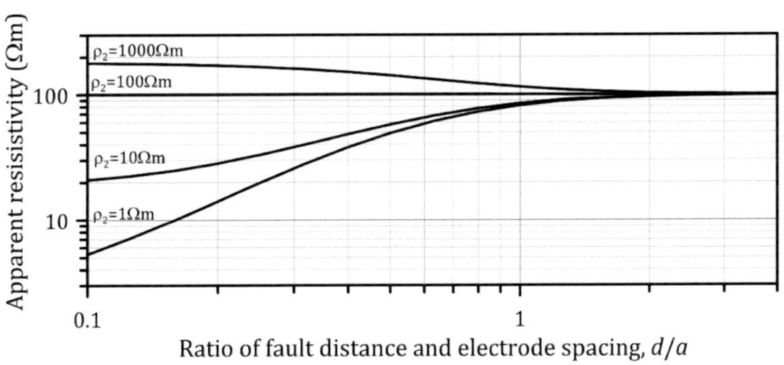

Figure 4.11 Apparent resistivity of a Wenner quadrupole in a 100 Ωm unit adjacent to a vertical fault of resistivity ρ_2. The quadrupole is parallel to the fault as in Figure 4.10.

sensitivity of voltage sensors, limiting the type of measurements that can be made. For example, if we consider the geometric factor for an array in Table 4.1, we can compute estimated voltage across potential electrodes for a given current injection, subsurface resistivity and electrode spacing. Figure 4.12 shows how the measured voltage decays with depth of investigation (based on Roy and Apparao, 1971) for a dipole–dipole ($a = 5$ m) and Wenner array ($a = 5n$, $n = 1, 2, 3, \ldots$) with 0.2 A current injection in a uniform resistivity of 100 Ωm. The graphs show clearly how the dipole–dipole signal weakens with increasing depth of investigation, limiting its use under such conditions. In contrast, the Wenner array retains high signal strength despite low current injection. Such characteristics of the latter configuration make it a popular choice for low-power/-sensitivity instruments.

As discussed in Chapter 3, most DC resistivity surveys in the field are carried out using metallic rod electrodes, typically made of stainless steel to avoid corrosion and normally about 10 mm in diameter. The potential difference measurement is influenced by the contact resistance between the electrode and the ground (see Box 3.2). Electrodes need to have good electrical contact to ensure that the contact resistance does not dominate the measurement of potential difference and most problems associated with errors are attributed to poor electrical contact.

The contact resistance can be reduced by increasing the surface area of the electrode (e.g. a larger diameter electrode, or deeper insertion of the electrode). The addition of small quantities of saline fluid around the electrode can also assist. In some extreme cases (very dry soil cover, for example), wire mesh electrodes are used to enhance the surface area, and hence reduce contact resistance. A potential problem with these remedies is that the electrode then no longer acts as a point sensor or source. So far in this chapter we have assumed in our calculations that the voltage difference is made between two points. The same assumption applies to the current electrodes. We can model non-point electrode effects (see later and in Chapter 5) but this is rarely done for field surveys. If the electrode spacing is relatively short (say less than 1 m), then the impact of non-point conditions may be significant (see Section 4.2.2.8). Failure to achieve point electrode conditions (or to treat them appropriately) thus contributes to the error in the measurement. Measurement errors

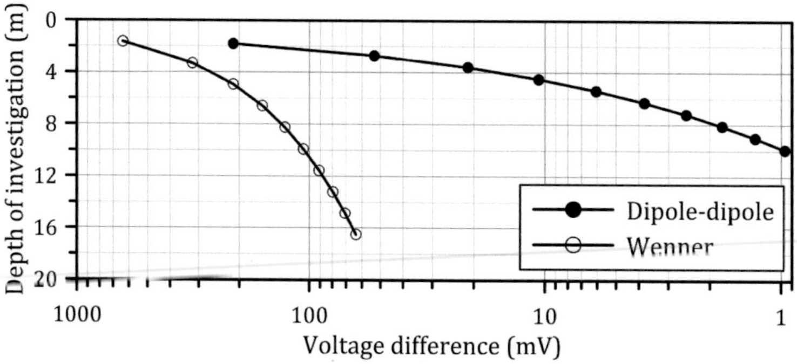

Figure 4.12 Variation in measured potential difference with depth of investigation for two quadrupoles (see text for parameters of the survey).

will also exist due to inaccurate placement of electrodes: positioning errors of the order of a few per cent are not unrealistic and may be higher in rugged terrain. Such geometrical errors will be systematic, not random.

Polarization will also occur at the current electrodes, despite the low-frequency alternating current used, and if these electrodes are used soon after (within several minutes) as potential electrodes, then further contact resistance problems may exist (Dahlin, 2000; LaBrecque and Daily, 2008). Degradation of contact can also occur with corrosion of the electrode surface, although such issues are more likely to be relevant for long-term monitoring arrays. Deterioration of contacts on electrode-cable connectors can also degrade quality of measurements. Intrinsic voltages of natural or anthropogenic origin exist within the ground. The sampling of voltage through a square wave cycle (see Chapter 3) makes assessment of spikes or drift, due to such effects, possible, to some degree, through appropriate filtering within the instrument. Other problems may exist due to high polarizability of the subsurface and inadequate switch on/off times of the signal.

Measurements are, therefore, subject to sources of systematic and random errors. As we show in Chapter 5, measurement errors can have a significant impact on modelling (inversion) of electrical measurements. Interestingly, this was not widely appreciated until relatively recently (but see earlier works of Binley et al., 1995; LaBrecque et al., 1996b). Assessment of data quality is, therefore, an important step of any DC resistivity survey (and, as we show later, even more significant for IP surveys). Errors due to sporadic and persistent degradation of voltage signals can be assessed through repeatability and reciprocity checks. The operator will normally request a number of cycles of the alternating (normally square wave) current signal and thus the standard deviation of a measurement will be recorded. This is a stacking error, not a true repeatability error. Measurements may be repeated but this is rarely done. Alternatively a reciprocity check can be carried out.

A total of 24 configurations of electrodes A, B, M and N are possible on a given set of four-electrode positions. Of these 24, three arrangements will have a different geometric factor (and hence a different transfer resistance in the absence of noise and assuming the ground is homogenous). The configurations are A-M-N-B, A-B-M-N and A-M-B-N (i.e. Wenner α,β,γ if the electrodes are collinear and equally spaced). Carpenter (1955) demonstrated with the Wenner configuration (which was subsequently extended to the more general case by Carpenter and Habberjam, 1956) how these three measurements could be used to assess measurement quality. What such a comparison does is determine the effect of lateral and vertical variability in resistivity, since each of the three quadrupoles has different sensitivity patterns.

The reciprocal measurement of configuration A-M-N-B is M-A-B-N, i.e. source and receiver dipoles are switched. Interestingly, the work of Parasnis (1988) is often cited as defining reciprocity in electrical geophysics; however, much earlier Searle (1911) and Wenner (1912) highlighted the principle in relation to electrical measurements. It is easy to see from Equation 4.6 (for the homogenous case) that swapping current electrodes A, B with potential electrodes M, N results in the same geometric factor and, consequently, the same measured voltage for a given magnitude of current injection. This demonstrates the principle of reciprocity, which also applies for a heterogeneous system, although Wenner (1912) highlight constraints when using alternating current sources.

The problem with reciprocity checking is that it requires an additional measurement, thus increasing the survey time. Inefficiency of configuring certain reciprocal measurements with multichannel instruments without full array switching capability (see Chapter 3) may result in significantly extended survey periods: Wenner and Schlumberger arrays are particularly prone to such inefficiency. A further problem (for multi-electrode cable surveys) is that potentials measured on electrodes recently used for current injection may be subject to residual voltage (e.g. Dahlin, 2000) and thus care must be taken to ensure that the reciprocal check does not introduce additional error sources. Figure 4.13 illustrates a measurement sequence for a dipole–dipole survey using a multi-electrode array that includes a full reciprocity check whilst minimizing the effect of residual voltages.

Figure 4.14 illustrates how stacking errors tend to be much smaller than reciprocal (or repeatability) errors. The data for this illustration are taken from surveys at a riparian wetland site at which an array with 32 electrodes at 0.6 m spacing was deployed for long-term monitoring of changes in resistivity in the wetland. A dipole–dipole array was used with a = 0.6 m and a = 2.4 m, with n = 1, 2, 3, …, 8. Errors in the transfer resistance $R = \Delta V/I$, expressed as ε_R, were computed based on stacking, reciprocity and repeatability. Reciprocal measurements were carried out during a given survey of normal quadrupoles, and repeats of the entire survey were done at 30-minute intervals, allowing repeatability checks. For further details, see Tso et al. (2017).

The histograms in Figure 4.14 show clearly how errors based on stacking underestimate those based on reciprocity, and that the range of errors is much smaller for the former. Also shown in Figure 4.14 are repeatability histograms for surveys conducted at 30-minute and 120-minute intervals. Note that the distribution of errors for reciprocity and short-term repeatability are similar; however, longer-term repeatability (in this case 120 minutes) reveals higher errors. This is not necessarily due to changes in error but due to variation in near surface resistivity caused by hydrological processes (surface soil wetting and drying, warming and cooling). We include these observations to highlight that the process of error checking can lead to an overestimation of errors if the system under investigation is

Measurement	●	●	●	●	●	●	●
1	A	B	M	N			
2		A	B	M	N		
3			A	B	M	N	
4				A	B	M	N
5				M	N	A	B
6			M	N	A	B	
7		M	N	A	B		
8	M	N	A	B			

Figure 4.13 Sequence of normal and reciprocal measurements for a dipole–dipole configuration with a multi-electrode array.

changing over time. This is a particular challenge for time-lapse monitoring systems, as discussed later.

Measurement errors that are due to the voltage sensing, not the geometrical effect of electrode placement, will tend to increase with increases in transfer resistance. Figure 4.15a illustrates this from a dipole–dipole survey with 96 electrodes placed at 1 m spacing using $a = 2$ m; $n = 1, 2, 3, \ldots, 10$. Note that in this figure we show absolute values of error and transfer resistance to illustrate the trend better. Although plots like Figure 4.15a are useful for initial checks in the field, they are limited in the assessment of a quantitative error model. Slater et al. (2000) show how such plots can be used to estimate trends in errors; the problem with doing this, however, is that, for each measurement we have only two samples. The error we require should equate to the standard deviation of the measurement. To overcome this we can subdivide our measurement set into groups, each group covering a range of transfer resistances, that is, we bin the samples. By doing this, provided we have an adequate number of samples in each bin, we can assess the standard deviation of the transfer resistance assigned to each bin (see e.g. Koestel et al., 2008). Figure 4.15b shows the result of this process for the dataset shown in Figure 4.15a. A near linear trend in error with transfer resistance is seen in this case, which tends to be typical for DC resistivity measurements. The plot in Figure 4.15b shows that reciprocal errors in this case are very low ($\ll$1% of the transfer resistance), indicating high-quality voltage measurements. Surveys with different quadrupole geometry should follow the same trend given that the likely sources of error are independent of the configuration. Note, however, that this does not account for geometric placement errors, which, for a field survey, will be inevitably higher. For most surface-based DC resistivity surveys with good electrode contact, reciprocal errors should be 1 per cent or better. Poor electrode contact can lead to significantly higher errors. It is important to note that the error analysis described applies to the treatment

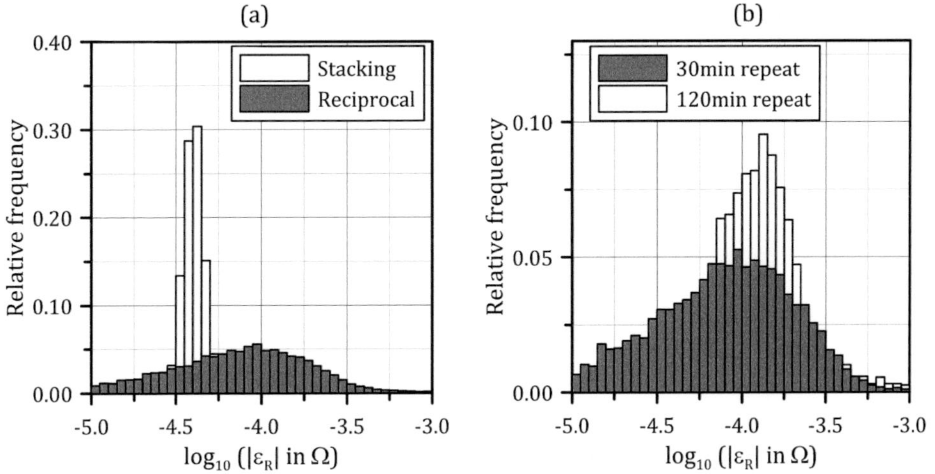

Figure 4.14 Comparison of measurement errors. (a) Stacking and reciprocal errors. (b) Repeatability errors. For details of the dataset used, see Tso et al. (2017).

of transfer resistances, not apparent resistivity, since the former is the fundamental measurement. Scaling with the geometric factor will not reveal such error structure, highlighting the value of analysing transfer resistance, rather than apparent resistivity, values. Finally, it is also important to note that transfer resistances can have different polarity and so the polarity should be retained with the measurement.

The quantification of an error model, as in Figure 4.15b, requires a sufficiently large enough dataset. Ideally, error checks (reciprocal or repeat, but not only stacking) should be made on all quadrupole measurements. If any inverse modelling of the data (see Chapter 5) is to be carried out (as is normally the case), then confidence in the error model is necessary. If survey time is constrained, then a subset of the quadrupoles could be used for error checking, provided it covers the measurement range. Quadrupoles with high geometric factors will clearly result in low transfer resistance measurements. Some operators tend to filter out such measurements in a survey; however, if an error model (as in Figure 4.15b) can be established, such a process may not be necessary.

So far we have assumed that errors are uncorrelated; however, it is intuitive to expect that in multi-electrode surveys electrodes subject to particularly bad contact will lead to higher

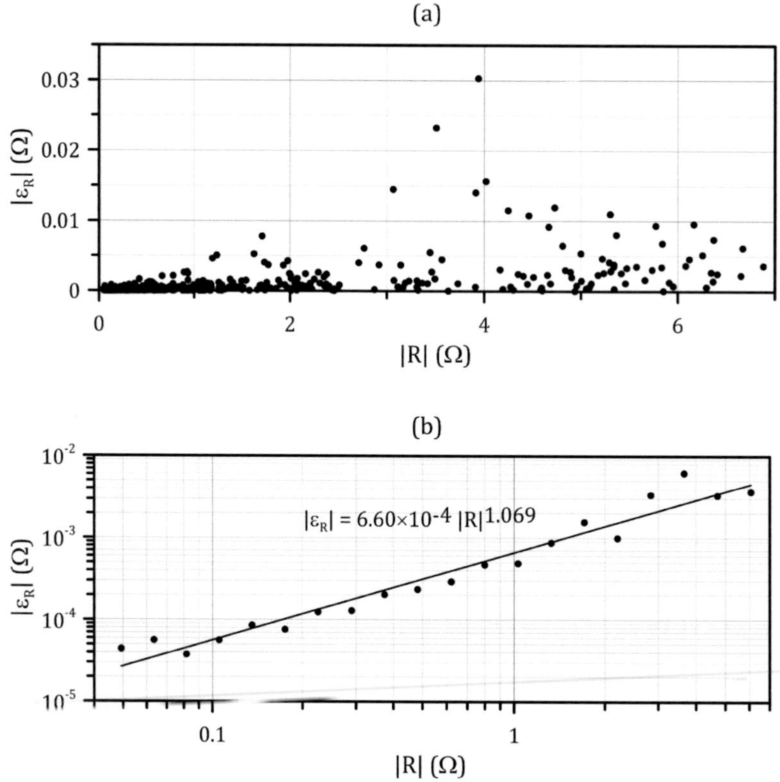

Figure 4.15 Example trend in reciprocal error with transfer resistance. (a) All measurements plotted. (b) Binned (aggregated) measurements to show trend.

measurement errors for quadrupoles utilizing these electrodes, i.e. the errors may be correlated to some degree. Tso et al. (2017) investigated this and illustrated the effect of *memory* in the error model. They then proposed an error modelling strategy, which is widely used in other disciplines, based on the linear mixed effects (LME) approach. Such an approach groups measurement errors by the electrodes associated with each measurement. In cases where significant variation in voltage measurement error may exist, e.g. sections of a survey subject to particularly bad electrode contact, such an approach may be effective. Oldenborger et al. (2005) examined the impact of electrode positioning errors on resistivity imaging, demonstrating that such effects can be comparable to measurement errors. Later we also discuss errors due to the assumption of a point electrode, which can be significant for small-scale imaging if not correctly accounted for.

4.2.2.2 Profiling

Profiling is typically carried out using a particular quadrupole (Wenner, dipole–dipole, etc.) along transects along the ground surface in order to assess lateral variation in resistivity. By maintaining a constant electrode spacing, the results are typically reported at the centre of the quadrupole, as an apparent resistivity. As shown in Figure 4.6 and Figure 4.7, the way in which lateral variation in resistivity will be revealed will depend on the quadrupole geometry. Profiling is now considered somewhat dated given the wide availability of computer controlled multi-electrode resistivity systems. Mobile units can be easily fabricated for small electrode spacing (1 m or less), typically in a wooden frame. The operator then moves along the transect with the frame and instrument and records apparent resistivity (or transfer resistance) at specific intervals. Although such surveys are relatively rapid to make, frequency domain electromagnetic induction (EMI) *terrain conductivity* surveys (e.g. Everett, 2013) are much easier to carry out given that there is no need for galvanic contact with the ground. Furthermore, the relatively recent availability of multi-coil EMI instruments (e.g. Mester et al., 2011) also permits apparent conductivity measurements over several depths. EMI measurements are, however, unable to differentiate contrasts in resistive areas (and also cannot measure IP). For larger spaced (greater depth of investigation) quadrupoles, profiling is labour intensive since each of the four electrodes needs to be moved for each measurement, often requiring multiple operator assistants. For surveys with electrode spacing of 20 m, measuring at 50 m intervals, it may take an experienced crew of three operators one day to complete a 2 km profile. Bernard and Valla (1991) illustrate a number of profiling case studies using large space arrays for mapping fracture zones in bedrock. The advantage of profiling is that relatively basic instrumentation is needed: a four-electrode instrument, four electrodes and four cables.

Profiling is routinely used by archaeological geophysicists with the twin array (mobile pole–pole) configuration (e.g. Clark, 1990). By fixing two of the electrodes (the remote electrodes, B and M), the measurement frame is relatively mobile, enhanced by the fact that surveys are necessarily shallow; thus, the spacing of the two mobile electrodes is short (typically 0.5 m or 1 m). The remote electrodes should be located at least 30 times the mobile electrode pair spacing away from the mobile frame. Given the shallow depth of investigation, relatively low-power (10 mA current injection or lower) DC resistivity instruments are used, weighing around 1 kg or less, and thus may be mounted on the survey frame. Mobility is further enhanced by using current injection frequencies of around 100 Hz, allowing rapid measurements at each location. Variants of the twin-array exist which utilize six electrodes (three pairs of electrodes and different spacing), allowing three depths of investigation in the profile (e.g. Gafney and Gater, 2003). Archaeological geophysicists often refer to profile surveys as 'resistance surveys', since the data are often presented as raw transfer resistances. Profiles are made on parallel transects, forming a grid. According to Gaffney and Gater (2003), a 20 m by 20 m grid with 1 m spacing can be surveyed in about 15 minutes.

Other mobile profiling configurations have been developed for a range of shallow subsurface investigations. Sørensen (1996) developed the Pulled Array Continuous Electrical Profiling (PACEP) system, which consists of two mobile current electrodes and two pairs of potential electrodes (ensuring two depths of investigation), towed by an all-terrain vehicle (Figure 4.16). The array is 90 m long with a 10 m spaced Wenner and 30 m spaced Wenner (in reciprocal configuration). The current and potential electrodes are heavy steel cylinders (10–20 kg for current electrodes, 10 kg for potential electrodes) to ensure continuous contact with the ground. Sørensen (1996) claims that a crew of one or two operators can complete 10 to 15 km of profiling in one day. Clearly, topography and access may reduce mobility of such surveys.

Mobile profiling systems have also been developed for applications in agriculture (Allred et al., 2008). Panissod et al. (1998) describes an arrangement using coulters (cutting discs) as electrodes, towed behind a vehicle. The configuration consists of a pair of current electrodes (1 m apart) and three pairs of potential electrodes (0.5, 1 and 2 m apart) parallel to the current electrode pair (similar to a square array quadrupole). Panissod et al. (1998) refer to the array as 'Vol-de-canards' ('flight of ducks'). The configuration (Figure 4.17) provides three depths of investigation at each location: 0.3 m, 0.52 m, 0.97 m (Gebbers et al., 2009). André et al. (2012) illustrate the use of the system for mapping soil characteristics in a vineyard. The commercial instrument ARP-03 (Geocarta, France) is based on this system. The Veris 3100 Soil EC Mapping System (Veris Technologies, USA) also uses coulters as electrodes in an array consisting of two current and four potential electrodes, towed behind a vehicle (Figure 3.12c) (Lund et al., 1999). Like the PACEP system, this uses a Wenner and reciprocal Wenner array, in this case with a spacing of ~0.22 m and 0.74 m (Figure 4.17) giving a depth of investigation of 0.12 m and 0.37 m (Gebbers et al., 2009).

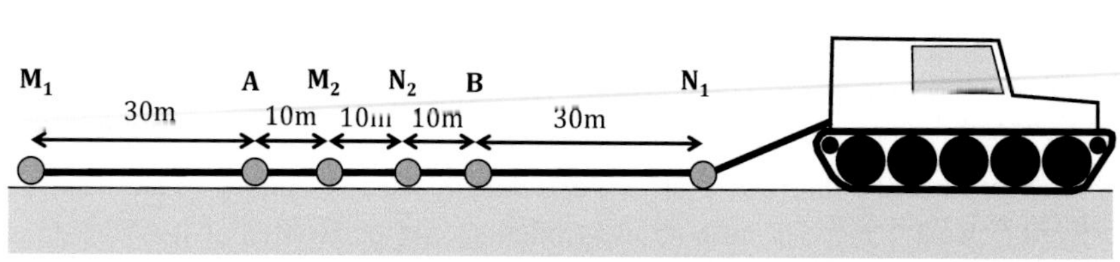

Figure 4.16 Schematic of the Pulled Array Continuous Electrical Profiling (PACEP) system.

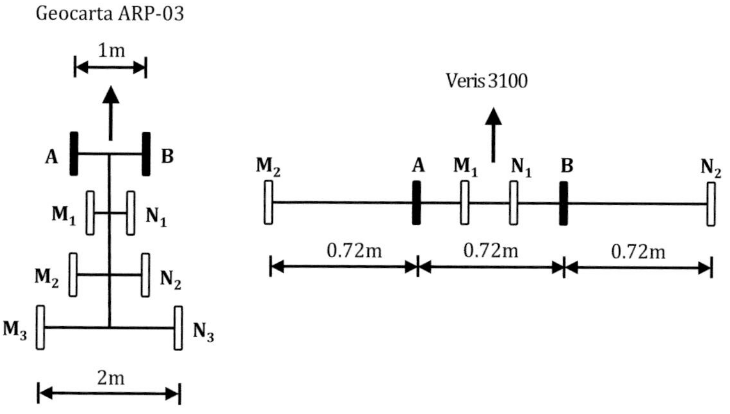

Figure 4.17 Schematic plan view of the Geocarta ARP-03 and Veris 3100 profiling systems. The arrow indicates the direction of travel of the frame.

Each coulter is a 4 mm thick steel disc, 43 cm in diameter. The manufacturers claim an operating survey speed of 25 km/hr. Although designed to map soil textural variation across a field site to aid agricultural management and increase crop yield, applications have also examined variability of soil strength (Cho et al., 2016) and soil water (Nagy et al., 2013). Gebbers et al. (2009) compare the ARP-03 and Veris 3100 systems alongside EMI methods for field-scale soil mapping (see also Sudduth et al., 2003). Luek and Ruehlmann (2013) outline the Geophilus Electricus, which is similar to the Geocarta ARP-03 but also measures induced polarization. Recognizing the mobility advantages of coulter-based electrode systems, a number of attempts have been made to deploy them for archaeological studies (e.g. Terron et al., 2015).

4.2.2.3 Anisotropy and Azimuthal Surveys

Many geological media are anisotropic at the microscale due to the microstructure of the porous media, although we rarely consider such anisotropy at the field scale. At the larger scale, macroanisotropy can exist due to the presence of layers of different media. We may treat this as a layered isotropic system (as in the next section) or we may view it as a homogenous anisotropic system. Consider a vertical profile of layered media with resistivities ρ_i, $i = 1, 2, 3, \ldots, N$, where N is the number of layers of equal thickness. If we consider current flow parallel to the bedding plane, then the effective resistivity of the N layers, $\rho_{||}$, is the harmonic mean of the resistivities. However, for current flow orthogonal to the bedding plane, the effective resistivity, $\rho_{\perp}$, is the arithmetic mean of the resistivities. $\rho_{\perp}$ will always be greater than $\rho_{||}$. Note that if the layers are not of equal thickness, then they need to be incorporated in the calculation of the (harmonic or arithmetic) mean.

A coefficient of anisotropy of resistivity is typically defined as

$$\lambda_A = \sqrt{\frac{\rho_\perp}{\rho_{||}}}. \tag{4.19}$$

As shown by Keller and Frischknecht (1966), for measurements on the ground surface in a macroanisotropic system with layers parallel to the ground surface, the apparent resistivity will be equal to $\lambda_A\rho_{||}$. For vertically dipping beds, the apparent resistivity from the ground surface array crossing (orthogonal to) the bedding planes will, perhaps non-intuitively, be $\rho_{||}$ (which is referred to as the *paradox of anisotropy*), and if the array is parallel to the bedding (i.e. in the strike direction), then the apparent resistivity will be $\lambda_A\rho_{||}$. Measurements of the apparent resistivity along these minor and major axes thus allow an assessment of λ_A from the ratio of the major and minor apparent resistivity. In the field, the orientation of the major (or minor) axis will not be known and so a series of measurements need to be made at different azimuths, resulting in an elliptical pattern of apparent resistivity (Figure 4.18). As the dip angle gets smaller, the effect of anisotropy is less pronounced, i.e. the elliptical pattern will become more circular.

An azimuthal survey consists of a series of co-linear four-electrode measurements (typically Wenner or Schlumberger) made at different angles, centred on a single point, thus assessing any change in the apparent resistivity with orientation, and hence the strike direction in the example in Figure 4.18. Measurements are made between 0° and 180° (ideally 360°) in angular steps of 10° to 20°. Such surveys are commonly used in fractured media, since the orientation of the fracture planes can be relatively easily determined (see e.g. Taylor and Flemming, 1988; Nunn et al., 1983). Watson and Barker (1999) show how the offset array of Barker (1981) enhances the sensitivity to anisotropy over a conventional Wenner configuration; Lane et al. (1995) illustrate the effectiveness of the square array for azimuthal surveys, due to its higher anisotropic sensitivity (Habberjam, 1972) and the reduced surface area (ground access) requirements in comparison to a co-linear array. In all cases, the apparent resistivities are normally be presented in a polar plot (Figure 4.18).

4.2.2.4 Vertical Sounding for a 1D Layered Media

In Section 4.2.1.3, we illustrated how 1D layering of resistivity influences the measured apparent resistivity. Early applications of DC resistivity at the field scale focused on using such knowledge to determine a 1D structure of the shallow subsurface (e.g. Gish and Rooney, 1925). The approach is commonly referred to as a VES. Apparent resistivity measurements are made at different electrode spacing, centred about a common point. As shown in, for example, Figure 4.8, as the electrode array size increases, measurements become sensitive to resistivity at greater depths. The Wenner and Schlumberger configurations (Figure 4.3) are most commonly used arrays for VES. With the Schlumberger array, the potential electrode dipole (M, N) is fixed and the current electrode dipole (A, B) is extended (maintaining distance $AB > 5MN$). This particular configuration is effective because of its limited sensitivity to lateral variation in resistivity, as discussed earlier. The Schlumberger array is also logistically easier to operate in sounding mode since only one of the dipoles is moved for each measurement. For a small MN spacing the measured potential differences can be relatively

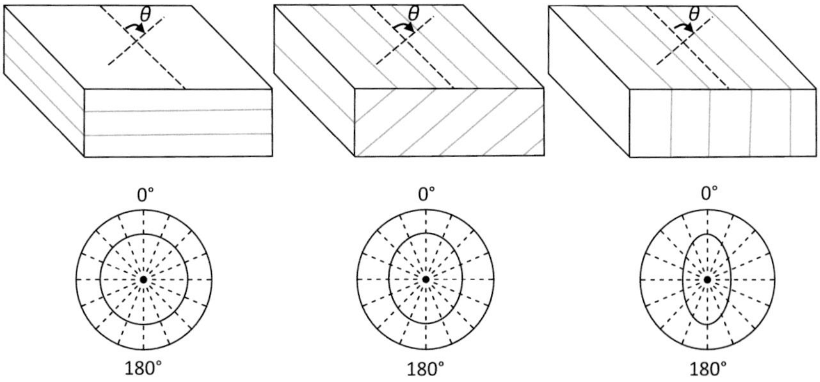

Figure 4.18 The anisotropic resistivity ellipse, showing orientation of minimum and maximum apparent resistivity of arrays measured on the surface above horizontal and dipping beds. For each case, the polar plot of apparent resistivity is shown.

weak and thus high-sensitivity instrumentation may be required. As the current electrodes are expanded, the measured voltage difference between the M and N electrodes will become too small; when this happens, *MN* is increased and the process continued until *AB* is too large again, which leads to expansion of *MN* again. The process, therefore, is a segmented series of measurements. Ideally, measurements will be taken that overlap each segment (i.e. *AB* constant and *MN* at the original and expanded position) to ensure continuity of the sounding data. Lack of continuity in the overlapping segments, as illustrated in Figure 4.19, may be caused by an inappropriate assumption in the calculation of the geometric factor. Most analysis software assumes $n \gg 1$, and from Table 4.1 it can be seen that the geometric factor for such an ideal case is smaller than the correct geometric factor. Such effects are, however, easy to remedy. Lateral heterogeneity in the near surface cover may also cause such failure to overlap. If such errors occur, then the segmented curves should be shifted to align the composite curve (see Koefoed, 1979). Many modelling tools adopt such filtering.

Use of the Wenner array for sounding does not require checks on continuity because the measured voltages do not suffer from the same constraints as in the case of the Schlumberger configuration. However, the array suffers more from near-surface heterogeneity. To overcome this, Barker (1981) proposed the 'offset Wenner' array. This configuration uses five equally spaced electrode sites for one recorded value. A Wenner measurement is taken with the left-most four electrodes and with the right-most four electrodes. The average is used as a record of the measurement for that spacing, and the difference between the two is recorded as a measure of the effect of near-surface variability.

An electrical sounding requires four electrodes, each with a suitable cable to connect to the instrument. Cable reels typically contain several hundred metres length of cable. When carrying out a VES with the Schlumberger array, the current electrode spacing (*AB*) is normally increased in a logarithmic (or approximately logarithmic) manner. A sounding curve is presented as a plot of the logarithm of apparent resistivity versus the logarithm of *AB*/2, as shown in Figure 4.8 and Figure 4.9. In Chapter 5, we describe data modelling tools

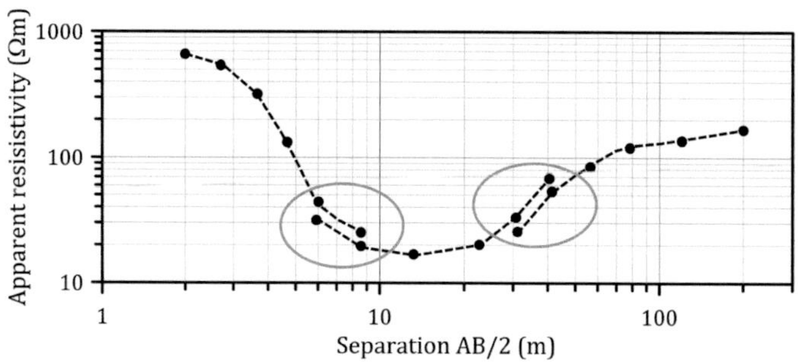

Figure 4.19 Example sounding curve showing build-up of segments, each with constant *MN* spacing. The two circled areas show overlapping sections that should match.

that are used to recover a 1D resistivity structure that is consistent with the measured sounding of apparent resistivity.

Vertical electrical soundings have been used widely used in hydrogeology to differentiate units (e.g. Kosinski and Kelly, 1981). The method suffers from the necessary assumption of 1D variation in electrical resistivity, and as most modern DC resistivity instruments are equipped with multi-electrode capability, information about the lateral variability of resistivity can be easily determined using combined sounding and profiling (see next section). As discussed in Chapter 5, VES survey interpretation can also be ambiguous or non-unique (see Simms and Morgan, 1992).The limited equipment demands continue, however, to make VES surveys a popular choice in some areas. Vertical electrical soundings are widely used, e.g. in Africa for siting water wells (e.g. MacDonald et al., 2001), and, in fact, in parts of Africa today public water wells cannot be drilled unless a VES survey has been carried out. Alle et al. (2018) highlight the problems with assuming one-dimensionality for electrical investigations in hard rock aquifers in Benin, Africa, noting that typical rate of success in identifying a successful well with VES is only 60%.

4.2.2.5 2D Imaging

We have already shown in Section 4.2.2.2 how profiling is sometimes carried out using more than one electrode spacing, thus sensing more than one depth of investigation. Two-dimensional imaging is a logical combination of sounding and profiling. Four-electrode measurements are made along a transect (as in profiling) but quadrupole spacing is incrementally changed in order to achieve sensitivity over different depths. The approach has been used for several decades, although early applications will have been labour intensive. As discussed in Chapter 3, many modern multi-electrode instruments permit connection to an array of electrodes through a multicore (or series of multicore) cable(s). Griffiths and Turnbull (1985) outlined one of the earliest field geophysics prototypes of this approach, using an eight-core cable; similar developments in electrical imaging were being made at the same time in other fields (Lytle and Dines, 1978; Wexler et al., 1985).

Two-dimensional imaging of resistivity with a surface arrangement of electrodes is sometimes referred to as electrical resistivity (or resistance) tomography (ERT) or electrical resistivity imaging (ERI). Griffiths et al. (1990) referred to the technique, as it was emerging, as microprocessor-controlled resistivity traversing (MRT). We discuss resistivity imaging with electrodes bounding a region later (Section 4.2.2.8).

The choice of quadrupole configuration will be influenced by a number of factors. If the objective of the survey is to delineate lateral variability, then a dipole–dipole configuration may be appropriate. However, such an array (as discussed earlier) can suffer from signal-to-noise constraints, which will depend on the resistivity of the study area and the instrumentation available. The Wenner configuration is a popular choice when using low-power instruments, but suffers from a weak sensitivity to lateral variability. The gradient array (see Figure 4.3) may be an attractive compromise (see e.g. Dahlin and Zhou, 2006). In practice, with multi-cable capability, multiple array types should be considered rather than limiting a survey to one configuration. Much of the survey time is spent installing the series of electrodes, and so running through sweeps of different array configurations often adds little survey time overhead.

Survey time can impose constraints when run in a basic four-cable mode (i.e. without multicore cable connectivity); the gradient and dipole–dipole arrays are efficient to use as one pair of electrodes is moved at a time. With multi-channel capability, these arrays may also be very efficient in terms of survey time as (for some manufacturers) multi-channel switching can be constrained to particular combinations of measurements. Early single channel multi-electrode systems did not suffer from this constraint and the Wenner array was a popular choice. However, use of such an array cannot exploit multi-channel capability efficiently.

Measurements from a 2D imaging survey are often presented (in the field) as a pseudosection of apparent resistivity. Edwards (1977) attributes the pseudosection concept to Hallof (1957) (see also Seigel et al., 2007). The pseudosection is a graphical presentation of the apparent resistivity – it does not represent an image of the subsurface, as shown later. A pseudosection has value for displaying raw measurements in resistivity units, allowing the operator to assess the range of apparent resistivity at the site, and note any outliers and anomalies. Some modern instruments have the capability to display a pseudosection as the survey progresses. In Chapter 5, we show how the measured data can be modelled to determine a 2D image of the resistivity. Prior to the availability of such tools, interpretation of field data relied on the use of pseudosections. However, modelling of 2D imaging data is now easily achieved and preliminary modelling can be done in the field, making pseudosections redundant to some degree, although they can be useful for identification of anomalous readings.

Figure 4.20 illustrates how a pseudosection is constructed, in this case for a dipole–dipole array. Each measurement of apparent resistivity is positioned graphically at the intersection of two lines dipping 45° from the centre of the electrode pairs. In some cases, the level axis is shown as a pseudo-depth using an appropriate assignment of a depth of maximum sensitivity (e.g. Roy and Apparao (1971); see also, Edwards (1977)). However, assigning a single depth may be misleading: the pseudosection is simply a means of displaying apparent resistivity data. In fact, a difficulty arises with pseudosections when we use a combination of array types (e.g. combining Wenner, dipole–dipole, etc., or even

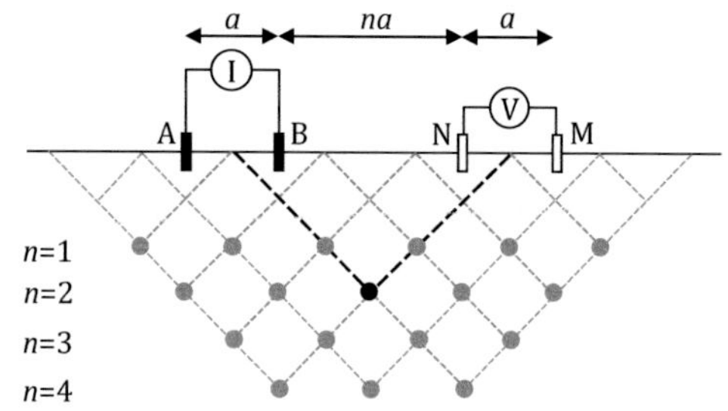

Figure 4.20 Construction of a pseudosection for dipole–dipole survey. Modified after Edwards (1977).

combining dipole–dipole surveys with different *a* values, since the level, *n*, will equate to a different depth of investigation). Consequently, separate pseudosections are often displayed for each array type.

Figure 4.21 shows pseudosections for dipole–dipole ($a = 5$ m) and Wenner measurements computed for a simple 2D structure of resistivity. A 2 m thick, 100 Ωm layer overlies a 500 Ωm formation, in which a 10 m wide 50 Ωm dike protrudes vertically. Twenty-five electrodes are placed on the ground surface at 5 m spacing. The pseudosections were computed using a solution to Equation 4.1 for the given resistivity distribution, which extends in the strike direction (the methods used to do this are covered in Chapter 5). The pattern of apparent resistivity in the pseudosection differs significantly for the two array geometries, both in terms of pattern and range, which is a result of the different sensitivity patterns (Figure 4.5). The vertical dike unit manifests itself in the dipole–dipole pseudosection as an upturned 'V' feature. In the Wenner survey, lateral sensitivity is limited and the pseudosection shows limited effect of the vertical feature for $n > 2$ (in this example). Note also that the dipole–dipole layout offers potentially greater coverage at the left and right margins of the survey line.

As discussed above, 2D imaging surveys are now commonly performed with instruments capable of addressing multiple electrodes. Even with relatively modest multi-electrode capability, long survey lines can be carried out using 'roll along' techniques (e.g. Dahlin, 1995). With such an approach, the survey is progressively developed through segments (Figure 4.22). In the first segment, the full set of measurements are made, whereas for subsequent sections a reduced dataset is acquired to avoid replicating those collected in the previous survey. A roll-along survey may involve disconnecting electrodes and moving all cables along for each segment, or, with dual-ended cables, it may be more efficient to move just one cable at a time, e.g. in a dual-cable set-up the leftmost cable in segment 1 becomes the rightmost cable in segment 2, and the rightmost cable in segment 1 remains in place as the leftmost cable for segment 2. The choice of approach will depend on the depth of investigation required and the quadrupole configuration adopted. As shown in Figure 4.22,

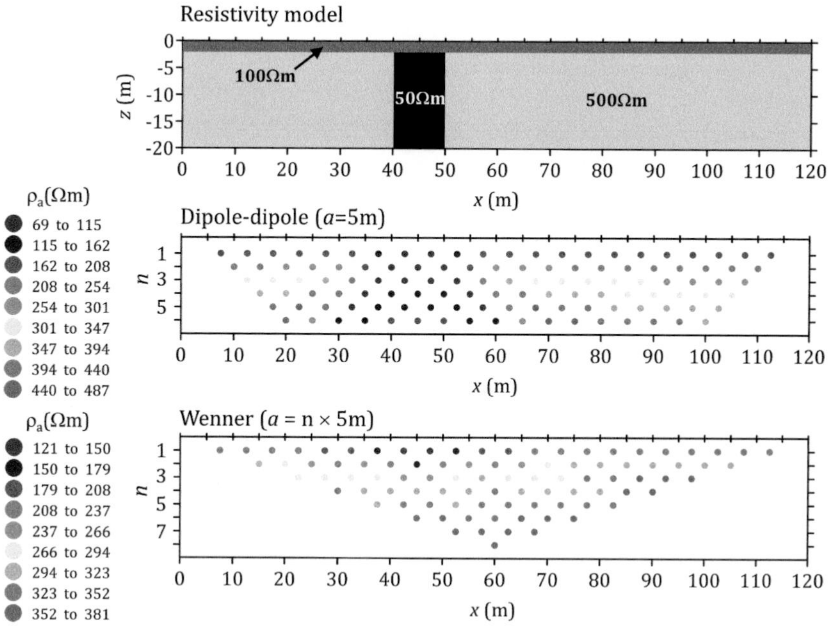

Figure 4.21 Comparison of dipole–dipole and Wenner pseudosections for a synthetic resistivity model. Note the different apparent resistivity scales. (A black and white version of this figure will appear in some formats. For the colour version, please refer to the plate section.).

a dipole–dipole configuration allows much greater progression of the roll-along survey, in comparison to a Wenner arrangement.

Towed electrode systems have been developed to allow coverage over long transects with high lateral resolution. These are often referred to as continuous vertical soundings (CVES) as datasets are typically treated as a series of 1D electrical soundings. Christensen and Sørensen (2001) illustrate the Pulled Array Continuous Electrical Sounding (PACES) system, which was a development of the PACEP continuous profiling method shown in Figure 4.16. PACES uses a 30 m spaced current dipole and eight potential electrodes in a variety of configurations with a maximum cable length of 90 m. Thomsen et al. (2004) show the effectiveness of the PACES system for mapping clay overburden as part of an aquifer vulnerability assessment. Other continuous sounding systems have been devised. Simpson et al. (2010) report on the tractor towed 'Geophilus electricus' system, which is similar to the Geocarta ARP-03 shown in Figure 4.19 and discussed in Section 4.2.2.2. The arrangement was devised for relatively shallow investigations and incorporates a 1 m current dipole and five pairs of 1 m spaced potential dipoles positioned parallel to the current dipole at distances of 0.5 m, 1 m, 1.5 m, 2 m and 2.5 m. Electrical contact with the ground is achieved through spiked wheels. Simpson et al. (2010) illustrate the system applied to an archaeological investigation.

The CVES concept for 2D imaging has also been applied for mapping river/lake bed variation in resistivity using a waterborne array. The earliest record of waterborne resistivity

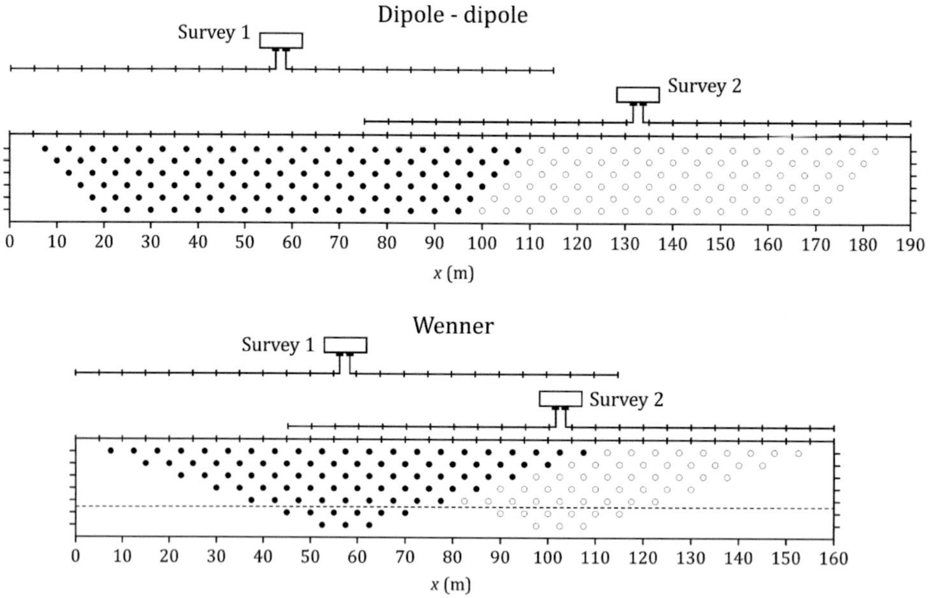

Figure 4.22 Roll-along surveys, shown schematically with pseudosection levels for a dual twelve-core electrode array applied at 5 m electrode spacing for dipole–dipole ($a = 5$ m) and Wenner quadrupoles. The cable pair is moved from Survey 1 (solid circles) to Survey 2 (open circles). The dashed line for the Wenner survey shows the limit of investigation with this set-up (and measurements for levels beyond this would probably not be acquired in the field).

surveys (not imaging, but profiling and soundings) is that of Schlumberger et al. (1934). More recent applications include: Bradbury and Taylor (1984); Allen and Merrick (2005); Sambuelli et al. (2011). Butler (2009) offers a useful review of waterborne electrical methods. Several manufacturers offer floating electrode cable (commonly referred to as a streamer) extensions to traditional systems (Figure 4.23). Rucker et al. (2011) document results from an impressive total of 660 line kilometres surveyed along the Panama Canal, using a 170 m long floating array of 11 electrodes, capable of measuring eight dipole–dipole measurements at each location, in 3.75 m intervals along the transects.

Waterborne surveys are typically carried out with a GPS unit recording position and a water column depth recording device (e.g. using ultrasound). Knowledge of the electrical conductivity of the water column is also valuable if data inversion is to be carried out (see Chapter 5). The electrode spacing in the electrode streamer will need to be selected according to the water column depth. Clearly, as either depth or conductivity of the water column increases, the sensitivity of the quadrupoles to the river/lake bed will be reduced (e.g. Lagabrielle, 1983; see also Box 4.2). Sensitivity can be improved by using an electrode array that sits on the river/lake bed (e.g. Crook et al., 2008; Orlando, 2013), although this can create logistical challenges due to the heightened risk of cable snagging. For deepwater column applications, information about vertical variation in the electrical conductivity is also required for interpretation of results. Baumgartner and Christensen (1998) discusses the use of a more elaborate arrangement of electrodes for application in Lake Geneva with water depths over 100 m. Note that a number of other logistical challenges exist with streamer-type surveys, in particular: (i) determining the location of the electrodes, since often the line is not straight; (ii) difficultly in making data error checks since reciprocal measurements are not achievable and stacking is often limited because of the speed of the survey.

DC resistivity methods have also been applied in marine environments. As Chave et al. (1991) discuss, such applications may at first appear unsuitable for the DC methods given the high conductivity and depth of the water column. Chave et al. (1991) state that in order to detect 10% change in marine bed resistivity a precision of the order of 0.3% is needed in measured signal (assuming a water column with ten times the electrical conductivity of the marine bed). However, working in such an environment has the advantage of ideal coupling of electrodes and low noise. Shallow marine surveys have been conducted, e.g. to identify submarine freshwater discharge (e.g. Henderson et al., 2010) and archaeological prospecting (e.g. Passaro, 2010), using similar arrangements to those used in freshwater studies. For deeper marine surveys electrode arrays located on the water surface are unlikely to be effective, and yet placement on the marine bed is not practical and likely to result in cable snagging. A compromise is therefore an array of electrodes elevated above the marine bed, high enough to avoid obstructions but not too high to ensure some penetration of signal into the bed. Ishizu et al. (2019) illustrate the approach for deepwater prospecting for sulphide deposits. Box 4.2 illustrates how the sensitivity pattern of two different measurement configurations is affected by positioning of the electrode array.

4.2.2.6 3D Imaging

3D imaging with an array of electrodes on the ground surface is a logical extension of 2D imaging with the advent of computer-controlled multi-electrodes instruments and data inversion tools (discussed in Chapter 5). Even with current instrumentation 3D imaging is significantly constrained by the hardware required and thus remains a reasonably specialized application method, typically on relatively small plots. For example, 50 electrodes is a fairly typical lower limit for the number of electrodes used on a 2D imaging transect. If the same coverage were to be achieved in both directions on the ground surface, then a system capable of measuring 2,500 electrodes would be required (well beyond the means of most instruments available today) along with several hundred kilograms of cables and electrodes, and immense labour for installation. Consequently, compromises need to be made for 3D imaging to be a practical option.

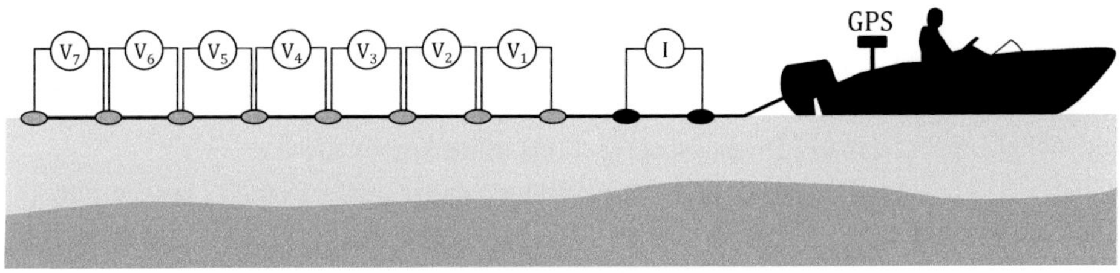

Figure 4.23 Schematic of a typical waterborne electrical resistivity streamer.

Box 4.2

Marine resistivity sensitivity patterns

The figure below illustrates the sensitivity pattern (Equation 4.10) of a quadrupole placed 5 m deep on a marine bed and 1 m above the bed. The arrangement is shown in (a). In (b) and (c), the sensitivity for a Wenner measurement is shown, illustrating the limited penetration of signal into the bed due to the low-resistivity water column. In (d) and (e), the same is shown for a dipole–dipole configuration. For marine surveys, the electrode separation may need to be considerably greater than that for terrestrial surveys, and to ensure satisfactory signal strength, larger currents are used.

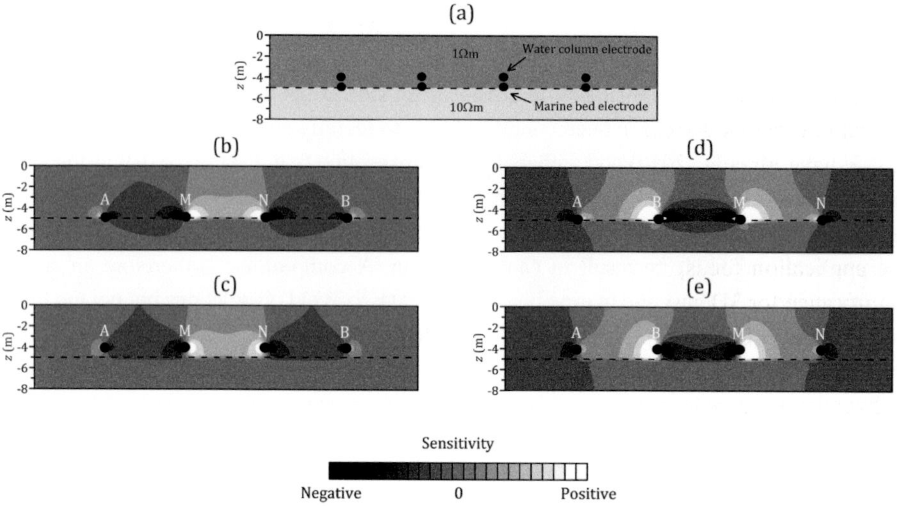

Quasi-3D imaging combines multiple 2D imaging surveys that are typically run as a series of parallel transects (although this is not a requirement for data inversion – see Cheng et al., 2019a; Dahlin and Loke; 1997; and Section 5.2.2.6). Chambers et al. (2002) refer to quasi-3D imaging as the 3D interpolation (rather than modelling) of 2D imaging results. Whilst not truly 3D since current is not injected (or potential gradients measured) in the strike direction, i.e. orthogonal to the transects, such surveys can still be effective provided the distance between transects is not great, relative to the scale of variation in resistivity (Aizebeokhai et al. (2011) recommend a line separation no greater than four times the electrode spacing along a line). The advantages of such surveys are that (i) standard 2D imaging approaches can still be applied to visualize and model the data (allowing rapid checks in the field, for example); (ii) standard multi-electrode hardware is required; (iii) field effort is simply a multiple of that required for 2D surveying. In addition, the operator can always resort back to 2D analysis if sections of the survey are corrupted.

True 3D imaging offers much greater sensitivity to horizontal variability in resistivity (as illustrated by the study of Chambers et al. (2002) – see also Section 5.2.2.6). To make fully 3D measurements surveys can still be made in linear arrays by connecting pairs (or sets) of transects in sweeps across the region of interest. Figure 4.24 illustrates this approach. The survey starts by addressing electrodes in lines 1 and 2, then lines 1 and 3, lines 1 and 4, and lines 1 and 5. The survey continues by connecting lines 2 and 3, then 2 and 4, etc. A total of ten combinations of line pairs are then made (for the five-line case in Figure 4.24). Such a survey can be implemented with standard quadrupoles discussed earlier, using survey schedules (the list of quadrupoles to be measured) identical to those used for 2D imaging. However, Wenner and dipole–dipole quadrupoles are likely to be least effective as the coverage will be limited in the orthogonal direction to the lines, and thus careful design of the survey schedule is needed. Figure 4.25 shows an alternative approach for connecting electrodes for a 3D imaging survey (see e.g. Chambers et al., 2012). In this case, lines are connected in two orthogonal directions to ensure unbiased sensitivity to horizontal variability. The schematic in Figure 4.25 is shown with all electrodes connected for illustration purposes. For large electrode arrays, subsets may be adopted in a similar manner to that shown in Figure 4.24 (e.g. Dahlin et al., 2002a).

We have already shown how quadrupole configurations (Wenner, dipole–dipole, etc.) have different depths of investigation and sensitivity to horizontal and vertical variation in resistivity, and thus the choice of quadrupole (or combination) must be selected according to the application focus, instrumentation and environmental conditions. This is of greater significance for 3D surveys since the spacing of quadrupoles will be highly variable.

For many 3D imaging surveys, the pole–pole quadrupole has proved to be popular. Park and Van (1991) present one of the earliest 3D resistivity imaging applications, and utilize the pole–pole geometry. The pole–pole configuration is potentially attractive because horizontal coverage will be achieved in all directions in an unbiased (symmetric) manner and there is no requirement for complex survey design and customization of

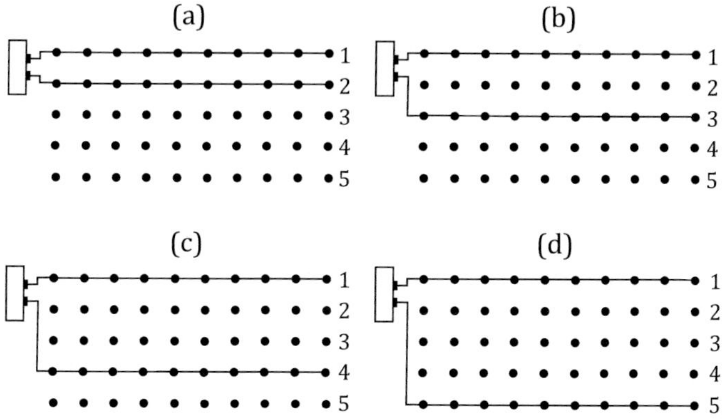

Figure 4.24 Schematic showing a sequence of line combinations for a 3D survey. (a) Lines 1 and 2 are connected. (b) Lines 1 and 3 are connected. (c) Lines 1 and 4 are connected. (d) Lines 1 and 5 are connected. The survey continues with six further combinations of pairs of lines. The solid circles indicate electrodes (note that only ten are shown for each line for illustration purposes).

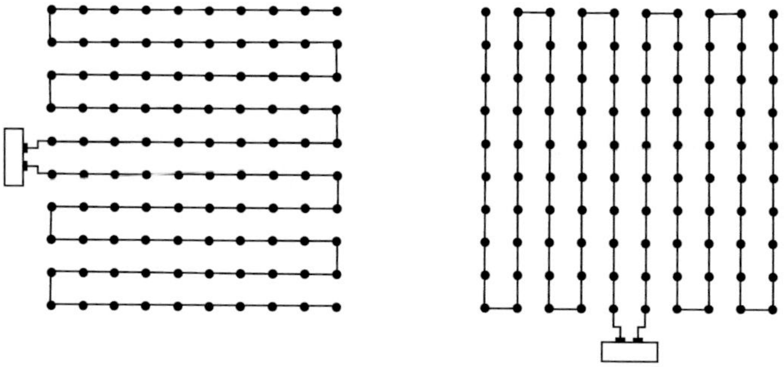

Figure 4.25 Orthogonal cable connectivity for 3D image data acquisition.

measurement schedules. Measurements from pole–pole surveys in a 3D configuration can also be visualized easily as pseudo-volumes, as illustrated by Dahlin et al. (2002a). For other quadrupoles (other than the square array), the construction of a 3D equivalent of a pseudosection is challenging as a nominal vertical position would need to be assigned to each measurement. In addition, a full pole–pole survey (i.e. all electrodes are used as current A electrode, with the remaining electrodes used as potential M electrode) is a truly complete independent set of measurements (Xu and Noel, 1993). For N electrodes (in addition to the two remote electrodes), there will be $N(N\text{–}1)$ possible combinations, half of which are reciprocals. All other quadrupole measurements can, in theory at least, be generated from combinations of these measurements, by superposition. However, in the presence of noise (as will always be the case), care must be taken in adopting such an assumption. Dahlin et al. (2002a) also comment on logistical challenges of pole–pole surveys due to rodents damaging remote electrode cables (i.e. outside the main survey area).

The pole–dipole configuration can overcome some of the constraints due to noise in pole–pole surveys. As noted by Nyquist and Roth (2005), using orthogonal potential dipoles in a pole–dipole survey is inefficient as voltage differences for many dipole combinations will be lower than noise levels. They propose dipole measurements along lines fanning out from the current (A) electrode, as shown in Figure 4.26 (see also Loke and Barker, 1996a). Nyquist and Roth (2005) show in a field study with a 10 by 11 grid of electrodes the radial arrangement results in a measurement set 20% smaller than a Cartesian arrangement, in addition to showing significant increase in measurement quality.

Samouelian et al. (2004) show the effectiveness of the square array for 3D resistivity imaging in a very small-scale study of soil cracks, using a grid of 8 by 8 electrodes. By considering two orientations of the square array, they were also able to examine anisotropic effects, which were assumed to be related to cracking with the soil sample. The square array allows easy visualization of a pseudo-volume and, by moving the square configuration with the electrode grid, gives unbiased horizontal sensitivity. Such an approach could also be replicated with a rectangular array, giving a larger measurement set and enhanced spatial coverage (see Figure 4.27).

4.2.2.7 Borehole-Based Measurements

So far we have considered DC resistivity measurements made on the ground surface. Electrodes placed within boreholes allow greater sensitivity at depth and in some cases are the only means of obtaining reliable information about resistivity variation, either because a surface-based approach has inadequate depth of investigation and resolution or because of access constraints.

4.2.2.7.1 Borehole Logging

DC resistivity measurements are widely used in geophysical logging of boreholes (wireline logging). The first electrical log in a borehole was run by Schlumberger in 1927 in France. The original log was run using four electrodes with fixed spacing lowered in a borehole and measurements recorded at 0.5 to 1 m intervals. Johnson (1962) shows the originally recorded log, which, although coarse in resolution, shows clear variation in resistivity throughout the profile. The approach became known as 'electric coring'. The approach developed with revisions to the configuration of electrode. Wireline electrical logs have proved invaluable for oil reservoir investigations for estimating properties of the formation

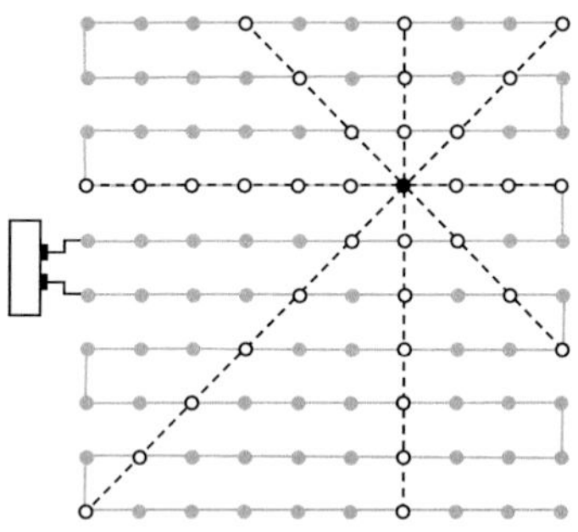

Figure 4.26 Selected quadrupole electrodes for pole–dipole surveys for improved measurements over the connectivity shown in Figure 4.27. The solid black circle is the current A electrode; open circles are potential electrodes.

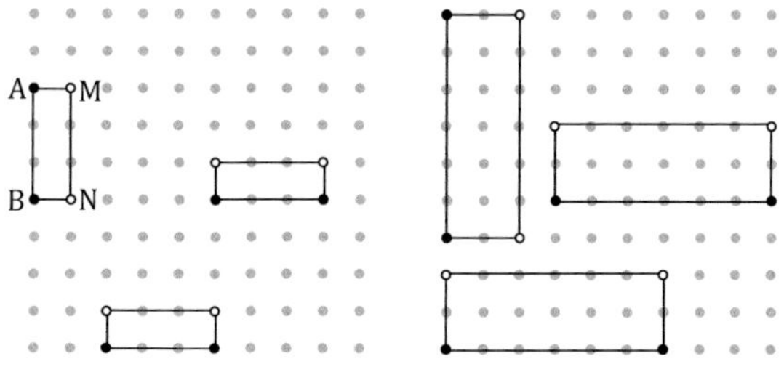

Figure 4.27 Alternative 3D imaging geometry using a rectangular array. Two different quadrupole spacings are shown for illustration.

(porosity, permeability) or the salinity of the formation fluid, and more recently in groundwater investigations for delineation of hydrogeological units (e.g. Keys et al., 1989).

The resistivity measurement is made using the same four-electrode principle covered earlier in this chapter. A sonde containing electrodes in a particular configuration is connected to a logging device on the surface. Measurements can be made in a water- (or mud-) filled section of the borehole, which must clearly be uncased to allow connectivity to the formation (dry uncased wells can be logged electrically with induction tools). The configuration of the electrodes within the sonde controls the vertical resolution and depth of investigation (laterally into the formation). Typical configurations are: short normal, long normal and lateral (Figure 4.28). The short normal will have high resolution of vertical variation in resistivity but will be influenced by the fluid within the well (which may vary along the well). Using short and long normal, measurements are made with the same sonde. The lateral log will have, as the name suggests, even greater lateral penetration but suffer from a relatively weak vertical resolution.

In the 1950s, the microlog (microresistivity log) was developed (see Segesman, 1980) to allow measurements of resistivity of the borehole wall using an expanding calliper arrangement (see Figure 4.28). The focus of the tool, specifically for oil field exploration, is the detection of 'mud cakes' (originating from the drilling mud that is circulated in a well) that form against permeable formations. Further developments include the focused current resistivity sonde in which guard (current) electrodes are fitted above a central current electrode to enhance lateral channelling of current into the formation from the main current electrode. Keller and Frischknecht (1966) provide detailed analysis of geometrical factors for the various resistivity sonde configurations.

Wirelogging tools are effective for measurements in conventionally drilled boreholes in consolidated media. For unconsolidated investigations, direct push tools, based on cone penetrometer technology (CPT), have become popular, particularly in contaminated land

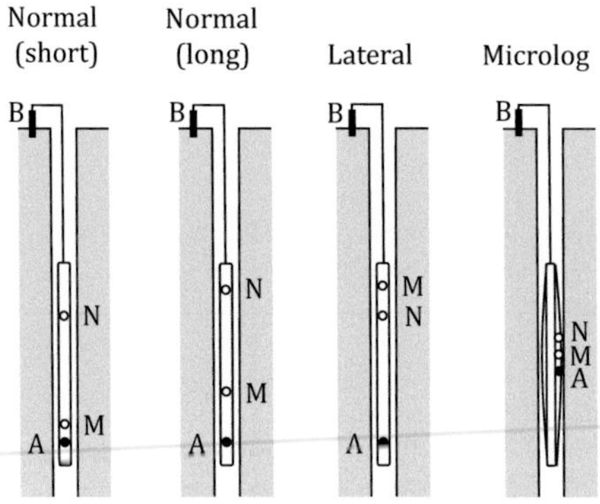

Figure 4.28 DC resistivity wireline logging configurations (not to scale). Standard electrode spacings are (i) normal (short) $AM = 41$ cm (16"); (ii) normal (long) $AM = 163$ cm (64"); (iii) lateral $AO = 569$ cm (18'8") (where 'O' is the midpoint of electrodes M and N).

studies. The technology has advanced to include a range of sensor techniques, including DC resistivity. Because the electrodes used are in direct contact with the formation and there is less concern about drilling disturbance of the formation (as in conventional drilling), a small electrode spacing can be used, and thus resolution is exceptionally high. Schulmeister et al. (2003) show several examples using a Wenner configuration with 2 cm electrode spacing on a probe, with measurement sampling at 0.015 m along the drive point. According to Schulmeister et al. (2003), a two-person crew can complete a single log of 20 to 30 m in about two hours. Figure 4.29 illustrates results from a direct push resistivity log carried out by UFZ-Leipzig, Germany, as part of a study of contamination at the Trecate site in northern Italy (see Cassiani et al. (2014) for more details). Figure 4.29 also shows a 2D resistivity image obtained from dipole–dipole measurements using 2 m electrode spacing, illustrating clearly the contrast in resolution between the two approaches.

4.2.2.7.2 Imaging Using Borehole Electrodes

Schlumberger's early developments of electrical methods included the use of a deep electrode as a current source (see discussion on the mise-à-la-masse method later). Subsequent studies explored the sensitivity of borehole-based methods in a mineral exploration context. Clark and Salt (1951) and Snyder and Merkel (1973) show how a borehole-installed current electrode can improve the sensitivity of resistivity measurements to an ore body. Alfano (1962), in contrast, shows that electrical sounding can be enhanced by using a current electrode in a borehole. Daniels (1977) was probably the first to consider borehole to borehole quadrupole measurements. However, it was only with the advent of suitable data inversion tools in the late 1980s/early 1990s that imaging methods based on borehole-based electrode measurements emerged (Sasaki, 1989; LaBrecque and Ward, 1990; Shima, 1990; Shima, 1992), enhanced by the emerging availability of multi-electrode measurement systems.

Imaging using borehole electrodes can significantly enhance resolution and, in some cases, is the only viable approach if site access restricts installation of surface electrodes

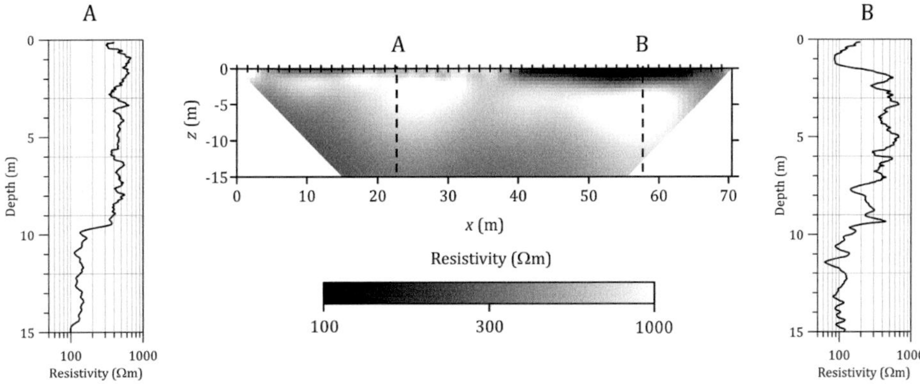

Figure 4.29 Example direct push resistivity logs at the Trecate site in Italy. Logs A and B are shown alongside a DC resistivity imaging survey using 1.5 m spaced electrodes.

(e.g. imaging underneath a building). As discussed in Chapter 3, a range of approaches for borehole electrode placement are available. Electrode arrays in a single borehole typically consist of a few tens of electrodes. When electrodes are placed within the borehole water column, the effect of short-circuiting of current can be significant (Osiensky et al., 2004; Nimmer et al., 2008; Doetsch et al., 2010a) and should ideally be accounted for in any modelling. Wagner et al. (2015) also highlight the impact of non-point electrode effects and non-verticality of boreholes in cross-borehole resistivity applications (see case study in Section 6.1.8).

Figure 4.30 shows a range of configurations for borehole electrode arrays. Resistivity profiling along a single borehole can be effective (although will suffer from poor resolution in comparison to wireline logging methods). A short-spaced dipole–dipole quadrupole is an ideal configuration to enhance resolution to vertical changes along the profile. Such an approach may be useful for monitoring changes in resistivity over time (e.g. Binley et al., 2002a). A single borehole profile is also an important first step in quality control checks before embarking on cross-hole surveys. By adopting a range of electrode spacing (as in the 2D imaging approaches discussed earlier), it is possible to image the formation surrounding the borehole, assuming radial symmetry (e.g. Tsourlos et al., 2003). Pseudosections can be derived, as before, but using an appropriate geometric factor that accounts for the electrode location in the vertical and the proximity of the (insulating) ground surface boundary (Equation 4.9).

Surface electrodes may also be adopted (Figure 4.30) to enhance resolution (e.g. Marescot et al., 2002; Tsourlos et al., 2011). Ideally, these are placed in at least two orthogonal directions, to allow checks of symmetry to be made, or in multiple directions in order to assess anisotropy. The apparent resistivity is easily computed for such configurations but production of a pseudosection is challenging unless assumptions are made about an appropriate 'pseudo-position' for each measurement. Bergmann et al. (2012) document results from a number of borehole-surface resistivity surveys in which the potential to monitor deep CO_2 storage is explored. They used surface dipoles of length 150 m in two concentric rings 800 m and 1500 m from a borehole, in which a string of 15 electrodes was installed over a monitored interval 600 m to 750 m from the ground surface.

Cross-borehole surveys utilize at least two boreholes for electrode sites. Three-dimensional imaging is possible with three or more borehole arrays. Lytle and Dines (1978) were one of the first to propose cross-borehole resistivity imaging (they referred to the technique as the 'impedance camera'). Daily and Owen (1991) demonstrated the concept, referring to it as cross-borehole resistivity tomography. William Daily and Abelardo Ramirez (Lawrence Livermore National Laboratory) continued to develop the technique with Douglas LaBrecque (University of Arizona), demonstrating on a wide range of applications primarily focused on remediation of contaminated groundwater (e.g. Daily et al., 1992; Ramirez et al., 1993). Further developments and numerous applications of this 2D imaging technique followed in the 1990s (e.g. Morelli and LaBrecque, 1996a; Schima et al., 1993; Slater et al., 1996; Bing and Greenhalgh, 1997; Slater et al., 1997a, 1997b).

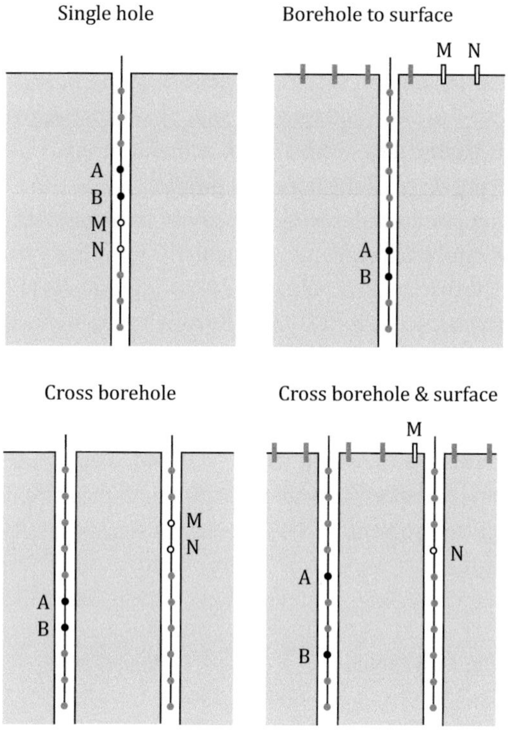

Figure 4.30 Electrode arrangements for borehole-based 2D imaging. The ABMN quadrupoles shown are possible configurations.

Three-dimensional cross-borehole imaging applications are now relatively common (e.g. Binley et al., 2002b; Wilkinson et al., 2006a; Doetsch et al., 2012a; Binley et al., 2016).

Cross-borehole measurements can be made in a variety of electrode configurations (Figure 4.31). A number of authors have explored sensitivity patterns for a range of quadrupole geometries (e.g. Bing and Greenhalgh, 2000). Inline dipole–dipole measurements will result in higher resolution, particularly with a short dipole spacing but can suffer poor signal-to-noise. In contrast, cross-well dipole–dipole measurements will lead to higher measurement signals but at the cost of weaker resolution. A combination of cross-well and inline quadrupoles may, in many cases, be a useful compromise. Performance of the choice of quadrupole geometry will depend on: (1) spacing of boreholes; (2) subsurface resistivity; (3) access to supplementary surface electrode sites; (4) the depth of the borehole electrode array. In cases where existing boreholes are utilized, boreholes may differ in length (and inclination), making optimization of quadrupole choice site specific. In Section 4.2.2.9, we discuss formal methods for optimizing measurement schedules and array geometry.

Visualizing measured cross-borehole datasets in a pseudosection equivalent is challenging because a spatial proxy for each measurement is difficult to define. However, for pole–pole surveys measurements can be graphically presented easily (as shown by Herwanger et al., 2004a) using a Cartesian plot of apparent resistivity for a given current pole location vs potential pole location. Another notable feature of the survey methodology of Herwanger et al. (2004a) is the combination of different electrode string positions. They utilized two 32-electrode arrays with 1 m electrode spacing, and, by sequentially positioning in a pair of boreholes at different depths, they were able to achieve an equivalent survey of two 96-electrode strings covering 95 m of the borehole profile.

A particular challenge of cross-borehole resistivity measurements is the wide range of geometric factors that can be encountered, particularly for inline quadrupoles, resulting in a number of extremely low measured voltages. Following Daniels (1977), we illustrate this in Figure 4.32. The geometric factor (K from Equation 4.9) was calculated for measurements in an adjacent borehole at varying distance from the borehole containing current electrode. Two cases are shown in Figure 4.32: (i) a 1 m dipole; (ii) a 3 m dipole. The figure shows clearly a zone of very high geometric factors; in fact, as Daniels (1977) shows, there is a diagonal line that marks a transition from $-\infty$ to $+\infty$. The figure illustrates that cross-borehole measurement schedules (particularly with inline measurements) should be carefully selected for the given problem.

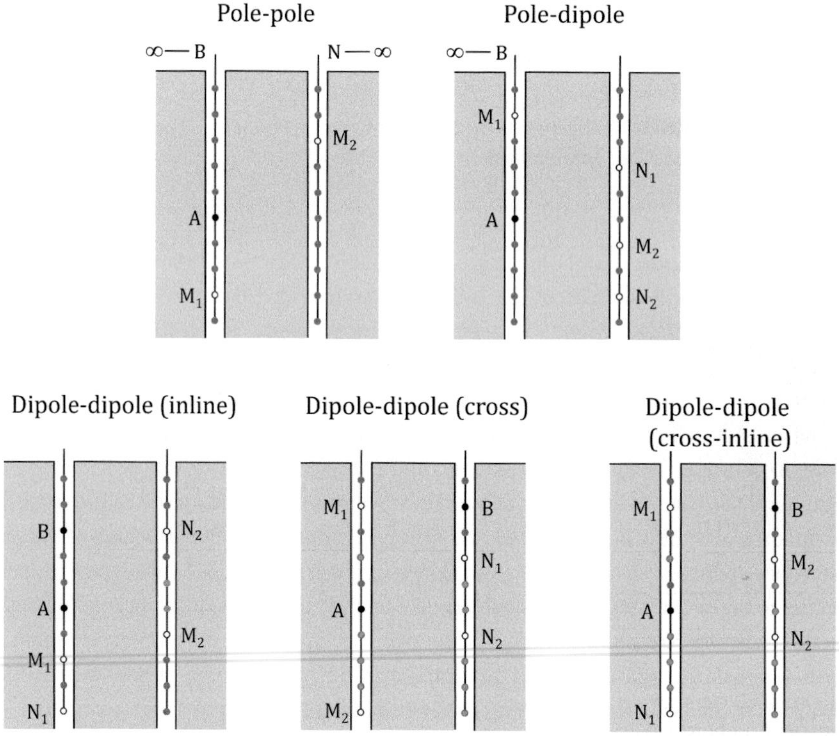

Figure 4.31 Cross-borehole measurement configurations. M_1, M_2, N_1 and N_2 illustrate possible potential electrode locations.

It is important to recognize that sensitivity (and resolution) will diminish away from the electrodes. As a result, cross-borehole imaging is limited to relatively small survey areas; the distance between boreholes should be less than the shortest length of an electrode array in a borehole, ideally half this length. Consequently, cross-borehole resistivity imaging is often limited to local site investigations. It is ideally suited to assessing technologies for groundwater remediation (e.g. Ramirez et al., 1993; Daily and Ramirez, 1995; Lundegard and LaBrecque, 1995; LaBrecque et al., 1996a) and protection (e.g. Daily and Ramirez, 2000; Slater and Binley, 2003, 2006). Cross-borehole imaging has also been successfully deployed for investigating CO_2 injection (e.g. Dafflon et al., 2012; Schmidt-Hattenberger et al., 2014; Wagner et al., 2015; Bergmann et al., 2017); see also case study in Section 6.1.8. In Bergmann et al. (2017), the focus was an injection zone approximately 700 m below ground level. Borehole-based resistivity imaging has also been deployed for mine/tunnel investigations (e.g. Sasaki and Matsuo, 1993; Kruschwitz and Yaramanci, 2004; Van Schoor and Binley, 2010; Simyrdanis et al., 2016).

There has been interest in the oil industry on the measurement of formation resistivity from electrical measurements along the metal casing of a borehole. Early studies include Schenkel and Morrison (1990), Kaufman and Wightman (1993) and Schenkel (1991), with recent developments on methods of analysis by, for example, Qing et al. (2017). Schenkel and Morrison (1990) first recognized the potential for borehole to borehole resistivity imaging using metal-cased boreholes. Ramirez et al. (1996) explored the use of metal-cased boreholes as individual long (current or potential) electrodes, in order to determine a horizontal image of electrical resistivity, integrated over the length of the borehole. Daily and Ramirez (1999) document the US patent, and Rucker et al. (2010) illustrate its use.

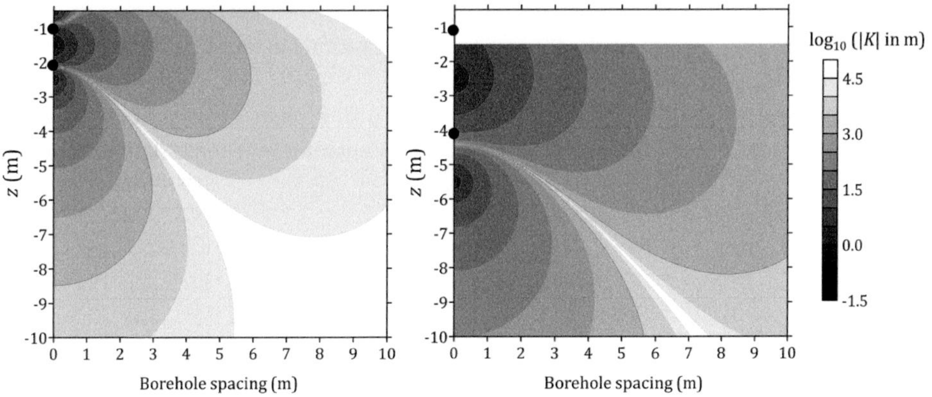

Figure 4.32 Geometric factor, K, for two inline current dipoles for inline potential dipoles measured in boreholes at different spacing from the current injection borehole. Left-hand image: electrode A at $z = 1$ m, B at $z = -2$ m, potential dipole spacing 1 m. Right-hand image: electrode A at $z = 1$ m, B at $z = -4$ m, potential dipole spacing 3 m. Note that the plot shows the logarithm of absolute values of K.

4.2.2.8 Small-Scale Imaging: Tanks and Columns

The electrode configurations discussed earlier are adopted for the majority of near surface applications of electrical methods; however, measurements can be made in a wider range of arrangements and scales. When excitation fields are created and measured using a full perimeter set of sensors to obtain an image of an object, the term 'tomography' is often used (although, as stated earlier, electrical resistivity tomography (ERT) is routinely used to refer to 2D and 3D resistivity imaging). Figure 4.33 shows example of electrode configurations for resistivity measurements in bounded systems, e.g. the imaging of soil and rock cores (Figure 4.33a) or soil-filled tanks in the laboratory (Figure 4.33b). Although more traditional resistivity measurements originate from exploration geophysics, tomographic imaging of resistivity has been adopted (and further developed) in other fields of study.

In medical physics, the term 'applied potential tomography' was coined by Barber and Brown (1984) for their Sheffield (UK) resistivity imaging system, although the term 'electrical impedance tomography' (EIT) appears to be universally adopted in medical physics (Webster, 1990; Brown, 2001). Early medical physics applications focused on imaging a 2D plane in vivo, allowing the study of electrical resistivity changes due to, for example, gastric or respiratory function (e.g. Barber, 1989). 3D imaging approaches evolved (e.g. Metherall et al., 1996) as did further application areas, for example, in the study of brain activity and breast imaging (see review by Bayford, 2006). Interestingly, medical physics researchers have also explored the value of linear arrays (i.e. traditional 2D geophysical imaging) due to the advantages of mobility and object target (e.g. Powell et al., 1987).

In parallel, EIT evolved in process engineering for the study of fluid-fluid and fluid-particle mixing in pipelines and vessels (e.g. Dickin and Wang, 1996; Wang, 2015), with a particular need for imaging highly dynamic processes. Resistivity imaging has also been successfully used for non-destructive evaluation of building materials (e.g. Karhunen et al., 2010; Zhou et al., 2017). In contrast to most geophysical applications, medical physics and process engineering require relatively low-powered instrumentation but much faster data acquisition.

In the pioneering study of Lytle and Dynes (1978), an *impedance camera* concept was proposed. They proposed a multiple electrode configuration for imaging resistivity, arguing

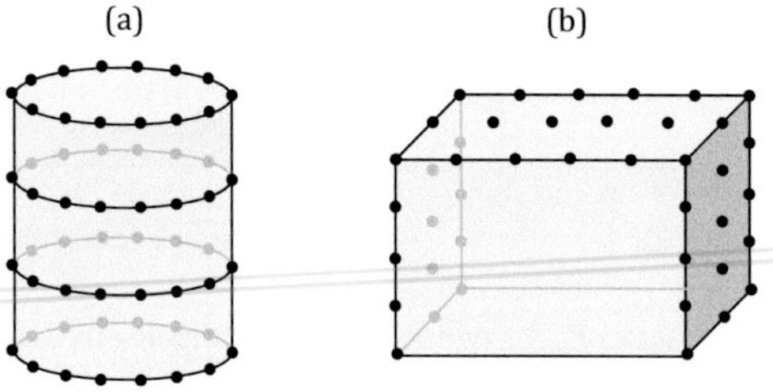

Figure 4.33 Example electrode geometries for small-scale imaging. (a) Cylindrical (column) set-up. (b) Tank-style set-up.

that by injecting current through multiple electrodes the potential field can be enhanced to better distinguish anomalies within the study object of study. Such an approach has also been developed for biomedical applications (e.g. Gisser et al., 1987) and has been referred to as an adaptive method or optimal current method (see Webster, 1990).

Daily et al. (1987) documents one of the first applications of ERT in geophysics using a circular array of 14 electrodes around an 8 cm diameter rock core in a study of changes in resistivity due to wetting and drying. Binley et al. (1996a, 1996b) examined preferential flow of solutes in 30 cm diameter soil cores, also using circular arrays (see case study in Section 6.1.5). For their study, Binley et al. (1996a, 1996b) adopted a process engineering EIT system, allowing rapid data collection, and in fact such data capture has been rarely replicated since. Other cylindrical applications of resistivity tomography in geophysics include Koestel et al. (2008). Block or tank-style applications include the study of moisture content changes in a 3 m × 3 m × 4.5 m block of welded tuff by Ramirez and Daily (2001) and the cross-borehole imaging of solute transport in a soil tank by Slater et al. (2002). Wehrer and Slater (2015) used a small tank study to investigate nitrate transport in soils. Fernandez et al. (2019) investigated the degradation of de-icing chemicals using soil tanks equipped with electrode arrays at the base, top and sides of the tank. Small-scale imaging can be effective for evaluating processes under controlled conditions, but it is important to remember that they will not necessarily represent real field conditions, particularly given different sensitivity patterns which result from the bounded, rather than semi-infinite domain. Consequently, their use as demonstrators of the potential for field implementation of techniques can be questionable, although this is often overlooked (e.g. Ts et al., 2016).

For small-scale resistivity imaging, all non-pole quadrupole geometries can be used. Often, these are dipole–dipole or gradient-based, or a combination (in process tomography the term *adjacent* is often used to describe a dipole–dipole array and *opposite* is used for the gradient array). As discussed before, increasing the dipole spacing will increase signal strength and depth of investigation at the expense of reduced resolution. Figure 4.34 shows example sensitivity patterns for two 2D geometries (c.f. Figure 4.5 for surface arrays). De Donno and Cardarelli (2011), amongst others, illustrate the effect of increasing the number of electrodes on the resolution of anomalies within a vessel.

Unlike conventional (half-space) applications, the derivation of an apparent resistivity for a general quadrupole and bounded volume geometry is more challenging. Zhou (2007) provides analytical expressions for cylindrical and block geometries. For most applications, data are managed as transfer resistances and pseudosections or pseudo-volumes are not used.

Small-scale resistivity imaging presents additional challenges. The electrode size required for reasonably low contact resistances often means that the electrode size – electrode spacing ratio is high. To overcome contact resistance problems, electrode size (contact area) may be increased either by using a plate geometry or by penetrating rod shaped electrodes deeper into the object. In both cases, modelling approaches based on point electrode assumptions may lead to erroneous results. When using plate electrodes, the passive electrodes can shunt electrical current and hence reduce resolution (e.g. Pinheiro et al., 1998) in addition to creating measurement errors (Rücker and Günther, 2011). For rod-shaped electrodes that are inserted deep relative to the electrode spacing, artefacts in images can result, if not

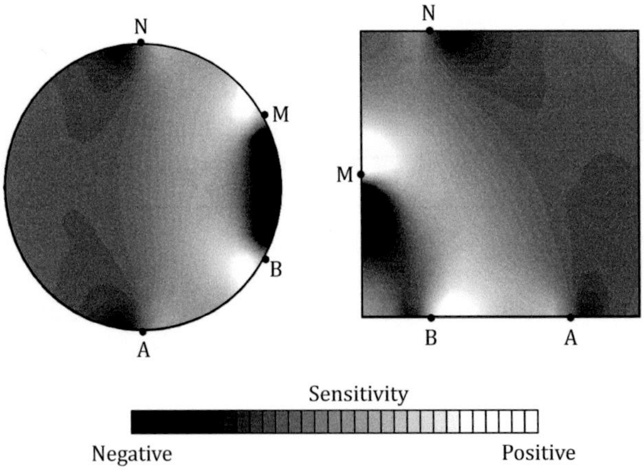

Figure 4.34 Sensitivity patterns, e.g. quadrupoles for circular and square geometries.

properly accounted for. Rücker and Günther (2011) modelled the effect of such electrodes and concluded that if the depth of penetration (embedment) is less than 20% of the spacing, the effect is minimal in half-space applications. For bounded applications, the criteria for neglecting such effects are more limiting. To mitigate such effects, Rücker and Günther (2011) suggest that an equivalent point electrode at 60% of the embedment leads to satisfactory results (Figure 4.35). This is also supported, in part, by the more recent study of Verdet et al. (2018) (who state that an equivalent point electrode depth of 73% of the embedment is appropriate). As such, studies are based on a limited number of numerical modelling scenarios (including assumptions of homogeneity); it is impossible to generalize findings, and consequently problem-specific modelling of such effects may be beneficial.

4.2.2.9 Optimal Measurement Schemes

2D resistivity imaging surveys with surface installed electrodes are usually carried out with *standard* quadrupole geometries (Figure 4.3), ideally in some combination; however, it will be apparent from the discussions earlier that the choice of measurement scheme forgeneralized imaging (3D, borehole-based, bounded, etc.) is less straightforward and often impossible to prescribe without accounting for problem specific conditions (survey aims, resistivity variability, instrumentation available, survey time constraints, etc.). Modern resistivity instruments with multi-channel capability are often constrained in multichannel configuration flexibility (see e.g. Stummer et al., 2002), and typically limited to addressing less than 100 electrodes without significant hardware additions. The choice of measurement scheme is then an optimization problem: we wish to determine the best set of quadrupoles to measure given the hardware (instrument, cables and electrodes), environment and labour constraints.

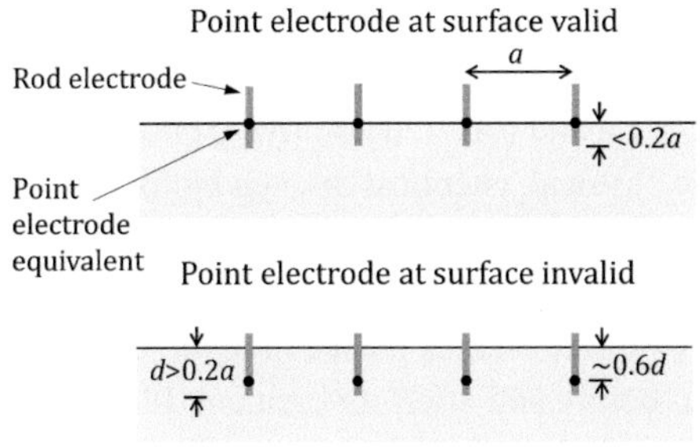

Figure 4.35 Effect and mitigation of electrode length.

We can approach such a search for an optimum set using sensitivity patterns (e.g. Figures 4.5, 4.34), which can be computed (see Section 5.2.2.3) for any quadrupole in any problem geometry. Furman et al. (2007) illustrate such an approach. However, the sensitivity patterns are also affected by the resistivity structure, which is unknown. The process thus needs to be done with recognition of resistivity variation and so the task becomes sequential experimental design: given a set of electrode positions, a trial set of measurements are used to determine an approximation of the variation in resistivity, which is then used to assess sensitivity patterns, which are used to refine the survey scheme and enhance resolution (see Stummer et al., 2004; Wilkinson et al., 2006b; Hennig et al., 2008; Loke et al., 2014a). Loke et al. (2014b) illustrate the approach for 3D surface imaging and 2D cross-borehole imaging. For application in real conditions, appropriate noise levels also need to be included (see Blome et al., 2011; Wilkinson et al., 2012), also recognizing that delays between the use of an electrode as a current and potential electrode should be accounted for to minimize charge-up effects. An alternative approach is to determine the optimal set of electrode positions (e.g. Wagner et al., 2015), or, in a more generalized sense, both locations and measurement sequence (Uhlemann, 2018).

Such methods are still highly specialized and, due to significant computational demands, remain research-focused and often limited to 2D imaging applications, despite the ease at which they can be adapted to parallel computation (e.g. Loke et al., 2010). Despite such constraints, the selection of quadrupoles should be assessed for every survey, ideally using forward and inverse modelling (see Chapter 5) to, at least, confirm that the quadrupole set choice is capable of addressing the aims of the survey. The use of such modelling for survey design cannot be overstated.

4.2.2.10 Time-Lapse Data Acquisition Considerations

Although originally developed for static characterization of the subsurface, electrical methods have immense power for the investigation of dynamic subsurface processes, e.g. temporal changes in fluid content (e.g. Binley et al., 2002b), solute concentration (e.g. Kemna et al., 2002), geochemical reactions (e.g. Kiessling et al., 2010) and temperature (e.g. Musgrave and Binley, 2011). Such capability has significant potential applications, e.g. monitoring landslide processes (e.g. Wilkinson et al., 2010), crop water uptake (e.g. Whalley et al., 2017), the effectiveness of remediation technologies (e.g. LaBrecque et al., 1996a) and engineered hydraulic barriers (e.g. Daily and Ramirez, 2000), leakage from underground storage tanks (e.g. Daily et al., 2004), and numerous others (see also case studies in Sections 6.1.4, 6.1.6 and 6.1.7). By studying changes in electrical signatures over time, we can focus on the physical, chemical or even biological state and remove the effect of other (static) factors that influence the electrical property.

Time-lapse surveys simply involve the repeat of an electrical survey a number of times over which the process of interest takes place. In some cases, only two surveys may be conducted, to establish a 'before and after' assessment. Ideally, electrodes remain in place (i.e. they are semi-permanently installed) throughout the monitoring. In some cases, this is not a practical option, and for each survey electrodes need to be reinstalled, in their original location. As discussed in Section 3.2.4, some instrument manufacturers now offer monitoring capability as a standard feature, and in some cases data can be acquired remotely, although clearly health and safety matters need to be thoroughly addressed for such applications. Some researchers claim the use of autonomous systems for monitoring, although, strictly speaking, they are generally automated, not autonomous, since self-learning capability (e.g. accounting for environmental condition changes) is rarely included, although the adaptive time-lapse survey strategy of Wilkinson et al. (2015) attempts to evolve the measurement sequence over time to reflect the temporal changes in resistivity.

There are factors that need to be considered when planning a time-lapse survey: the capability to leave electrode arrays installed; the choice of electrode array and measurement sequence; the time interval of measurements; the choice of a reference (starting) condition; the duration of the entire monitoring process; other environmental factors that may influence results. The choice of electrode array and measurement sequence will reflect the objectives of the survey (i.e. the dynamic process which is to be observed) but will also be constrained by the speed of both the process and data capture. It is important to recognize that during the collection of one dataset, which represents one point in time, the process may evolve significantly, which can result in problems in data inversion and interpretation. For highly dynamic processes, sacrifices in spatial resolution may be necessary in order to meet a required short data capture period. In many time-lapse studies, comparisons are made to a reference (baseline) case. If this is the case, then it is critical that a reliable reference survey is undertaken. Ideally, this will be done over a period of time to determine changes that are a result of other environmental factors. For example, for a study of a process over 30 days using daily surveys, it may be useful to obtain daily surveys for five days prior to the onset of the process under investigation.

Typically, the same measurement sequence is used throughout the time-lapse survey to avoid any biasing; however, the adaptive optimization strategy of Wilkinson et al. (2015) explicitly revises the measurement sequence to address changes in the evolving resistivity structure. Such an approach is ideally suited to long-term monitoring installations.

Assessment of measurement errors throughout the time-lapse survey is important, particularly for long-term monitoring as electrode contact can degrade over time, resulting in high errors. A sequence consisting of normal and reciprocal measurements (see Section 4.2.2.1) is ideally made; however, this can sometimes be impractical for all datasets given the survey time burden it imposes. In such cases, a subset of measurements may be suitable for development of error models (e.g. Figure 4.15). Such an error model may change through time. Figure 4.36 illustrates this using data from a time-lapse 3D cross-borehole resistivity survey tracking the movement of a solute plume in the unsaturated zone of an aquifer (for details of the experiment, see Winship et al., 2006; see also case study in Section 6.1.4). Figure 4.36a shows the median reciprocal error (expressed as a percentage) of the 3,188 measurement set during the experiment; even though the median error is low, there is clearly a degradation of data quality during the experiment. The error model will impact on inversion of measured data, and for time-lapse inversion (see Chapter 5) we need to assess a single error model that represents two datasets. Lesparre et al. (2017) recommend that this be done by examining the difference in the change in both normal, R_N, and reciprocal, R_R, transfer resistance measurements by

$$\Delta\log|R_N| - \Delta\log|R_R|, \tag{4.20}$$

where absolute values of transfer resistances have been used to reflect that transfer resistances can have both positive and negative polarity. Lesparre et al. (2017) suggest that the absolute value of the term in Equation 4.20 should show an inverse relationship with the transfer resistance. This is illustrated in Figure 4.36b for the dataset used in Figure 4.36a, comparing the dataset from 16 April 2003 to that of 6 March 2003. In this case, for resistance magnitudes less than 1 Ω, the trend suggested by Lesparre et al. (2017) exists but above 1 Ω the error is reasonably constant. Such approaches for investigating data errors should help with quality control of time-lapse surveys and will inevitably have immense value for data inversion.

4.2.2.11 Current Source Methods

4.2.2.11.1 Mise-à-la-masse

The mise-à-la-masse (excitation of mass) method is a potential mapping method that has been widely used in mineral exploration (e.g. Parasnis, 1967; Mansinha and Mwenifumbo, 1983; Bhattacharya et al., 2001), dating back to the early Schlumberger trials in 1920. In its original form, boreholes are not necessarily used. A current source is installed in a mineralized zone, either at an outcrop on the surface or in a borehole. A remote current electrode is then sited and either pole–pole or pole–dipole measurements are made on the ground surface. The objective of the approach is to map the potential field and deduce the orientation of the conductive ore body. Interpretation of the measured potential field can be challenging when used in isolation but the method can be valuable as an early-stage reconnaissance tool to help design subsequent geophysical surveys or borehole deployment (e.g. Ketola, 1972).

By using the steel casing of a borehole for current excitation, the method has been used in a number of geothermal studies (e.g. Kauahikaua et al., 1980; Mustopa et al., 2011). In hydrogeology, the method has been utilized for tracing the migration of tracers injected in boreholes (e.g. Bevc and Morrison, 1991; Nimmer and Osiensky, 2002; Perri et al., 2018). Gan et al. (2017) used the outlet of a karstic channel to place the excitation electrode, and then mapped the potential field on the surrounding ground surface in order to assess the orientation of the channel in the subsurface. Recently, Mary et al. (2018) used the mise-à-la-

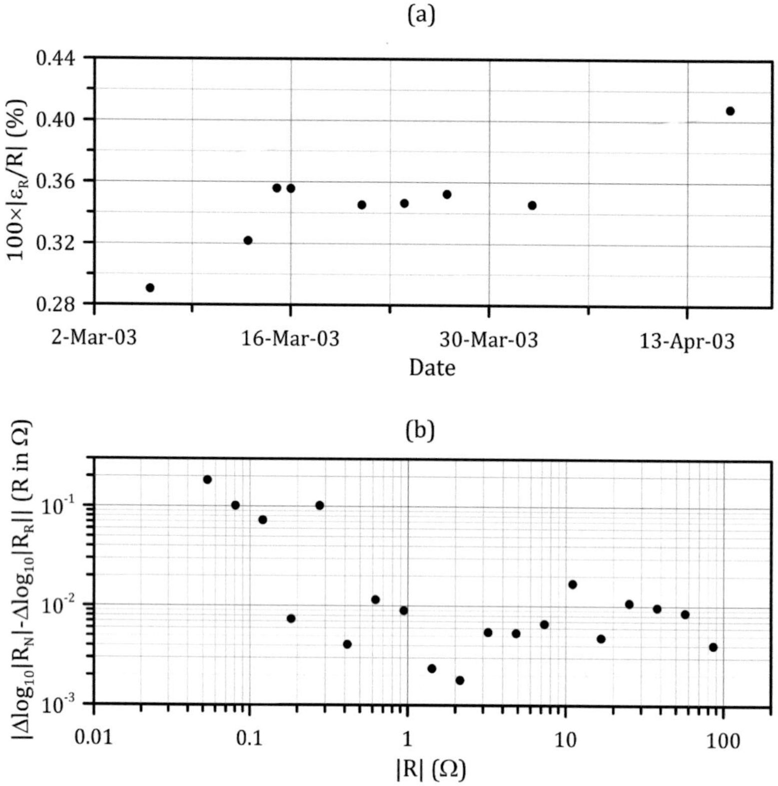

Figure 4.36 Time-lapse errors. (a) Variation in median reciprocal error over time for cross-borehole survey (details in text). (b) Time-lapse error.

masse method, in a small-scale study, to map plant root structure, using the plant stem as the excitation source and an array of shallow borehole electrodes to detect the potential field.

4.2.2.11.2 Current Leak Location

The mass excitation concept has an interesting (and successful) application for detection of fluid leaks through (electrically insulating) membranes. This was proposed in the US patent by Boryta and Nabighian (1985) and outlined in detail by Parra (1988). Key (1977) also discusses methods based on the same concept for tank and pipelines. Although this is not a true resistivity method, we include it here as it adopts a similar measurement approach.

If a current dipole is placed between a remote electrode and a conducting material (liquid or solid) impounded by an electrically insulating membrane or liner, current will flow through any holes in the liner. Then, from potential measurements across dipoles close to the liner it is possible to determine the location(s) of the leaks. Such an approach is now routinely used for quality control after installation of a liner, prior to waste filling. The mass excitation is created and a short portable potential electrode dipole is traversed over the liner, using a thin layer of sand above the liner (added to protect the liner from physical damage) for contact.

The approach described is effective if the potential dipole is positioned close to the leak site. For active waste storage sites, several metres of waste may exist above the liner. Frangos (1997) documents a modification to the method, using an array of electrodes beneath the liner (see Figure 4.37a), i.e. installed prior to fitting the liner. Frangos (1997) used a pole–pole electrode array, which allows easy interpretation of the leak source, although dipole potential electrodes are equally viable (as shown, for illustration, in Figure 4.37a). White and Barker (1997) present a similar application to a UK landfill site, also exploiting the electrode array for imaging the resistivity beneath the landfill liner. Binley et al. (1997) demonstrates an alternative approach for investigation of existing waste sites (given that retrofitting an electrode array beneath a liner is not possible). In their approach, an array is installed around the perimeter of the waste site (see Figure 4.37a), which requires inverse modelling to determine the (unknown) location of the current source(s). The array can also be installed within the boundary of the waste, although as Binley and Daily (2003) illustrate, in both cases the ability to differentiate multiple leaks is challenging. Binley et al. (1997) also illustrate how the method can be utilized for conducting containers, e.g. underground steel shell tanks (see also Key, 1977).

4.3 Induced Polarization

The objective of an IP survey is to determine the spatial variation of polarizability of the subsurface. IP measurements are made using four electrodes in a similar way to DC resistivity (see Chapter 3). In fact, many of the principles already discussed in this chapter can be applied to IP problems.

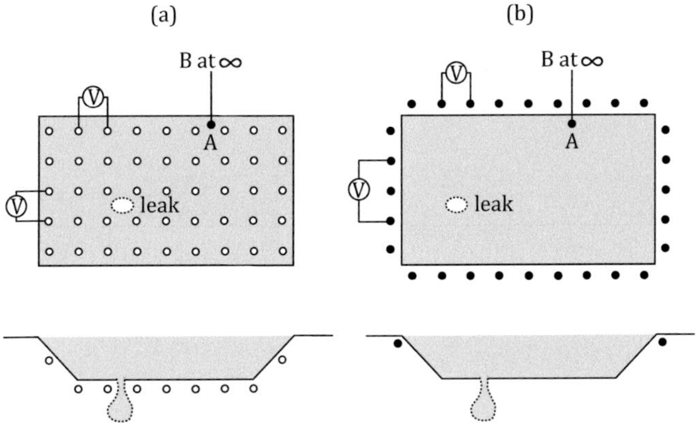

Figure 4.37 Current excitation applied to leak location. Plan and vertical cross-sections shown for (a) electrode array beneath a liner and (b) external electrode array.

4.3.1 Characteristics of a Polarizable Subsurface

Seigel (1959) first introduced the concept of chargeability as a property and illustrated how the apparent chargeability M_a is affected by subsurface variation in the intrinsic chargeability $\hat{m}$. Following Seigel (1959), the apparent chargeability of a homogenous subsurface with resistivity ρ_0 and chargeability $\hat{m}$, can be written as

$$M_a = \frac{V(\rho_{IP}) - V(\rho_0)}{V(\rho_{IP})}, \tag{4.21}$$

where $\rho_{IP} = \rho_0/(1 - \hat{m})$ and $V(\rho)$ represents the measured voltage. Since V is linearly related to ρ, the apparent chargeability is equal to the true intrinsic chargeability, $\hat{m}$, of the homogenous earth.

Although Seigel's definition of apparent chargeability has limited practical value (in terms of measurement) because the ratio of secondary to primary voltage is impossible to measure, it does have immense value in understanding how variation in polarization of the subsurface propagates to the measurement. Such an approach has been widely used by others (e.g. Patella, 1972; Oldenburg and Li, 1994).

Seigel (1959) illustrates how the apparent chargeability varies as a function of properties of a layered subsurface. For a system with N units, each with resistivity ρ_i and intrinsic chargeability $\hat{m}_i$, $i = 1, 2, \ldots, N$, Siegel (1959) shows that the apparent chargeability can be approximated by

$$M_a = \sum_{i=1}^{N} \hat{m}_i \frac{\partial \log \rho_a}{\partial \log \rho_i} = \sum_{i=1}^{N} \hat{m}_i \frac{\rho_i}{\rho_a} \frac{\partial \rho_a}{\partial \rho_i}, \tag{4.22}$$

where ρ_a is the apparent resistivity. Equation 4.22 reveals how variation in resistivity propagates through to the observed chargeability. We will discuss this later in the context of a negative IP effect.

We can use analytical expressions, such as those presented in Section 4.2.1.3, to model the apparent chargeability for a layered system. Patella (1972) presents an apparent chargeability model for a two-layered Schlumberger sounding configuration. An upper layer of resistivity ρ_1 and intrinsic chargeability $\hat{m}_1$ and thickness h overlies a unit with resistivity ρ_2 and chargeability $\hat{m}_2$. The model can be written as

$$M_a = \hat{m}_1 \left(1 + 2\sum_{n=1}^{\infty} \frac{k_{1,2}^{'}}{\left(1 + (4nh/AB)^2\right)^{3/2}} \right) \Bigg/ \left(1 + 2\sum_{n=1}^{\infty} \frac{k_{1,2}}{\left(1 + (4nh/AB)^2\right)^{3/2}} \right), \tag{4.23}$$

where $k_{1,2} = (\rho_2 - \rho_1)/(\rho_2 + \rho_1)$, $k_{1,2}^{'} = (\hat{m}_2\rho_2 - \hat{m}_1\rho_1)/(\hat{m}_2\rho_2 + \hat{m}_1\rho_1)$, and AB is the current electrode spacing. Adopting a similar configuration to that in Figure 4.8, Figure 4.38 illustrates the effect of the lower-layer resistivity on the observed chargeability.

The chargeability definition of Seigel (1959), as expressed in Equation 4.22, can lead to an interesting resistivity paradox resulting in negative apparent chargeability. For

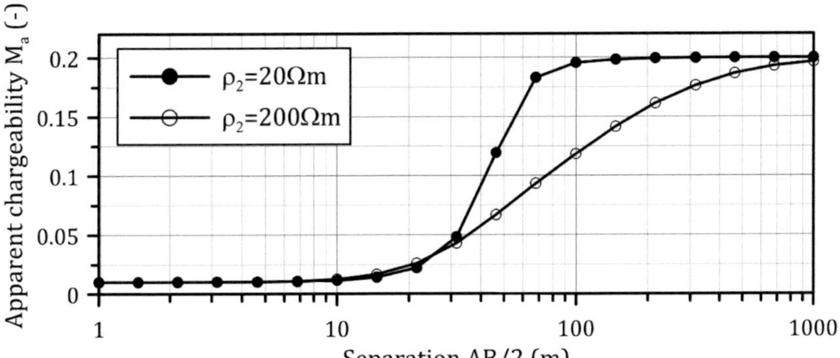

Figure 4.38 Effect of lower-layer resistivity on apparent chargeability $\rho_1 = 500\ \Omega\text{m}$, $\hat{m}_1 = 0.01, \hat{m}_2 = 0.2, h = 10\ \text{m}$.

a particular resistivity structure, the term $\partial\rho_a/\partial\rho_i$ in Equation 4.22 can be negative, and if $\hat{m}_i$ is relatively large (in comparison to other values of chargeability), a negative apparent chargeability can result. This 'negative IP effect' has been recognized for some time (e.g. Sumner, 1976). Nabighian and Elliot (1976) analysed various synthetic layered models and demonstrated that when a conductive layer overlying a resistive layer exists, negative contributions to the apparent chargeability can occur. In fact, as illustrated by Dahlin and Loke (2015), the problem is easily illustrated by considering the sensitivity function in Equation 4.10 (examples of which are shown in Figure 4.5). Areas of the region of investigation will exhibit negative sensitivity (the darker areas in Figure 4.5). If the intrinsic chargeability in such regions is relatively high, say due to a shallow polarizable layer, or localized lateral variation, then a negative apparent chargeability will be observed. As Dahlin and Loke (2015) note, such negative IP observations should not be discounted as poor measurements since they contain information about the subsurface (in terms of resistivity and chargeability).

The apparent chargeability concept is easily applied to the DC resistivity approaches discussed earlier in this chapter. For example, a pseudosection of apparent chargeability can be constructed by assigning the measured chargeability in the same graphical representation as apparent resistivity. As in the case of DC resistivity, such an image is not a true pictorial representation of the true intrinsic chargeability (magnitude and variation) but simply serves as a means of displaying field data. Furthermore, given the comments on negative IP effects, regions of the pseudosection may display negative apparent chargeability.

Representation of polarizability by complex resistivity is more straightforward conceptually, although somewhat more complex mathematically. We can describe any volume of the subsurface by its complex resistivity, ρ^*, which is commonly expressed as a magnitude, $|\rho|$, and phase angle, φ. At suitably low current injection frequency, given likely low phase angles, the magnitude is effectively equivalent to the DC resistivity. A non-zero (and negative) phase angle represents a polarization

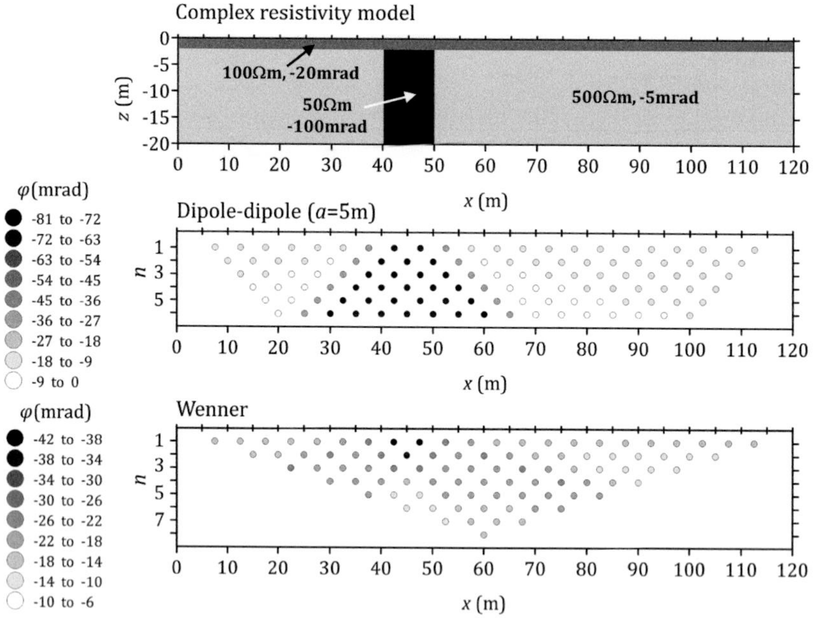

Figure 4.39 Comparison of dipole–dipole and Wenner phase angle pseudosections for a synthetic complex resistivity model. Note the different phase angle scales.

effect but it is important to recognize that the phase angle is also influenced by the real component of resistivity since, by definition, $\varphi = \tan^{-1}(-\rho''/\rho')$. Again, the concepts discussed earlier for DC resistivity can be applied. For a given quadrupole geometry, the geometric factor can be determined as before and used to convert the measured impedance to an apparent complex resistivity, i.e. $\rho_a(= kZ)$ and φ. This is illustrated in Figure 4.39 using the same structure as in Figure 4.21, but in this case polarizability is added in terms of resistivity phase angle. Note how poorly resolved the vertical polarizable feature is with the Wenner quadrupole arrangement. However, it is important to recognize that the inevitable weaker signal strength with the dipole–dipole configuration will impact on the resolution of any polarizable body. The nature of measurement errors are discussed next.

4.3.2 Measurement Errors

In Section 4.2.2.1, we illustrated how measurement errors in a DC resistivity survey typically show an increase in reciprocity or repeatability error of the transfer resistance with increasing transfer resistance. Since chargeability is a measure of a secondary potential, it follows that IP errors will also exhibit a similar trend. Figure 4.40 shows error plots for a time domain IP survey carried out along the same survey line used to illustrate DC resistivity errors in Figure 4.15. In this case, the survey used the same 96 electrodes placed at 1 m spacing, again with a dipole–dipole configuration but using

a = 1 m; n = 1, 2, 3, ..., 10. Measurements were made with a Syscal Pro (Iris Instruments) with a 4s cycle (1s on, 1s off). The DC resistivity transfer resistance error plot is shown in Figure 4.40a revealing consistent behaviour to that in Figure 4.15; in Figure 4.40b, the chargeability errors reveal non-linear behaviour: high errors occur for low apparent chargeability measurements due to the resolution of the secondary potentials; high errors also tend to increase for high transfer resistances given the implicit link between apparent chargeability and transfer resistance. Figure 4.40c shows even clearer systematic behaviour of errors in apparent chargeability. If a simple linear translation of apparent chargeability to phase angle applies (as discussed earlier and in Section 3.3.3), then identical behaviour in inferred phase angle would result. Mwakanyamale et al. (2012) illustrate similar behaviour in time domain IP measurements carried out at the Hanford 300 Area adjacent to the Columbia River (Washington, USA) (see also case study in Section 6.2.2). Irrespective of whether plots in Figure 4.40a or Figure 4.40c are used, the operator must ensure that some assessment of measurement error is carried out. When surveying targets of low polarizability (as in the example used in Figure 4.40) errors may be similar (or even exceed) the magnitude of the measurement and thus some filtering of data is advisable. Furthermore, as we show in Chapter 5, incorporation of an appropriate error model for DC resistivity and IP surveys is crucial for satisfactory inversion of data.

Flores Orozco et al. (2012) studied the nature of frequency domain IP errors and argued that the error in measured phase angle should progressively decrease with increasing transfer resistance (magnitude of the measured transfer impedance), i.e. not showing the behaviour seen with apparent chargeability errors at high transfer resistances (as in Figure 4.40b). In Figure 4.40, computed errors from a frequency domain IP survey are shown. The survey used 20 surface electrodes, 1.5 m apart in a dipole–dipole configuration with a = 1.5 m. Measurements were made with a SIP256 (Radic-Research) over a range of current injection frequencies; the data used in Figure 4.41 were conducted with an injection frequency of 0.156 Hz. Unlike the time domain error behaviour in Figure 4.40, low phase angle errors are seen at high transfer resistances, which is consistent with that reported by Flores Orozco et al. (2012). For multi-frequency (i.e. SIP) measurements, measurement error behaviour needs to be assessed for each injected frequency, such analysis may reveal extremely high errors at high injected frequencies (see Flores Orozco et al., 2012), which may limit their value in subsequent modelling (inversion).

Gazoty et al. (2013) studied the repeatability of time domain IP measurements and observed a relatively weak relationship between stacking IP errors (i.e. differences in apparent chargeability over repeated injection cycles) and repeatability. Such observations reinforce the problem of relying on stacking errors (as was noted in DC resistivity surveys earlier, e.g. Figure 4.14). Gazoty et al. (2013) highlight that stacking errors in IP surveys can be particularly problematic if the time on/off period in a cycle is not long enough for charge/discharge in a cycle. In time domain IP surveys, it is, therefore, advisable to examine the decay curves measured (not just the measured chargeability) to ensure that the cycle settings are appropriate.

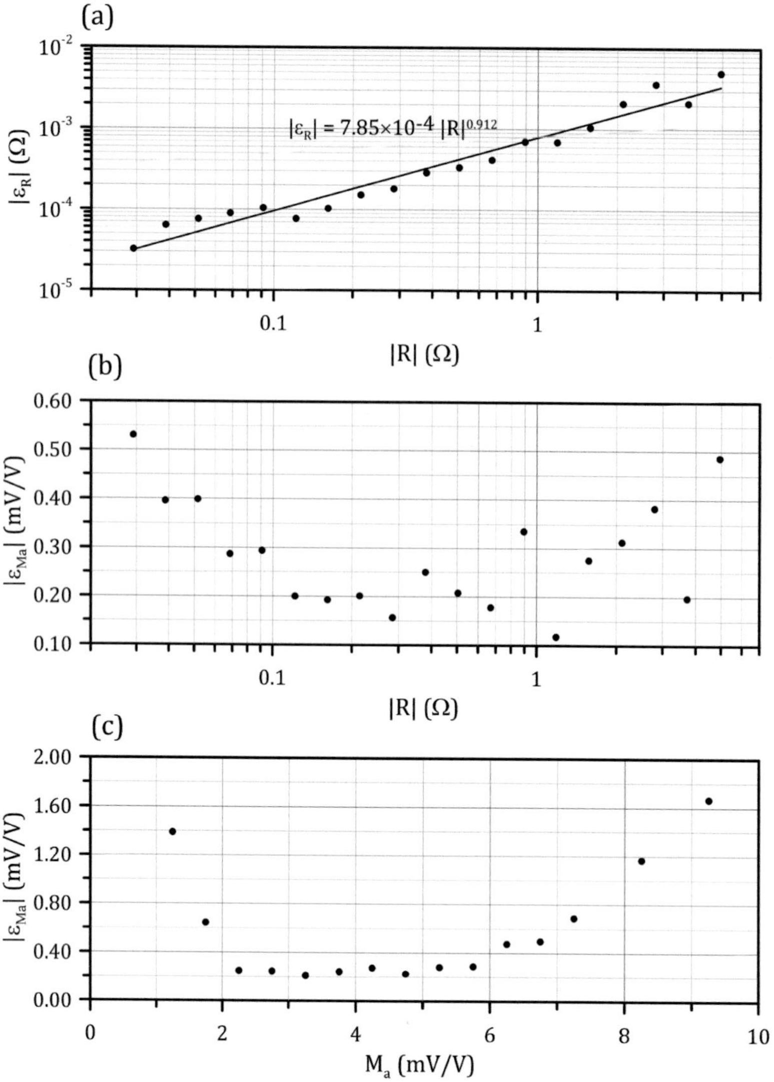

Figure 4.40 Aggregated measurement errors in a time domain IP survey. (a) Transfer resistance error and (b) chargeability error, as a function of transfer resistance. (c) Chargeability error as a function of chargeability. For details of the survey, see text.

4.3.3 Electrode Geometries

In Chapter 3, we discussed the choice of electrodes for IP measurement. Despite several investigations of optimum IP electrodes, most IP surveys use standard stainless steel electrodes (e.g. Dahlin et al., 2002b; LaBrecque and Daily, 2008). Care must be taken within a survey to avoid measurement on electrodes that have recently been used for current

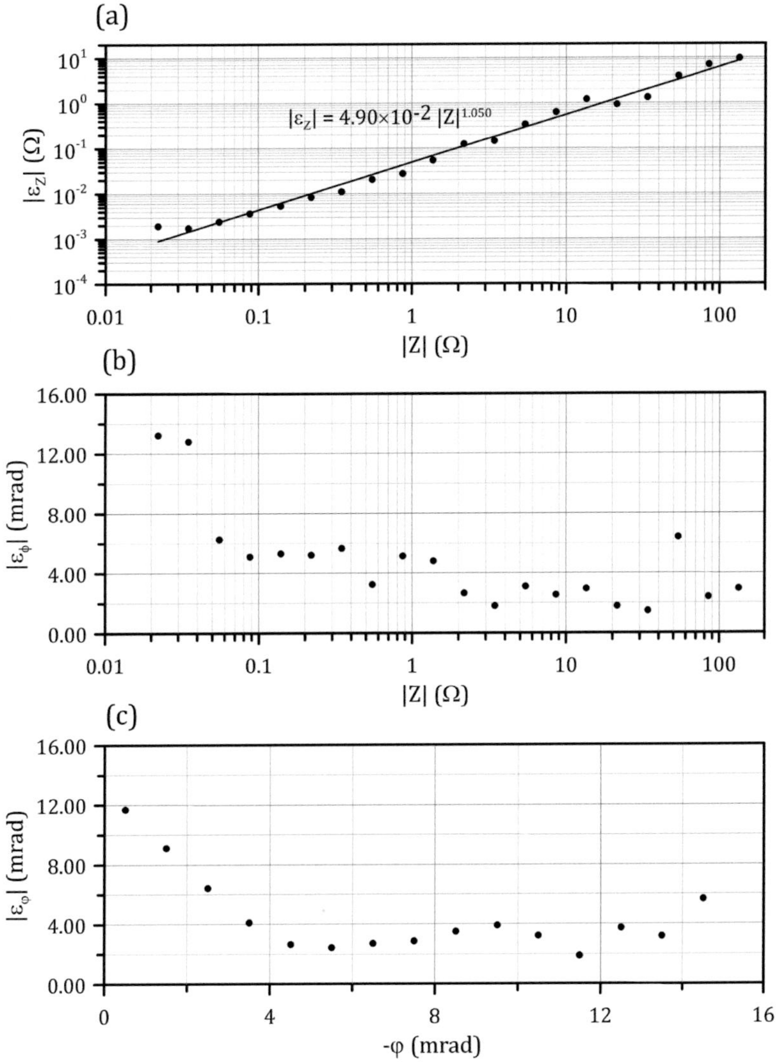

Figure 4.41 Aggregated measurement errors in a frequency domain IP survey. (a) Transfer resistance error and (b) phase angle error, as a function of transfer resistance. (c) Phase angle error as a function of phase angle. Note the different phase angle error behaviour to that in Figure 4.40. For details of the survey, see text.

injection since any residual polarization at the electrode can deteriorate detection of secondary potentials due to polarization of the subsurface. Dahlin (2000) reports on this phenomenon in DC resistivity surveys; for IP surveys, the effect can be even more significant.

In theory, any quadrupole used for DC resistivity surveying can be used for IP measurement; however, a number of factors need to be considered. The key consideration is the strength of the measured secondary voltage, which will be dictated by the resistivity and polarizability of the subsurface, the current injected and the geometric factor for a given quadrupole. Gazoty et al. (2013) illustrate the sensitivity of the secondary voltage to the geometric factor. For example, if we assume an apparent chargeability of $M_a = 20$ mV/V and $\rho = 100$ Ωm, then the minimum geometric factor is 400 m in order to achieve a 2 mV secondary voltage. For a dipole–dipole configuration using $a = 5$ m, then following Table 4.1 such a survey will be limited to $n < 3$, which clearly limits depth of investigation. For comparison, a Wenner array will satisfy the same constraint with a spacing of $a = 63$ m, although we are now limited by lateral resolution, as shown in Figure 4.39. Careful attention to survey design is, therefore, even more critical for IP surveys, particularly with relatively low-power instrumentation.

One way to overcome the problem of satisfactory secondary voltage signal strength is to increase the current injection. Many modern DC resistivity/IP units have the capability of replacing a typical basic 200 W transmitter to a higher-power (>1 kW) unit, allowing current injection of several amperes, although clearly this only increase signal strength by a factor of 5 or so. The use of such high currents requires even more stringent health and safety considerations; furthermore, appropriate cabling capable of operating at high current and voltage is also required.

The choice of electrode geometry will also be influenced by the availability and suitability of multi-core cabling and computer controlled switching. Traditionally, in mineral resource exploration IP surveys were conducted with high-power transmitters and manual movement of electrode quadrupoles. A dipole–dipole was commonly used and has significant characteristics for such manual placement: by keeping cables used for current injection away from those assigned for potential measurement, interference due to inductive coupling between cables was minimized. There are also significant health and safety benefits from such a configuration as the 'active area' with the current dipole can be separated from the potential dipole pair.

Inductive coupling effects (see Section 3.3.2.4) must be considered in field surveys. Zonge et al. (2005) notes that when using multicore cables the effect of inductive coupling in complex resistivity measurements can be severe, and in highly conductive (e.g. marine) environments such effects are even greater given the low resistivity. Cultural coupling may also occur from, for example, power lines. In the frequency domain, it may be possible to identify such effects at specific frequencies. In the time domain, examination of the decay curves may reveal anomalous behaviour of the decay at specific IP measurement gates. Good survey design can reduce inductive coupling effects in field data (Zonge et al., 2005), thereby minimizing the need to rely on imperfect models for inductive coupling removal. As discussed in Section 3.3.2.4, capacitive coupling can also exist between transmitter and receiver cables, between receiver cables and also between the subsurface and cables. It will be prominent at high frequencies (or early times in a time domain measurement) in resistive environments, particularly when long cables are used. For frequency domain measurements, shielded cables are used to minimize such effects (see Section 3.3.2.4). Conventional time domain IP measurements are less susceptible to coupling effects due to the delay time (t_d) in recording after current switch off (Figure 3.20).

It should be clear from this discussion that selection of a suitable electrode geometry for IP surveys is more challenging than for DC resistivity surveys. Surveys will greatly benefit from basic modelling prior to any measurement. The chosen electrode geometry should ensure good strength of secondary voltages, whilst also achieving adequate resolution of the target of interest. In addition, during surveys checks should be made of voltage decays (time domain) to identify any anomalous behaviour. Data quality should also be assessed, ideally through reciprocity checks; however, in doing so it is important to note that a reciprocal configuration may be susceptible to greater coupling effects, and therefore any systematic differences should be assessed. High-quality IP surveys will also require considerably more time for planning and execution in the field (e.g. longer time windows, more stacking, etc.). It is sometimes assumed that as a DC resistivity instrument has IP measurement capability the operator will conduct an IP survey for no additional effort – this is not true.

4.3.4 Borehole Measurements

As in the DC resistivity method, IP surveys can also be carried out using electrodes in boreholes in order to obtain better resolution at depth. Snyder and Merkel (1973) illustrate how a conventional surface electrode survey for mineral exploration can be enhanced using buried electrodes. Daniels (1977) first proposed the use of cross-borehole IP measurements. Some of the earliest cross-borehole IP applications are reported by Iseki and Shima (1992) and Schima et al. (1993). Kemna et al. (2004) illustrate the value of such measurements for 2D cross-borehole imaging in a number of case studies (see also Sections 6.2.3, 6.2.4 and 6.2.6). Other 2D examples include Slater and Binley (2003) and Slater and Glaser (2003); Binley et al. (2016) show 3D cross-borehole IP imaging results.

Quadrupole geometries discussed earlier (Figure 4.31) can be used, although cross-borehole IP measurements are more challenging than DC resistivity measurements. Zhao et al. (2013, 2014) discuss methods for correction of coupling effects in cross-borehole complex resistivity measurements (see also Kelter et al., 2018). From the previous discussion on signal strength, cross-borehole IP measurements are more susceptible to poor data quality. For example, the geometric factor plot in Figure 4.32 reveals that many quadrupole combinations are likely to suffer from poor signal strength, highlighting the need for careful survey design.

Borehole logging using IP is also a straightforward extension of DC resistivity wireline logging (see e.g. Freedman and Vogiatzis,1986) and more recently spectral IP logging tools have been developed (e.g. LoCoco, 2018).

4.3.5 Small-Scale Imaging

IP imaging can also be carried out in small-scale imaging studies using electrode geometries discussed in Section 4.2.2.8. Clearly, for small-scale applications signal strength is less of a challenge than at the field scale, which is enhanced by the bounded nature of tanks and vessels. Furthermore, shielded cabling is more of a practical option, allowing minimization of coupling effects. Recent instrumentation development has led to the availability of devices capable of measuring complex impedance at multiple frequencies, permitting imaging of complex resistivity over a broad range of frequencies (Zimmerman et al., 2008b). Kelter et al. (2015) use complex resistivity imaging in an artificially packed soil column to assess relaxation model characteristics under drying conditions. Weigand and Kemna (2017, 2019) show changes in images of multi-frequency complex resistivity of a plant root system undergoing stress; see also the case study in Section 6.2.9 showing the use of IP imaging to assess tree health. As in the case of DC resistivity, small-scale imaging of polarization has also been adopted in other fields, e.g. in human respiration (Brown et al., 1994) and brain imaging (Yerworth et al., 2003).

4.4 Closing Remarks

We have illustrated in this chapter how the four-electrode array can be used in practice to measure the apparent resistivity and polarizability of the subsurface and how the sensitivity of such a measurement varies for different quadrupole configurations. We have illustrated, using analytical models, how relatively simple non-uniformity in resistivity affects the measured apparent resistivity and introduced the concept of a pseudosection. Methods to assess the quality of measurements have also been outlined. The four-electrode measurement can be used to map lateral variability of electrical properties, to conduct vertical soundings or to image in 2D and 3D. Although four-electrode resistivity and IP measurements are commonly applied using electrode arrays deployed on the ground surface, they are easily applied to a range of more specialized surveys, e.g. using boreholes or for small-scale imaging. Armed with a series of measurements and an assessment of measurement quality, we can now explore the use of forward and inverse modelling to allow the determination of resistivity and IP properties of the subsurface. This is the focus of the following chapter.

5

Forward and Inverse Modelling

5.1 Introduction

In Chapter 4, we explained how a four-electrode DC resistivity and induced polarization (IP) measurement is made for different types of surveys, illustrated the wide range of electrode configurations possible and showed how a given measurement is sensitive to electrical properties within the region of interest. We also introduced the concept of apparent resistivity and how a pseudosection can be used for graphical presentation of such values. The calculation of the 'measurement' that would be observed for a given geometry is referred to as *forward* modelling. We showed in Chapter 4 how, for some relatively simple cases, the apparent resistivity can be computed analytically. For more complex variation of resistivity (and polarization), different forward modelling methods are needed. In this chapter, we present such methods. For most applications of electrical methods (and geophysical methods in general), we are interested in determining the spatial (or spatio-temporal) variation of a property (e.g. an image of resistivity) given a set of measurements, i.e. the reverse action of a forward model. This is referred to as an *inverse* model (Figure 5.1), which is underpinned by a forward model.

In basic science, we routinely fit equations to measurements, e.g. the determination of two parameters of a linear equation given a set of x, y values. In electrical geophysics, most of the problems we need to solve are non-linear, requiring more advanced treatment than linear regression. Furthermore, we are often seeking the solution of an under-determined problem, i.e. more unknowns (grid cells with unknown resistivity) than equations, which also require specific treatment.

The availability of robust methods for inverse modelling and expansion of personal computer power in the 1980s transformed our ability to image the variation of electrical properties in the near surface. Prior to this, type curves were used to graphically match (1D) sounding data (e.g. Slichter, 1933; Flathe, 1955; Keller and Frischknecht, 1966; Bhattacharya and Patra, 1968), resulting in inevitable subjectivity. For 2D problems, pseudosections were used to image electrical data (e.g. Edwards, 1977), recognizing that different electrode configurations give different pseudosection responses for a given structure of electrical properties (as shown in Chapter 4, Section 4.2.2.5). Nowadays, the 3D structure of resistivity can be computed automatically on a laptop computer, avoiding inherent operator subjectivity of former methods. However, we will

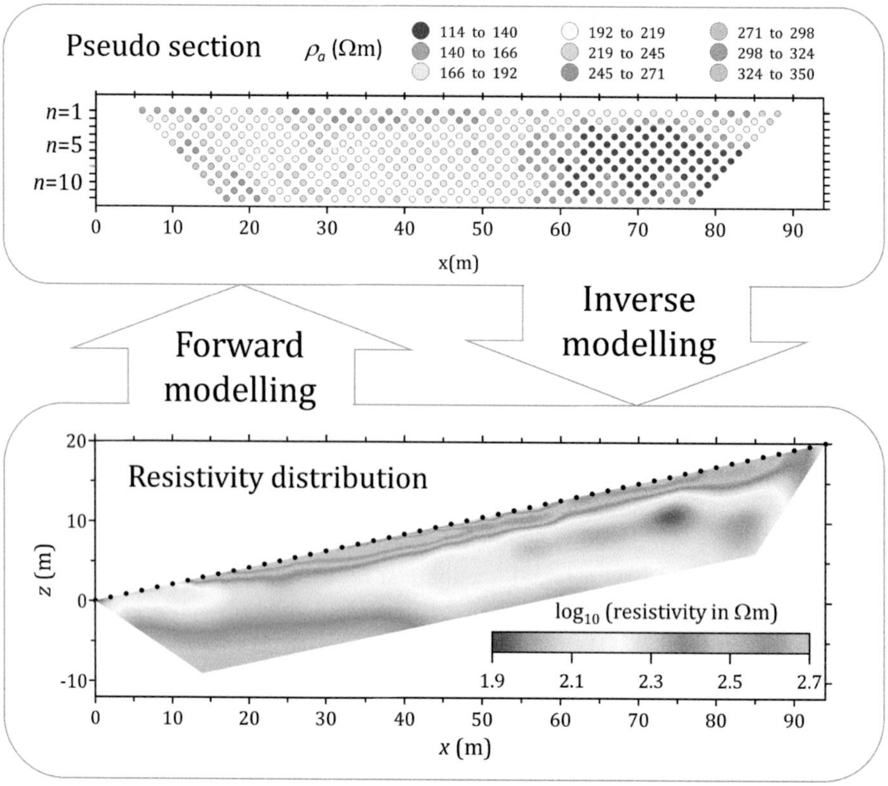

Figure 5.1 Forward and inverse modelling. The example is a dipole–dipole DC resistivity survey carried out on a hillslope (see de Sosa et al., 2018). Electrode positions are marked in the lower figure with solid circles. (A black and white version of this figure will appear in some formats. For the colour version, please refer to the plate section.)

show that even though the process is now much easier to automate, the user should be aware of the basic assumptions of such inverse methods. Unfortunately, such assumptions are frequently neglected, limiting the validity of any interpretation based on the resulting image.

In their pioneering study of the *impedance camera*, Lytle and Dines (1978) noted: '*Items worthy of future research include an assessment of the influence of noise in the data, a study of the accuracy of the reconstruction and its spatial dependence, an evaluation of the degree of dependence of various measurement configurations, an analytic study of the resolution limit, and a determination of the extent to which the use of a priori knowledge affects the interpretation*'. We address most of these issues in this chapter in the context of inverse modelling.

There are numerous publications on forward and inverse modelling in electrical geophysics. To the new user, this can be overwhelming. However, the vast majority of approaches are surprisingly similar. In this chapter, we provide details of all stages of the modelling process with the intention of giving the reader an understanding of all elements.

We provide a modelling platform (see Appendix A) to give the reader the capability of analysing several examples used for illustration in this chapter, and to provide tools needed to carry out their own forward and inverse modelling.

5.2 DC Resistivity

5.2.1 Forward Modelling

Equation 4.3 is a forward model for the DC resistivity problem with uniform resistivity. The objective of the general forward model is to determine the transfer resistance (or apparent resistivity if preferred) that satisfies Equation 4.1 for a given spatial distribution of resistivity and specific geometry. To do this, the variation in resistivity is represented by discretizing the region into layers (in 1D) or cells (in 2D and 3D), with a uniform value assigned to each discretization unit. Since electrodes are placed at discrete positions, the fundamental task of a forward model is the determination of voltage at each measurement electrode which results from current excitation. As shown in Chapter 4 (Section 4.2.1), we can use the principle of superposition to compute the resulting quadrupole measurement. A forward model thus typically involves a solution of the governing equations N_c times, where N_c is the number of current electrode locations.

Forward modelling is a necessary stage of any inversion approach since we require a means to evaluating the goodness of fit of our model to observations. Forward modelling is also extremely powerful for survey design. Sensitivity patterns of measurements (see Figure 4.5) can be computed from forward modelling, allowing the user to assess the suitability of electrode geometry (configuration, spacing, etc.) for the specific problem under investigation.

5.2.1.1 1D Modelling

If we represent the subsurface as a sequence of horizontal layers with resistivity ρ_i and thickness d_i ($i = 1, 2, \ldots N$, with $i = 1$ the upper layer), then, as stated in Chapter 4, the solution to Equation 4.1 giving the voltage V at distance r from a current source I, can be written in an integral form as Equation 4.13. In Chapter 4, we showed how the *Stefanesco kernel function* K_s is formed for a two-layer case.

For a Wenner array (with spacing a), using Equation 4.13, we can obtain

$$\rho_a = a\int_0^\infty T_s(\lambda, k, d)[J_0(\lambda a) - J_0(\lambda 2a)]d\lambda, \tag{5.1}$$

where $J_0(x)$ is the zero-order Bessel function, λ is the integration variable and $T_s(\lambda, k, d)$ is the *resistivity transform function* (or sometimes called the *Schlichter kernel function*), which is governed by the reflection coefficients k and thicknesses d and related to kernel K_s by

$$T_s(\lambda, k, d) = \rho_1[1 + 2K_s(\lambda, k, d)], \tag{5.2}$$

where ρ_1 is the resistivity of the upper layer.

For a Schlumberger array, we can write (e.g. Zohdy, 1975):

$$\rho_a = s^2 \int_0^\infty T_s(\lambda, k, d) J_1(\lambda s) d\lambda, \tag{5.3}$$

where s is $AB/2$ (see Figure 4.3) and J_1 is the first-order Bessel function. The integral in Equation 5.3 is commonly referred to as the *Hankel Transform* of the kernel T_s.

In order to compute the integrals in Equation 5.1 (or Equation 5.2), a log transformation is applied, which allows the use of a standard digital linear filter method (e.g. Ghosh, 1971a, 1971b), resulting in an expression of the form:

$$\rho_a = \sum_{j=1}^{M} b_j T_{sj}, \tag{5.4}$$

where b_j are filter coefficients and T_{sj} are discrete values of the resistivity transform. Methods for computation of a series of filter coefficients have been developed by many authors (e.g. Ghosh, 1971a, 1971b; O'Neil, 1975; Johansen, 1975; Koefeod, 1979; Anderson, 1979; Guptasarma, 1982). The M values of T_{sj} are computed based on the resistivity and thickness of the N layers. Sheriff (1992) provides a useful spreadsheet for computation of a Schlumberger apparent resistivity sounding curve using the O'Neil (1975) filter coefficients. Alternative approaches to the digital filter method have also been proposed (e.g. Santini and Zambrano, 1981; Niwas and Israil, 1986) to minimize the computational effort of linear filter design, although fast Hankel transform filters have also been proposed (Johansen and Sorensen, 1979; Christensen, 1990).

5.2.1.2 2D and 3D Modelling

We first consider the 2D forward modelling problem. We can write Equation 4.1 in a 2D form (with x horizontal and z vertical) and resistivity, ρ, varying in x and z but constant in y, as

$$\frac{\partial}{\partial x}\left(\frac{1}{\rho}\frac{\partial V}{\partial x}\right) + \frac{\partial}{\partial z}\left(\frac{1}{\rho}\frac{\partial V}{\partial z}\right) = -I\delta(x)\delta(z); \tag{5.5}$$

however, this assumes that the current source is infinitely long in the y direction, and so of limited practical value. To account for point electrodes, we require a 2.5D solution: resistivity varies in 2D but we must recognize the 3D current flow. Following Hohmann (1988), we can we can use the Fourier cosine transformation:

$$v(x, k_w, z) = 2\int_0^\infty V(x, y, z) \cos(k_w y) dy, \tag{5.6}$$

where k_w is the wave number. This allows us to form a 2D equation in terms of the transformed variable, v as

$$\frac{\partial}{\partial x}\left(\frac{1}{\rho}\frac{\partial v}{\partial x}\right)+\frac{\partial}{\partial z}\left(\frac{1}{\rho}\frac{\partial v}{\partial z}\right)-\frac{vk_w^2}{\rho}=-I\delta(x)\delta(z)\cdot \tag{5.7}$$

Equation 5.7 can then be solved for v, for a given value of k_w, in a 2D manner. To determine the voltage V, the inverse Fourier transform is needed. This can be stated as

$$V(x,y,z)=\frac{1}{\pi}\int_0^\infty v(x,k_w,z)\cos(k_w y)dk, \tag{5.8}$$

which can be approximated with numerical integration. LaBrecque et al. (1996b) recommend a combination of Gaussian quadrature and Laguerre integration (see also Kemna, 2000). Typically, this will involve approximately 10 values of k_w for satisfactory accuracy. To determine the voltage at the potential electrodes we, therefore, solve the 2D problem in Equation 5.7 for a given number of values of k_w and then approximate V through Equation 5.8. This approach is the basis of the majority of 2D resistivity modelling.

Equation 4.1 may also be expressed in spherical coordinates (see e.g. van Nostrand and Cook, 1966), allowing us to formulate an alternative 2.5D model based on resistivity variation in Cartesian and radial coordinates (e.g. depth and radial distance from a borehole array).

For most applications, we require the solution of Equation 4.3 or 5.7 for a given distribution of resistivity $\rho(x, y, z)$ or $\rho(x, z)$, respectively. The solution will provide estimates of voltages at potential electrode locations, due to injection of current at given electrodes. As analytical solutions are not available for the general case, grid-based methods (commonly either finite difference or finite element based) are used to approximate the solution. The resistivity distribution is assigned within a grid of cells or elements, and voltage is computed at node points (defined at either cell corners or centres). Finite difference methods are conceptually very simple and date back to the early twentieth century (or earlier). They are based on a discrete approximation to the partial derivatives, as illustrated in Figure 5.2. The result is a set of N linear equations of the form $\boldsymbol{A}\,\boldsymbol{V}=\boldsymbol{b}$, where $\boldsymbol{A}$ is a very structured (banded) sparse square matrix (the *conductance matrix*); $\boldsymbol{b}$ is a vector, which includes the current source terms; $\boldsymbol{V}$ is the vector of unknown voltages; and N is the number of cells/node points. An alternative formulation based on nodes at cell corners is easily made, as is an extension accounting for variable cell sizes.

In contrast, finite element methods (which were developed in branches of engineering in the 1960s) are based on integral approximations using variational calculus. In their simplest form (and that which is most widely used for electrical methods), linear variation of the variable of interest (i.e. voltage) is assumed within an individual cell (finite element); consequently, for the same grid shape there is no advantage of finite elements over finite differences based on accuracy of solution, and, in fact, finite difference methods are

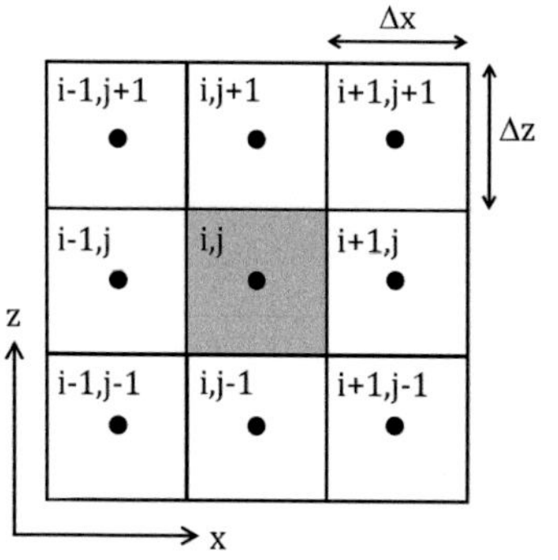

$$\frac{\partial}{\partial x}\left(\frac{1}{\rho}\frac{\partial V}{\partial x}\right) \approx \frac{1}{\Delta x}\left[\frac{2}{(\rho_{i,j}+\rho_{i+1,j})}\left(\frac{V_{i+1,j}-V_{i,j}}{\Delta x}\right) - \frac{2}{(\rho_{i-1,j}+\rho_{i,j})}\left(\frac{V_{i,j}-V_{i-1,j}}{\Delta x}\right)\right]$$

$$\frac{\partial}{\partial z}\left(\frac{1}{\rho}\frac{\partial V}{\partial z}\right) \approx \frac{1}{\Delta z}\left[\frac{2}{(\rho_{i,j}+\rho_{i,j+1})}\left(\frac{V_{i,j+1}-V_{i,j}}{\Delta z}\right) - \frac{2}{(\rho_{i,j-1}+\rho_{i,j})}\left(\frac{V_{i,j}-V_{i,j-1}}{\Delta z}\right)\right]$$

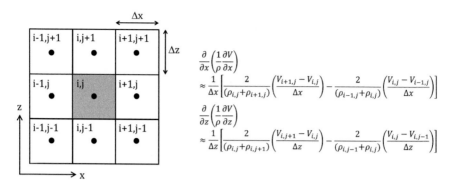

Figure 5.2 2D finite difference grid centred at cell i, j. Resistivity is assigned to each cell and voltages at each node point (cell centre). Extension to a 3D grid is straightforward.

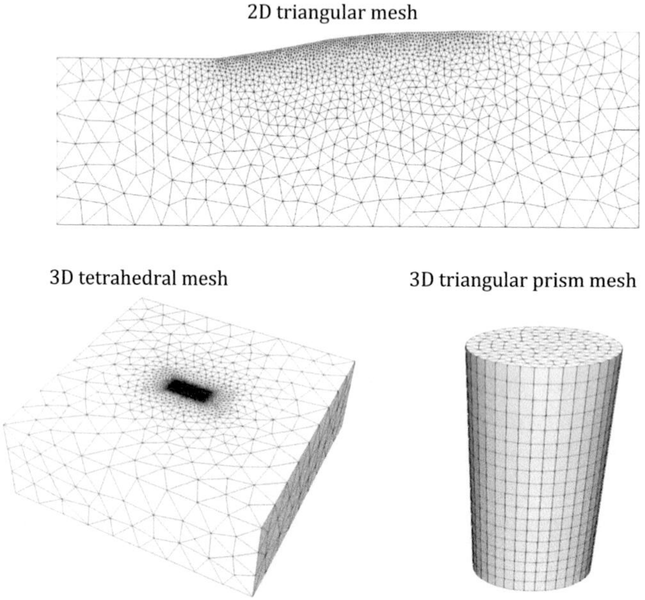

Figure 5.3 Example of finite element mesh discretization.

computationally more efficient. Discretization using finite difference is inherently structured, whereas the use of triangles (2D) or tetrahedra (3D) in finite element modelling allows unstructured meshing (Figure 5.3). The finite element method allows a wide range of element shapes: the basic shapes are triangle (2D) and tetrahedron (3D), although other shapes can be combined in a grid. Consequently, the modelled region can be discretized more efficiently: a higher density of cells can be placed in regions where greatest potential differences occur (i.e. close to current electrode sources). Furthermore, complex geometry (e.g. topography) is easily represented in the finite element mesh. In contrast, for finite difference solutions, resistive cells need to be embedded in a structured grid to account for topographic effects; however, as noted by Wilkinson et al. (2001), significant errors can result in such treatment of boundary conditions. Figure 5.3 illustrates a number of finite element meshes for 2D and 3D problems.

In the finite element method, a basis function is used to approximate variation of the variable (voltage in this case) within the element (the simplest basis function is linear). The method involves integration of the differential equation across each element. As the variable is represented by a product of the basis function and nodal (unknown) values, integration is easily achieved, either analytically (for simple element shapes) or by approximation (e.g. using Gauss quadrature). The basic linear shapes (triangle in 2D, tetrahedron in 3D) allow analytical integration, which adds to computational efficiency. More importantly, their tessellation (the fitting together of a collection of shapes) properties allow efficient mapping of complex geometry.

As in the case of finite difference methods, application of finite element methods results in a system of linear equations of form $\boldsymbol{A}\,\boldsymbol{V} = \boldsymbol{b}$; a local form of the equations is developed and added to the global (composite) system of equations. Whereas finite difference methods result in a very structured conductance matrix, $\boldsymbol{A}$, a sparse matrix is formed in the application of finite element methods. The choice of linear equation solver can have a significant effect on the efficiency of the method, and capability to extend to very large problems.

Note that the grid-based method provides values of the potential field over the entire mesh (vector $\boldsymbol{V}$ mentioned earlier), even though only values at the measurement electrodes are needed. As noted in Chapter 4, we can use the method of superposition to compute the voltage difference at dipole (M,N) due to current injected between dipole (A,B); thus, computation of the potential field is carried out for every current electrode and computational effort increases linearly with the number of electrodes (for a given mesh, and assuming all electrodes are used for current).

Hohman (1975) presented the first 3D numerical solution of the forward problem (Equation 4.3), which was followed by Dey and Morrison (1979). At the time, applications of such models required access to large supercomputer facilities: solutions to problems with 10,000 unknowns (voltages) were considered large, whereas today such solutions are easily achievable with modest personal computing resources. Park and Fitterman (1990) used the Dey and Morrison (1979) code to model a 3D volume using 27 × 21 × 10 finite difference cells, which was the maximum problem they could solve with their 4Mb computer core storage (equivalent to the term 'RAM' on a modern computer) memory storage limit. At the time, attempts to manually match 3D model responses to field data required several months of effort (Park and Van, 1991).

Coggon (1971) developed a finite element solution of the 2D resistivity problem, which was later extended to 3D by Pridmore et al. (1981). The modelling by Pridmore et al. (1981) allowed the use of tetrahedral finite elements (permitting unstructured meshes), although in their application they adopted hexahedral (brick) elements; such structured discretization offered limited advantages to the finite difference solution of Dey and Morrison (1979), and, in fact, was likely to be relatively inefficient, computationally. Pridmore et al. (1981)

highlighted the computational challenges of their approach, remarking: '*In all probability, an efficient solution to this class of problem will await the development of the next generation of computers*'.

The 1990s saw significant advances in computing hardware, some of which translated to desktop computers. Forward modelling algorithms continued to develop (see e.g. Zhang et al., 1995; Bing and Greenhalgh, 2001), although structured meshes remained the norm, thus limiting the size of 3D problems that could be solved. A major constraint of unstructured meshing is the complexity of mesh generation. Fortunately, such tools are now widely available (e.g. the open source *Gmsh* code of Geuzaine and Remacle, 2009), which, coupled with significantly reduced costs of computer (RAM) storage, now enables large-scale forward modelling on complex meshes. Rücker et al. (2006) formulated a forward modelling approach in a finite element unstructured mesh, allowing (at the time) the computation of voltages on a mesh with several hundred thousand nodes. Solutions of problems with 10^6 unknowns (voltages) are now achievable with non-specialized computer hardware. Exploitation of parallel computing permits application to even larger problems (e.g. Johnson et al., 2010). The approach of Rücker et al. (2006) is probably the current benchmark in 3D resistivity modelling, not only offering accuracy (due to the ability to discretize finely near current sources) but also the capability of discretizing complex geometry (e.g. Udphuay et al., 2011). Other advances in forward modelling include accounting for metallic infrastructure (underground pipes, tanks, etc.) (Johnson and Wellman, 2015), allowing application to industrial sites, particularly for groundwater quality assessment.

Note that for field-based applications, infinite boundary conditions must be recognized within the model. The simplest, and probably the most common, approach is to extend the grid to a reasonable distance from the electrode array, increasing cell sizes towards the pseudo-infinite boundaries. Unstructured meshing allows this to be done relatively efficiently (see e.g. Figure 5.3); however, for structured meshes a significant computational overhead results, which is particularly problematic for 3D problems. At the pseudo-infinite boundaries, Neumann (second type) boundary conditions can be applied, i.e. the potential gradient normal to all boundary nodes is set to zero. Alternatively, Dirichlet (first type) boundary conditions may be applied, i.e. the potential is fixed. If the pseudo-infinite boundary is not extended sufficiently from the electrode array, then if Neumann or Dirichlet conditions are imposed, the computed potentials will typically be over-estimated or under-estimated (Coggon, 1971). An alternative approach, first proposed by Dey and Morrison (1979), is to apply a mixed boundary condition at some distance from the electrode array. In this approach, the condition

$$\frac{\partial V}{\partial n} + \frac{V}{r} \cos\theta = 0 \tag{5.9}$$

is applied at all nodes along the pseudo-infinite boundary, where n is the outward normal direction, r is the distance from the source electrode and θ is the angle between n and r. Such an approach allows reduction of the extended mesh. Kemna (2000) provides an equivalent formulation for the 2.5D form of the governing equation (Equation 5.7).

In most cases, linear approximations of the potential field are implicit in grid-based methods, and so most model computations are subject to discretization errors, particularly close to current electrodes, where gradients will be high. With unstructured meshes, errors can be reduced in an efficient manner by dense gridding near electrodes. However, as shown in Figure 5.4, even for reasonably fine grids, errors can be high for specific quadrupole configurations. In the example shown, elements roughly one quarter of the electrode spacing are needed to ensure sub 1 per cent voltage difference estimates for a Wenner configuration. In this case, errors were computed by comparing against an analytical solution. For non-trivial problems (variable resistivity, topographic effects, etc.), such analytical solutions do not exist, making model error checking challenging.

In finite element models, accuracy can be improved by adopting quadratic or cubic shape functions (e.g. Rucker et al., 2006), although these come with significant computational overheads. Sophisticated adaptive meshing techniques may also be used to enhance accuracy of the model (e.g. Ren and Tang, 2010), particularly for complex 3D modelling. For triangle (2D) and tetrahedral (3D) elements, integration of the local differential equations can be performed analytically and is, therefore, computationally efficient. For other shapes (e.g. quadrilaterals in 2D), numerical integration (e.g. using Gauss quadrature) is required.

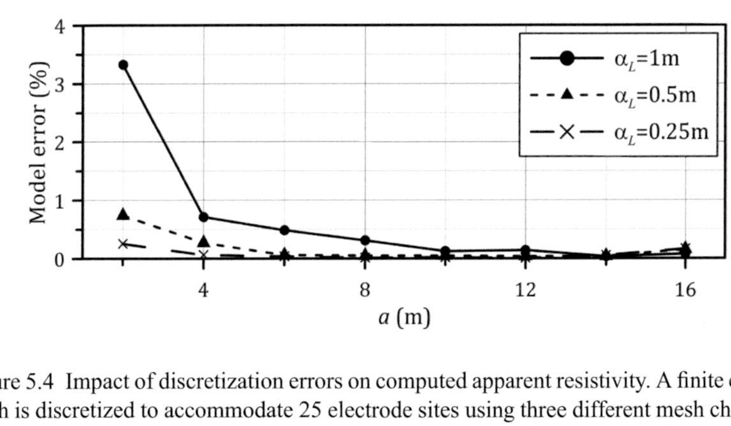

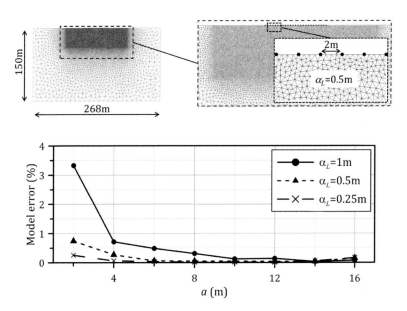

Figure 5.4 Impact of discretization errors on computed apparent resistivity. A finite element mesh is discretized to accommodate 25 electrode sites using three different mesh characteristic lengths (α_L). The extended region of the mesh (shown in the upper-left diagram) is used to account for infinite boundaries. The graph shows errors in Wenner computed apparent resistivity with electrode spacing (a) for three different characteristic lengths.

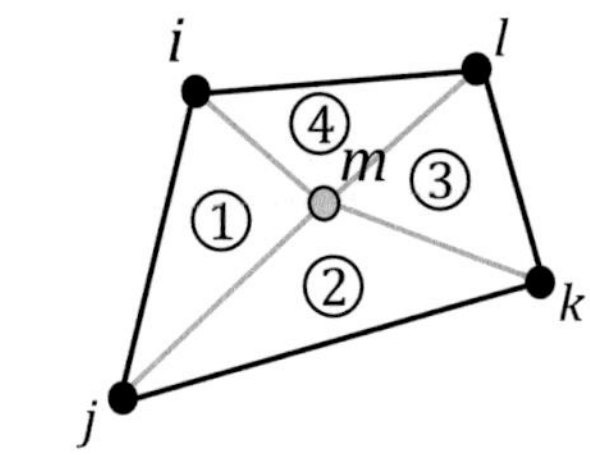

Original equation for triangle ①

$$\begin{bmatrix} A_{i,i} & A_{i,j} & A_{i,m} \\ A_{j,i} & A_{j,j} & A_{j,m} \\ A_{m,i} & A_{m,j} & A_{m,m} \end{bmatrix} \begin{bmatrix} x_i \\ x_j \\ x_m \end{bmatrix} = \begin{bmatrix} b_i \\ b_j \\ b_m(=0) \end{bmatrix}$$

Eliminating node m

$$\begin{bmatrix} A_{i,i} - \dfrac{A_{i,m}A_{m,i}}{A_{m,m}} & A_{i,j} - \dfrac{A_{i,m}A_{m,j}}{A_{m,m}} \\ A_{j,i} - \dfrac{A_{j,m}A_{m,i}}{A_{m,m}} & A_{j,j} - \dfrac{A_{j,m}A_{m,j}}{A_{m,m}} \end{bmatrix} \begin{bmatrix} x_i \\ x_j \end{bmatrix} = \begin{bmatrix} b_i \\ b_j \end{bmatrix}$$

Figure 5.5 Enhanced discretization of quadrilateral finite elements and elimination of an equation of the form $\boldsymbol{A}\,\boldsymbol{x} = \boldsymbol{b}$. The grey lines show sub-division of the element. An example sub-element local equation is shown.

However, embedding a dummy node internally within such an element, and elimination of corresponding equations, allows the use of more efficient triangle integration formulae for a quadrilateral element (see Figure 5.5).

Accuracy can be substantially improved with relatively modest computational overhead by the removal of singularities in the potential field, as outlined by Coggon (1971), Lowry et al. (1989) and Zhao and Yedlin (1996). This method considers the total potential, V, as a combination of a primary, V_a, and secondary potential, V_b:

$$V = V_a + V_b. \tag{5.10}$$

Note that the use of primary and secondary voltage terms here is unrelated to primary and secondary voltage in the context of time domain induced polarization measurements.

The primary potential is computed from the solution of the governing equations with a uniform resistivity, ρ_0; the secondary potential is then obtained from the solution of

$$\nabla.\left(\frac{1}{\rho}\nabla V_b\right) = \nabla.\left(\left[\left(\frac{1}{\rho_0}\right) - \left(\frac{1}{\rho}\right)\right]\nabla V_a\right) \tag{5.11}$$

with boundary conditions as before (in terms of V_b). Note that the current source (Dirac delta) condition is not applied at the electrode location, since it is accounted for in the right-hand side of Equation 5.11. Also note that the right-hand side of Equation 5.11 is of same form as the original equation, allowing computation through the same grid-based approximation.

The primary voltage, V_a, is easily computed for the given homogenous resistivity, ρ_0, using Equation 4.3, for a semi-infinite region without topographic variation. However, if such conditions are violated the primary potential field must be approximated. This can be achieved using a finer grid, although this is only required once for each source electrode – secondary voltages may be computed on a coarser grid. Lowry et al. (1989) propose volumetrically averaging the cell values of resistivity in order to specify the resistivity,

ρ_0; Zhao and Yedlin (1996) suggest that a value close to that near the source electrode is a more appropriate value.

From the previous discussion, it will be apparent that a range of forward modelling approaches exists for dealing with 2D (2.5D) and 3D problems. Structured meshes are conceptually simple to implement and have some computational advantages; however, for large-scale problems, particularly with complex geometry, unstructured meshes are far more efficient, although each problem requires initial overheads in designing a suitable mesh. Accuracy of the chosen method can be improved through singularity removal (but note the complexity when topographic variation exists), mesh refinement and use of non-linear approximating functions. Computational overheads can also be removed through appropriate boundary condition assignment for treatment of infinite boundary conditions. Whatever the choice of method, the user should ensure that an assessment of accuracy of the solution is made. This is often overlooked, and yet forward modelling errors can potentially exceed measurement errors discussed in Section 4.2.2.1. The combination of model and measurement errors can have a significant effect on the application of inverse methods, as we show later.

5.2.1.3 Anisotropy

It is normally assumed in potential field models that resistivity is isotropic, and yet under some conditions (e.g. distinct layering or fracturing), the electrical properties may be anisotropic. The distinction between anisotropy and heterogeneity can be challenging, since one can view macroscale anisotropy as small-scale heterogeneity (e.g. layering or individual fractures). Accounting for anisotropy in the governing equations requires the tensor representation of electrical resistivity or conductivity (e.g. Bibby, 1977), i.e. the flux in one direction is not only related to the potential gradient in that direction but also the potential gradient in the orthogonal directions. For a 2D problem, this equates to a total of three parameters: the resistivity (or conductivity) along principal axes and the orientation of this axis relative to the Cartesian axes. For a 3D problem, a total of five parameters are required. Incorporation of anisotropy in a forward model is relatively straightforward (e.g. Herwanger et al., 2004b) and, in fact, finite element solutions are easily adapted for this. However, such principal directions are likely to vary over some scale, making practical application of such approaches challenging. Sensitivity patterns (like the isotropic case illustrated in Figure 4.5) can be computed for anisotropic models (e.g. Greenhalgh et al., 2009), which can be used to assess the impact of anisotropy on DC resistivity measurements (e.g. Greenhalgh et al., 2010).

5.2.2 Inverse Modelling

5.2.2.1 General Concepts

If we state the forward model in terms of the vector of observations, $\boldsymbol{d}$, and parameters, $\boldsymbol{m}$, using operator F, as: $\boldsymbol{d} = F(\boldsymbol{m})$, then the inverse model can be expressed as $\boldsymbol{m} = F^{-1}(\boldsymbol{d})$. As the problem is non-linear (F is a function of $\boldsymbol{m}$), the inverse solution is obtained in an iterative manner (unlike, for example, linear regression).

Inverse modelling of DC resistivity problems determines the set of M spatial electrical parameters, $\boldsymbol{m}$, that is consistent with the N observations, $\boldsymbol{d}$ (apparent resistivity or transfer resistance). For 1D problems, the parameters are either a sequence of resistivities of given layers or a set of layer thicknesses and associated resistivities. For 2D/3D problems, the parameters are typically the resistivities of a given set of cells. For most problems, log-transformed resistivities are used for parametrization since (i) resistivity can vary over orders of magnitude and (ii) the log transform ensures a positive resistivity in the inverse model. Similarly, log-transformed data are commonly used. However, if working with transfer resistances, a measure of goodness of fit for a particular measurement can only be computed if the polarity of both data and model are the same (positive or negative).

Typically for 2D/3D problems, the discretization of parameter cells is aligned to the mesh used for forward modelling, the simplest case being one parameter per forward model grid cell/element. In order to minimize the number of parameters (hence improving computational efficiency), groups of adjacent cells/elements may be clustered to form parameter cells. Advances in inverse methods over the past few decades have led to the availability of flexible, robust and computationally efficient algorithms for 1D, 2D and 3D problems.

Unconstrained inverse modelling of DC resistivity data is inherently non-unique, i.e. a large number of resistivity distributions are consistent with the observed data. To address this, the inverse model is constrained, as discussed later. Such constraints also reduce the likelihood of instabilities in the iterative process by damping the effect of the propagation of errors (e.g. due to numerical rounding).

Most resistivity inverse models used today are based on a least squares fit between data and model parameters. We can express the data-model misfit as

$$\Phi_d = \left(\boldsymbol{d} - F(\boldsymbol{m})\right)^T \boldsymbol{W}_d^T \boldsymbol{W}_d \left(\boldsymbol{d} - F(\boldsymbol{m})\right), \tag{5.12}$$

where $\boldsymbol{W}_d$ is a data weight matrix, which, assuming uncorrelated errors, is a diagonal matrix with entries equal to the reciprocal of the standard deviation of each measurement. This ensures that misfit between model and observations is weighted according to their quality. Note that this should include the error due to measurement and modelling, although the latter is normally (and incorrectly) ignored. For 3D imaging, modelling errors are often likely to dominate the total error.

An inverse modelling process may then be adopted that seeks the vector $\boldsymbol{m}$ that minimizes Φ_d. One challenge with this is how small should Φ_d be? This is sometimes considered in terms of the chi-squared statistic $\chi^2 = \Phi_d/N$, where N is the number of measurements. A satisfactory solution can be considered if $\chi^2 = 1$ (although Günther et al. (2006) advocate a range $1 \leq \chi^2 \leq 5$). Note also that we can express the misfit as a root mean square (RMS) error, χ.

Since the forward model is a function of parameters $\boldsymbol{m}$, a linearization process is required for inverse modelling. This is normally achieved using the Gauss–Newton approach (see Box 5.1), which results in the following iterative sequence:

$$\begin{aligned}\left(\boldsymbol{J}^T \boldsymbol{W}_d^T \boldsymbol{W}_d \boldsymbol{J}\right)\Delta \boldsymbol{m} &= \boldsymbol{J}^T \boldsymbol{W}_d^T \left(\boldsymbol{d} - F(\boldsymbol{m}_k)\right), \\ \boldsymbol{m}_{k+1} &= \boldsymbol{m}_k + \Delta \boldsymbol{m},\end{aligned} \tag{5.13}$$

where $\boldsymbol{J}$ is the Jacobian (or sensitivity) matrix, given by $J_{i,j} = \partial F(\boldsymbol{m}_k)_i / \partial m_j$, $i = 1, 2, \ldots N$, $j = 1, 2, \ldots M$; $\boldsymbol{m}_k$ is the parameter set at iteration k; $\Delta \boldsymbol{m}$ is the parameter update at iteration k. Note that Equation 5.13 is a linear system of equations of the form $\boldsymbol{Ax} = \boldsymbol{b}$, where A is a full matrix of size $M \times M$, $\boldsymbol{x}$ and $\boldsymbol{b}$ are column matrices of size $M \times 1$.

In practice, Equation 5.13 is of limited practical value. Problems can occur resulting in convergence to a local minimum or instability of the solution, thus failing to converge.

5.2.2.2 Damping and Regularisation

The Levenberg–Marquardt method (sometimes called the Marquardt method) includes a damping parameter λ_{LM} in a revised form of Equation 5.13:

Box 5.1

Derivation of the Gauss–Newton solution

We can write the objective function using Taylor expansion as

$$\Phi_d(\boldsymbol{m}_k + \Delta \boldsymbol{m}) \approx \Phi_d(\boldsymbol{m}_k) + \frac{\partial \Phi_d(\boldsymbol{m}_k)}{\partial \boldsymbol{m}} \Delta \boldsymbol{m} + \frac{\partial^2 \Phi_d(\boldsymbol{m}_k)}{\partial \boldsymbol{m}^2} \Delta \boldsymbol{m}^2, \tag{5.14}$$

where higher-order terms have been neglected.

At the solution, the derivative of Equation 5.14 is zero, which, ignoring high-order terms, gives

$$\frac{\partial \Phi_d(\boldsymbol{m}_k + \Delta \boldsymbol{m})}{\partial \boldsymbol{m}} \approx \frac{\partial \Phi_d(\boldsymbol{m}_k)}{\partial \boldsymbol{m}} + \frac{\partial^2 \Phi_d(\boldsymbol{m}_k)}{\partial \boldsymbol{m}^2} \Delta \boldsymbol{m} = 0. \tag{5.15}$$

Which can be rearranged as

$$\frac{\partial^2 \Phi_d(\boldsymbol{m}_k)}{\partial \boldsymbol{m}^2} \Delta \boldsymbol{m} = \frac{-\partial \Phi_d(\boldsymbol{m}_k)}{\partial \boldsymbol{m}}. \tag{5.16}$$

Given the definition in Equation 5.14, from the chain rule and ignoring higher-order terms $2\left(\nabla \boldsymbol{J}^T\right) \boldsymbol{W}_d^T \boldsymbol{W}_d \left(\boldsymbol{d} - F(\boldsymbol{m}_k)\right)$, etc., we can write:

$$\frac{\partial \Phi_d(\boldsymbol{m}_k)}{\partial \boldsymbol{m}} = -2\boldsymbol{J}^T \boldsymbol{W}_d^T \boldsymbol{W}_d \left(\boldsymbol{d} - F(\boldsymbol{m}_k)\right), \text{ and} \tag{5.17}$$

$$\frac{\partial^2 \Phi_d(\boldsymbol{m}_k)}{\partial \boldsymbol{m}^2} = 2\boldsymbol{J}^T \boldsymbol{W}_d^T \boldsymbol{W}_d \boldsymbol{J}. \tag{5.18}$$

Using Equations 5.17 and 5.18, Equation 5.16 can be expressed in the form shown in Equation 5.13.

$$\left(\boldsymbol{J}^T\boldsymbol{W}_d^T\boldsymbol{W}_d\boldsymbol{J}+\lambda_{LM}\boldsymbol{I}\right)\Delta\boldsymbol{m}=\boldsymbol{J}^T\boldsymbol{W}_d^T\boldsymbol{W}_d\left(\boldsymbol{d}-F(\boldsymbol{m}_k)\right), \tag{5.19}$$

where $\boldsymbol{I}$ is the identity matrix. λ_{LM} is a positive value that is adjusted during the iteration process. The normal procedure is to reduce the value of λ_{LM} at each iteration if misfit reduces, or increase λ_{LM} if there is no improvement. The $\lambda_{LM}\boldsymbol{I}$ damping term helps to stabilize the solution.

The damping approach above may prove effective if the number of parameters is small relative to the number of measurements (which can be the case for 1D sounding), but as the number of parameters increases (as in 2D and 3D imaging), then the problem becomes ill-conditioned. To address this, Tikhonov regularization (e.g. Tikhonov and Arsenin, 1977) is widely adopted. In this approach, the solution is constrained by a penalty function based on the parameter values, e.g.:

$$\Phi_m=\boldsymbol{m}^T\boldsymbol{R}\boldsymbol{m}, \tag{5.20}$$

where $\boldsymbol{R}$ is a roughness matrix that describes the spatial connectivity of the parameter values.

The objective function to be minimized is then

$$\Phi_{total}=\Phi_d+\alpha\Phi_m, \tag{5.21}$$

where α is a scalar that controls the balance of model smoothing relative to data misfit.

The most common form of regularization in electrical imaging is a minimum structure function based on the sum of squared differences between adjacent parameter values. To illustrate this, consider a 1D arrangement of parameters m_1, m_2 and m_3. The roughness penalty term is then $\Phi_m=(m_1-m_2)^2+(m_2-m_3)^2$ and so matrix $\boldsymbol{R}$ can be written as

$$\boldsymbol{R}=\begin{bmatrix}1 & -1 & 0\\ -1 & 2 & -1\\ 0 & -1 & 1\end{bmatrix}. \tag{5.22}$$

From Figure 5.2, it can be seen that this form of $\boldsymbol{R}$ is equivalent to a second derivative operator. If the parameter cells are of different sizes, then, as pointed out by Oldenburg et al. (1993), the elements in $\boldsymbol{R}$ should be inversely scaled by the distances between centres of adjacent cells. For example, if $\Delta z_{1,2}$ and $\Delta z_{2,3}$ are the distances between centres of cells 1 and 2, and cells 2 and 3, respectively, then

$$\boldsymbol{R}=\begin{bmatrix}\frac{1}{\Delta z_{1,2}} & \frac{-1}{\Delta z_{1,2}} & 0\\ \frac{-1}{\Delta z_{1,2}} & \left(\frac{1}{\Delta z_{1,2}}+\frac{1}{\Delta z_{2,3}}\right) & \frac{-1}{\Delta z_{2,3}}\\ 0 & \frac{-1}{\Delta z_{2,3}} & \frac{1}{\Delta z_{2,3}}\end{bmatrix}. \tag{5.23}$$

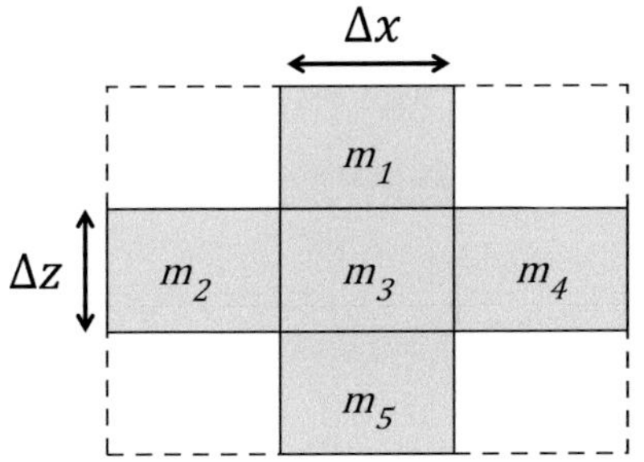

$$
\boldsymbol{R} = \begin{bmatrix} 0 & 0 & 0 & 0 & 0 \\ 0 & \Delta z/\Delta x & -\Delta z/\Delta x & 0 & 0 \\ 0 & -\Delta z/\Delta x & 2\Delta z/\Delta x & -\Delta z/\Delta x & 0 \\ 0 & 0 & -\Delta z/\Delta x & \Delta z/\Delta x & 0 \\ 0 & 0 & 0 & 0 & 0 \end{bmatrix} + \begin{bmatrix} \Delta x/\Delta z & 0 & -\Delta x/\Delta z & 0 & 0 \\ 0 & 0 & 0 & 0 & 0 \\ -\Delta x/\Delta z & 0 & 2\Delta x/\Delta z & 0 & -\Delta x/\Delta z \\ 0 & 0 & 0 & 0 & 0 \\ 0 & 0 & -\Delta x/\Delta z & 0 & \Delta x/\Delta z \end{bmatrix}
$$

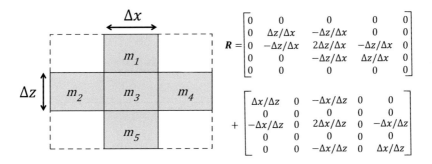

Figure 5.6 2D regularization template. Note that the cell size is assumed to be constant in this example and so cell dimensions can be used to represent cell centre spacing.

Figure 5.6 shows how the same approach is applied to a rectangular 2D mesh. Extension to other parameter meshes (e.g. 2D triangular and 3D tetrahedral) is straightforward.

The spatial regularization in the objective function (Equation 5.21) may be incorporated in a number of ways within the optimization process. Constable et al. (1987) first coined the (now widely used) term 'Occam's inversion' (to emphasize the search for the simplest model, after Occam's razor, attributed to the fourteenth-century English philosopher William of Ockham) in which the parameter set that provides the minimum value of Φ_m is solved for subject to satisfying Φ_d meeting the target misfit, Φ_d^* (see comments earlier on χ^2). Constable et al. (1987) tackled the optimization using the method of Lagrange multipliers, minimizing

$$\Phi_{total} = \Phi_m + \mu(\Phi_d - \Phi_d^*), \tag{5.24}$$

where the Lagrange multiplier, μ, is effectively equivalent to $1/\alpha$ in Equation 5.21. An iterative process can be then be devised in which Φ_d^* is adjusted at each iteration.

Alternatively (and effectively similar) is the formulation of a Gauss–Newton optimization, which results in the following iterative equations (see Box 5.2):

$$
\begin{aligned}
\left(\boldsymbol{J}^T \boldsymbol{W}_d^T \boldsymbol{W}_d J + \alpha \boldsymbol{R}\right)\Delta \boldsymbol{m} &= \boldsymbol{J}^T \boldsymbol{W}_d^T \boldsymbol{W}_d \left(\boldsymbol{d} - F(\boldsymbol{m}_k)\right) - \alpha \boldsymbol{R} \boldsymbol{m}_k, \\
\boldsymbol{m}_{k+1} &= \boldsymbol{m}_k + \Delta \boldsymbol{m},
\end{aligned} \tag{5.25}
$$

which is repeated until a satisfactory value of Φ_d is achieved. The general sequence of steps for an inversion is shown in Box 5.3.

The regularization scalar α is normally adjusted at each iteration. By starting with a suitably large value in the first iteration, the optimization process is dominated by finding the best very smooth model; as α reduces during subsequent iterations, the data misfit Φ_d begins to dominate: the model structure thus roughens to fit the data. One may adopt a target data misfit Φ_d^* at each iteration to help in a gradual transition to the solution and avoid being trapped in a local minimum. Note that the adoption of a large value of α in the early stage of

Box 5.2

Derivation of the Gauss–Newton solution with regularization

Following the procedure in Box 5.1, the equations for first and second derivatives of the total objective function are

$$\frac{\partial \Phi_{total}(\boldsymbol{m}_k)}{\partial \boldsymbol{m}} = -2\boldsymbol{J}^T \boldsymbol{W}_d^T \boldsymbol{W}_d \left(\boldsymbol{d} - F(\boldsymbol{m}_k)\right) - 2\alpha \boldsymbol{R}\boldsymbol{m}_k \text{ and} \tag{5.26}$$

$$\frac{\partial^2 \Phi_{total}(\boldsymbol{m}_k)}{\partial \boldsymbol{m}^2} = 2\boldsymbol{J}^T \boldsymbol{W}_d^T \boldsymbol{W}_d \boldsymbol{J} + 2\alpha \boldsymbol{R}. \tag{5.27}$$

Using Equations 5.26 and 5.27, Equation 5.16 can be expressed in the form shown in Equation 5.25.

Box 5.3

General inversion sequence

1. Define mesh for forward model.
2. Define discretization of parameters, which will be aligned to the mesh used for forward modelling – the simplest case being one parameter per mesh cell/element.
3. Compute roughness matrix, $\boldsymbol{R}$.
4. Given starting resistivity model (usually homogenous), compute forward model and hence data misfit by comparing with observations.
5. Determine starting value of α and select range of values of α for line search.
6. Compute Jacobian $\boldsymbol{J}$ for given resistivity model (see later).
7. Select target misfit for the iteration, e.g. 10 per cent reduction of initial misfit.
8. Solve Equation 5.25 for parameter updates for each value of α, starting with the largest value in the range.
9. For each update in step 8, compute the forward model and hence data misfit.
10. Terminate line search if target misfit is reached or if reducing α leads to an increase in data misfit (see Figure 5.7a).
11. If $\Phi_d = N$, then convergence is reached. Otherwise, select the starting α in the line search based on optimum α from previous iteration (see Kemna, 2000) and return to step 6.

the process ensures independence of the starting model. Kemna (2000) offers an effective means of estimating an initial value of α based on the work of Newman and Alumbaugh (1997).

At each iteration, an optimum value of α (or μ in the Lagrange multiplier method) may be determined. This is usually done with a line search (e.g. deGroot-Hedlin and Constable, 1990; LaBrecque et al., 1996b). In the approach adopted by Kemna (2000), Equation 5.25 is solved for several trial values of α (spanning several orders of magnitude). For each trial α, the data misfit is computed and the value of α that is adopted is that which results in the minimum value of Φ_d or the target misfit at that iteration. Figure 5.7a shows example

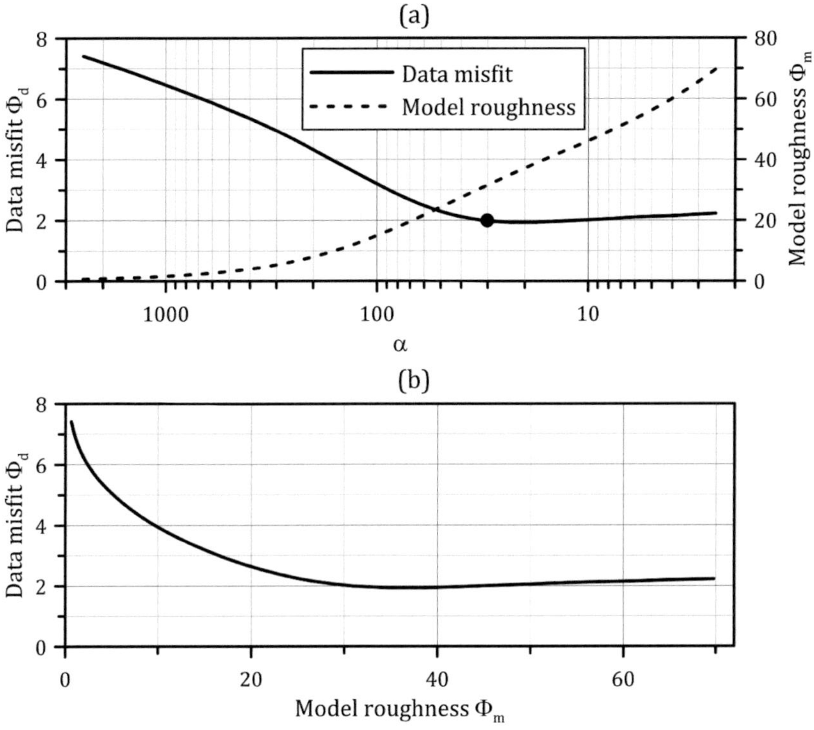

Figure 5.7 Example behaviour of data misfit and model roughness. (a) Data misfit and model roughness as a function of regularization scalar α. The black circle identifies the selected value of α. (b) L-curve.

behaviour of such a line search. Note for α less than 20, in this example, the data misfit remains at minimum, whereas roughness continues to rise as α decreases. In this case an α value greater than 20 (say, 30) is assigned (Figure 5.7a).

Some authors advocate the use of the L-curve method (e.g. Hansen, 1992) to determine the optimum regularization parameter. Examples applied to DC resistivity problems include Li and Oldenburg (1999) and Günther et al. (2006). In such an approach, the relationship between Φ_d and Φ_m is examined (e.g. Figure 5.7b) to determine the point of maximum curvature. The challenges with this are that (i) L-curve behaviour is not guaranteed, and (ii) a sufficient number of solutions for a range of α values is required in order to estimate, with sufficient accuracy and resolution, the curvature of the relationship. In fact, Li and Oldenburg (1999) recommend that an approximate solution is used to carry out such a line search in order to reduce computational overheads. In the first author's experience, a simple line search for the minimum data misfit (or target value) over a range of three orders of magnitude of α, discretized in ten steps, is adequate for this process. The key requirement is a suitable initial estimate of α – the method proposed by Kemna (2000) appears to be extremely robust for this.

The second-order regularization operator in Equation 5.22 applied to the model misfit definition in Equation 5.20 is referred to as an L_2 norm (as is the least squares data misfit). This is an extremely robust and reliable approach (and the most widely used approach in 2D and 3D electrical imaging). The assumption implicit in their use is that a smooth model is consistent with *a priori* knowledge. Such smooth models can be attractive in many applications; however, for situations where known contrasts in electrical properties exist, they can be inappropriate. Furthermore, they often result in some oscillatory behaviour outside regions of sharp contrast (Figure 5.8). To address this, other model misfit objective functions may be used. The L_1 norm penalizes absolute differences in adjacent values and thus results in a flatter model, except where data drive changes in the parameters. Consequently, this results in a 'blocky' model, which may be preferred in some applications. Incorporation of an L_1 model norm in the Gauss–Newton formulation discussed earlier is not possible but alternative optimization methods (e.g. linear programming, Dosso and Oldenburg, 1989) can be utilized. However, incorporation of pseudo L_1 norm approaches in the formulation in Equation 5.25 is possible. Farquharson and Oldenburg (1998) present a method for general non-L_2 norm model structure which involves iteratively reweighting the roughness matrix $\boldsymbol{R}$ at each iteration of the Gauss–Newton process. Their approach effectively adds an additional term to the diagonal of the roughness matrix, which is a function of the parameter set at the given iteration. Farquharson and Oldenburg (1998) illustrate the effectiveness of the approach in 1D inversions; Loke et al. (2003) demonstrate the same technique applied to 2D resistivity imaging. We discuss other forms for regularization in Section 5.2.5.

5.2.2.3 Computation of the Sensitivity Matrix

Calculation of the sensitivity (Jacobian) matrix, $\boldsymbol{J}$, in Equation 5.13, can present a significant computational burden to the inversion process. As stated earlier, the elements are given by $J_{i,j} = \partial F(\boldsymbol{m}_k)_i / \partial m_j, i = 1, 2, \ldots N, j = 1, 2, \ldots M$, where $\boldsymbol{m}_k$ is the parameter set at iteration k. For 1D problems defined by a set of layers, the parameters are the thickness and resistivity of each layer. For general problems based on a model norm (e.g. the Occam's approach), the parameters are the resistivities of the model layers (1D) or cells (2D and 3D). As stated earlier, it is normal to

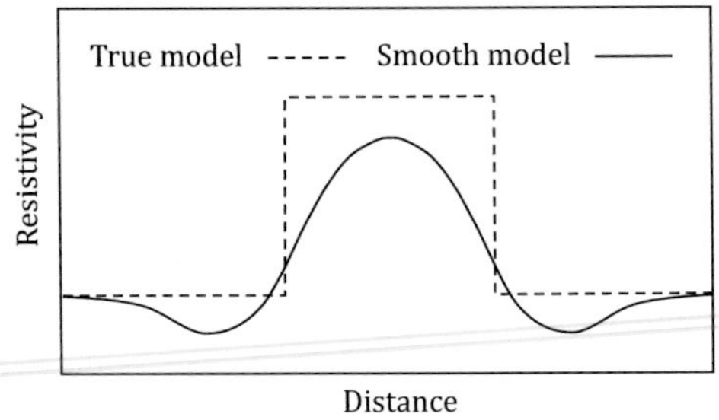

Figure 5.8 Over-smoothing of a regularized 1D model using the L_2 model norm.

use a logarithm transformation of the resistivity for parameterization given the potential variation of resistivity over orders of magnitude. This has the added advantage of constraining the parameters to ensure non-negative resistivity estimation.

For vertical electrical soundings, the sensitivities can be computed based on Equation 5.4 (see e.g. Koefoed, 1979; Inman et al., 1973; Constable et al., 1987). For the general problem, parameterized in terms of a 1D, 2D or 3D distribution of log resistivities, if the forward model for measurement *i* is expressed as a logarithm of the apparent resistivity

$$F(\boldsymbol{m}_k)_i = \ln\rho_{a,i} = \ln\left(K_i(V/I)_i\right), \tag{5.28}$$

where K_i is the geometric factor for transfer resistance measurement $(V/I)_i$, and parameter $m_j = \ln\rho_j$, where ρ_j is the resistivity of cell *j*, then, from the chain rule, the Jacobian can be written as

$$\partial F(\boldsymbol{m}_k)_i/\partial m_j = \left(\rho_j/V_i\right)\partial V_i/\partial\rho_j\cdot,\ i = 1,\ 2,\ldots N,\ j = 1,\ 2,\ldots M. \tag{5.29}$$

Conceptually, the simplest way of computing the terms in Equation 5.29 is to apply a finite difference operation, referred to as the influence coefficient method. The forward model is computed for a given resistivity distribution and then each parameter is perturbed, one at a time, and a forward model is recomputed. Thus, for *M* parameters the forward model must be computed *M*+1 times, each time providing the *N* transfer resistances. Such a method can be applied to any form of parameterization.

A more attractive method, computationally, is the adjoint method that uses the principle of reciprocity (Geselowitz, 1971) to determine the derivatives on the right-hand side of Equation 5.29 (e.g. Sasaki, 1989; Kemna, 2000). When applied with the finite element method, this approach incorporates a matrix of the same form as the conductance matrix, $\boldsymbol{A}$, used for computation of the potential field and, effectively, only requires one forward model calculation, reducing computation cost significantly for large problems.

Note that if the data are represented by the logarithm of the transfer resistances, then, for Φ_d to be definable, both measured and modelled transfer resistances must have the same polarity. For surface-based surveys, the polarity is independent of the spatial variability of resistivity; however, for subsurface electrodes (e.g. in a cross-borehole setup) the resistivity distribution can influence the polarity of the measurement. Consequently, for such surveys, some measurements may have a different polarity to that computed by the forward model, and thus cannot be used in the calculation of the log-based data misfit. This usually only occurs early on in the inversion process as the model evolves from an initially homogenous case. Such measurements need to be ignored at this stage in the inversion process, but can later be incorporated as the inversion progresses.

5.2.2.4 Inverse Models for Vertical Soundings

Early methods for interpretation of vertical electrical soundings were based on type (or master) curve matching (e.g. Flathe, 1955). In the 1970s, a number of automated methods evolved based on Gauss–Newton-type solutions (e.g. Inman et al., 1973; Inman, 1975; Johansen, 1977). In these methods, the target is the solution of a number of layers, for which the thickness and resistivity is determined. The Marquardt method (sometimes called *ridge regression*) (see Equation 5.19) has proved popular for such an approach. Zohdy (1989) proposed a fast iterative method for inverse modelling of VES data without the need for sensitivity calculations. Gupta et al. (1997) offer a non-iterative approach based on solving the set of resistivities of a layered model, but comment on the sensitivity to the choice of layer thicknesses. There now exist a large number of commercial and academic codes for applying such methods. In fact, as computation of a theoretical sounding curve for a given layered model is readily achievable on a spreadsheet (e.g. Sheriff, 1972), solving the inverse problem on a spreadsheet using built-in optimization routines is now easily done.

Rather than forming the parameter set in terms of a small number of layers, the problem can be cast as series of many layers of fixed thickness, using regularization (as discussed earlier) to constrain the solution (e.g. Constable et al., 1987), ultimately resulting in a smoother transition of resistivity through the profile. When considering inversion based on an unknown set of layers, it is important to select a minimum number of layers that fits the data adequately. Increasing the number of layers will inevitably lead to better fitting but this can enhance non-uniqueness in a model. Figure 5.9 shows results from an inversion of data for the three-layer Schlumberger sounding in Figure 4.9. For this example, the synthetic dataset was perturbed with 10 per cent Gaussian noise (i.e. for each apparent resistivity, noise with a mean of zero and standard deviation equal to 10 per cent of the value of apparent resistivity was added). Note that such an error level will probably exceed that observed in the field. The code *IPI2WIN* (Bobachev, 2003; see also Appendix A) was used for the inversion. A three-layer model was deemed satisfactory for the inversion, which is consistent with the true case. *IPI2WIN* provides minimum and maximum models (see Figure 5.9) to allow the user to assess the uniqueness of the final model. As can be seen in the example in Figure 5.9, for this inversion, the true resistivity lies within the range. The thickness of the upper layer is, however, underestimated slightly. Electrical sounding models, derived in this way, are prone to non-uniqueness and the nature of equivalence often needs to be assessed (e.g. Simms and Morgan, 1992). A useful way of doing this is to examine the correlation of parameters (see also Section 5.2.4.). Table 5.1 reports the correlation matrix for the model in in Figure 5.9, showing generally low correlation, except the negative correlation for the upper layer, implying that increasing the resistivity has a similar effect to decreasing the thickness, i.e. it is the conductance h/ρ that is resolved by the model, where h is the layer thickness. The problem of equivalence is more pronounced in cases where a resistive layer is embedded between two conducting layers, or when a conductor lies between two resistive beds. In the former case, the resistance $h\rho$ is resolved, whereas in the latter case the conductance h/ρ is resolved. In both cases, the intermediate layer needs to be thick enough for the resistivity to be resolved with an electrical sounding.

5.2.2.5 Generalized 2D Inverse Modelling

Vertical electrical soundings have been immensely powerful, and continue to be so. Because of the typically large datasets that result from mobile arrays (see Chapter 4), 1D inverse

Table 5.1 Correlation matrix for inverse model in Figure 5.9. ρ_i and h_i are the resistivity and thickness of layer i. Layer 1 is the upper layer

	ρ_1	h_1	ρ_2	h_2	ρ_3
ρ_1	1	−0.53	0.00	0.06	0.03
h_1	−0.53	1	0.00	−0.54	−0.15
ρ_2	0.00	0.00	1	0.00	0.00
h_2	0.06	−0.54	0.00	1	−0.35
ρ_3	0.03	−0.15	0.00	−0.35	1

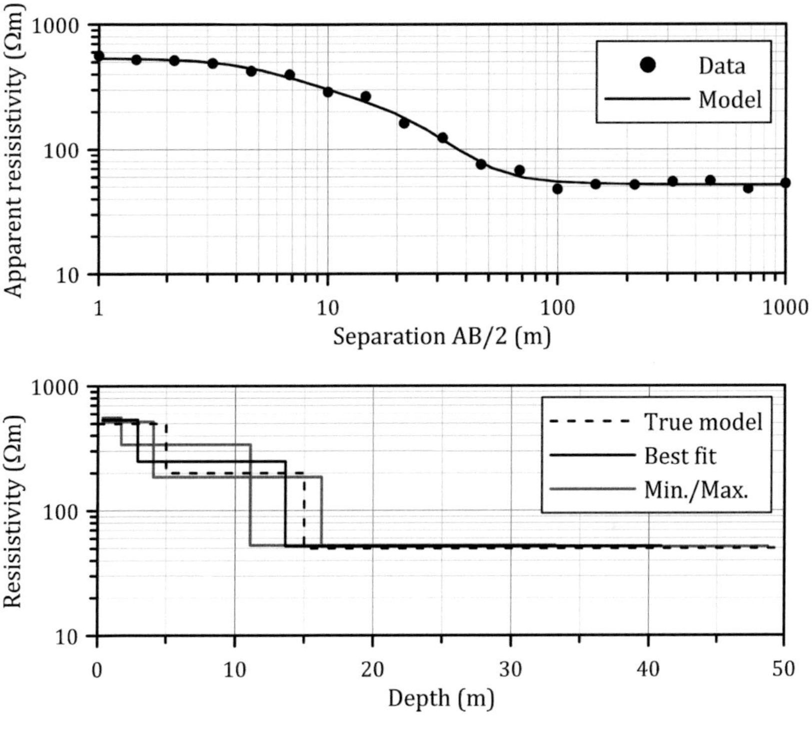

Figure 5.9 Inversion of sounding data from Figure 4.9. 10% Gaussian noise added prior to inversion. Inversions carried out with the *IPI2WIN* code.

modelling has also been used successfully for quasi-2D and quasi-3D imaging of data from such systems (e.g. Auken and Christiansen, 2004; Auken et al., 2005; Guillemoteau et al., 2017), in some cases using lateral smoothing. Figure 5.10 shows an example inversion of continuous vertical electrical sounding (CVES) data from the Aarhus PACES system.

The development of multi-electrode instrumentation (see Chapter 3) in the late 1980s stimulated parallel advances in algorithms for 2D imaging of resistivity. The Occam's approach of Constable et al. (1987) for 1D problems led to the logical development of

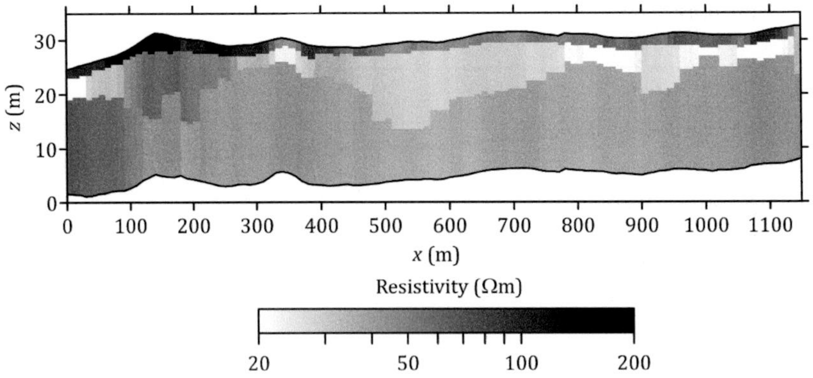

Figure 5.10 Example inversion of CVES data using a 1D three-layer model with lateral constraints. (Inversion provided by Nikolaj Foged, Aarhus University.)

a number of smoothness-based 2D solutions (e.g. Sasaki, 1989, 1992; deGroot-Hedlin and Constable, 1990; Loke and Barker, 1995). Loke and Barker (1995) focused efforts on computationally efficient methods, using finite difference based forward modelling, with a view to offering tools that could ultimately be used in the field (considered at the time to be computationally prohibitive). This earlier work, for example, assumed that the Jacobian matrix during iterations does not change from that for a homogenous model, allowing precalculation and storage of the Jacobian matrix for specific array geometries. Their work led to the most widely used 2D DC resistivity inversion software (*RES2DINV*; see Appendix A), which permitted (for the first time) relatively rapid interpretation of field data. This code has no doubt contributed significantly to the wide success of 2D imaging, particularly for near-surface geophysics problems.

Two areas of focus emerged. Loke and Barker (1995), and subsequent studies, addressed surface-based electrode applications, targeting more widely used field investigations. In contrast, methods addressing more generalized application, in particular cross-borehole resistivity imaging, evolved. Daily and Owen (1991) and Shima (1992) proposed inverse methods for early cross-borehole applications, but it was the incorporation of the regularized solution (e.g. Sasaki, 1992) that led to more robust and reliable techniques (see also, Lesur et al., 1999; Zhou and Greenhalgh, 2002), many of which adopted finite element based forward modelling.

Figure 5.11b shows an example 2D inversion based on the L_2 model norm regularization discussed earlier. For this example, dipole–dipole data were generated based on the resistivity model in Figure 5.11a for a twenty-five–surface electrode array at 5 m spacing (a = 5 m, n = 1 to 6). Forward modelled data were perturbed using a Gaussian error of 2 per cent. The inversion weighted data according to the same error level and convergence to an RMS of 1.0 was achieved in two iterations. Clearly, given the smoothness-based regularization, sharp interfaces are not resolved; however, a clear demarcation between

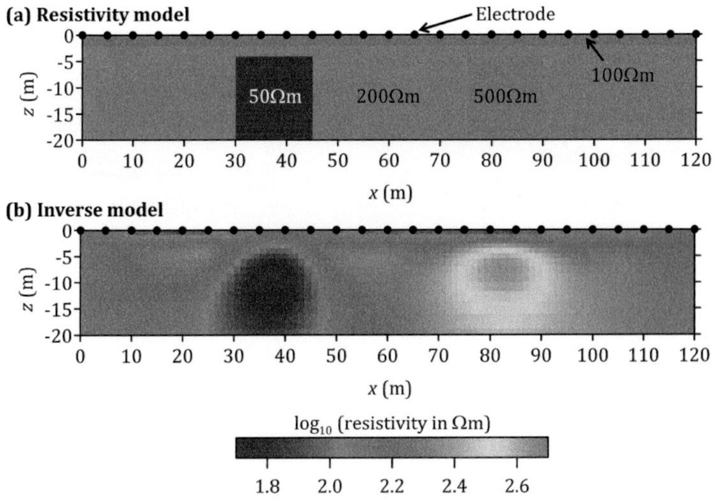

Figure 5.11 Inversion of synthetic 2D surface electrode resistivity model. (a) Synthetic model. (b) Inverted model. (A black and white version of this figure will appear in some formats. For the colour version, please refer to the plate section.)

anomalies is evident. Also note that the two localized anomalies are truncated at depth due to the limited sensitivity at depth.

As discussed in Section 5.2.2.8, the regularization operator can be modified to enhance the inverse model (see also below). Figure 5.12 shows a comparison of L_1 and L_2 norm inversions of a field dataset from Wilkinson et al. (2012). In this example, the L_1 norm model enhances contrasts in resistivity, although both models reveal the three (roughly horizontal) geological units at the site (see Wilkinson et al., 2012).

As explained in Chapter 4, greater resolving power at depth can be achieved using cross-borehole imaging, although the required maximum spacing of the borehole pair can be limiting in many applications (see Figure 4.32). Figure 5.13 shows an example inversion for a synthetic model using two boreholes containing 16 electrodes at 1 m spacing. For this example, a dipole–dipole configuration was used with 1 m spaced dipoles, incorporating all possible combinations. In practice, such a range of measurements is likely to result in poor reciprocity for large dipole separations given the inevitable low-voltage measurements that would result. For this example, forward modelled data were perturbed with a noise model $\varepsilon_R = 0.001 + 0.02R$, where R is the absolute value of the transfer resistance (see Figure 4.15). This is equivalent to 2 per cent Gaussian noise with an offset of 0.001 Ω. The offset component was added to account for more realistic field noise conditions, which can impact on the performance of surveys with such large dipole separations. Convergence of the inversion was achieved in one iteration.

The same principles can be applied for treatment of small scale resistivity imaging problems. Figure 5.14 shows an example in which a resistive target is analysed using a series of 103 dipole–dipole measurements. For this problem, the generated forward model was perturbed with a realistic noise, $\varepsilon_R = 0.001 + 0.02R$, and then inverted on a mesh containing 13,654 parameter cells. Convergence was achieved in two iterations. The inverse model in Figure 5.14 shows

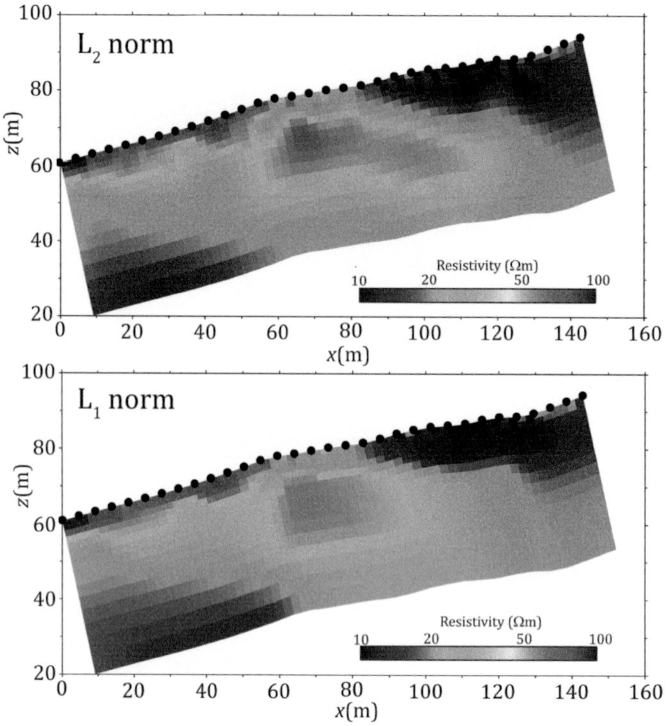

Figure 5.12 Comparison of L_1 and L_2 norm inversions of a 32-electrode dipole–dipole 2D dataset discussed in Wilkinson et al. (2012). Electrode locations are marked by solid circles. The field site is discussed in more detail in the case study in Section 6.1.7. (A black and white version of this figure will appear in some formats. For the colour version, please refer to the plate section.)

recovery of the resistive target, albeit with inevitable smoothing and some overshoot around the perimeter of the resistive target. The case study in Section 6.1.5 illustrates the use of small-scale resistivity imaging in the study of solute transport in soil cores.

Although convergence of the inversion indicates satisfactory misfit overall, it is often useful to examine misfit in other ways. A simple way of doing this is to plot observed and modelled apparent resistivity, as shown in Figure 5.15a for the inversion in Figure 5.11. An alternative approach is to examine the misfit for each measurement relative to the prescribed error. From Equation 5.12, we can express the relative misfit as $(d_i - F(\boldsymbol{m})_i)/\varepsilon_i$, where ε_i is the standard deviation assigned to measurement *i*. The relative misfit should, in theory, follow a Gaussian distribution with a mean of zero and unit standard deviation, e.g. we expect 95 per cent of the measurements to be fit with a relative misfit between −2 and 2. Figure 5.15b illustrates such a distribution for the inversion in Figure 5.11.

5.2.2.6 3D Inverse Modelling

During the 1990s, 3D DC resistivity inversion algorithms evolved, based on finite difference techniques (e.g. Park and Van, 1991; Zhang et al., 1995; Loke and Barker, 1996a) and

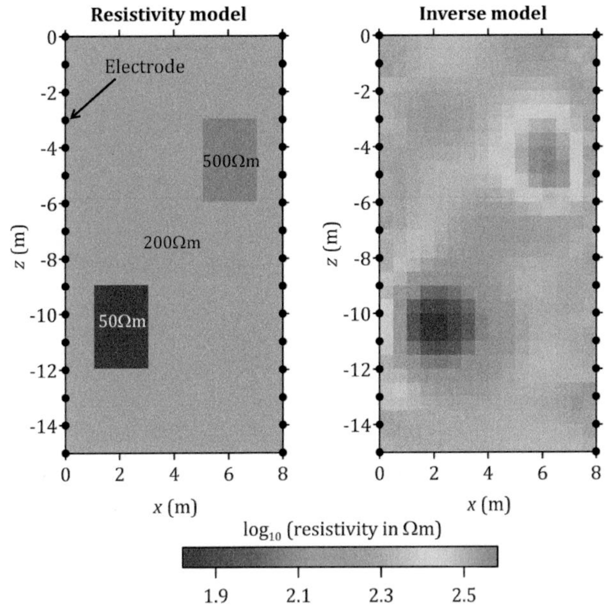

Figure 5.13 Inversion of synthetic 2D cross-borehole resistivity model. (A black and white version of this figure will appear in some formats. For the colour version, please refer to the plate section.)

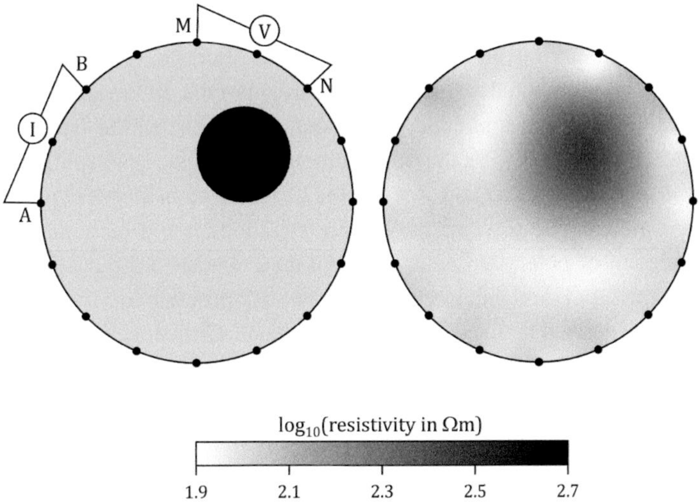

Figure 5.14 Inversion of 2D resistivity in a circular geometry. The left-hand image shows the synthetic model. Measurements are modelled in a dipole–dipole geometry, as illustrated. The right-hand image shows the inversion. Note that for the inversion a different mesh (with similar discretization) was used to avoid biasing of the inverse model by the boundary of the resistive anomaly.

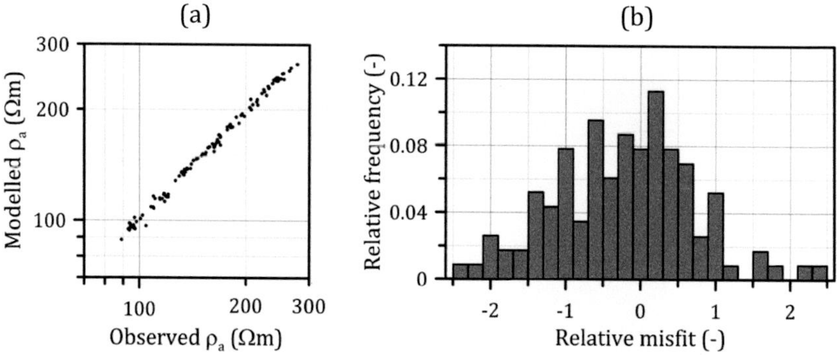

Figure 5.15 Misfit plots for inverted model in Figure 5.11. (a) Comparison of modelled and measured apparent resistivity. (b) Histogram of relative misfits.

(computationally more expensive) finite element solutions (e.g. Sasaki, 1994; Binley et al., 1996c; LaBrecque et al., 1999). Early applications were constrained by computational resources, particularly for calculation and storage of the Jacobian matrix, as $\boldsymbol{J}$ is a full matrix and must be computed at each iteration of Equation 5.13. Note that $\boldsymbol{J}$ is of size $N \times M$, and so for a parameter mesh with $100 \times 100 \times 50$ parameter cells, and a modest dataset of 1,000 measurements, the required storage of $\boldsymbol{J}$, using the necessary 16-byte precision, is 8Gb. Such demands were unrealistic for core storage on personal computers in the 1990s. Efforts addressed at reducing storage of the Jacobian include Zhang et al. (1995). Other attempts to make 3D imaging viable include the use of quasi-Newton simplifications (e.g. Loke and Barker, 1996b) in which the Jacobian at subsequent iterations was approximated based on that from the starting model (although such an approach can be unreliable). Even if the Jacobian could be computed and stored, the solution of the full linear matrix equation in Equation 5.13 also restrained applications to modest grid sizes, and was typically carried out using iterative techniques, such as the preconditioned conjugate gradient method.

Note that the parameter mesh does not need to be the same as the mesh used for solution of the forward model. The forward model discretization should be chosen to ensure an accurate calculation of potential fields; by adopting a coarser discretization for the parameter mesh, the computational demands on the inverse problem are reduced. Thus, parameter cells may be defined by the boundary of groups of adjacent cells/elements within the forward modelling mesh. Günther et al. (2006) illustrate three levels of discretization: coarse parameterization, finer discretization for the potential field and a very fine discretization for secondary voltage computation allowing use of singularity removal techniques (see earlier section on forward modelling).

Most of the early finite element based inverse solutions used structured meshes, offering limited (if any) advantage over computationally simpler finite difference methods. Finite element solutions on unstructured grids were well established from the 1970s but it was the availability of 3D meshing tools, along with a significant increase in in-core computer storage on personal computers, which permitted wider application. This increase in in-core storage resulted in a re-emergence of linear equation solvers that proved effective on modern computers for solving large problems. These advances were first exploited by Günther et al. (2006). Johnson et al. (2010) and Johnson and Wellman (2015) demonstrated that by exploiting the natural parallel computational elements of the inverse problem 3D resistivity inversions can be applied to very large-scale problems (e.g. 4,850 electrode positions, 208,000 measurements and over 10^6 parameter values, computed on an array of 1,024 processors).

For many large 3D imaging applications, true 3D acquisition of data is rarely carried out since few available instruments have the capability to address large electrode arrays (or, if they do, the resources to achieve this are often confined to specialized applications). A common approach is to conduct quasi-3D imaging, whereby data from multiple 2D acquisition surveys are combined and inverted in a single 3D model. Figure 5.16 illustrates such an approach using an inversion of data reported in Chambers et al. (2012). In this example, data were collected in a dipole–dipole configuration on 32 2D lines, each with 32 electrodes at 3 m spacing, resulting in a total of over 23,000 measurements (including reciprocals). Data were then combined and inverted in a single 3D model. In this case, the 3D model clearly shows the demarcation of river terrace deposits overlying a conductive clay bedrock. The case study in Section 6.1.7 illustrates how 3D resistivity imaging can be used to assess slope stability.

The density of measurements on a regular grid shown in the example in Figure 5.16 is rarely applied. Site access constraints often limit the coverage. However, by adopting an unstructured grid for modelling, more complex configurations of 2D arrays can still be combined, as illustrated in Figure 5.17. In this example, 15 2D resistivity surveys (Figure 5.17a) were carried out in a karstic region of southwest China in order to assess 3D

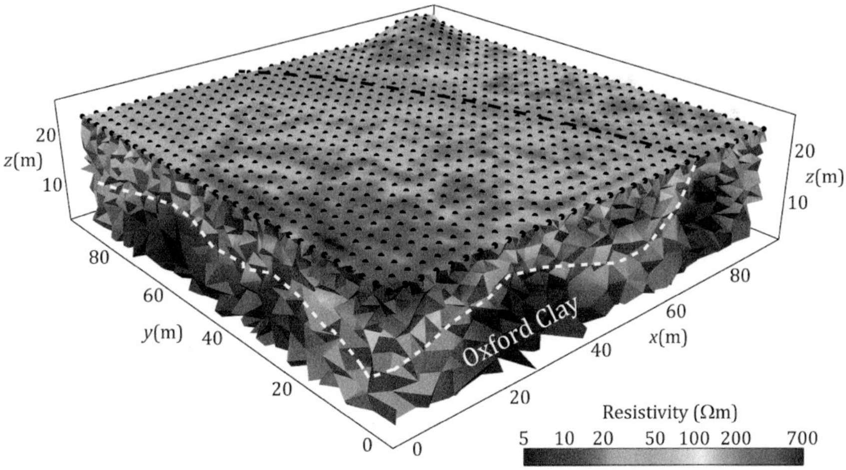

Figure 5.16 3D resistivity imaging of river terrace deposits after Chambers et al. (2012). Electrodes are shown by black circles. The white dashed line marks the interpreted interface of resistive river terrace deposits overlying conductive Oxford Clay. The black dashed line marks a horizontal boundary of deposition. (A black and white version of this figure will appear in some formats. For the colour version, please refer to the plate section.)

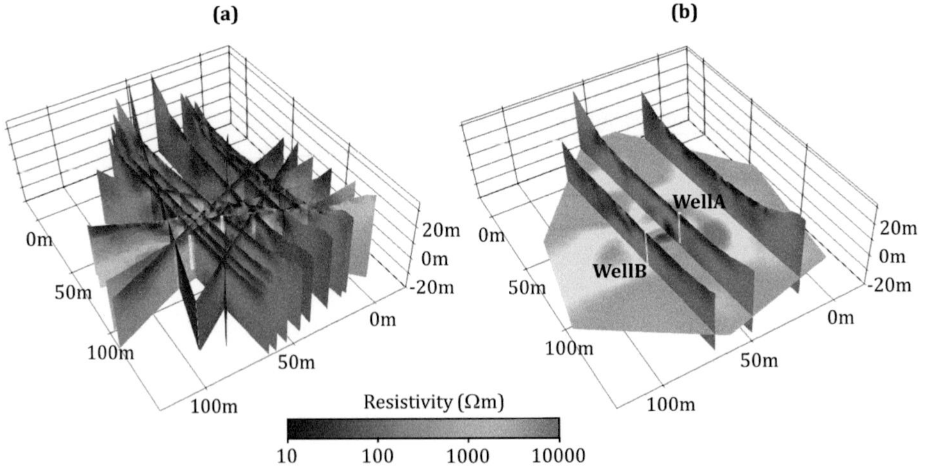

Figure 5.17 Quasi-3D inversion of resistivity data at a karst site in southwest China. (a) Location of 2D survey lines and sections through the 3D resistivity model. (b) Simplified extract of the 3D model showing localized resistivity variation consistent with observed artesian conditions at well A (which were not observed at well B). For more information, see Cheng et al. (2019a). (A black and white version of this figure will appear in some formats. For the colour version, please refer to the plate section.)

variability of resistivity that could help hydrologists understand why two closely spaced wells revealed completely different hydraulic responses (Cheng et al., 2019a). Data from the 2D surveys were combined, resulting in approximately 7,000 measurements for the set of 559 unique electrode sites. 3D inversion was then performed on a mesh of over 700,000 parameters. Figure 5.17b shows sections of the inverse model, which reveals the localized nature of the (hydraulically conductive) low-resistivity zones within the study area.

The ability to image 3D resistivity structures in complex topography using large unstructured meshes stimulated interest in imaging volcanoes (e.g. Revil et al., 2010). The resistivity variation in such applications is large (Soueid Ahmed et al., 2018); 3D imaging offers a means of enhancing knowledge of geological and tectonic features within these complex systems.

Figure 5.18 illustrates application of the same inverse modelling approach to a simple column experiment. For this demonstration, 96 stainless steel electrodes were installed around the perimeter of a water (85 Ωm resistivity) filled acrylic 6.4 cm diameter cylinder (Figure 5.18a). A 2 cm diameter plastic tube was installed within the cylinder, serving as a resistive target. Measurements were then made using a short dipole configuration, arranged in a radial geometry (Figure 5.18c). Using a 3D triangular prism mesh (Figure 5.18b), smoothness-based inversions were carried out on the dataset. Despite using a smoothness-based inversion, the inverse model resolves the resistive target well (Figure 5.18c).

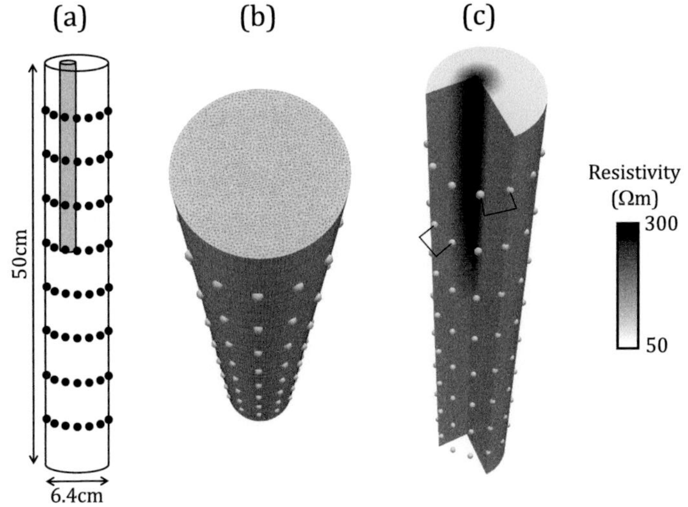

Figure 5.18 3D inverse modelling of a laboratory column experiment. (a) Arrangement of electrodes around a cylindrical vessel and position of resistive target within the water filled column. (b) Mesh discretization. (c) Inversion of measurements with example quadrupole configuration shown.

5.2.2.7 Accounting for Electrical Anisotropy

As mentioned earlier, the electrical properties of the subsurface may be considered anisotropic at a particular scale. In applications of cross-borehole resistivity imaging, LaBrecque and Yang (2001a) noted instances of horizontal banding and attributed this to anisotropy in the electrical properties. Rödder and Junge (2016) illustrate how data from an anisotropic system impacts the interpretation of an inverse model based on isotropic assumptions. Some researchers have attempted to formulate the inverse problem in terms of anisotropic resistivity (e.g. Herwanger et al., 2004b; Kim et al., 2006; Zhou et al., 2009). However, such methods are inevitably fraught with non-uniqueness given the additional parameter degrees of freedom. In some cases, sufficient parameter discretization may remove the need to formulate in this way, since anisotropy can be considered as a different scale of heterogeneity. In studies of fractured rock environments, anisotropy is often an inevitable consequence of small-scale macroporosity, which will often exist at a smaller scale than the model parameter discretization (e.g. Herwanger et al., 2004b), necessitating the use of anisotropic solutions to the inverse problem. In cases where anisotropic conditions could exist, measurements made on quadrupoles of different orientations (e.g. borehole and surface) may help diagnose the level of anisotropy present (Greenhalgh et al., 2010).

5.2.2.8 Enhancing the Regularisation

We have focused much of the discussion and examples so far on regularization that ensures isotropic smoothing, as this is the most commonly used approach. In Section 5.2.2.5, we explained how more abrupt changes in resistivity can be accounted for using an L_1 norm based equivalent inversion (see also Figure 5.12). However, a range of other approaches are available.

Attempts have been made to refine regularization spatially within a grid of parameters. Morelli and LaBrecque (1996) argued that the extent of regularization should be reduced in areas of lower data sensitivity. In contrast, Yi et al. (2003) proposed a method called active constraint balancing, which essentially enhances regularization in such areas, using the resolution matrix (see Section 5.2.4.2). The preferred approach of many, however, is to apply regularization uniformly within the grid of parameter cells, accounting for grid cell sizes as shown in Figure 5.6.

The regularization used should reflect *a priori* information about the region under investigation. Isotropic smooth models may be consistent with prior knowledge; however, in sedimentary environments, anisotropic models may be more consistent. This is easily accounted for using the methods discussed so far. For a 2D regularization, as shown in Figure 5.6, the horizontal and vertical smoothing can be separated and consequently enhancement to smoothing can be applied in different orientations. Figure 5.19 illustrates the effect of anisotropy in regularization, which can be compared with the isotropic case in Figure 5.11. In this case, the roughness operator, $\alpha \boldsymbol{R}$, is expressed as $\alpha_x \boldsymbol{R}_x + \alpha_z \boldsymbol{R}_z$.

In addition to L_1 norm-type approaches, a number of methods have been developed to enhance regularization, often in an iterative manner. The minimum gradient support approach of Portniaguine and Zhdanov (1999) minimizes the area where significant model parameter variations and discontinuities occur, thus resulting in sharp contrasting electrical images (see also Blaschek et al., 2008; Nguyen et al., 2016). Barboza et al. (2019) offer an adaptive method that adjusts regularization within different parts of the modelled region; Bouchedda et al. (2012) describe an approach using edge detection (see also Section 5.2.2.9) iteratively within an inversion.

The regularization operator can be easily modified to account for known discontinuities by removing smoothing completely at such locations. Such conditions may exist, for example, at the positions of a water table, or known geological boundaries. Slater and Binley (2006) adopted this approach for electrical imaging of permeable reactive barriers, where, in this case, the engineered structure boundaries were known (see case study in Section 6.2.6). In some cases, additional data from other geophysical surveys (e.g. ground-penetrating radar or seismic methods) can provide locations to constrain regularization (Doetsch et al., 2012b; Zhou et al., 2014). Clearly, inappropriate designation of disconnected regions is likely to lead to erroneous inverse model results and may even result in failure to converge, and thus must be carefully applied.

Figure 5.20 shows the effect of applying disconnection in the regularization operator. In this case, the model shown in Figure 5.11 is inverted using a roughness matrix that removes smoothing along the boundary of the two anomalies. Thus, three zones with normal smoothing exist. Note that a similar boundary could have been applied to the shallow overburden layer. The effect of disconnecting the regularization in this case is striking, demonstrating how effective such *a priori* information can be. However, as stated earlier, the application of such constraints must be evidence-based and used with caution.

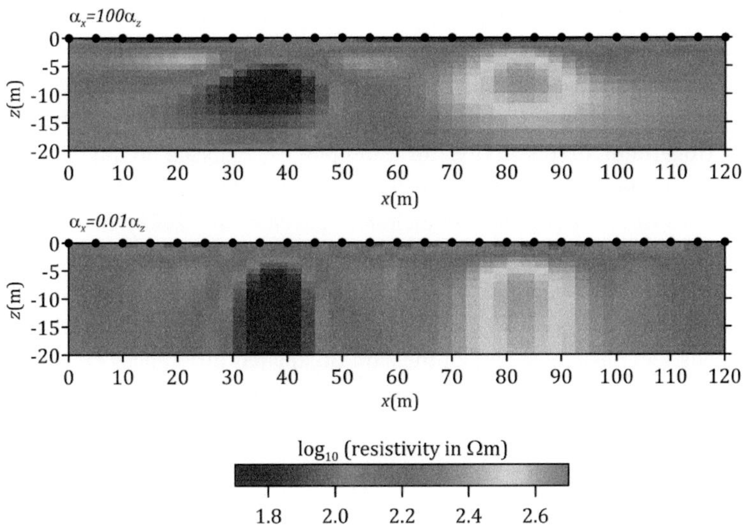

Figure 5.19 Effect of anisotropy in regularization. The upper image shows the case for $\alpha_x = 100\alpha_z$; the lower image shows the case for $\alpha_x = 0.01\alpha_z$. The true model is shown in Figure 5.11a. (A black and white version of this figure will appear in some formats. For the colour version, please refer to the plate section.)

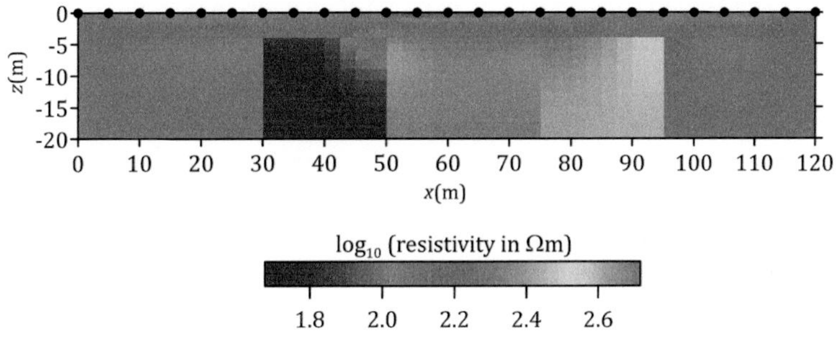

Figure 5.20 Effect of regularization disconnection. The true model is shown in Figure 5.11. (A black and white version of this figure will appear in some formats. For the colour version, please refer to the plate section.)

More sophisticated use of *a priori* knowledge includes the addition of geostatistical functions as a regularization operator. Linde et al. (2006) demonstrate such an approach in their inversion of cross-borehole resistivity data, using a geostatistical model based on electromagnetic induction well logs. Johnson et al. (2012b) adopted a similar approach, constraining resistivity values at boreholes, based on well-logged values.

5.2.2.9 Post-Processing of Inverse Models

Rather than enhancing the regularization operator to create greater contrasts in an inverse model, one can apply edge detection methods to standard L_2 norm inverse models, i.e. the task of sharpening an image is done as a post-processing step. Given the advancement in image processing techniques, a number of approaches are available and have been applied to electrical geophysical problems. Perhaps the simplest approaches, conceptually, are those based on maximum gradient analysis of the inverse model. In such methods, the grid of parameters (logarithm of resistivity, for example) is analysed to find lines or planes demarking points of greatest change in the parameter. Examples of the application of such approaches to resistivity models include Nguyen et al. (2005) and Chambers et al. (2012). An alternative approach is cluster analysis. The k-means method is one such approach that has been widely used in a range of geospatial analysis problems. In this approach, a number of clusters are defined by the user and an algorithm seeks to group the parameter cells into clusters such that the sum of squared differences between parameter cell location and cluster centre is minimized. Increasing the number of clusters will reduce such a misfit and thus the user has to determine an adequate level of clustering. Ramirez et al. (2005), Melo and Li (2016) and Binley et al. (2016) illustrate the use of k-means cluster analysis of electrical inverse models. Chambers et al. (2014) compare a number of post-processing analysis techniques (including cluster analysis) in an application to shallow hydrostratigraphy.

5.2.2.10 Time-Lapse Inversion

Over the past few decades electrical methods have been widely used to study dynamic processes (as discussed in Chapter 4), e.g. to monitor changes in pore fluid conductivity due to alteration of solute concentration or changes in temperature. From a series of datasets $\boldsymbol{d}_t$, $t = 0, 1, 2, \ldots N_t$, the objective is to determine parameter sets $\boldsymbol{m}_t$, $t = 0, 1, 2, \ldots N_t$, or changes in the parameters $\delta \boldsymbol{m}_t = \boldsymbol{m}_t - \boldsymbol{m}_0, t = 1, 2, \ldots N_t$. By analysing changes in resistivity, we can remove the effect of static variation in resistivity (e.g. due to lithological effects).

The problem can be approached by inverting each dataset individually, although this can be problematic as often the spatial variation of the parameter (e.g. resistivity) within each inverse model is significantly greater than the temporal variation of interest. Each inverted dataset will be affected by the iterative inverse sequence, e.g. level of regularization. Consequently, any temporal variation (the signal of interest) may be too subtle to be resolved from comparison of two independent inversions. Ideally, we would invert a change in measurement to determine the change in parameter. Daily et al. (1992) first approached this using a 'ratio method' (although they don't actually refer to the method explicitly). The method has proven to be extremely robust and effective for a wide range of resistivity imaging applications (e.g. Daily and Ramirez, 1995; Binley et al., 1996a; Slater et al., 1996; Ramirez et al., 1996; Zaidman et al., 1999; Daily and Ramirez, 2000). A number of cases studies in Chapter 6 further illustrate the use of time-lapse imaging.

The approach taken is to create a new dataset, $\boldsymbol{d}_{rat}$, from the ratio of a pair of datasets, scaled by the forward model for a uniform model, $\boldsymbol{m}_{hom}$, i.e.

$$\boldsymbol{d}_{rat} = \frac{\boldsymbol{d}_t}{\boldsymbol{d}_o} F(\boldsymbol{m}_{hom}) \cdot \tag{5.30}$$

The dataset, d_{rat}, can then be inverted using Equation 5.25. The choice of m_{hom} is arbitrary but typically a uniform resistivity of 100 Ωm is used. Increases in resistivity between the two times are reflected by values above m_{hom}, and decreases are revealed by values lower than m_{hom}. While this may be considered to be rather qualitative, the method has proven to be effective in showing relative changes in resistivity from a reference case. Implicit in the approach is that the Jacobian matrix computed by the parameter set is a valid approximation to the true sensitivity matrix. For relatively uniform resistivity cases, this will be justified. Note also that the two datasets must be the same size, although even if analysis of individual inversions is considered, bias from analysing different electrode configurations in the two datasets should be avoided.

One aspect often overlooked in time-lapse imaging is the choice of data errors, i.e. the data weight matrix W_d. Since datasets d_0 and d_t are subject to error, which we commonly assume are random and Gaussian, then since the data variances compound in an operation like Equation 5.30, the data errors for d_{rat} should reflect this. However, it is typically observed that such increased error levels result in significant under-fitting. This is likely to be the result of a systematic error component in data errors that is effectively removed by the transformation in Equation 5.30.

An alternative approach is to cast the regularization in Equation 5.20 in terms of changes in the parameters,

$$\Phi_m(\boldsymbol{m} - \boldsymbol{m}_0)^T \boldsymbol{R}(\boldsymbol{m} - \boldsymbol{m}_0). \tag{5.31}$$

LaBrecque and Yang (2001b) utilized this in their difference inversion scheme, which adopts the following modification to Equation 5.25:

$$(\boldsymbol{J}^T \boldsymbol{W}_d^T \boldsymbol{W}_d \boldsymbol{J} + \alpha \boldsymbol{R})\Delta \boldsymbol{m} = \boldsymbol{J}^T \boldsymbol{W}_d^T \boldsymbol{W}_d \Big((\boldsymbol{d} - \boldsymbol{d}_0) - (F(\boldsymbol{m}_k) - F(\boldsymbol{m}_0)) \Big) - \alpha \boldsymbol{R}(\boldsymbol{m}_k - \boldsymbol{m}_0). \tag{5.32}$$

This approach effectively removes the effect of systematic data errors, and has proved effective in a number of time-lapse imaging studies (e.g. LaBrecque et al., 2004; Doetsch et al., 2012a; Yang et al., 2015). In order to apply Equation 5.32, the parameter set m_0 is first evaluated from the inversion of dataset d_0. The inversion of dataset changes is then carried out in order to derive the parameter set m_t (note that Equation 5.32 does not determine the change in parameter set, m_t, directly, although it is easily computed).

Figure 5.21 shows the result from a difference inversion applied to data collected using a 64-electrode permanent resistivity array installed within a riparian wetland. Data were collected at monthly intervals, using a Wenner configuration (see Musgrave and Binley, 2011). Figure 5.21 shows the reference inversion based on the dataset collected at the start of the monitoring. This inversion was then used as the reference model, m_0 and successive datasets were inverted using the difference inversion approach discussed above (Equation 5.32). Example time-lapse changes in resistivity are shown in Figure 5.21; convergence was achieved in one iteration in all time steps. The reference inversion shows low resistivity in the upper metre due to peat soils, which overlies a more variable resistive zone due to chalk/

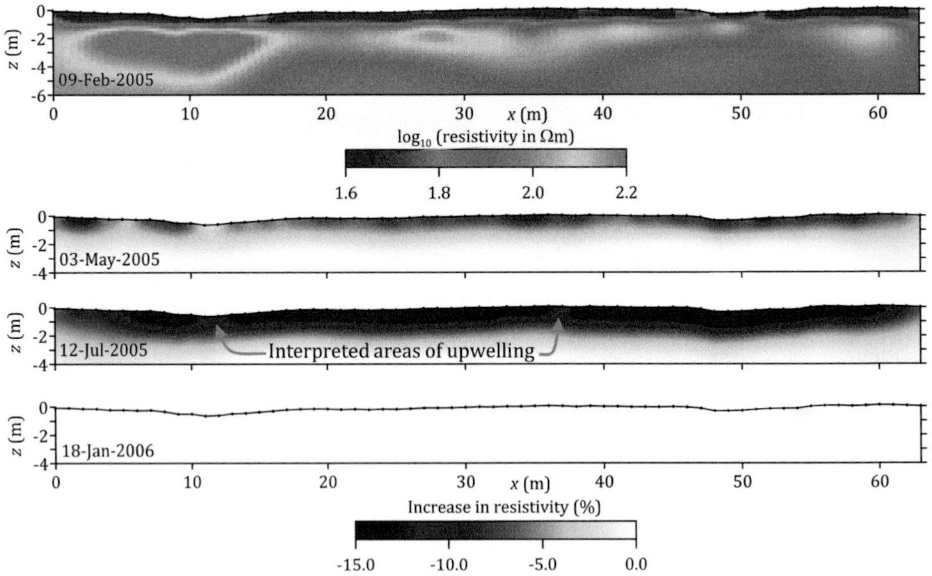

Figure 5.21 Time-lapse imaging of changes in resistivity in a riparian wetland. The upper figure shows the reference inversion. The lower figures show changes in resistivity along the transect, computed using a difference inversion. The symbols show the position of electrodes. Inversions carried out using data reported in Musgrave and Binley (2011). (A black and white version of this figure will appear in some formats. For the colour version, please refer to the plate section.)

flint gravels, which in turn overly chalk of intermediate resistivity. Musgrave and Binley (2011) provide observations from shallow drilling that support this interpretation. The time-lapse results reveal a decrease in resistivity during summer months, followed by a return to conditions similar to the reference case. Independent sampling from dip wells installed along the transect revealed that pore water conductivity and water table levels remained reasonably static over the 12-month period (see Musgrave and Binley, 2011); however, temperature increases of 7°C were recorded, which explains the decrease in resistivity (a 7°C change in temperature can result in 14 per cent change in resistivity; see Chapter 2). Of significance in the context of the study, localized zones of suppression of temperature (and hence resistivity) change (e.g. 11 m and 37 m along the 12 July 2015 transect in Figure 5.21) can be interpreted as regions of (reasonably constant temperature) groundwater upwelling.

A further alternative time-lapse approach (e.g. Oldenburg et al., 1993; Oldenborger et al., 2007) is to adopt an additional penalty function by modifying the objective function in Equation 5.21 to

$$\Phi_{total} = \Phi_d + \alpha\Phi_m + \alpha_t\Phi_t, \tag{5.33}$$

where α_t is a scalar that weights a penalty relating to the change in parameter values from a reference, e.g.

$$\Phi_t = (\boldsymbol{m} - \boldsymbol{m}_0)^T(\boldsymbol{m} - \boldsymbol{m}_0). \tag{5.34}$$

A logical extension is to form a series of equations that allow spatial regularization and temporal regularization across a number of datasets (Kim et al., 2009). For example, given reference dataset $\boldsymbol{d}_0$ and two subsequent datasets $\boldsymbol{d}_1$ and $\boldsymbol{d}_2$, we can formulate the time-lapse problem as the solution of (cf Equation 5.25):

$$(\boldsymbol{J}^T\boldsymbol{W}_d^T\boldsymbol{W}_d\boldsymbol{J} + \alpha\boldsymbol{R})\Delta\boldsymbol{m} = \boldsymbol{J}^T\boldsymbol{W}_d^T\boldsymbol{W}_d(\boldsymbol{d} - F(\boldsymbol{m}_k)) - \alpha\boldsymbol{R}(\boldsymbol{m}_k - \boldsymbol{m}_0) \tag{5.35}$$

with data vector $\boldsymbol{d} = [\boldsymbol{d}_1,\ \boldsymbol{d}_2]$, parameter vector $\boldsymbol{m} = [\boldsymbol{m}_1,\ \boldsymbol{m}_2]$, and reference parameter set $\boldsymbol{m}_0$, which contains two duplicates of the reference model. The Jacobian in Equation 5.35 is

$$\boldsymbol{J} = \begin{bmatrix} \boldsymbol{J}_1 & 0 \\ 0 & \boldsymbol{J}_2 \end{bmatrix}, \tag{5.36}$$

where $\boldsymbol{J}_1$ and $\boldsymbol{J}_2$ are Jacobians computed for $\boldsymbol{d}_1,\ \boldsymbol{m}_1$ and $\boldsymbol{d}_2,\ \boldsymbol{m}_2$, respectively, and the roughness matrix is given by

$$\boldsymbol{R} = \boldsymbol{R}_{x,y,z} + \frac{\alpha_t}{\alpha}\boldsymbol{R}_t, \tag{5.37}$$

where $\boldsymbol{R}_{x,y,z}$ is a block diagonal matrix containing spatial smoothing coefficients (as in the standard inversion), $\boldsymbol{R}_t$ contains –1 and 1 (see e.g. Hayley et al., 2011) to link corresponding elements of $\boldsymbol{m}_1$ and $\boldsymbol{m}_2$ (all other elements are 0), and α_t is a scalar that weights the temporal regularization.

The problem is now formulated as a joint spatiotemporal inversion, which can be extended to multiple time-lapse datasets, although the computational demands clearly grow. Authors have argued improvements in performance of such combined spatiotemporal regularization, although the optimum number of datasets to consider at once is not clear, and the choice of temporal regularization α_t appears to be somewhat subjective (e.g. Kim et al., 2009) or subject to further method development (e.g. Karaoulis et al., 2011, 2014). Nevertheless, there has been a recent growth in applications of true 4D imaging of resistivity (e.g. Zhang and Revil, 2015; Uhlemann, et al., 2017) and further demonstrations will no doubt emerge.

5.2.3 The Impact of Measurement and Model Errors

In Chapter 4, we discussed methods for estimation of measurement errors. The matrix $\boldsymbol{W}_d$ in Equation 5.12 was included to allow measurements to be weighted according to their reliability. As stated earlier, since we assume uncorrelated data errors, $\boldsymbol{W}_d$ is a diagonal matrix with each entry equal to the reciprocal of the standard deviation of a measurement. This also assumes that the model is error free. Applying weights to measurements allows differential weighting of poor/good data and also allows a definition of convergence:

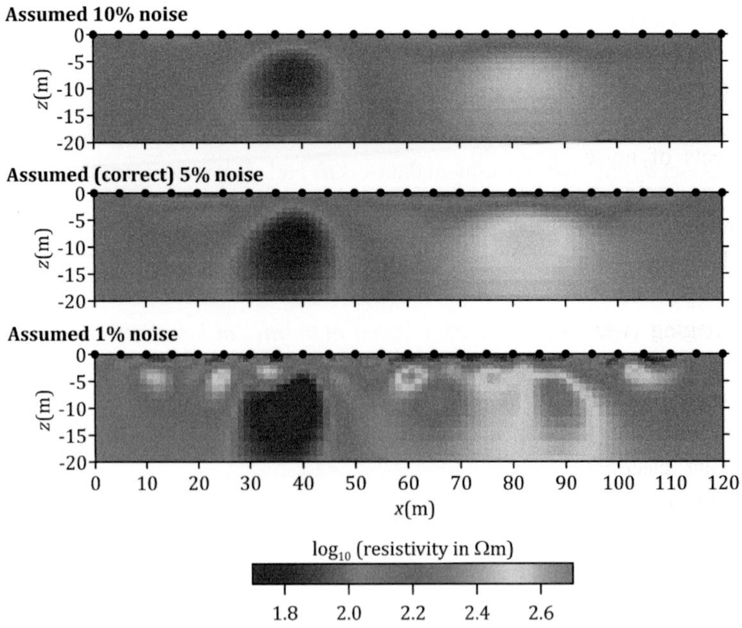

Figure 5.22 The impact of incorrect error estimation on an inverse model. The true model is shown in Figure 5.11. (A black and white version of this figure will appear in some formats. For the colour version, please refer to the plate section.)

$\Phi_d = N$, where N is the number of measurements. However, such a criterion is often not adopted and typically the user reports an equivalent uniform data error that the final model represents.

Despite the significance of data errors in an inversion, very few studies have examined their impact on inverse models. Binley et al. (1995) and LaBrecque et al. (1996b) are rare examples, both revealing how incorrect data error estimation can affect the final model. To illustrate the significance of correct data weighting, the model in Figure 5.11 was simulated using a dipole–dipole configuration (as before) and perturbed with 5 per cent Gaussian noise. Then, inversions were carried out assuming (i) 10 per cent noise (i.e. the values in $\boldsymbol{W}_d$ are set smaller than they should be; measurements are assumed to be lower quality); (ii) 5 per cent noise (i.e. the entries in $\boldsymbol{W}_d$ are correct); and (iii) 1 per cent noise (i.e. data weights are larger than they should be). Figure 5.22 shows the inversion results. When the correct noise level is assumed, the model is reasonably recovered, albeit, as expected, with slightly less contrast than the 2 per cent noise case in Figure 5.11. If we assume greater noise (in this case 10 per cent) than the true level, then not all the information in the data is recovered by the inversion. In contrast, if the noise level is underestimated (in this case 1 per cent), then higher variability results in the final model due to 'over-fitting', i.e. the inversion attempts to fit variation in signal due to noise. This illustrates the importance of (i) assessing error levels (as discussed in Chapter 4), and (ii) accounting for such error levels in the inversion process. Clearly, in this example the true model is known and thus the impact of errors is clearly seen;

however, in a real dataset, over-fitting may lead to incorrect interpretation of the subsurface electrical structure. In contrast, under-fitting from over-estimation of errors can result in failure to exploit all the information in the data. The above comments apply not only to imaging but also electrical sounding.

The effects of noise have probably been widely overlooked in resistivity imaging because many applications have been carried out with surface arrays, with good electrical contact, and measurement configurations less prone to noise (e.g. Wenner, Schlumberger, gradient). However, for configurations susceptible to elevated noise (e.g. dipole–dipole), in conditions of poor electrical contact and/or in more specialized applications, such as cross-borehole imaging (where a greater range of measured voltages and contact resistances commonly exist), understanding data quality is often critical for successful application.

So far we have concentrated on measurement quality, but we should also recognize that the models are subject to errors (e.g. discretization errors or conceptual errors, such as incorrectly assumed point electrode sources, positioning errors, inappropriate assumptions of two-dimensionality (in 2D models), or borehole effects (see Chapter 3 and case study in Section 6.1.8)). In the example above, the inversion employed the same forward modelling approach used to synthesize the data (committing what is commonly referred to as an 'inverse crime' (Colton and Kress, 1992)) and thus forward modelling errors are insignificant since any discretization error present in the generation of the dataset is reproduced in the forward modelling stage of the inversion. This will not be the case for real data. In fact, as instruments and operational procedures (e.g. quality assurance checks) improve and data quality is enhanced, the role of model errors can begin to dominate. We may find that errors in the forward model need to be recognized and correctly accounted for in the data misfit objective function. This is particularly the case for 3D problems, where, due to computational constraints on discretization, forward modelling errors can be high unless appropriate steps are taken (as discussed earlier).

Assessment of forward modelling errors due to discretization for half space problems with no topography is easily done using analytical solutions for homogenous problems (see e.g. Figure 5.4). For more complex problems mesh refinement can be used in order to determine a reference 'accurate' solution for forward modelling error computation. Computation of the modelling error in a measurement can then be accounted for in the data weight matrix $\boldsymbol{W}_d$ in Equation 5.12 and subsequent equations, using

$$\boldsymbol{W}_{d,i} = 1 \Big/ \sqrt{\varepsilon_{R,i}^2 + \varepsilon_{M,i}^2}, \tag{5.38}$$

where $\varepsilon_{R,i}$ is the measurement error (see Section 4.2.2.1) and $\varepsilon_{M,i}$ the modelling error.

5.2.3.1 Robust Inversion

In an effort to address the influence of data errors, attempts have been made to develop schemes to refine the data weight matrix $\boldsymbol{W}_d$ during the inversion. As noted by Farquharson and Oldenburg (1998), the least squares data objective function (Equation 5.12) is susceptible to outliers and non-Gaussian noise, whereas an L_1 formulation, which minimizes the absolute values of data misfit, can be more attractive in such situations. The latter has been referred to as robust inversion, following Claerbout and Muir (1973). Morelli and LaBrecque (1996) developed an adaptive data weighting scheme, based on Mostellar and Tukey (1977), that allows the inversion process to adjust weights throughout the inversion process to achieve an L_1-like data misfit function. Farquharson and Oldenburg (1998) propose a similar iterative reweighting scheme (and also include a means of refining the model misfit to evolve into a pseudo L_1 formulation). Such reweighting schemes can be effective in guaranteeing convergence, since data outliers (which cause the problem with L_2 schemes) become weighted less, but they must be used with care. Outliers may be present because the forward model assumptions are incorrect (e.g. the assumption of two-dimensionality), and thus it is advisable to analyse final data weights after the inversion is completed, as this may reveal systematic errors, or incorrect assumptions.

5.2.4 Inverse Model Appraisal

5.2.4.1 General Concepts

As shown in Figure 4.5, a sensitivity pattern exists for each DC resistivity measurement, and this pattern is relatively complex and complicated further if there is significant variation in resistivity (the patterns shown in Figure 4.5 were computed for a uniform resistivity). The sensitivity pattern helps us address two important tasks. First, we can use it to help design a survey, e.g. select the appropriate electrode configuration and electrode spacing given the objectives of a survey. This was discussed in Chapter 4. The second task is to assess the reliability of the inverse model. We discussed earlier equivalence in 1D inversions (sounding); here we focus on appraisal of 2D and 3D images.

It is common practice to present inverse models from 2D surface electrode surveys using a trapezium boundary, i.e. we clip a left- and right-most triangle of the image of resistivity (e.g. Figure 5.1). For example, in Figure 5.11 the resistivity is presented in some areas where we clearly have little or no sensitivity (the pseudosection plots in Figure 4.21 illustrate some of the spatial coverage, albeit in a qualitative manner). Although such clipping of the images is useful, it presents a rather binary perspective of the inverse model, implying, perhaps, that the region of the model that is not clipped is perfectly known. The user may be fully appreciative that this is not the case, but must remember that others (non-specialists), who use the results, may not. The experienced geophysicist will be used to 'filtering' parts of the image visually in their interpretation. However, rarely are any uncertainties in the computed model presented. Furthermore, clipping 2D images from surface electrode data is relatively trivial, but for other configurations (e.g. cross-borehole, surface-borehole, 3D, etc.) such filtering is not so straightforward. A number of approaches are available, and are discussed below.

5.2.4.2 Model Resolution Matrix Approaches

One approach for model appraisal, which is widely appreciated in general inverse theory (e.g. Menke, 2015), is the model resolution matrix, $\boldsymbol{R_m}$, which describes the mapping of data and model space, and is defined by

$$\boldsymbol{m} = \boldsymbol{R}_m \boldsymbol{m}_{true}, \quad (5.39)$$

where $\boldsymbol{m}$ is the inverted parameter set and $\boldsymbol{m}_{true}$ is the (unknown) true parameter set. Clearly, ideally $\boldsymbol{R}_m = \boldsymbol{I}$, any deviation reveals the lack of sensitivity of the parameter values to the measured data, manifested by regularization.

From the formulation in Equation 5.25, $\boldsymbol{R}_m$ can be approximated by the solution of

$$(\boldsymbol{J}^T \boldsymbol{W}_d^T \boldsymbol{W}_d \boldsymbol{J} + \alpha \boldsymbol{R})\boldsymbol{R}_m = \boldsymbol{J}^T \boldsymbol{W}_d^T \boldsymbol{W}_d \boldsymbol{J}, \quad (5.40)$$

where the Jacobian, $\boldsymbol{J}$, has been computed using the final (inverted) parameter set and the regularization scalar, α, is the value at the end of the inversion. The determination of $\boldsymbol{R}_m$ using Equation 5.40 requires significant computational effort: the formation and solution of M sets of equations, each of size $M \times M$, where M is the number of parameters. Note that the definition of the model resolution matrix following Equation 5.40, strictly speaking, only applies to a linear inverse problem; however, it is commonly assumed that it can be adopted for the linearization of the non-linear problem here.

The simplest way to use the resolution matrix is to display the diagonal value for each parameter (Figure 5.23), which should be unity for perfectly resolved parameters. Stummer et al. (2004) recommend 0.05 as a cut-off value, although this is somewhat subjective. Ramirez et al. (1993) report one of the earliest illustrations of the model resolution matrix, expressed as a 'resolution radius', which they define as a distance over which parameters are smoothed. Alumbaugh and Newman (2000) illustrate a similar concept using the Backus and Gilbert (1970) point spread function, which is the row (or column) of $\boldsymbol{R}_m$ for a given parameter. This can be shown for all parameters as a distance, by assessing, for example, the spatial extent of 50 per cent of the diagonal entry of $\boldsymbol{R}_m$ for each parameter (Alumbaugh and Newman, 2000). Oldenborger and Routh (2009) report more detailed analysis of the point spread function for a 3D resistivity imaging problem. Day-Lewis et al. (2005) used the model resolution matrix to examine how well the geostatistical properties of the subsurface are recovered through inversion of DC resistivity data.

Because of the computational burden of computing $\boldsymbol{R}_m$, particularly for 3D problems, it is rarely reported. Park and Van (1991) and Kemna (2000) offer an alternative cumulative sensitivity matrix, which is much easier to compute and is given by

$$\boldsymbol{S} = \boldsymbol{J}^T \boldsymbol{W}_d^T \boldsymbol{W}_d \boldsymbol{J}. \quad (5.41)$$

Figure 5.23 shows a comparison of the diagonal of $\boldsymbol{R}_m$ and $\boldsymbol{S}$ for the 2D imaging problem in Figure 5.21, showing similarity in pattern. The cumulative sensitivity map is extremely useful as a qualitative guide, particularly as it is easy to compute. Kemna (2000) recommends that a value of $\boldsymbol{S}$ equal to 10^{-3} of the maximum value of the diagonal of $\boldsymbol{S}$ is a useful guide for demarcation of sensitivity (see Figure 5.23b). As shown in Figure 5.23c, the sensitivity map (or resolution matrix) can be effective at highlighting uncertainty in the inverse model through an opaqueness filter. In this example, a graded opaqueness is applied for all parameter cells with $\boldsymbol{S}$ less than 10^{-3} of the maximum value.

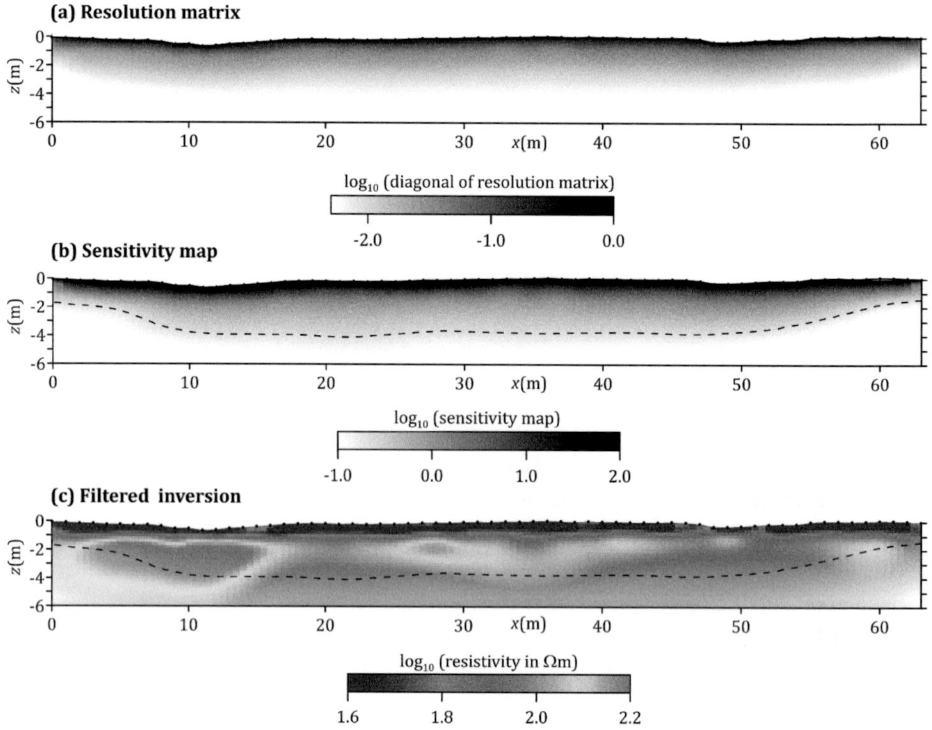

Figure 5.23 (a) Model resolution matrix (Equation 5.40). (b) Cumulative sensitivity map (Equation 5.41) for inversion shown in Figure 5.21. (c) Replotted image in Figure 5.21 using the sensitivity map to change opaqueness of the image (c.f. Figure 5.21). The dashed line in (b) and (c) shows the sensitivity map equal to 10^{-3} of its maximum value (see text). (A black and white version of this figure will appear in some formats. For the colour version, please refer to the plate section.)

5.2.4.3 Depth and Volume of Investigation

An alternative model appraisal approach was developed by Oldenburg and Li (1999), which uses a formulation similar to Equation 5.35 with the additional α_s term that constrains the inversion to the reference model:

$$(\boldsymbol{J}^T \boldsymbol{W}_d^T \boldsymbol{W}_d \boldsymbol{J} + \alpha \boldsymbol{R} + \alpha_s \boldsymbol{I})\Delta \boldsymbol{m} = \boldsymbol{J}^T \boldsymbol{W}_d^T \boldsymbol{W}_d(\boldsymbol{d} - F(\boldsymbol{m}_k)) - \alpha \boldsymbol{R} \boldsymbol{m}_k - \alpha_s \boldsymbol{m}_0. \quad (5.42)$$

The problem is solved for two reference models $\boldsymbol{m}_0^{(a)}$ and $\boldsymbol{m}_0^{(b)}$, and computes a depth of investigation, DOI, for each parameter cell, i, from

$$DOI_i = \frac{m_i^{(a)} - m_i^{(b)}}{m_{0,i}^{(a)} - m_{0,i}^{(b)}}. \quad (5.43)$$

Oldenburg and Li (1999) recommend that the reference models vary by a factor of 5 to 10. Marescot et al. (2003), in contrast, recommend reference models that span the resistivity by two orders of magnitude. Such a range could be computed from the geometric mean of the measured apparent resistivities. For parameters which are well resolved by the data, the DOI value will be close to zero since the numerator should be independent of the reference model, whereas values of DOI close to unity indicate little sensitivity to the measurement.

As noted by Miller and Routh (2007), unlike the model resolution matrix approach, the DOI method does not rely on the linearization assumption. Oldenburg and Li (1999) suggest that a value of DOI equal to 0.1 to 0.2 represents a reasonable upper threshold for satisfactory sensitivity, although, as noted by Oldenborger et al. (2007), the gradient of the DOI function may also provide valuable insight for model appraisal (see also Caterina et al., 2013). Oldenborger et al. (2007) offer a straightforward extension of the DOI method to 3D problems, expressed as a volume of investigation, VOI, but highlight the subjectivity of the DOI/VOI approach. The choice of α_s can be particularly problematic – too high a value may lead to failure to converge, whereas too low a value results in limited constraint. Marescot et al. (2003), in a similar formulation to Equation 5.42, recommend $\alpha_s = 0.01\alpha$ (see also Hilbich et al., 2009). Trials of different reference models and regularization parameters are recommended when such an approach is adopted. In an attempt to overcome some of the subjectivity, Deceuster et al. (2014) suggest an extension to the DOI approach based on the computation of a scaled probability density function. Caterina et al. (2013) compared the DOI approach to the model resolution matrix, recommending the latter for quantitative assessment of inverse models, although, as noted in Section 5.2.4.2, such calculations can be computationally demanding for large problems.

5.2.4.4 Model Covariance Matrix and Parameter Uncertainty

The model covariance matrix (e.g. Menke, 2015) provides insight into how data and model errors propagate in the inverse model. Following Alumbaugh and Newman (2000), a linearized approximation can be written as

$$\boldsymbol{C_m} = \left(\boldsymbol{J}^T \boldsymbol{W}_d^T \boldsymbol{W}_d \boldsymbol{J} + \alpha \boldsymbol{R}\right)^{-1}. \tag{5.44}$$

The diagonal entries of $\boldsymbol{C_m}$ quantify the uncertainty (variance) of parameters estimates, whereas off-diagonal terms reveal the degree of correlation between parameters (which we previously discussed in relation to vertical electrical sounding models; see Table 5.1). As in the case of the model resolution matrix, estimation of $\boldsymbol{C_m}$ requires significant computational effort.

An alternative approach is to generate realizations of the assumed noise model and perturb the dataset with such realizations in a Monte Carlo framework, inverting each perturbed dataset in the normal way (e.g. Aster et al., 2018; Tso et al., 2017). Bootstrap methods (e.g. Efron and Tibshirani, 1994) can also be used for sampling, in which case some data are resampled and some are retained in each realization (e.g. Schnaidt and Heinson, 2015). If $\boldsymbol{M_{mc}}$ is the matrix of N_{mc} realizations of parameter sets $\boldsymbol{m}_i$, I = 1, 2, . . ., N_{mc}, each of size M, then $\boldsymbol{C_m}$ can be estimated by

$$C_m = \frac{(M_{mc} - \overline{m})^T (M_{mc} - \overline{m})}{N_{mc}}, \tag{5.45}$$

where $\overline{m}$ is the mean of the model realizations. Sampling realizations in a Monte Carlo framework can be computationally demanding, although they are inherently parallelizable.

Figure 5.24b illustrates the model covariance matrix for the problem in Figure 5.13. For this example, 500 realizations of the (noisy) dataset used for inversion in Figure 5.13 were perturbed with the same level of noise. Equation 5.44 was then applied to the 500 inverted model parameters (expressed as $\log_{10}$ resistivity). The diagonal of C_m is shown in Figure 5.24b, expressed as a standard deviation of the logarithm of resistivity. Higher values indicate greater impact of the data noise on the inverted model. In this case, they show how the low-resistivity zone in the lower left of the image has greater variability within the realizations. For comparison, Figure 5.24a shows the error in log resistivity between true and inverted models in Figure 5.13. The suite of realizations produced can also be explored to examine, for example, minimum and maximum parameters for each cell. Schnaidt and Heinson (2015) also illustrate how the spatial gradient of a parameter can be effective at highlighting uncertainty in anomaly detection. For example, the low-resistivity target in the lower left of Figure 5.13 may be subject to higher uncertainty (Figure 5.24b) but all realizations may confirm presence of an anomaly from the computed gradients.

Fernández-Muñiz et al. (2019) offer an alternative approach to model uncertainty assessment. Their method, named data kit inversion, selects realizations of 'random data bags' from the field dataset, each bag consisting of between 25 per cent and 75 per cent of the full

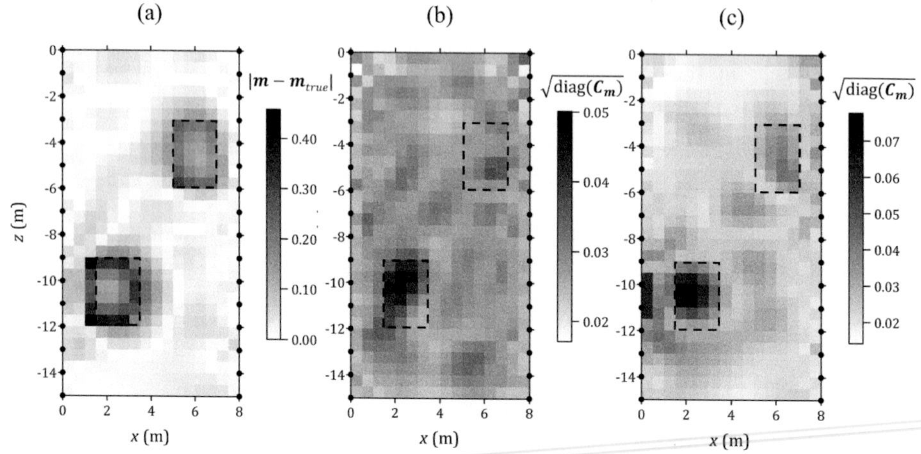

Figure 5.24 Model uncertainty for the problem in Figure 5.13. (a) Difference in $\log_{10}$ resistivity of true and inverted models. (b) Diagonal of the model covariance matrix using Monte Carlo simulations of data noise. (c) Diagonal of the model covariance matrix using data kit modelling. In (a), (b) and (c), the two rectangles with broken lines show the position of the two anomalies in the synthetic model.

dataset (the fraction used is selected at random). Fernández-Muñiz et al. (2019) claim that such an approach requires fewer sets of realizations compared to a standard Monte Carlo approach. Figure 5.24c shows the performance of such an approach for the problem in Figure 5.13. Five hundred realizations were used in this example, although almost identical behaviour was noted for a 100-realization set. The overall pattern (and magnitude of uncertainty) in Figure 5.24c is similar to that in Figure 5.24b except for the region close to the electrodes adjacent to the low left conductive anomaly. In this zone, the apparent uncertainties are high, and yet this is inconsistent with the model error (Figure 5.24a). It appears that removing some measurements in this region leads to an over-estimation of model uncertainty.

Note that the model covariance matrix discussed so far only reflects the propagation of data errors in the inversion. In the highly regularized problems discussed throughout much of this chapter, regions of low data sensitivity will be strongly influenced by the regularization operator and consequently show low uncertainty due to data errors. This should clearly not be interpreted as low overall uncertainty. In fact, in the extreme case, regions with no sensitivity to measurements will clearly have high uncertainty but this will not be reflected in the model covariance matrix as discussed so far. The total uncertainty should also reveal the impact of choice of regularization. For time-lapse inversions, the effect of uncertainty in the reference model ($\boldsymbol{m}_0$ in Equation 5.32) can also be easily accounted for by sampling realizations of the model in the same manner. Yang et al. (2014) illustrate this in a study of monitoring CO_2 sequestration using a cross-borehole configuration. However, if the same spatial regularization (e.g. L_2 norm) is maintained for all realizations, then again the full uncertainty in the model is not assessed. We discuss uncertainty later in the context of global search methods.

5.2.5 Alternative Inverse Modelling Approaches

We have focused so far on gradient-based methods since they are by far the most commonly used techniques for analysis of electrical data. Such methods are classified as deterministic since they consider the solution to be the only one that is consistent with the data. They are also referred to as local methods as they converge to a local minimum of the objective function (Figure 5.25). In this section, we discuss alternative approaches to solving the inverse problem, including the ways in which the inversion can be constrained by additional data/information.

5.2.5.1 Bayesian Methods

Rather than considering the inverse problem in a deterministic manner, we can adopt a stochastic framework in which we start with a prior probability $P(\boldsymbol{m})$ of model parameter sets and use the data $\boldsymbol{d}$ to derive a posterior probability of the model parameters following Bayes' rule (e.g. Ulrych et al., 2001):

$$P(\boldsymbol{m}|\boldsymbol{d}) = \frac{P(\boldsymbol{m})P(\boldsymbol{d}|\boldsymbol{m})}{P(\boldsymbol{d})}, \tag{5.46}$$

where the notation $P(\boldsymbol{A}|\boldsymbol{B})$ refers to the probability of event $\boldsymbol{A}$ given event $\boldsymbol{B}$.

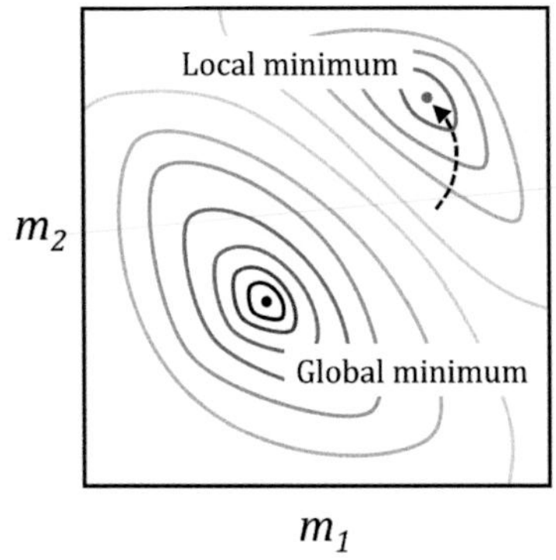

Figure 5.25 Global and local minima of an objective function. m_1 and m_2 are parameters. The contours show the objective function. The arrowed line shows the path of a gradient-based local method, missing the global minimum.

In Equation 5.46, $P(\boldsymbol{d})$ is the probability that data $\boldsymbol{d}$ are observed and is a constant given by

$$P(\boldsymbol{d}) = \int P(\boldsymbol{d}|\boldsymbol{m})P(\boldsymbol{m})d\boldsymbol{m}, \tag{5.47}$$

which ensures that $\int P(\boldsymbol{m}|\boldsymbol{d})d\boldsymbol{m} = 1$. $P(\boldsymbol{d}|\boldsymbol{m})$ in Equation 5.46 is referred to as the likelihood function, i.e. the likelihood of a dataset given a set of parameters.

Given the above, we can simplify Equation 5.46 to

$$P(\boldsymbol{m}|\boldsymbol{d}) = C\,P(\boldsymbol{m})\,L(\boldsymbol{m}), \tag{5.48}$$

where C is a scalar and the likelihood function is now expressed as $L(\boldsymbol{m})$.

The commonly used form of the likelihood function follows a least squares model (see Equation 5.12) and assumes a Gaussian distribution, allowing the expression:

$$L(\boldsymbol{m}) = \frac{1}{\left((2\pi)^N|\boldsymbol{W}_d|\right)^{1/2}}\exp\left[-\frac{1}{2}\left(\boldsymbol{d}-F(\boldsymbol{m})\right)^T\boldsymbol{W}_d^T\boldsymbol{W}_d\left(\boldsymbol{d}-F(\boldsymbol{m})\right)\right], \tag{5.49}$$

where $|\boldsymbol{W}_d|$ is the determinant of $\boldsymbol{W}_d$ and N is the number of measurements.

In the deterministic problem, we constrained the parameter model using regularization (Equation 5.20), which should be based on *a priori* information. In a stochastic inversion, we can include such constraints in the prior model, e.g. following Ulrych et al. (2001),

$$P(\boldsymbol{m}) = \left(\frac{\eta}{2\pi}\right)^{(M-1)/2}\exp\left[-\frac{\eta}{2}\boldsymbol{m}^T\boldsymbol{R}\boldsymbol{m}\right], \tag{5.50}$$

where M is the number of parameters and η is a parameter that controls the smoothness (c.f. α in the deterministic case).

The problem can then be solved to determine the parameter set corresponding to the maximum likelihood (e.g. Zhang et al., 1995); however, the real power of the stochastic approach is the flexibility of assigning a likelihood function and prior probability model, and the potential to determine the distribution of *a posterior* models, offering a truly global optimization approach. This is normally achieved following a Markov chain Monte Carlo (MCMC) search of the parameter space. A Monte Carlo search would be impractical (computationally); the Markov chain addition allows some memory in the search of parameter sets. Typically, the search consists of generating numerous plausible models (subject to $P(\boldsymbol{m})$) and evaluating their likelihood $(L(\boldsymbol{m}))$, based on a forward model calculation (i.e. no gradient (Jacobian) is needed). Multiple random walk Markov chains are constructed using the Metropolis candidate selection algorithm (Metropolis et al., 1953), which dictates acceptance or rejection of the proposed candidate at each step. After a 'burn in' period (a significantly large number of proposals to remove the memory of the starting model) for each Markov chain, convergence can be evaluated (e.g. Gelman and Rubin, 1992); following satisfactory convergence, proposed models can be used to evaluate the posterior distribution of parameters. Since multiple Markov chains are needed, adoption to parallel computing platforms is trivial. However, a very large number of parameter proposals are required, both for convergence and sampling the posterior distribution, making MCMC methods computationally expensive; consequently, applications to multi-dimensional resistivity problems are rare.

For application to 2D and 3D problems, rather than solving for a large number of parameters (as in the traditional gradient-based method), the problem needs to be posed in terms of a reduced set of parameters. Andersen et al. (2003) illustrate the approach for 2D resistivity CVES (see Section 5.2.2.5), using simple geometrical shapes as prior models of resistivity structure. Kaipio et al. (2000) apply a stochastic inverse method to resistivity imaging for biomedical imaging applications. Ramirez et al. (2005) report on a more sophisticated MCMC search, which utilizes prior models based on a wide range of data sources (geological, geophysical, hydrological). Their approach, referred to as the Stochastic Engine, offers unparalleled flexibility recognizing the immense value of a wide range of prior information, albeit at an immense computational cost. More recently, Galetti and Curtis (2018) describe a (parameter) transdimensional stochastic approach for solving electrical resistivity tomography problems. Their method parameterizes a region geometrically via tessellation of Voronoi cells (see Figure 5.26 and Bodin & Sambridge, 2009). The discretization is allowed to vary between model proposals (hence the number of parameters in each proposal can vary). Galetti and Curtis (2018) adopt a reversible-jump Markov chain Monte Carlo (Green, 1995) method to search the parameter space. They incorporated simulated annealing concepts (see next section) in their approach to avoid any of the chains being trapped in a local likelihood minima. Their method not only allows better definition of resistivity structure compared to a conventionally regularized solution but, through the stochastic framework, permits assessment of model uncertainty (Figure 5.27).

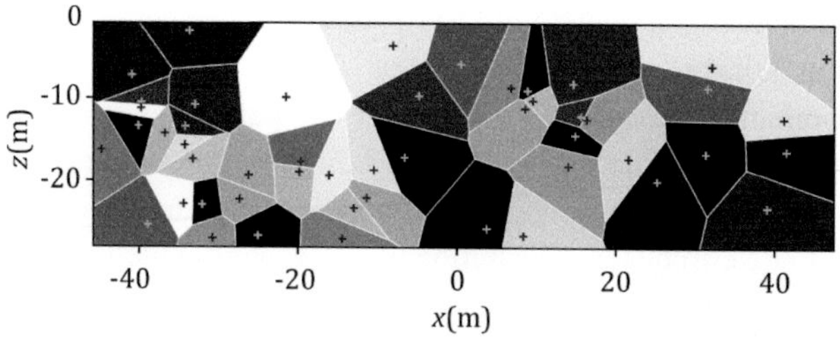

Figure 5.26 Illustration of discretization of a region using Voronoi cells. The centroid of each cell is marked with a cross.

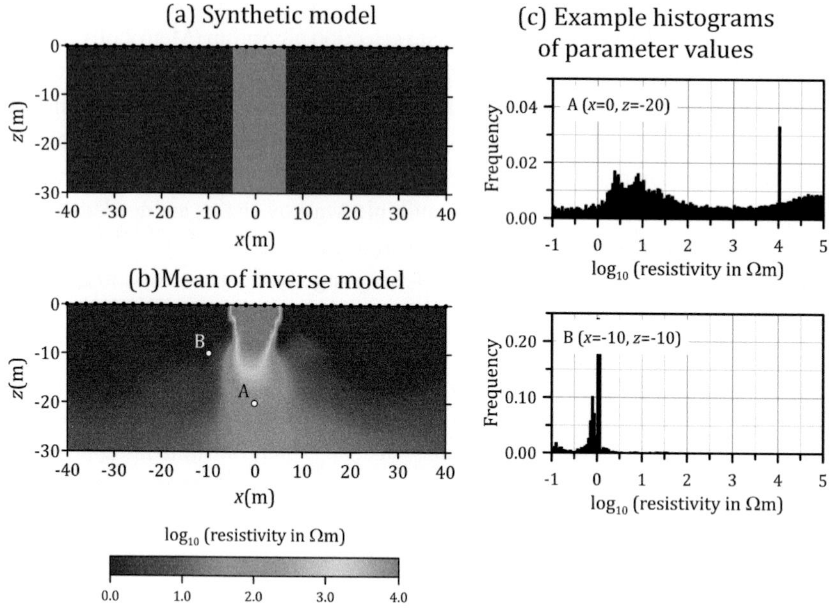

Figure 5.27 Example of transdimensional resistivity inversion based on Galetti and Curtis (2018). (a) Synthetic model. (b) Mean inverse model. (c) Parameter histograms from two locations within the region (marked in (b)). (A black and white version of this figure will appear in some formats. For the colour version, please refer to the plate section.)

5.2.5.2 Other Global Optimization Methods

A number of alternative global optimization methods have been developed, several of which have been established for some time. Sen and Stoffa (2013) give details on a range of such methods applied to geophysical inverse problems.

Simulated annealing (Kirkpatrick et al., 1983) takes its name from the annealing process in metallurgy when a metal is cooled slowly to enhance its final strength. The analogy in optimization refers to the slow progression to a global minimum of the objective function, rather than rapid arrival at a local minimum. Like MCMC, models are proposed within a search process and the probability of acceptance is influenced by the computed misfit and a temperature parameter, T, that is gradually reduced as the search progresses, e.g.,

$$P(\boldsymbol{m}) \propto \exp\left[-\frac{1}{T}\left(\boldsymbol{d}-F(\boldsymbol{m})\right)^T \boldsymbol{W}_d^T \boldsymbol{W}_d\left(\boldsymbol{d}-F(\boldsymbol{m})\right)\right]. \quad (5.51)$$

Thus, early on in the search process when T is large, the probability of accepting a model proposal is large even if the misfit is not small, to prevent being trapped in a local minimum. Sen et al. (1993) show application of simulated annealing for vertical electrical soundings. As in MCMC, application to problems with a large number of parameters is challenging because of the computational burden. Pessel and Gibert (2003) illustrate how simulated annealing can be used for 2D resistivity inversion by progressively increasing the level of parameterization of the problem.

The genetic algorithm, as the name suggests, follows an evolution analogy in the parameter search. In this case, parameter sets, within a population, are modified mimicking a genetic code, in a representation of reproduction from two parameter sets of the population. A probability is assigned to the progression of the members of the population to the next generation, allowing the strongest/fittest to survive. Jha et al. (2008) illustrate the use of the approach for 1D resistivity inversion. Schwarzbach et al. (2005) exploit parallel computation for tackling 2D resistivity inversion using a genetic algorithm, but still comment on the computational burden. Liu et al. (2012) apply a genetic algorithm for 3D resistivity inversion but include a gradient (Jacobian) calculation to assist the mutation process. It is not clear if such an approach results in a truly global optimization.

Another global search method that follows a biological analogy is particle swarm optimization. In this case, a population of parameter sets are represented by a swarm of particles. The particles change throughout the iterative process based on an analogy to each parameter position and velocity using a measure of cognition and social behaviour, mimicking the movement of a flock of birds or swarm of insects searching for food. Shaw and Srivastave (2007) illustrate the approach for 1D vertical electrical soundings.

The methods described so far are based on evolution of parameter sets following some analogy in natural sciences. Artificial neural networks (e.g. Lippmann, 1987) follow a different approach, using a network of processing operations (neurons) that map observations to models. The network is constructed through a training process and may include a feedback element. van der Baan and Jutten (2000) provide an overview of the approach in a geophysics context. Calderon-Macias et al. (2000) illustrate the approach for inversion of VES data, while Neyamadpour et al. (2010) apply neural networks to 3D resistivity problems. Neural networks are a class of machine learning models, and although they were first conceived in the 1940s (McCulloch and Pitts, 1943), it is only relatively recently that their potential for solving geophysical inverse problems has been widely appreciated. Russell (2019) offers some cautionary comments about such machine learning approaches in geophysics, and argues for a balance of physics and machine learning. Recent examples of machine learning combined with Markov chain Monte Carlo sampling include Laloy et al. (2018) and Ray and Myer (2019).

5.2.5.3 Joint and Coupled Inversion

We discussed earlier how additional data can be used to constrain a traditional gradient-based inversion of electrical data through adjustment of the roughness matrix. This can be very effective when, for example, lithological boundaries are known, e.g. from borehole logs, ground penetrating radar or seismic refraction surveys (e.g. Doetsch et al., 2012b). In some cases, we may have data from an additional geophysical survey which we can model in parallel with the electrical survey data, thus inverting the two (or more) datasets in a joint manner. The two surveys may be sensitive to resistivity (e.g. using DC resistivity and electromagnetic induction methods), and so the joint inversion is relatively straightforward as the joint objective function is simply an addition of the two individual terms. Note that the data weights are important to ensure appropriate balance of the likely different quality of the two datasets. Sasaki (1989) and Monteiro Santos et al. (2007) show examples of the joint inversion of DC resistivity and audio-magnetotelluric surveys. It should also be noted that occasionally the term 'joint inversion' has (incorrectly) been used by some to describe the inversion of multiple types of DC resistivity quadrupole geometries (as in Candansayar, 2008; Demirel and Candansayer, 2017).

The two surveys may sense different geophysical properties (e.g. resistivity and seismic velocity) but may have some dependence due to the geometry of layers within the subsurface (e.g. Hering et al., 1995; Misiek et al., 1997). Analysis of a joint inverse model can be examined to determine the information content in each geophysical modality (e.g. JafarGandomi and Binley, 2013). Alternatively, the link between the geophysical properties may be petrophysical, e.g. the link between fluid saturation and both resistivity and permittivity, allowing the joint inversion of cross-borehole radar and resistivity data through the common property or state. However, as noted by Tso et al. (2019) (see also Chapter 2), petrophysical models can be subject to high uncertainty, which should be accounted for in any inversion of this type.

An alternative approach to joint inversion is to consider the structural similarity of the two (or more) models that are consistent with data. Gallardo and Meju (2003) developed a simple, yet powerful, way of quantifying similarity using their cross-gradient penalty function, defined, for a dual modality problem, as the cross product:

$$\tau(x,y,z) = \nabla \boldsymbol{m}_1 \times \nabla \boldsymbol{m}_2, \tag{5.52}$$

where $\boldsymbol{m}_1$ and $\boldsymbol{m}_2$ are two model parameters.

In areas where the spatial trend of both parameters are in the same or opposite direction (or where one of the gradients is zero), the value of τ in Equation 5.52 will be zero and the parameter trends structurally similar. In areas where the trends are not parallel (e.g. orthogonal), the value of τ will be non-zero. In the formulation of Gallardo and Meju (2003), the two modality datasets are inverted, but coupled by minimizing the values of τ in Equation 5.52 for all cells. They illustrate the effectiveness of the approach for the joint inversion of 2D resistivity and seismic data, and show how plots of the relationships between the two inverted parameter sets can identify lithological units. Linde et al. (2006) used the same cross-gradient constraint to invert cross-borehole 3D resistivity and ground penetrating radar data. Doetsch et al. (2010b) extend the approach for the analysis of three datasets (resistivity, radar and seismic), again in a 3D cross-borehole arrangement. A review of the cross-gradient-based approach is presented in Gallardo and Meju (2011). Other examples of structurally constrained joint inversion of resistivity and other geophysical data include Bouchedda et al. (2012) and Hamdan and Vafidis (2013).

It is possible to formulate the same cross-gradient approach for treating time-lapse data. In their time-lapse study of cross-borehole resistivity and seismic monitoring of changes in fluid saturation, Karaoulis et al. (2012) include a cross-gradient control for both static and dynamic processes. The inversion of time-lapse data can also be coupled with an underlying process (e.g. hydraulic) model; in other words, rather than determining the geophysical behaviour to infer the hydraulic response, a hydraulic model is used to constrain the inversion of geophysical data. This concept was first proposed as part of the Stochastic Engine approach mentioned earlier (see Aines et al., 2002). With such an approach the geophysical data can be used to estimate directly properties of interest, e.g. permeability. Binley et al. (2002a) calibrated an unsaturated flow model by using time-lapse cross-borehole resistivity data to match the centre of mass of an injected tracer (see also Crestani et al., 2015). Kowalsky et al. (2004) presented a more sophisticated approach that is centred around the flow model parameterization, using the geophysical (radar) data to constrain it. Such an approach is often referred to as coupled hydrogeophysical inversion (e.g. Ferre et al., 2009). Looms et al. (2008), Hinnell et al. (2010), Huisman et al. (2010), Mboh et al. (2012), Phuong Tran et al. (2016) and Kang et al. (2019) are examples of the same underlying philosophy, in these cases using resistivity data to parameterize a hydraulic flow model. Such approaches rely on stochastic methods to explore the parameter space and, given the computational demands of flow simulation (particularly for highly non-linear unsaturated flow problems), they can often be challenging to solve. Oware et al. (2013) offer an alternative approach in which Monte Carlo simulations of flow and transport models are used to generate training images, which are subsequently used to constrain the inversion of geophysical data. It is important to note that most coupled hydrogeophysical inversion requires a reliable petrophysical model to link hydraulic properties or states to geophysical parameters. A geometrical transformation between model grids for geophysics and process (e.g. hydrology) model is required in order to link the two using petrophysical relationships. Furthermore, given the uncertainty of such petrophysical models in most applications, this requirement may further limit the use of such approaches (see Tso et al., 2019). The data–domain correlation approach of Johnson et al. (2009) attempts to overcome such a requirement. Finally, it is worth highlighting that coupled hydrogeophysical codes are now becoming available (e.g. Johnson et al., 2017), creating more opportunities for such applications.

5.2.6 Current Source Modelling and Inversion

In Section 4.2.2.11, we showed how DC resistivity measurements can be made to assess the integrity of hydraulic barriers (see Figure 4.37). For such applications, the objective is to determine the distribution of electrical current sources within the barrier, which may be interpreted as hydraulic leaks. We can formulate this as a similar inverse problem to that used for DC resistivity. If we first assume that the resistivity is homogenous and that the ground surface is flat ($z = 0$), then from Equation 4.5 we know that the voltage at coordinates $(x_p, y_p, 0)$, distance r from the source, is given by

$$V(x_p, y_p, 0) = \frac{I\rho}{2\pi r}, \tag{5.53}$$

and from superposition, the voltage due to M current sources is given by

$$V(x_p, y_p, 0) = \frac{\rho}{2\pi}\sum_{i=1}^{N}\left(\frac{I_i}{r_i}\right), \tag{5.54}$$

then the problem becomes the determination of unknowns I_i and r_i, $i = 1, 2, \ldots M$, given N measurements $V_j, j = 1, 2, \ldots N$, subject to $\sum_{i=1}^{N} I_i = I$, the total current applied.

For the more general problem we can assume that the resistivity varies spatially but is known (it could be assessed *a priori* from a resistivity imaging survey in the normal way), and the potential field (in 3D space) can be modelled (for a single current source) by

$$\frac{\partial}{\partial x}\left(\frac{1}{\rho}\frac{\partial V}{\partial x}\right) + \frac{\partial}{\partial y}\left(\frac{1}{\rho}\frac{\partial V}{\partial y}\right) + \frac{\partial}{\partial z}\left(\frac{1}{\rho}\frac{\partial V}{\partial z}\right) = -I\delta(x)\delta(z), \tag{5.55}$$

then using a mesh-based discretization we can assign a current source at each node within the region under investigation, i.e. the parameters, $\boldsymbol{m}$, to be solved for are the set of M values of I_i. We can use a least-squares formulation with a smoothness constraint as before, i.e.

$$\Phi_d = (\boldsymbol{d} - F(\boldsymbol{m}))^T \boldsymbol{W}_d^T \boldsymbol{W}_d(\boldsymbol{d} - F(\boldsymbol{m})) + \alpha \boldsymbol{m}^T \boldsymbol{R}\boldsymbol{m}, \tag{5.56}$$

where $F(\boldsymbol{m})$ is now the solution of Equation 5.55 for all current source nodes (via superposition). Note that because the problem is linear (scaling the current, scales the voltage at any node), the problem is easier to solve than the non-linear resistivity problem.

Binley et al. (1997) illustrate the application of the above approach for the computation of the distribution of current sources under a leaking metallic underground storage tank, with voltages recorded at electrodes in boreholes around the perimeter of the tank. In this case, the smoothness constraint (second term in Equation 5.56) is valid because current leakage occurs due to the conducting tank base in addition to the hydraulic leak. For leaks in insulating barriers, e.g. HDPE liners, a different regularization may be applied, or, as shown in Binley et al. (1997), a stochastic search approach may be adopted, although, as shown by Binley and Daily (2003), the ability to differentiate multiple leak sources is challenging. It is also worth noting that such methods have other potential applications, such as the analysis of self-potential signals (e.g. Minsley et al., 2007) and the electrical characterization of the plant root – soil interactions, as shown recently by Mary et al. (2018).

5.3 Induced Polarization

5.3.1 General Comments

As discussed in Chapters 3 and 4, we can approach the IP problem either in the time domain, using the concept of chargeability, or the frequency domain, using a complex resistivity formulation. We show below how these alternative approaches are formulated as an extension to the DC resistivity problem. Note that the use of a complex resistivity approach does not rely on measurements made in the frequency domain, since transformations between measured apparent chargeability and the equivalent complex resistivity phase angle can be applied (see Chapter 3).

The choice of method for analysing IP data is perhaps a personal one. Chargeability approaches have been established for some time and are computationally simpler (as shown below). However, the interpretation of an image of chargeability (which will result from inversion of time domain IP data) can be challenging in some cases since most petrophysical relationships (see Chapter 2) have been derived based on complex resistivity measurements. The complex resistivity approach is perhaps more elegant, mathematically, but can come at a greater computational cost.

As shown in Chapter 2, the frequency dependence of polarization, i.e. spectral IP, can provide insight into hydrological and biogeochemical properties and processes. We require a set of tools to be able to determine macroscopic spectral properties from measured electrical spectra; we present such approaches later. And finally, in some cases we may wish to model or interpret the spatial (and spatio-temporal) variation in such spectral properties from field data. A series of approaches are available, as outlined later.

5.3.2 Forward Modelling in the Time Domain

As discussed in Section 4.3.1, if we adopt the Seigel (1959) chargeability definition (Equation 4.21), then recalling Equation 4.42, we can express the apparent chargeability in a body defined as M cells of resistivity, ρ_j, and intrinsic chargeability, $\hat{m}_j$ $(0<\hat{m}_j<1)$, as

$$M_a = \sum_{j=1}^{M} \hat{m}_j \frac{\partial \log\rho_a}{\partial \log\rho_j}. \tag{5.57}$$

Given the implicit link between resistivity and apparent chargeability in Equation 5.57, we can use methods already established for the DC resistivity problem to model IP. In fact, the derivative in Equation 5.57 is the same as the Jacobian for the DC resistivity inverse problem (Equation 5.29) but expressed in terms of measurements of apparent resistivity and not transfer resistance. We have already shown (Section 4.3.1) how this formulation can be used to model the observed response for a horizontally layered system, each layer being represented by resistivity and chargeability. We can easily extend this for the general (2D and 3D cases).

5.3.3 Forward Modelling in the Frequency Domain

Weller et al. (1996a) first proposed an alternative approach to the IP problem by formulating the governing equations in terms of complex resistivity. In many ways, this is a straightforward extension of the DC resistivity problem, although there are a few key differences.

If we now express the resistivity as a complex variable ρ^*, which can be considered as a pair of real and imaginary values, or a magnitude $|\rho^*|$ and phase angle φ, then Equation 5.7 (for the 2D case) can now be written as

$$\frac{\partial}{\partial x}\left(\frac{1}{\rho^*}\frac{\partial v^*}{\partial x}\right) + \frac{\partial}{\partial z}\left(\frac{1}{\rho^*}\frac{\partial v^*}{\partial z}\right) - \frac{v^* k_w^2}{\rho^*} = -I\delta(x)\delta(z), \tag{5.58}$$

which can be solved for v^* and, as before, converted to (now complex) voltages $V^*(x,y,z)$ following Equation 5.8. In solving for v^*, the same methods can be used as in the DC resistivity problem, although an appropriate complex linear equation solver (e.g. Schwarz et al., 1991) is now necessary. Figure 4.39 shows an example complex resistivity phase angle pseudosection.

As before, singularity removal methods can be applied to increase accuracy and non-infinite mixed boundary conditions can be applied (see Kemna, 2000). Extension to 3D problems is equally straightforward, although it should be noted that (i) the formulation as a problem in terms of complex numbers leads to a doubling of the computer memory storage requirements, and (ii) arithmetic operations of complex variables are more time consuming than the same operations applied to real variables. Consequently, the overall computational demands for large 3D problems can be excessive. An alternative modelling strategy is to decouple the real and imaginary voltages (e.g. Commer et al., 2011; Johnson and Tholme, 2018), which can be solved jointly, thus avoiding the computational overheads of working with complex arithmetic. Farias et al. (2010) is a further example of complex resistivity forward modelling.

A significant attraction of defining the problem in terms of complex resistivity (rather than chargeability) is that the same underlying principles addressed in the DC resistivity case are relatively easily formulated for the IP problem, e.g. the impact of anisotropy (Kenkel et al., 2012). Furthermore, unlike the time domain chargeability formulation, the extension to address frequency dependence of polarization is straightforward. Moreover, as the real and imaginary parts are directly related to the transport mechanisms of interest (conduction and polarization), interpretation is less challenging.

5.3.3.1 Modelling Electromagnetic Coupling

Electromagnetic coupling between the transmitter, the receiver and the ground can have a significant effect in IP surveys. Its effect can increase with electrode separation, conductivity of the ground and frequency, as demonstrated in model studies (e.g. Millett, 1967; Dey and Morrison, 1973). In time domain surveys, it is usually ignored as it is assumed that the recording delay after current switch off is long enough to minimize such effects (although as there is a growing trend towards collection of, so called, full waveform data, then such effects may not be negligible (Section 3.3.2.3)). For frequency domain surveys coupling effects can be evident even at relatively low frequencies, limiting the usefulness of higher-frequency measurements in spectral IP surveys.

We can approach the problem in one of two ways if multi-frequency measurements are made. The simplest involves using a simple polynomial form for the coupling effect as a function of frequency, which can be parameterized given measurements at a number of frequencies (e.g. Song, 1984) and the effect subsequently removed. Alternatively, Cole–Cole-type relaxation models (mimicking the high-frequency effect) can be applied to measured spectra to remove the coupling component (e.g. Pelton et al., 1978). Although simple to implement, the validity of such approaches has been questioned (Major and Silic, 1981).

The second approach is a more physics-based approach to modelling of the coupling (e.g. Wait and Gruszka, 1986; Routh and Oldenburg, 2001; Ingeman-Nielsen and Baumgartner, 2006), inevitably leading to a Cole–Cole-type representation of the frequency effects. Such studies have been conducted using collinear arrays; for more complex geometries (in particular 3D arrays), the challenges in accurate modelling of the coupling effects may limit their application. Zhao et al. (2015) illustrate a physics-based modelling approach applied to a generalized electrode layout.

5.3.4 Inverse Modelling in the Time Domain

In general terms, inverse modelling of time domain data equates to the determination of a spatial distribution of intrinsic chargeabilities given a set of measured chargeabilities from a number of quadrupoles. The approach is implicitly coupled with the DC resistivity problem. Note that from an estimation of intrinsic chargeability and resistivity, a normalized chargeability can be computed. As discussed in Chapter 2, this normalized value is likely to offer more insight into polarizability rather than using just the intrinsic chargeability.

Adopting Equation 5.57, the procedure typically consists of first solving the DC resistivity problem (as before), which is then followed by a linearized inversion for intrinsic chargeabilities, which is carried out in one step. Pelton et al. (1978) first utilized the approach for 2D inversion of chargeability data, adopting a damped (see Section 5.2.2.2) solution. LaBrecque (1991) and Iseki and Shima (1992) both put forward regularized inverse schemes for 'IP tomography', focusing on the analysis of cross-borehole data.

Using an objective function of the same form as Equation 5.21, we aim to minimize:

$$\Phi_d = (\boldsymbol{d} - F(\boldsymbol{m}))^T \boldsymbol{W}_d^T \boldsymbol{W}_d (\boldsymbol{d} - F(\boldsymbol{m})) + \alpha \boldsymbol{m}^T \boldsymbol{R} \boldsymbol{m}, \tag{5.59}$$

where now the data, $\boldsymbol{d}$, are N apparent chargeabilities $M_{a,i}$ $(i = 1, 2, .., N)$; the parameters, $\boldsymbol{m}$, are intrinsic chargeabilities m_j $(j = 1, 2, .., M)$; the data weight matrix, $\boldsymbol{W}_d$, relates to uncertainties in measured apparent chargeability; the forward model operator, $F(\boldsymbol{m})$, is Equation 5.57; and α and $\boldsymbol{R}$ are defined as before.

In order to minimize Equation 5.59, the same procedure can be followed as before; however, note that given the linear forward model definition in Equation 5.57, the Jacobian is now independent of the chargeabilities, and given by

$$J_{i,j} = \sum\nolimits_{k=1}^{M} \frac{\partial \log\rho_{a,i}}{\partial \log\rho_k} \quad (i = 1, 2, .., N; j = 1, 2, \ldots, M), \tag{5.60}$$

i.e. the problem is linear and can be solved in one step.

The definition in Equation 5.57 does, however, rely on an appropriate resistivity distribution (derived from the DC resistivity data inversion). Oldenburg and Li (1994) present a similar chargeability inversion scheme and explored the impact of incorrectly defining the resistivity distribution on the chargeability inverse model. They concluded that the DC resistivity inverse model should provide an adequate representation for the chargeability modelling, but that an assumption of uniform resistivity can degrade the performance of the inversion. Oldenburg and Li (1994) also proposed a non-linear solution to the chargeability inverse problem, which does not suffer from the low-intrinsic chargeability assumption implicit in the above. They conclude that for most cases, the simpler, linearized, solution is satisfactory and clearly computationally simpler. Li and Oldenburg (2000) extend the approach to 3D problems. Beard et al. (1996) developed a computationally simpler approach based on a low-contrast (resistivity and IP) approximation; however, such an approach is now of limited value given subsequent computational advances.

5.3.5 Inverse Modelling in the Frequency Domain

We can formulate the IP inverse problem in terms of complex resistivity. In this case, IP data are provided in the form of complex impedances (transfer impedance magnitude and phase angle), which are inverted to produce a set of parameters defined in terms of complex resistivity (resistivity magnitude and phase angle). Weller et al. (1996b) first proposed the inversion of IP data in this way, although their approach proved to be limited because of inadequate treatment of data errors and no incorporation of regularization. Kemna (2000) proposed a more robust approach, which has subsequently been used in a wide range of applications to field data (e.g. Kemna et al., 2004).

In the formulation of Kemna (2000), the parameters and data are defined as the natural logarithm of complex conductivities and apparent conductivities, respectively (note that log conductivity is equivalent to negative log resistivity). Given that a complex number Z^* can be expressed as $|Z^*|e^{i\varphi}$, then, if based on resistivity, the real and imaginary parts of data are $\log_e(K_iZ_i)$ and φ_i, respectively, where K_i is the geometric factor, Z_i is the transfer impedance magnitude and φ_i is the phase angle for measurement i. Note that expressing the data in this way requires that the polarity of measurement and its forward modelled value are the same. For surveys conducted on boundary surfaces (e.g. ground surface), this is unlikely to be a problem since any negative apparent resistivities can be filtered prior to inversion. However, for surveys using buried electrodes (e.g. in a cross-borehole configuration) the modelled and measured apparent resistivities may differ in polarity during the inversion process (see e.g. Wilkinson et al., 2008). This can be addressed by ignoring data that do not match the polarity of modelled values. As the inversion progresses as the resistivity structure develops, such values may then be reintroduced.

As Kemna (2000) shows, the Jacobian can be formed in the same way as the DC resistivity problem, but is now a complex matrix. The data weight matrix $\boldsymbol{W}_d$ is, however, a real quantity, which includes error in both real and imaginary measurements. This can lead to over-emphasis in the inversion of the real component of the resistivity magnitude misfit relative to the phase angle. To address this, Kemna (2000) developed a two-stage process for inversion. First, the complex resistivity is solved for using a complex formulation equivalent to Equation 5.25 (terms $\boldsymbol{W}_d$, $\boldsymbol{R}$ and α remain as real terms). Once convergence is reached, a final phase improvement is introduced which involves maintaining the magnitude of the complex resistivity for all parameters but solving the same problem purely for the phase angle of all parameters. Kemna (2000) shows how the elements of the inversion in this step (e.g. the Jacobian) are similar to the full complex formulation. By performing this final phase improvement, the inversion can properly account for phase angle measurement errors.

The process can be viewed effectively as a decoupling of the real and imaginary parameter inversion, and from a computational perspective is less efficient than a true decoupling (e.g. Commer et al., 2011).

Figure 5.28 shows an example complex resistivity inversion using data from Mejus (2015). In this study, IP was used to help assess the vulnerability of a regional sandstone aquifer, the focus being an assessment of the thickness and hydraulic properties of the overburden. The survey shown in Figure 5.28 was carried out with

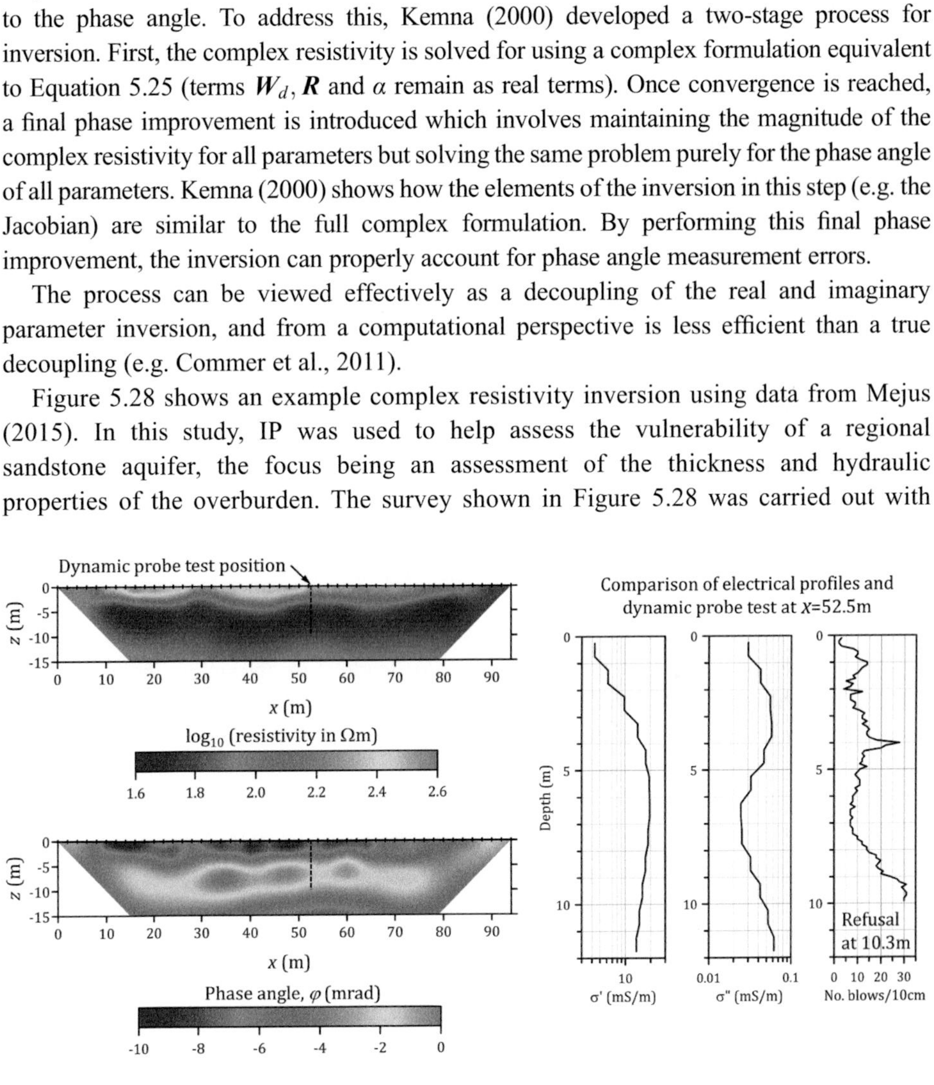

Figure 5.28 Complex resistivity inversion of surface electrode data. The survey was carried out with 48 electrodes at 2 m spacing (positions shown in the images). The three graphs show a comparison of inverted profiles (shown as real and imaginary conductivity) and a dynamic probe test. Refusal of the test was reached at 10.3 m (top of bedrock). (A black and white version of this figure will appear in some formats. For the colour version, please refer to the plate section.)

a time domain IP instrument (Syscal Pro, Iris Instruments), with apparent chargeabilities converted to an equivalent phase angle assuming 1 mrad ≡ 1 mV/V (Mejus, 2015; see also Section 3.3.3). The images show clear definition of the sandstone boundary (~10 m depth). A dynamic probe test carried out at one location along the transect reveals vertical variation in the strength of the overburden sediments (clays, sands, gravels), which mirrors more closely variation in electrical polarization (plotted as imaginary conductivity) than in electrical conductivity. Chapter 6 includes examples of IP inversion applied to a range of problems.

As in the DC resistivity case, IP inverse modelling can be enhanced by incorporating *a priori* information. In fact, given the (often) relatively weak signature of IP for many environmental applications, *a priori* constraints can improve significantly the recovered model, as illustrated by Slater and Binley (2006) for application to permeable reactive barrier imaging. Blaschek et al. (2008) propose a more sophisticated approach for complex resistivity imaging using a minimum gradient support function.

As discussed in Section 5.2.3, the appropriate assignment of measurement errors can have a significant effect on the inverse model. In IP modelling, this is an even greater issue since signal-to-noise in IP measurements is inevitably weaker. To illustrate the effect of noise, the resistivity model in Figure 5.11 was revised to include polarization anomalies as shown in Figure 5.29. Complex resistivity inversions were applied to data corrupted with Gaussian noise with a standard deviation of 1 mrad and 5 mrad (2 per cent Gaussian noise was applied to impedance magnitudes as before). The inverted phase angle models are shown in Figure 5.29 (the inverted resistivity magnitude is identical to that shown for the DC resistivity case in Figure 5.19). For the low-noise case, both polarizable bodies are recovered; however, for the larger-noise level, the resistive polarizable anomaly is only weakly resolved and, as expected, greater smearing of the two features is evident. Note also a slight over-shoot in the phase angle above the left-most polarizable body (a result of the regularization), leading to anomalous positive phase angles. It should also be noted that for this simple example the noise levels applied are not unlike those likely to be accounted in the field; however, the target anomalies are characterized by larger IP values (100 mrad) than those typical of environmental features (which may be of the order of one or two tens of mrad). Therefore, care must be exercised in quantitative analysis of IP images, particularly if they are to be utilized for determination of physical and chemical properties of the subsurface. Flores Orozco et al. (2012) further illustrate the role of measurement error on complex resistivity images.

Although illustrated above with 2D models, extension to 3D is straightforward, although computational costs can be high, necessitating parallel computation for large-scale problems. Binley et al. (2016), for example, illustrate the application of 3D complex resistivity imaging in a cross-borehole configuration. Similarly, extension to bounded problems (e.g. laboratory tanks, columns, etc.) is trivial (e.g. Kemna et al., 2000). In fact, biomedical applications of multi-frequency impedance imaging have been established for some time; examples include respiratory imaging (Brown et al., 1994) and neurology (Yerworth et al., 2003).

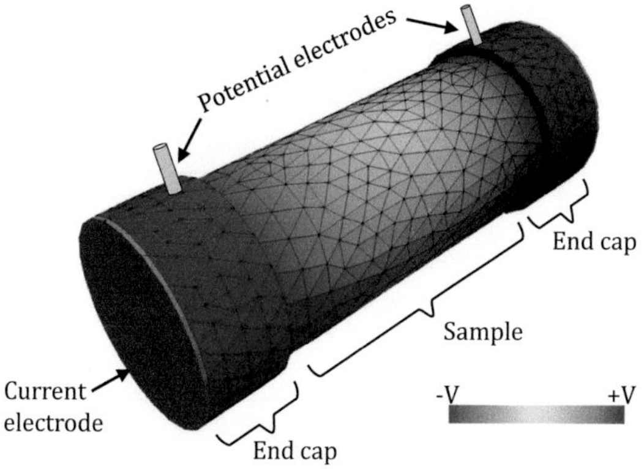

Figure 3.3 Example of a numerical model for the potential field in a sample holder and itsend caps. The sample (of homogeneous resistivity) is 2 cm long and 1 cm in diameter. The end caps are 5 mm long and 11 mm in diameter. The end caps have a resistivity of 20% of that of the sample, which results in lower potential gradients within the end caps.

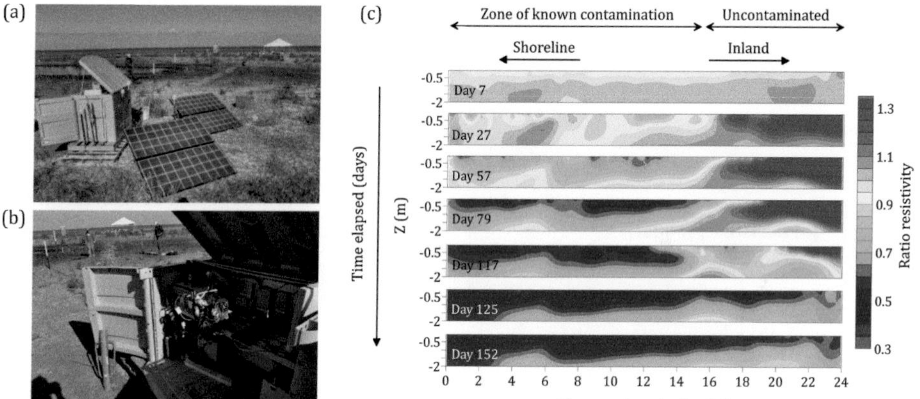

Figure 3.9 Long-term electrical resistivity monitoring system set up on GT1, just off Grand-Terre, Louisiana, USA, to monitor the degradation of oil-contaminated beach sediments: (a) site photo showing instrument storage, solar panels and electrode array; (b) close-up of resistivity meter being used for monitoring; (c) resulting time-lapse sequence of ratio changes in resistivity (unitless) recorded with this system. Data from Heenan et al. (2014). The system addressed a total of 96 electrodes (48 on the surface and 48 in boreholes) twice daily for an 18-month period.

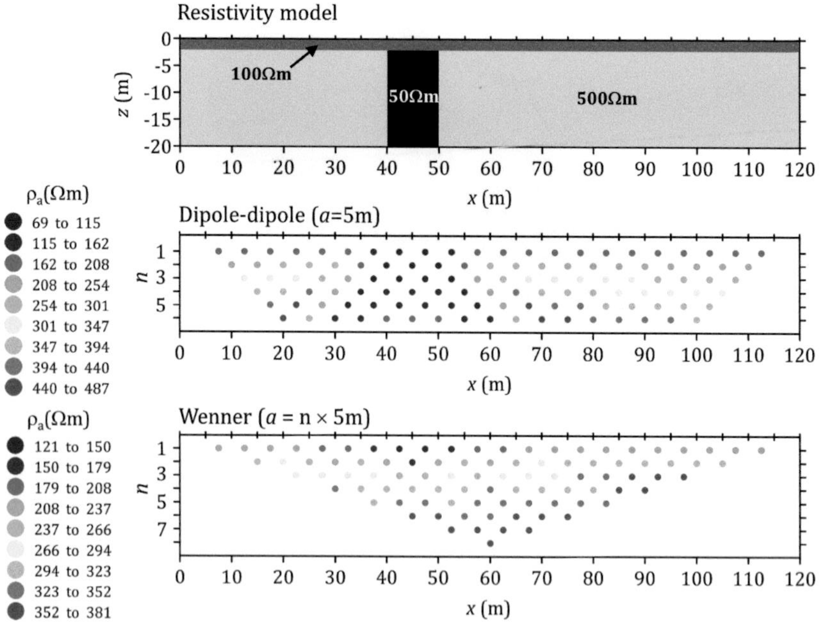

Figure 4.21 Comparison of dipole–dipole and Wenner pseudosections for a synthetic resistivity model. Note the different apparent resistivity scales.

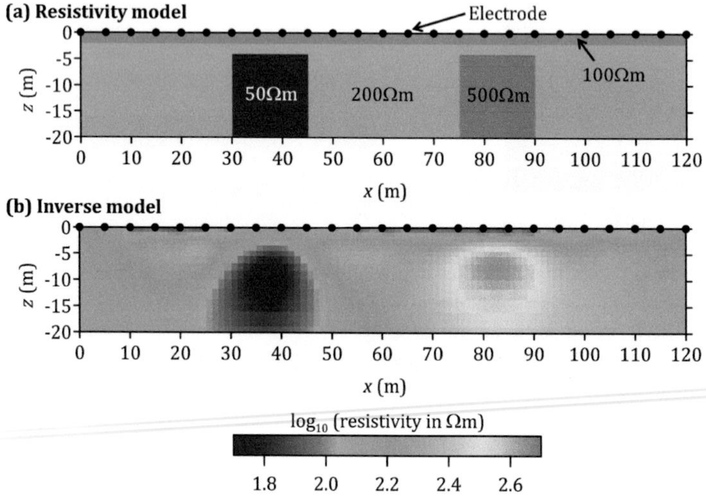

Figure 5.11 Inversion of synthetic 2D surface electrode resistivity model. (a) Synthetic model. (b) Inverted model.

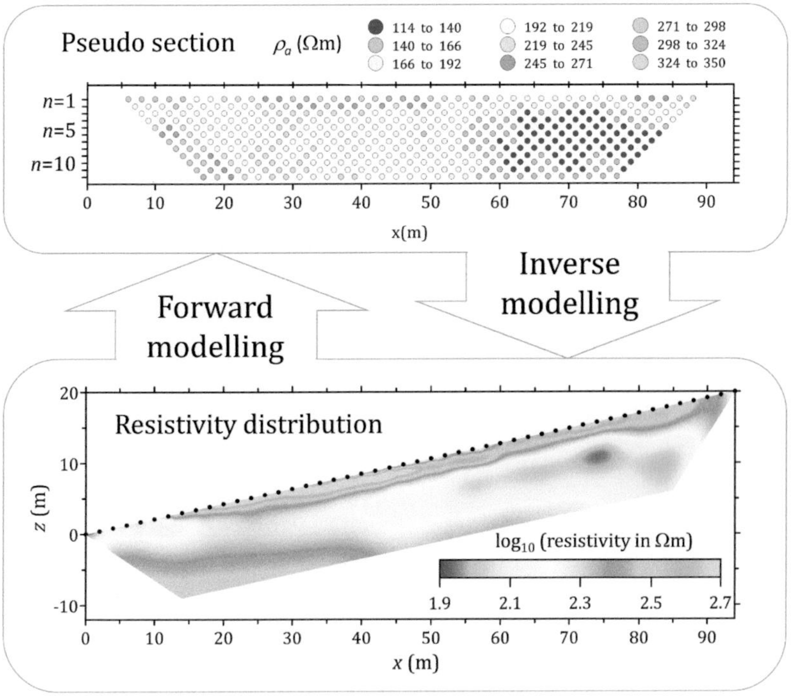

Figure 5.1 Forward and inverse modelling. The example is a dipole–dipole DC resistivity survey carried out on a hillslope (see de Sosa et al., 2018). Electrode positions are marked in the lower figure with solid circles.

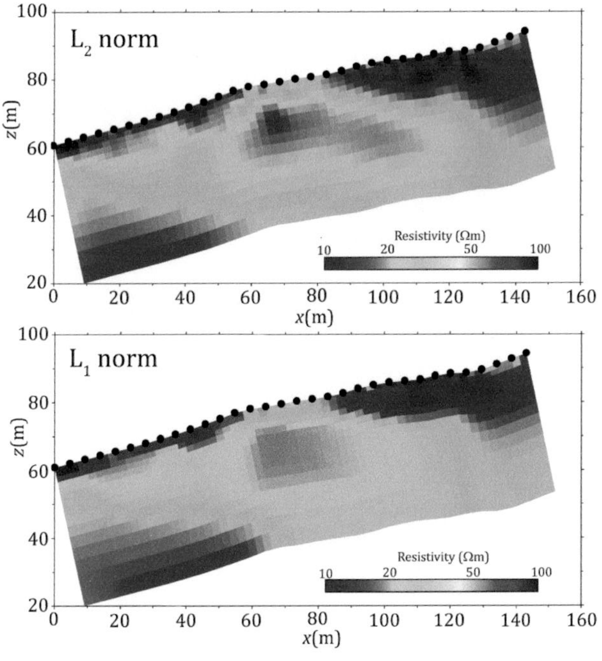

Figure 5.12 Comparison of L_1 and L_2 norm inversions of a 32-electrode dipole–dipole 2D dataset discussed in Wilkinson et al. (2012). Electrode locations are marked by solid circles. The field site is discussed in more detail in the case study in Section 6.1.7.

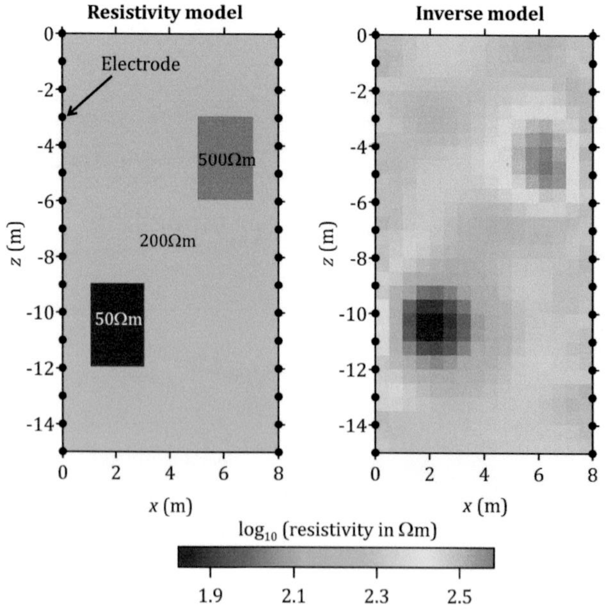

Figure 5.13 Inversion of synthetic 2D cross-borehole resistivity model.

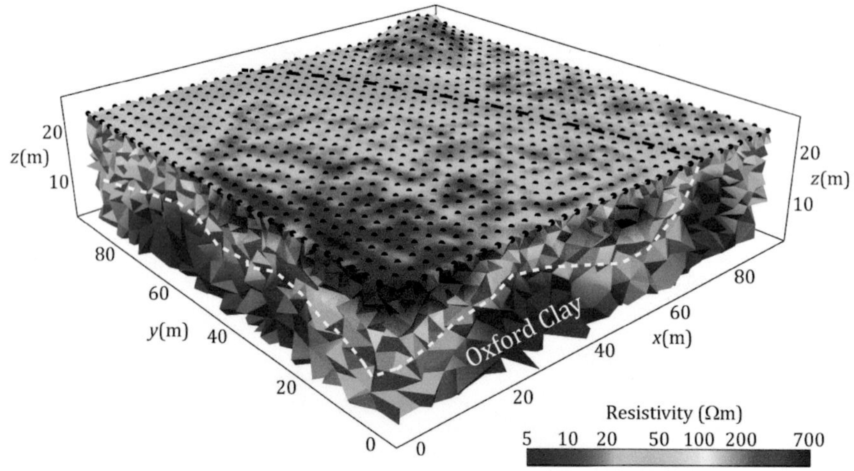

Figure 5.16 3D resistivity imaging of river terrace deposits after Chambers et al. (2012). Electrodes are shown by black circles. The white dashed line marks the interpreted interface of resistive river terrace deposits overlying conductive Oxford Clay. The black dashed line marks a horizontal boundary of deposition.

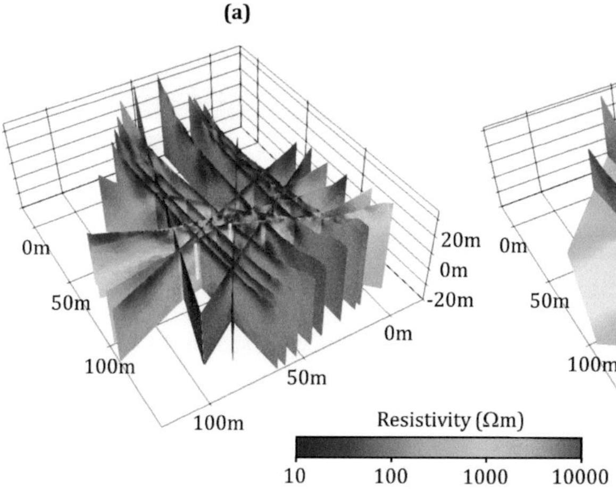

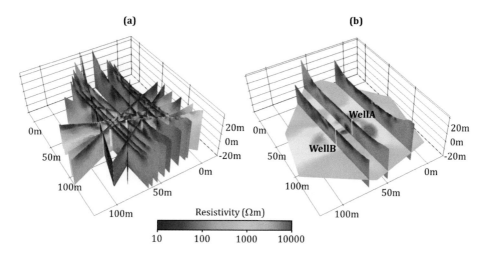

Figure 5.17 Quasi-3D inversion of resistivity data at a karst site in southwest China. (a) Location of 2D survey lines and sections through the 3D resistivity model. (b) Simplified extract of the 3D model showing localized resistivity variation consistent with observed artesian conditions at well A (which were not observed at well B). For more information, see Cheng et al. (2019a).

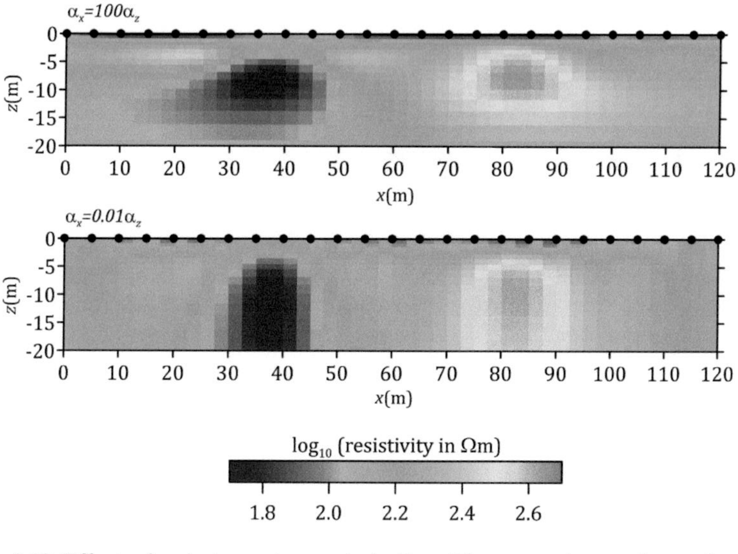

Figure 5.19 Effect of anisotropy in regularization. The upper image shows the case for α_x=100α_z; the lower image shows the case for α_x = 0.01α_z.The true model is shown in Figure 5.11a.

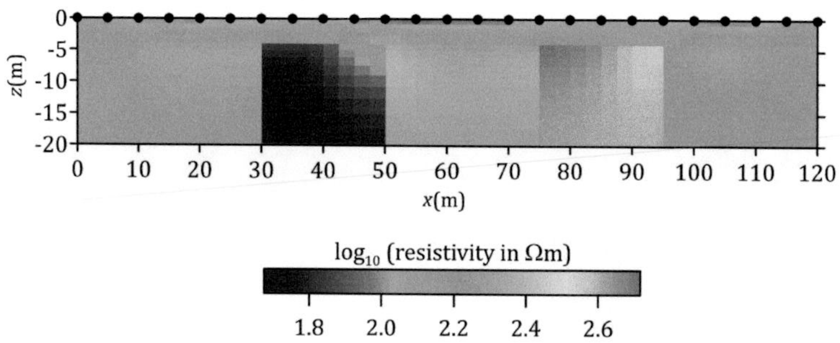

Figure 5.20 Effect of regularization disconnection. The true model is shown in Figure 5.11.

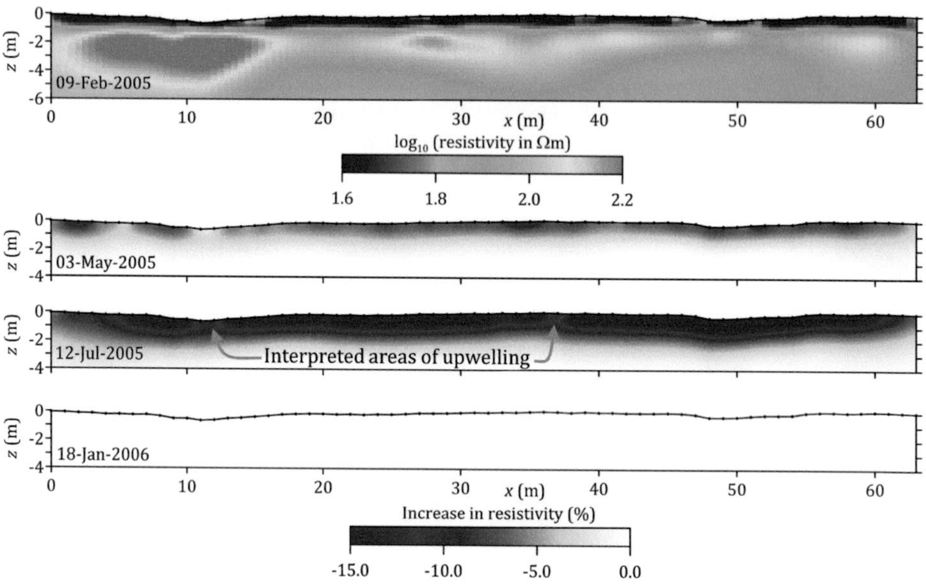

Figure 5.21 Time-lapse imaging of changes in resistivity in a riparian wetland. The upper figure shows the reference inversion. The lower figures show changes in resistivity along the transect, computed using a difference inversion. The symbols show the position of electrodes. Inversions carried out using data reported in Musgrave and Binley (2011).

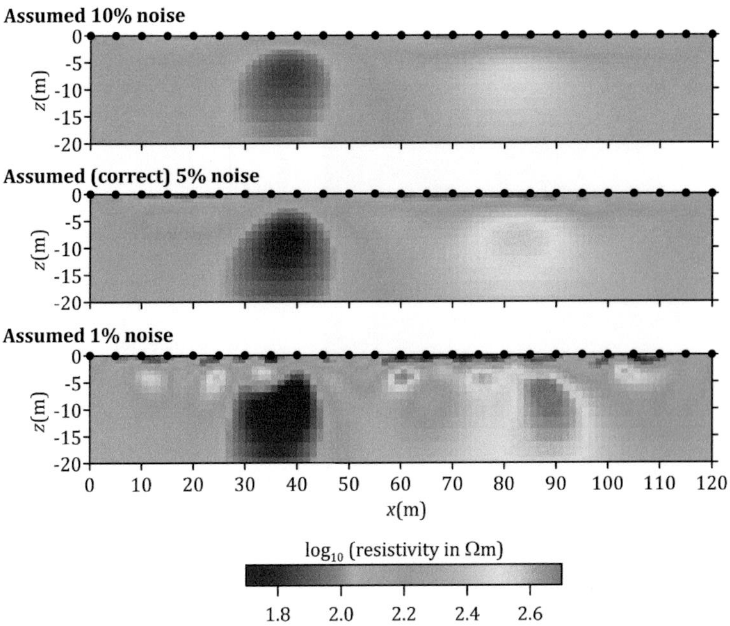

Figure 5.22 The impact of incorrect error estimation on an inverse model. The true model is shown in Figure 5.11.

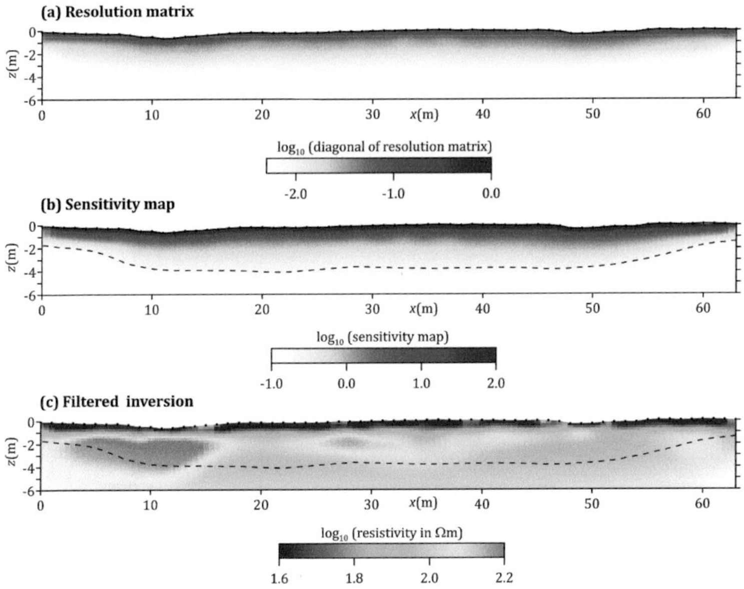

Figure 5.23 (a) Model resolution matrix (Equation 5.40). (b) Cumulative sensitivity map (Equation 5.41) for inversion shown in Figure 5.21. (c) Replotted image in Figure 5.21 using the sensitivity map to change opaqueness of the image (c.f. Figure 5.21). The dashed line in (b) and (c) shows the sensitivity map equal to 10^{-3} of its maximum value (see text).

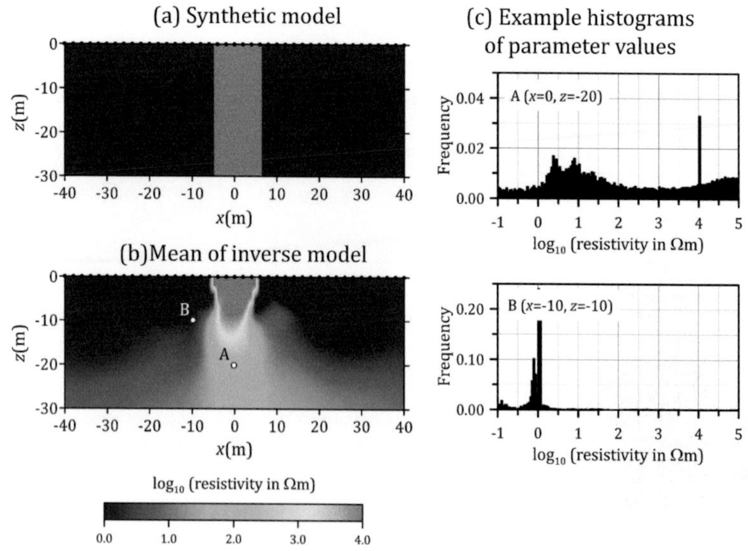

Figure 5.27 Example of transdimensional resistivity inversion based on Galetti and Curtis (2018). (a) Synthetic model. (b) Mean inverse model. (c) Parameter histograms from two locations within the region (marked in (b)).

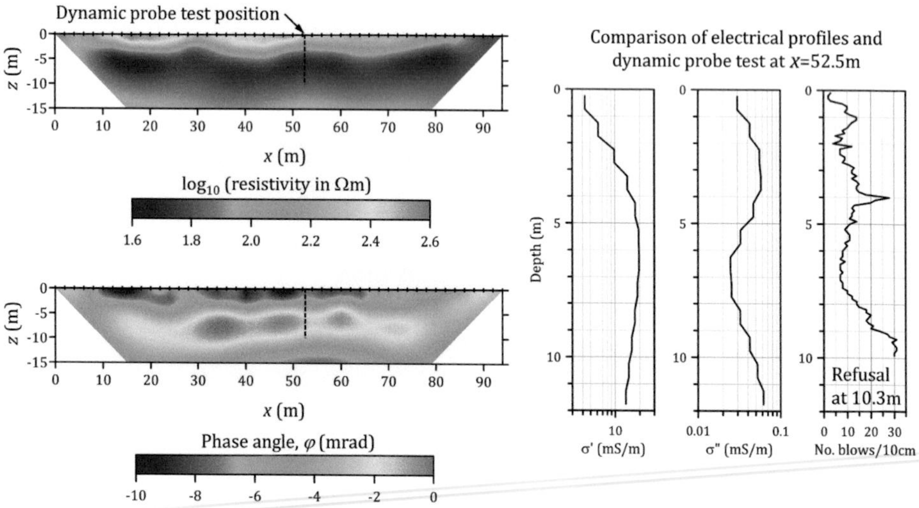

Figure 5.28 Complex resistivity inversion of surface electrode data. The survey was carried out with 48 electrodes at 2 m spacing (positions shown in the images). The three graphs show a comparison of inverted profiles (shown as real and imaginary conductivity) and a dynamic probe test. Refusal of the test was reached at 10.3 m (top of bedrock).

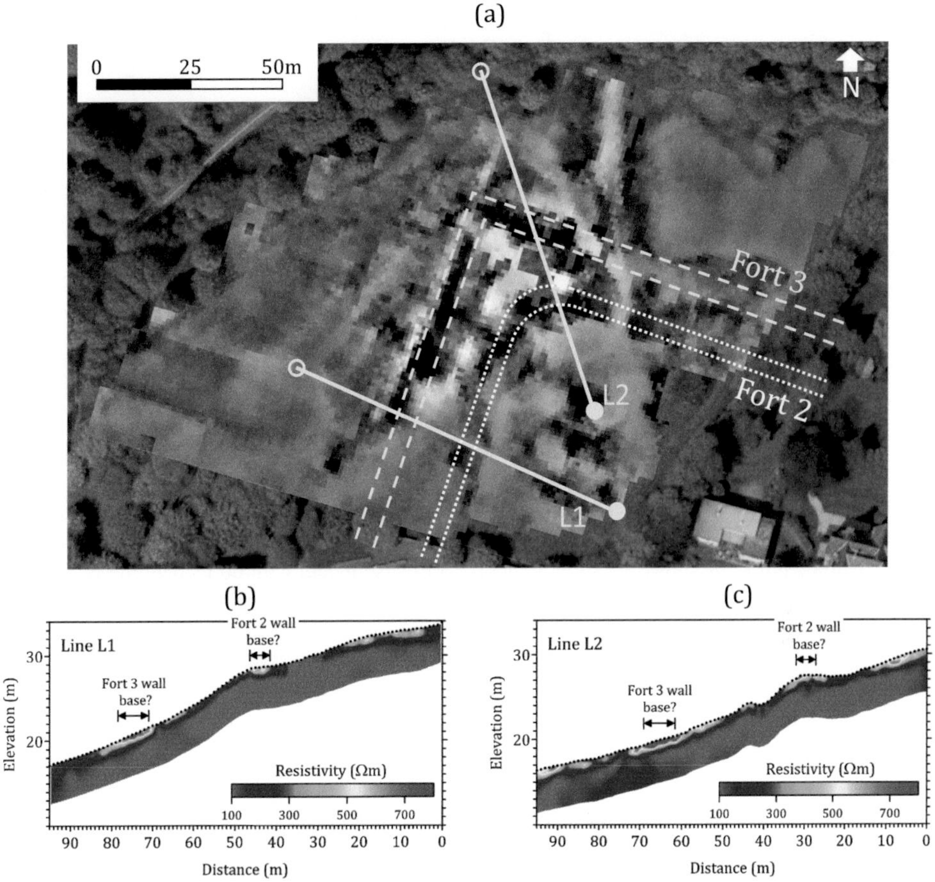

Figure 6.1 Geophysical surveys at the Lancaster Vicarage Fields site. (a) Composite twin array resistance survey (white: low resistance; black: high resistance), the interpreted position of the foundation of two Roman fort walls (from Wood, 2017) and the position of two 2D resistivity imaging lines (solid circle indicates zero distance on each line). 2D resistivity image for line L1 (b) and line L2 (c). Note distorted vertical scale in figures (b) and (c). The datum for the elevation scale in (b) and (c) is mean sea level.

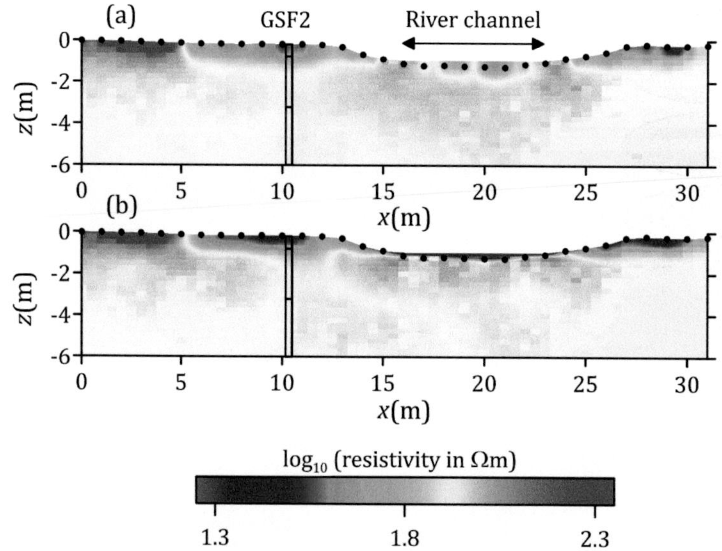

Figure 6.2 Resistivity image of a 32 electrode Wenner survey crossing the river Lambourn (electrode positions shown by solid circles). (a) Inverse model with no recognition of the water body. (b) Inverse model that fixes the resistivity of the cells in the water body to the measured value (20 Ωm). The images show the position of a shallow borehole adjacent to the river (log shown to the right of the figure). The cumulative sensitivity map (Equation 5.41) has been used to highlight opaqueness in the resistivity image in areas of low sensitivity.

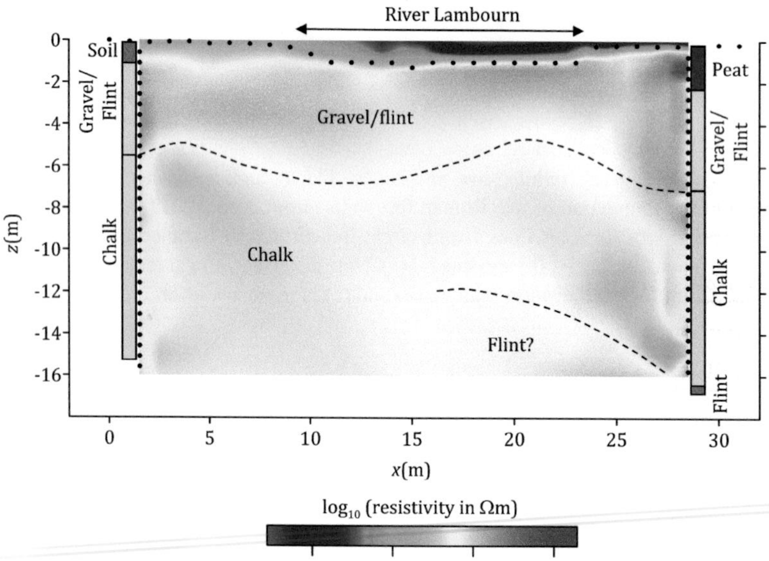

Figure 6.4 Cross-borehole resistivity imaging beneath the river Lambourn using data reported in Crook et al. (2008). The cumulative sensitivity map (Equation 5.41) has been used to highlight opaqueness in the resistivity image in areas of low sensitivity. Electrode positions are shown by solid circles. Borehole logs are shown for comparison.

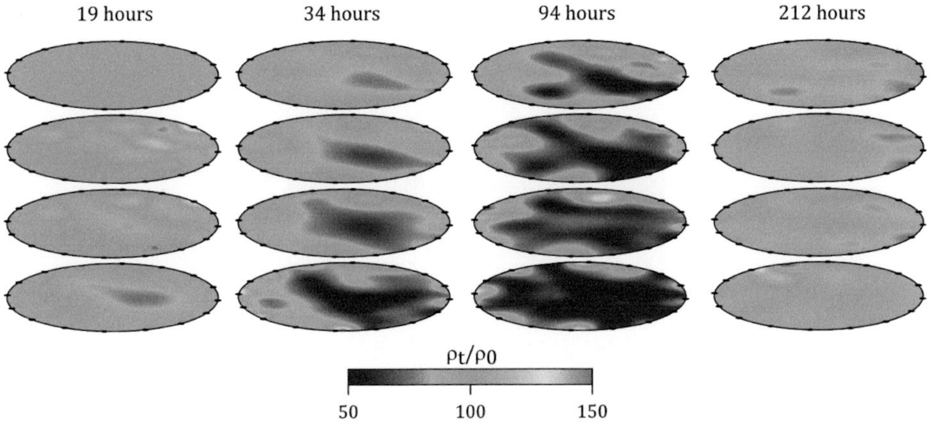

Figure 6.8 Time-lapse inversion of example datasets from the study of Binley et al. (1996a). The four planes are shown in the same layout as in Figure 6.7. The solute concentration is reduced 97 hours after the experiment starts.

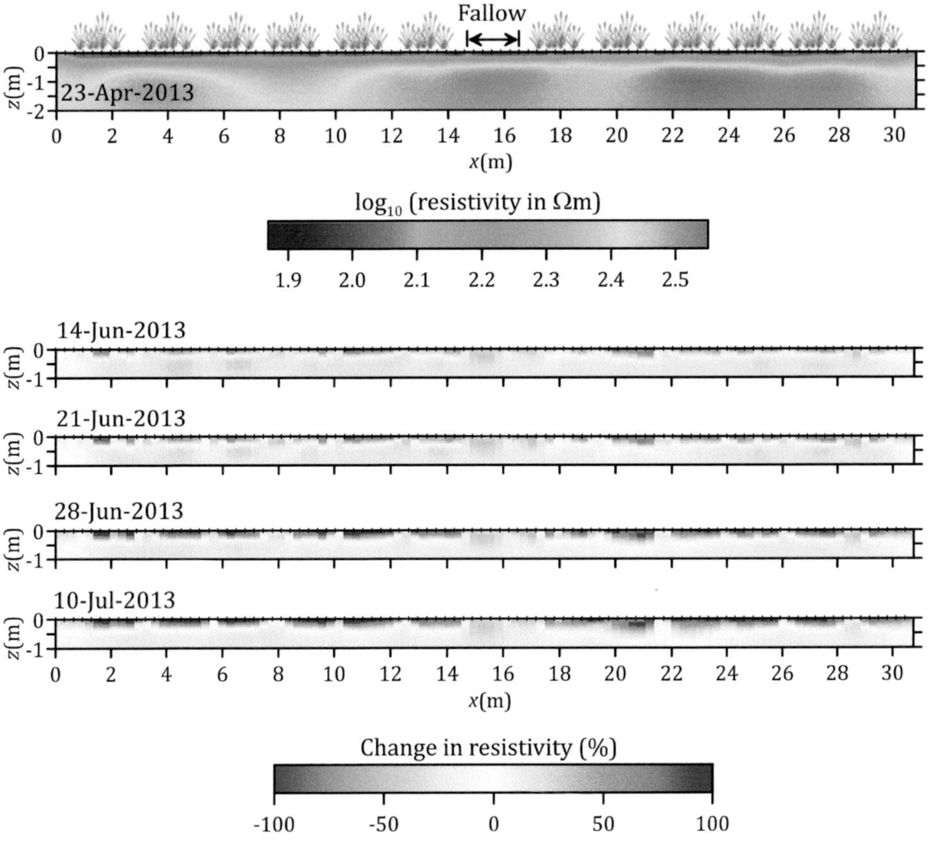

Figure 6.11 Monitoring changes in resistivity due to water uptake by winter wheat. Twelve different breeds of whcat are grown along the 30 m transect. The upper image shows the background resistivity image at early stages of crop growth. The lower images show changes in resistivity during crop growth The position of the fallow plot is shown in the upper image.

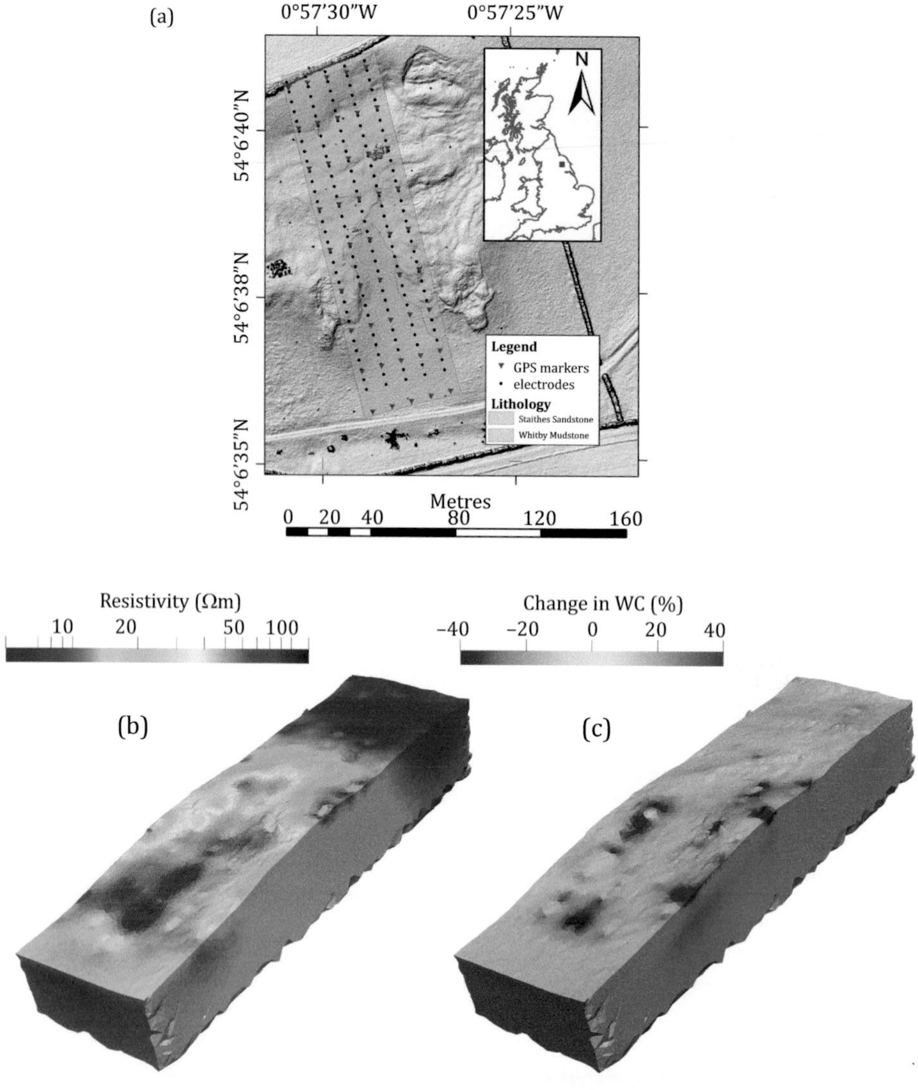

Figure 6.12 Example results from the Hollin Hill Landslide Observatory. (a) LiDAR image overlain by monitoring array positions and geological information. (b) 3D resistivity image from data collected 12 December 2012. (c) Example 3D image of change in gravimetric moisture content (between 12 March 2010 and 12 December 2012) inferred from temporal changes in resistivity. Inversions conducted by Jimmy Boyd (BGS, Lancaster University).

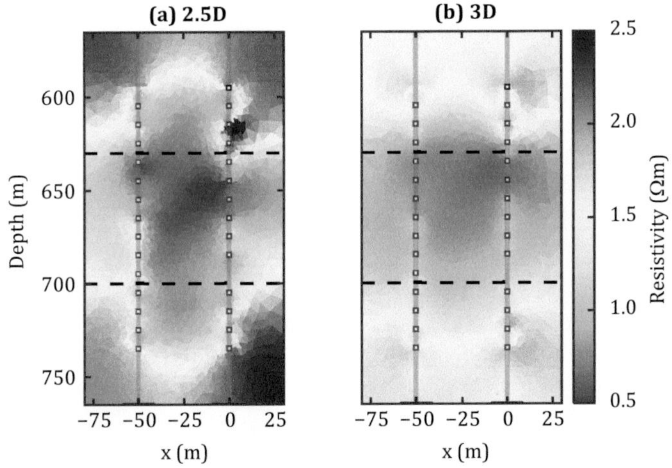

Figure 6.13 Inversion of Ketzin 21-June-2008 data. (a) 2.5D inversion. (b) 3D inversion accounting for borehole geometry. The horizontal broken lines show the formation into which CO_2 is injected. Figure based on Wagner et al. (2015).

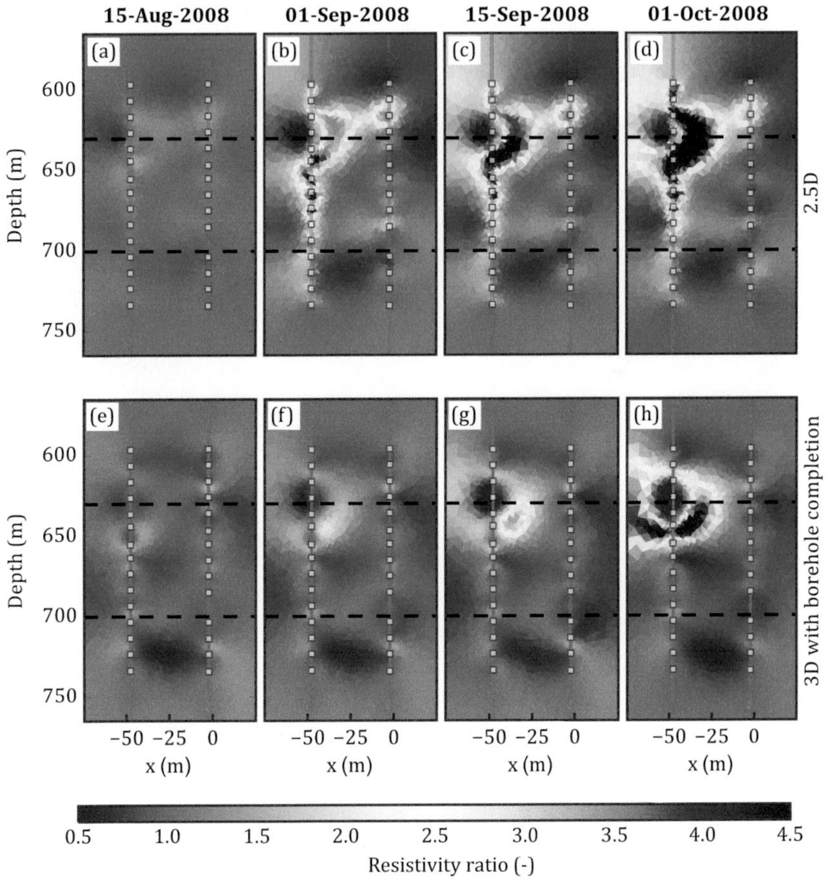

Figure 6.14 Time-lapse inversion of Ketzin data. (a–d) 2.5D inversion. (e–h) 3D inversion accounting for borehole geometry. The horizontal broken lines show the formation into which CO_2 is injected. The results are shown as a ratio of resistivity relative to the 21-June-2008 model. Figure based on Wagner et al. (2015).

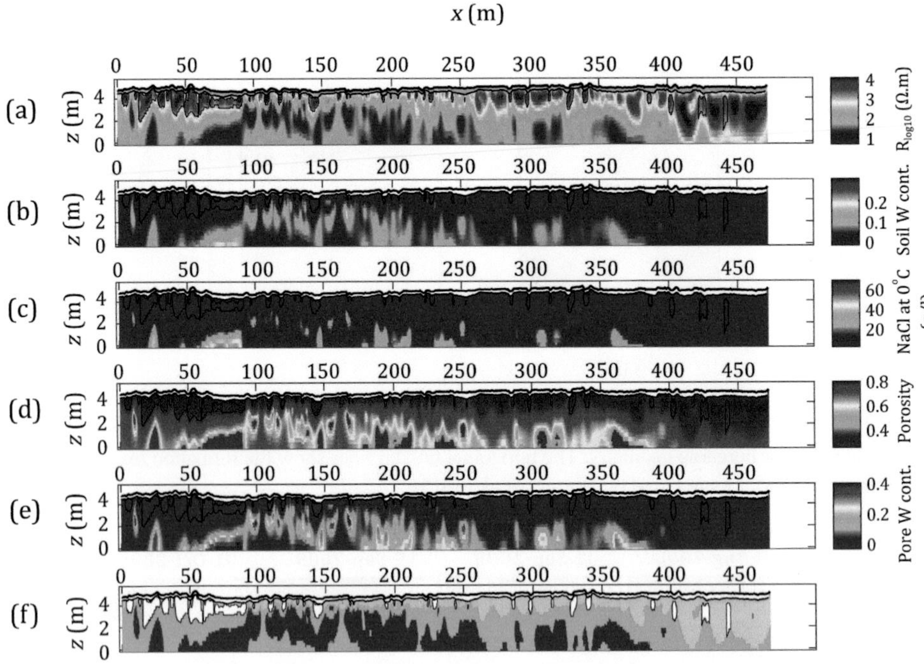

Figure 6.16 Results of a 500 m 2D ERI survey at an experimental research site near Barrow, Alaska. Black line shows depth of active layer from direct probing. (a) Resistivity distribution. Petrophysical interpretation: (b) soil water content, (c) initial salinity of thawed soil, (d) porosity, (e) fraction unfrozen water content and (f) interpretation including predicted location of ice wedges (white), as well as regions of permafrost with higher salinity (blue). After Dafflon et al. (2016).

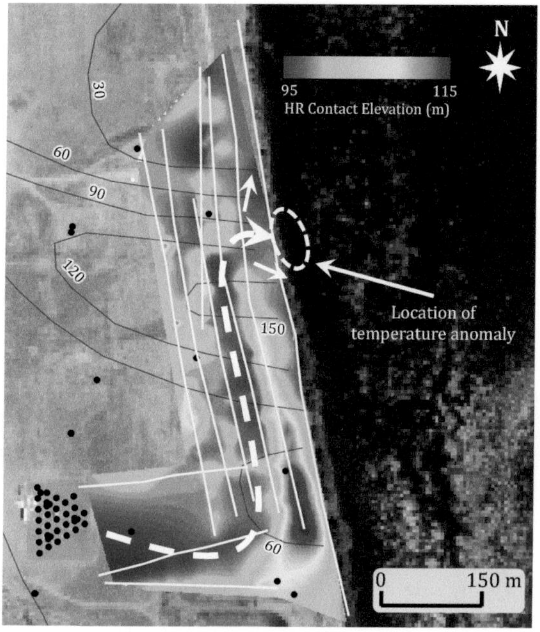

Figure 6.18 Elevation of the Hanford–Ringold contact determined from imaginary conductivity images compared to a prominent temperature anomaly (from distributed temperature sensing [DTS]) showing (1) groundwater–surface water exchange locations, and (2) contours of uranium concentrations (mg/L) in aquifer from boreholes (after Mwakanyamale et al., 2012).

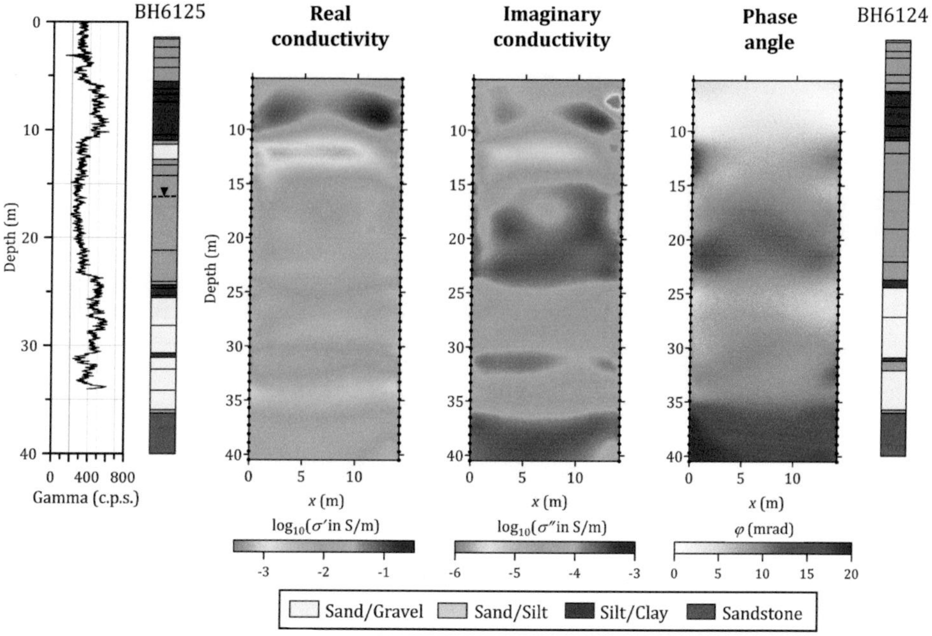

Figure 6.19 Cross-borehole IP imaging at the Low Level Waste Repository site, Cumbria, UK. Each image is between boreholes BH6125 and BH6124. Natural gamma log for BH6125 is shown, along with geological logs for both boreholes. Images based on data reported in Kemna (2004).

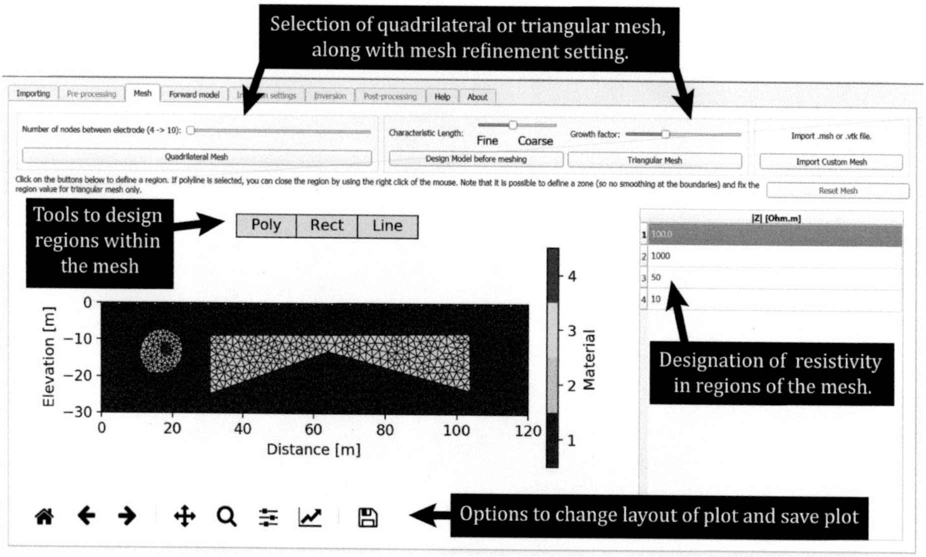

Figure A.1 Screenshot from *ResIPy* showing meshing options and designation of resistivity structure for forward modelling.

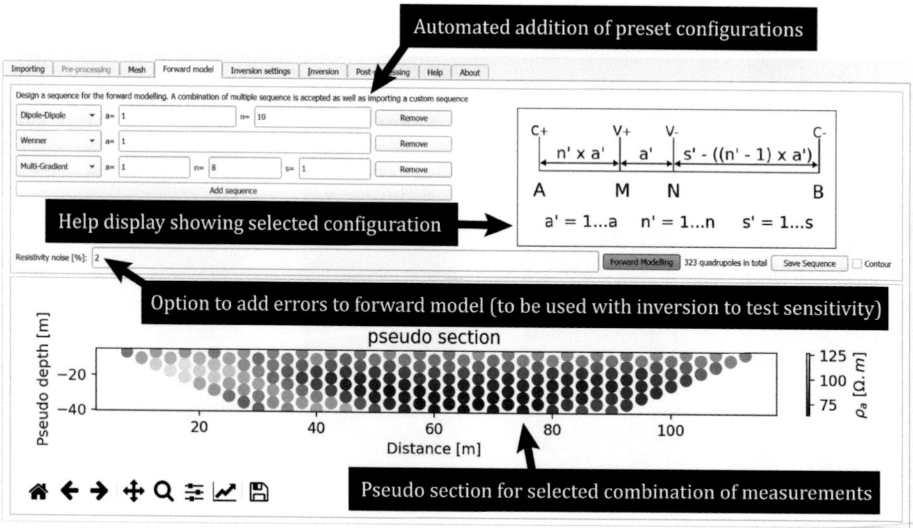

Figure A.2 Screenshot from *ResIPy* showing design of measurement sequence and pseudosection plotting following forward modelling stage.

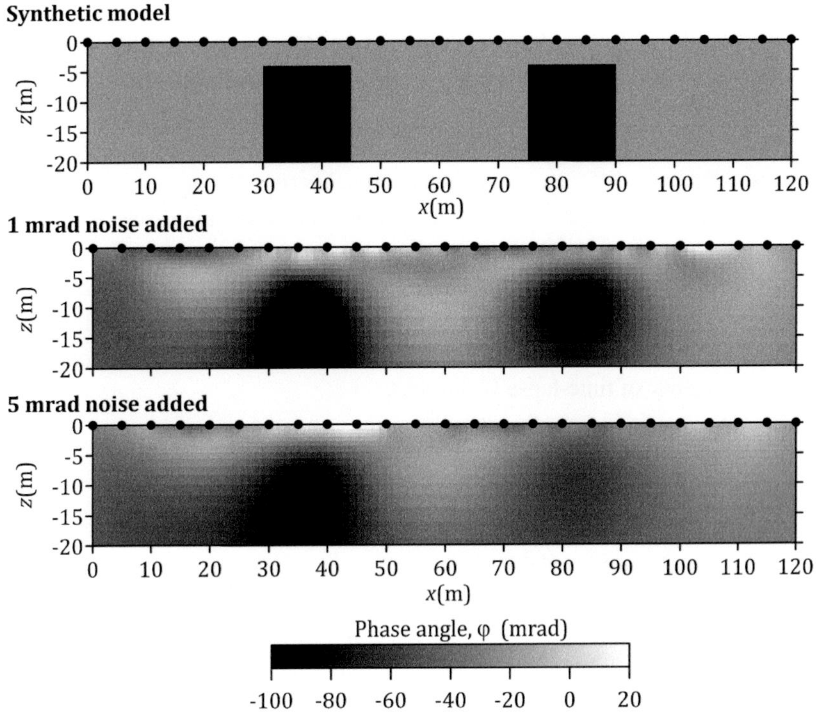

Figure 5.29 Complex resistivity phase angle inverse models for the problem in Figure 5.11 revised to include two polarizable bodies (as shown in the upper image). The two lower images show inverse models for different levels of noise applied to the synthetic datasets.

5.3.6 Time-Lapse Inverse Modelling

As in the case of DC resistivity (see Section 5.2.2.10), IP data can be modelled in a time-lapse framework. Although applications are rare, they have been driven by the appreciation that dynamic subsurface processes can be observed from changes in IP signatures. Examples include: fluid movement in hydraulic embankments (Abdulsamad et al., 2019); plant root-soil interactions (Weigand and Kemna, 2019); biogenic gas generation (Mendoca et al., 2015); remediation of contaminated groundwater (Williams et al., 2009; Flores Orozco et al., 2013; Sparrenbom et al., 2017). Example case studies are also described in Sections 6.2.6, 6.2.7 and 6.2.8.

The modelling of time-lapse IP data is a relatively straightforward extension to DC resistivity time-lapse analysis. For time domain IP data, if we adopt the formulation in Equation 5.57, then, assuming minimal change in the derivatives, a linear relationship between changes in observed apparent and intrinsic chargeabilities follows:

$$\Delta M_a = \sum_{j=1}^{M} \Delta \hat{m}_j \frac{\partial \log \rho_a}{\partial \log \rho_j}, \tag{5.61}$$

thus, the same linear inversion step used for static imaging can be applied to time-lapse data.

More sophisticated approaches that utilize true 4D inversion strategies have been proposed (e.g. Kim et al., 2018), which are extensions of those developed for DC resistivity time-lapse inversion.

IP problems formulated in terms of complex resistivity are equally extendable to time-lapse analysis: the same difference inversion approaches (e.g. Equation 5.32) can be applied to complex resistivity data, as can the simpler ratio inversion approach. Karaoulis et al. (2013) illustrate how the full 4D analysis of complex resistivity data can be applied.

When dealing with time-lapse IP data, measurement errors can have a strong influence on the quality of inverse models produced. Flores Orozco et al. (2019b) illustrate, with frequency domain data collected at a groundwater contaminated site, approaches for analysing data quality of time-lapse IP surveys.

5.3.7 Inversion of Frequency-Dependent Properties

As discussed in Chapter 2, electrical properties are frequency dependent and the spectral behaviour of polarization can offer insight into fundamental properties and states of subsurface materials. The intrinsic chargeability utilized so far in this chapter represents a cumulative measure of polarization, whereas complex resistivity represents one specific frequency. In some applications, we wish to determine spectral characteristics of the subsurface. In this section, we first show how specific properties can be determined from measurements in the time domain or frequency domain. We then illustrate how such concepts can be extended to field-based surveys.

5.3.7.1 Relaxation Modelling

Recalling Equation 2.113, the Pelton et al. (1978) form of the Cole–Cole relaxation model can be expressed as

$$\rho^*(\omega) = \rho_0\left[1 - \sum\nolimits_{i=1}^{n_r} \widetilde{m}_i\left(1 - \frac{1}{1 + \left(i\omega\tau_{0,i}\right)^{c_i}}\right)\right], \tag{5.62}$$

where n_r is the number of relaxation terms, ω is the angular frequency ($\omega = 2\pi f$), f is frequency, ρ_0 is the DC resistivity, $\widetilde{m}_i$, $\tau_{0,i}$ and c_i are the chargeability, time constant and Cole–Cole exponent for relaxation i. A number $n_r > 1$ is used to represent higher-frequency dispersion (e.g. associated with coupling) (Pelton et al., 1978).

Kemna (2000) outlines a Levenberg–Marquardt scheme to determine the $3n_r + 1$ parameters given a spectrum of $\rho^*(\omega)$ measurements. Measurements can be weighted in the same manner as discussed earlier. Figure 5.30 shows an example model fit to complex resistivity spectra measured on a sandstone sample (see Osterman et al. (2016) for the measurement setup). A dual relaxation model was used in this example, resulting in seven fitted parameters, although parameters, ρ_0, $\widetilde{m}_1$, $\tau_{0,1}$ and c_1 are the low-frequency properties of interest. The case study in Section 6.2.5 shows how Cole–Cole modelling can be used to assess links between spectral properties and permeability.

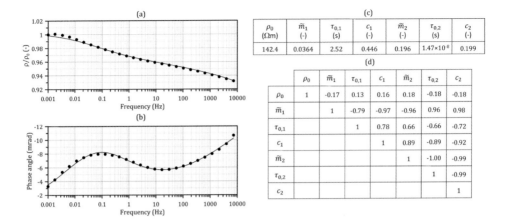

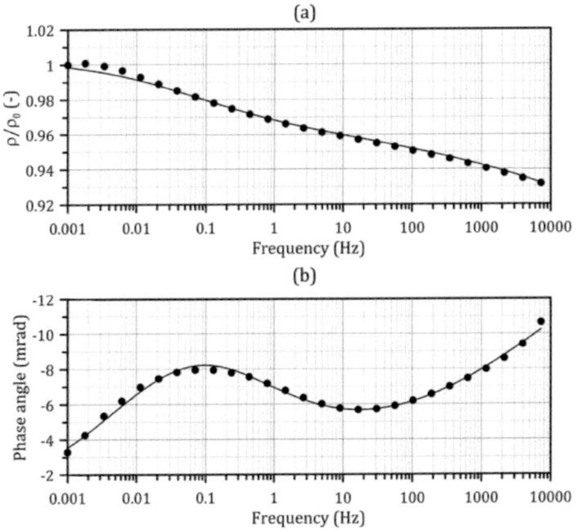

(c)

ρ_0 (Ωm)	$\widetilde{m}_1$ (-)	$\tau_{0,1}$ (s)	c_1 (-)	$\widetilde{m}_2$ (-)	$\tau_{0,2}$ (s)	c_2 (-)
142.4	0.0364	2.52	0.446	0.196	1.47×10^{-8}	0.199

(d)

	ρ_0	$\widetilde{m}_1$	$\tau_{0,1}$	c_1	$\widetilde{m}_2$	$\tau_{0,2}$	c_2
ρ_0	1	-0.17	0.13	0.16	0.18	-0.18	-0.18
$\widetilde{m}_1$		1	-0.79	-0.97	-0.96	0.96	0.98
$\tau_{0,1}$			1	0.78	0.66	-0.66	-0.72
c_1				1	0.89	-0.89	-0.92
$\widetilde{m}_2$					1	-1.00	-0.99
$\tau_{0,2}$						1	-0.99
c_2							1

Figure 5.30 Example Cole–Cole fit to measured spectral IP data. (a) Resistivity magnitude showing measurements (symbols) and modelled curve (line). (b) Phase angle showing measurements (symbols) and modelled curve (line). (c) Fitted parameters. (d) Correlation matrix of fitted parameters. The model fit was carried out using the *SpecFit* code written by Andreas Kemna.

To appraise the model, the model covariance matrix $\boldsymbol{C_m}$ (see Section 5.2.4.4) can be computed using Equation 5.44 (in this case α and $\boldsymbol{R}$ do not apply). From this, an assessment of correlation between model parameters i and j can be computed (as in Section 5.2.2.4 for 1D soundings) from

$$r_{i,j} = \frac{C_{m\ i,j}}{\sqrt{C_{m\ i,i} C_{m\ j,j}}}. \tag{5.63}$$

Figure 5.30b shows the correlation matrix for the fitted parameters in this example. The high correlations between high-frequency terms $\widetilde{m}_2$, τ_2 and c_2 is inevitable given minimal coverage of high frequency data, and of no real concern since these serve purely to remove high-frequency effects not related to the properties of interest. However, the low-frequency parameters $\widetilde{m}_1$, τ_1 and c_1 also show significant correlation, highlighting the non-uniqueness in determination of these parameters. Xiang et al. (2001) developed an approach for direct estimation of Cole–Cole parameters from complex resistivity spectra. Such an approach does not permit the assessment of the model covariance matrix (as above). The determination of relaxation model parameters, as above, using Gauss–Newton-based methods can also be highly dependent on the starting model, i.e. the inversion can get caught in a local minimum of the objective function. Ghorbani et al. (2007) and Chen et al. (2008) proposed Bayesian-based approaches as a means of quantifying posterior model uncertainty. If MCMC sampling is used (as in Chen et al. (2008)), then compared to the local-based Gauss–Newton method the computational costs are much higher, but not excessive given the small number of parameters to be determined.

The Pelton et al. (1978) Cole–Cole model, as used above, often requires at least two relaxation models in order to match observed spectra. An alternative relaxation model formulation is based on a distribution of n_r Debye relaxations:

$$\rho^*(\omega) = \rho_0 \left[1 - \sum\nolimits_{i=1}^{n_r} \widetilde{m}_i \left(1 - \frac{1}{1 + i\omega\widetilde{\tau}_{0,i}} \right) \right], \tag{5.64}$$

where n_r is now a sufficiently large number to allow the model to mirror the measured response. In fact, the Cole–Cole model is equivalent to a series of Deybe models, following a log-normal distribution of time constants $\widetilde{\tau}_{0,i}$.

Using Equation 5.64, the total chargeability $m_{tot} = \sum\nolimits_{i=1}^{n} \widetilde{m}_i$ is analogous to the chargeability $\widetilde{m}$ in the Cole–Cole model. Nordsiek and Weller (2008) present a method for the determination of such a distribution of Debye relaxation models (see also Zisser et al., 2010b), commonly deferred to as Debye decomposition. Keery et al. (2012) extended this approach using MCMC sampling in a Bayesian formulation, thus giving estimates of parameter uncertainty. Weigand and Kemna (2016b) developed a means of inverting time-lapse complex resistivity spectra using a regularized Debye decomposition approach.

So far in this section we have considered frequency domain measurements. The majority of field-based IP surveys are, however, carried out in a time domain mode. In Section 3.3.2.3, we showed how Fourier analysis of a suitably digitized current and measured voltage signal (over the entire pulse sequence) can be analysed to derive frequency domain equivalent measurements, although such analysis is unlikely to provide a broad enough spectral coverage to allow assessment of relaxation model parameters. Such an approach is now commonly referred to as full waveform analysis. Much earlier, however, there has been interest in using the measured voltage decay (typically measured in an IP survey) to derive relaxation parameters (Pelton et al., 1978; Tombs, 1981; Johnson, 1984). Early efforts used graphical solutions (master curves) (e.g. Tombs, 1981). More recently, stochastic inversion approaches have been developed (e.g. Ghorbani et al., 2007). The latter approach is particularly attractive given the computation of parameter uncertainty estimates. Vinciguerra et al. (2019) approached the inverse modelling using particle swarm optimization (see Section 5.2.5.2), analysing the full digitized current and voltage waveforms.

5.3.7.2 Imaging Relaxation Properties

In the previous section, we focused on the determination of spectral properties from a single measurement, e.g. for determination of the bulk properties of a homogenous sample in the laboratory. We can extend these approaches for analysis of imaging data. Treatment of complex resistivity (i.e. frequency domain) data is a logical extension of the analysis above. Data collected at individual frequencies can be inverted individually, after which parameter cell values can be analysed as spectra and relaxation model inversion applied. Kemna (2000) first illustrated this approach using a Cole–Cole model. Other examples include Williams et al. (2009). Weigand et al. (2017) compared the recovery of Cole–Cole and Deybe relaxation model parameters from multi-frequency imaging. Such an approach could, in theory, be extended by inverting data from all frequencies together (analogous to a time-lapse inversion), by regularizing across space and frequency dimensions.

Given the theoretical link between relaxation properties and time domain IP decay (as discussed in the previous section), it is also possible to image spectral properties using time domain data directly. Yuval and Oldenburg (1997) showed how an image of Cole–Cole parameters can be determined from analysis of individual gates in the voltage decay, building on the expression in Equation 3.10 (see also Hördt et al., 2006). Fiandaca et al. (2012) developed an alternative approach that utilizes a frequency domain transfer function to link Cole–Cole model parameters and voltage decay measurements. More significantly, perhaps, the method of Fiandaca et al. (2012) accounts for the superposition of the pulse response (i.e. the IP effect from the first cycle can exist in subsequent cycles, given the finite pulse length). Fiandaca et al. (2013) illustrate the approach in an imaging context; Fiandaca et al. (2018a) show how the method can be applied to different relaxation models; Bording et al. (2019) further demonstrate the method for a cross-borehole application. Recognizing the limitations of deterministic inversion approaches, as used by Fiandaca et al. (2013) and others, Madsen et al. (2017) used the forward modelling approach of Fiandaca et al. (2012) in a stochastic framework with MCMC sampling to illustrate how improved estimates of relaxation model parameter uncertainties can be determined from field-based time domain IP data. The computational constraints limited their analysis to 1D soundings but higher dimensionality problems will no doubt be soon realized.

5.3.8 Inverse Model Appraisal

The methods discussed in Section 5.2.4 for inverse model appraisal are equally applicable to the IP inverse problem, although their use is rarely reported. Kemna (2000) shows how the cumulative sensitivity map (illustrated in Figure 5.23 for DC resistivity) can be used for complex resistivity inversion. Weigand et al. (2017) examine the reliability of single and multi-frequency complex resistivity inversion and offer insight into the likely uncertainty in relaxation model parameters. As in the DC resistivity case, stochastic methods are ideally used to quantify the uncertainty in inverse models. Ghorbani et al. (2007) and Madsen et al. (2017) are rare examples of the use of such methods for IP model appraisal, in both cases focusing on 1D models. The quantification of such uncertainty is clearly important if petrophysical relationships (as discussed in Chapter 2) are to be extracted from IP inverse models to estimate properties of the subsurface.

5.4 Closing Remarks

For most applications, the interpretation of DC resistivity and IP data involves a degree of modelling, usually in the form of inversion. Developments in inverse modelling of electrical data over the past few decades have transformed our ability to interpret measurements. Such developments have been underpinned by advances in forward modelling. Fifty years ago, we relied on subjective curve matching approaches to interpret 1D sounding data; nowadays we can invert 3D datasets on personal computers, and even do this in the field for relatively small problems. For the majority of applications, the principles behind inverse modelling of electrical data have not really changed over recent decades: most utilize Gauss–Newton gradient-based methods in the search for an optimum model that is consistent with the data. The simplicity and robustness of such approaches make them particularly appealing, and their use has no doubt contributed immensely to the success of the DC resistivity and IP techniques. Appendix A includes a list of available codes for 1D, 2D and 3D modelling, along with details of *ResIPy* – an open source modelling environment developed to allow the reader to analyse datasets used throughout the book, and also model their own datasets.

We have detailed the main elements of forward and inverse modelling and attempted to illustrate how these have evolved to present day approaches. There is now a growing appreciation of uncertainty in inverse models, and advances are being made to formulate the inverse problem in a probabilistic manner. We have highlighted some of the advances in this area. Understanding the mathematical detail of forward and inverse models is not essential for practical use; however, the user needs to appreciate the underlying assumptions in order to make appropriate interpretation of their data.

In the next chapter we provide some example case studies, illustrating application of resistivity and IP modelling. In Chapter 7, we offer a perspective on future needs of modelling of DC resistivity and IP data, along with similar appraisals of instrumentation and petrophysics.

6
Case Studies

6.1 Resistivity Case Studies

6.1.1 Introduction

It will be apparent from the material covered so far that DC resistivity methods have been successfully applied to an immense range of applications and scales in environmental and Earth sciences. It is impossible (in any text) to showcase the full range of application areas; a number of case studies are presented in this section to highlight the breadth of use of DC resistivity, to showcase emerging areas and to illustrate some of the concepts covered earlier in the text. The case studies add support to material covered so far, showing specific context for the use of resistivity tools. A focus is the use of time-lapse measurements for monitoring dynamic processes. That said, we recognize that DC resistivity is widely used for more straightforward applications, including assessment of lithological boundaries. We include some examples of such applications but also refer the reader to the many established texts on applied geophysics.

For each case study, some context is provided and for most of the cases the reader is provided with references to allow further reading. Where possible, details of the approach used to measure and model data are given, and in a number of cases model files are provided in online supplementary information (www.cambridge.org/binley) to allow the interested reader to work with the data using the inversion codes provided with this text (see Appendix A).

6.1.2 Archaeology: Investigation of Roman Fort Remains in Lancaster

The castle in the UK city of Lancaster dates back to Norman times, although the grounds include evidence of earthworks and stone walls from Roman occupation. Situated close to the River Lune (hence the name Lancaster), it is believed that a Roman fortification at the site was a strategic base with easy access to the north west coast of England. Evidence of Roman occupation was revealed by archaeological excavations in the 1920s; further investigations took place between 1950 and 1970. In 2014, as part of Lancashire County Council and Lancaster City Council's 'Beyond the Castle' project, Oxford Archaeology North was commissioned to carry out a series of geophysical surveys within the site. These surveys, details of which can be found in Wood (2017), included magnetrometry, resistivity and ground penetrating radar, which helped direct trial excavations in the site. Two-dimensional resistivity imaging surveys performed in 2017 targeted areas of interest based on the original 2014 field campaign.

Figure 6.1a shows the results of the resistivity survey carried out in 2014 by Oxford Archaeology North using a twin array configuration (see Section 4.2.2.2) in Vicarage Fields: an area between the existing castle and the River Lune. For this survey a 0.5 m twin array was used, with a 1 m line spacing and 1 m spacing between measurements (Gaffney and Gater (2003) state that a 0.5 m twin array senses to depths of up to 0.75 m). Measurements were made using the RM15-D (Geoscan Research, UK) resistance meter. Surveys were conducted on 30 m by 30 m grids. Raw resistance measurements were de-spiked to remove anomalous readings and a filter was applied to remove large-scale trends due to geological variations within the site. Since remote electrode locations (for the twin array setup) changed for each 30 m grid square, in order to develop a composite resistance map a filter was applied to correct for the effect of remote position change. Note that, as explained in Section 4.2.2.2, archaeological geophysicists often report twin array surveys in terms of raw transfer resistances, not apparent resistivities.

The twin array resistance survey (Figure 6.1a) reveals significant variability within the site. However, a right-angled high resistance feature is clear. Wood (2017) interpreted this as the base of a substantial wall, the latest of three successive fort boundaries. Targeted excavations in 2016 confirmed the presence of the base of a 4 m wide stone and clay wall (photographic evidence is reported in Wood (2017)), along with several late Roman coins. 2D resistivity imaging surveys conducted in 2017 by Lancaster University used a Syscal Pro (Iris Instruments, France) with 96 electrodes at 1 m spacing; the dipole–dipole configuration was used with dipole spacing (a = 1 m, 2 m and 3 m) and levels (n = 1 to 11). Figure 6.1b, c show inverted resistivity images for the two lines marked in Figure 6.1a. Inversions were computed using *R2* (see Appendix A.2). Each image reveals a high resistivity (~1,000 Ωm) anomaly in the upper metre of the soil profile, with a width of several metres. The elevation of the anomaly on each line is similar and the positions are close to the interpreted Roman Fort 3 wall in Figure 6.1a, providing further geophysical evidence of historical earthworks at the site.

Other resistive anomalies along the survey lines suggest the presence of further localized features. As Wood (2017) argues, the site appears to have experienced a succession of Roman forts, expanding in size as occupation progressed. The resistive feature at ground elevation of approximately 28 m on lines L1 and L2 coincides with the location of the earlier Fort 2 wall, proposed by Wood (2017) and based on additional investigations at the site. In fact, subsequent shorter parallel surveys to line L1 indicate that the resistive feature approximately 45 m along line L1 (Figure 6.1b) extends to the north.

The geophysical results shown here are clearly susceptible to a range of interpretations and, as always, other observations must be integrated in order to further test hypotheses. At the Lancaster site, geophysical (and other) surveys continue, with the hope of establishing a clearer picture of the scale and nature of Roman occupation in the area, perhaps revealing how significant the site was, strategically, for the Roman Empire.

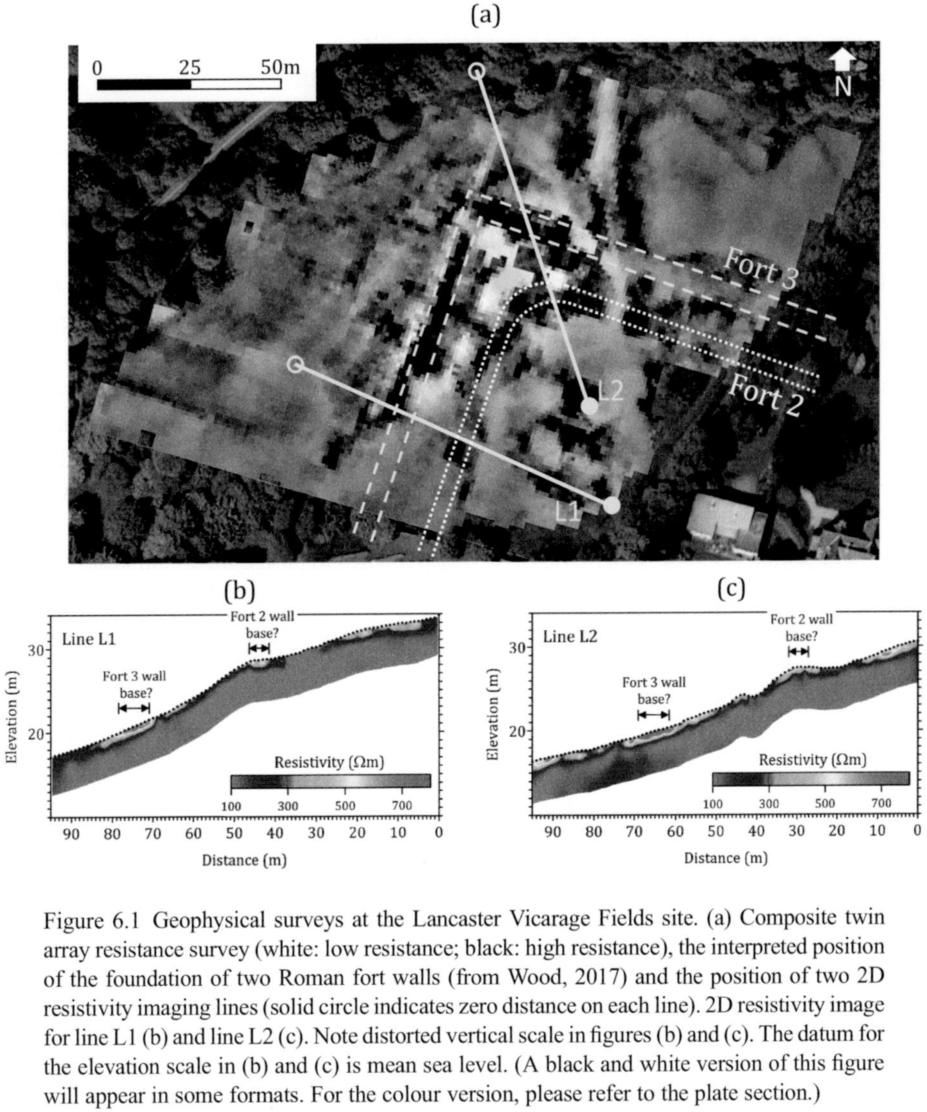

Figure 6.1 Geophysical surveys at the Lancaster Vicarage Fields site. (a) Composite twin array resistance survey (white: low resistance; black: high resistance), the interpreted position of the foundation of two Roman fort walls (from Wood, 2017) and the position of two 2D resistivity imaging lines (solid circle indicates zero distance on each line). 2D resistivity image for line L1 (b) and line L2 (c). Note distorted vertical scale in figures (b) and (c). The datum for the elevation scale in (b) and (c) is mean sea level. (A black and white version of this figure will appear in some formats. For the colour version, please refer to the plate section.)

6.1.3 Hydrogeology: Imaging at the Groundwater–Surface Water Interface

At the interface of aquifers and rivers (and other surface water bodies) the fabric of the subsurface can control exchange between groundwater and surface water. Such exchange controls the movement of water between the two systems and can influence the transfer of solutes. In groundwater-fed rivers, for example, the physical characteristics (e.g. permeability) of sediments beneath the river can impact on biogeochemical cycling, which can be critical for maintaining ecological health of the river.

Conventional drilling beneath water bodies is difficult and often prohibited because of the likely impact on the water course. Such methods are also localized and fail to give adequate spatial information. Electrical geophysical methods can offer additional insight into the structure of the deposits beneath a water body (Butler, 2009; McLachlan et al., 2017). Resistivity measurements can be made using streamer arrays floating on the water surface (see Section 4.2.2.5) or by placing electrodes on the riverbed. The former permits relatively long surveys (e.g. Rucker et al., 2011). Crook et al. (2008) illustrate the use of resistivity imaging in different settings and demonstrate the improved sensitivity to the subsurface when electrodes are placed at the bottom of the water column (see also Day-Lewis et al., 2006).

In this case study we show some results from resistivity surveys carried out in a lowland chalk catchment in the UK. The Chalk is the UK's major aquifer. In some areas, Chalk rivers are threatened by over abstraction of groundwater and deterioration of water quality, e.g. from excessive use of nitrogen-based fertilizers in the 1950s, which are slowly moving through the aquifer towards the river. In this study, resistivity surveys were conducted in the river Lambourn, within the Thames basin. The surveys were carried out in conjunction with other, more conventional, surveys to help develop a conceptual model of the Lambourn catchment, which could be used to improve management and protection of the water course.

Figure 6.2 shows the results of a resistivity imaging survey crossing the Lambourn near the village of Great Shefford. For this survey a Wenner configuration was used with 32 electrodes at 1 m spacing, 8 of which were placed within the channel (see Crook et al., 2006). Figure 6.2a shows an image of resistivity inverted in a conventional manner using *R2* (see Appendix A.2). Figure 6.2b shows the result of an inversion which honours the known resistivity (20 Ωm) of the cells representing the water body. In both resistivity images in Figure 6.2 opaqueness in areas of low sensitivity has been assigned based on the cumulative sensitivity map (Equation 5.41). A sharper definition of the subsurface structure is evident from the approach in which the surface water body is fixed, in particular the alluvial gravel unit. Comparison with a drilling log reveals some consistency, although the contrast between the gravel and Chalk bedrock is weak in places.

In this example, fixing the resistivity of the cells representing the water column appears to improve the inverse model, although Day-Lewis et al. (2006) show that fixing to an incorrect value can lead to artefacts. Defining an appropriate value of resistivity for the water column cells may appear trivial since measurement of fluid electrical conductivity is easy to make. However, it is important to recognize that the bulk resistivity of the water column may be affected by stratification (particularly in deep, slow-flowing rivers) and the presence of river vegetation. Furthermore, in high-energy upland water courses complex riverbed topography can exist, which if not accounted for can impact on the inverted model (McLachlan, 2020).

A cross-borehole survey configuration can also be used for investigating subsurface structures beneath a water course. Crook et al. (2008) reports results from a survey near the village of Boxford on the river Lambourn (several km downstream of the Great Shefford site). For this survey 32 electrodes were installed at 0.5 m spacing in two boreholes adjacent to the river, along with 32 electrodes at 1 m spacing crossing the river. A pole–pole configuration was used, with remote electrodes approximately 100 m from the watercourse. Note that the configuration does not allow the ideal borehole array length: separation aspect

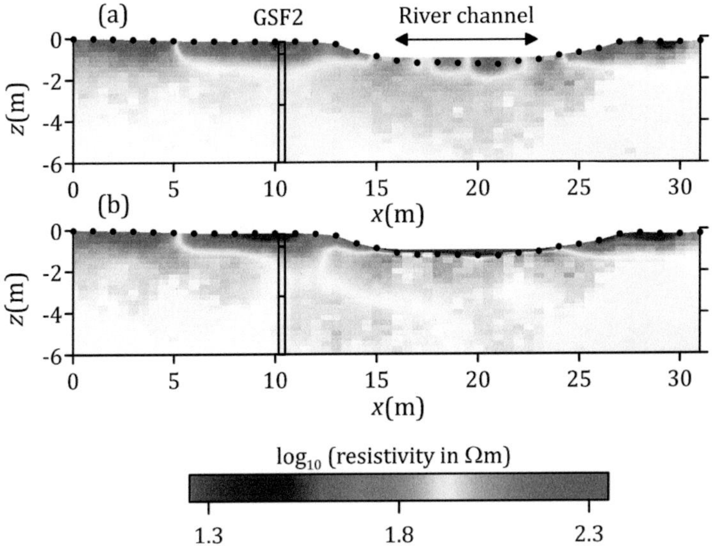

Figure 6.2 Resistivity image of a 32 electrode Wenner survey crossing the river Lambourn (electrode positions shown by solid circles). (a) Inverse model with no recognition of the water body. (b) Inverse model that fixes the resistivity of the cells in the water body to the measured value (20 Ωm). The images show the position of a shallow borehole adjacent to the river (log shown to the right of the figure). The cumulative sensitivity map (Equation 5.41) has been used to highlight opaqueness in the resistivity image in areas of low sensitivity. (A black and white version of this figure will appear in some formats. For the colour version, please refer to the plate section.)

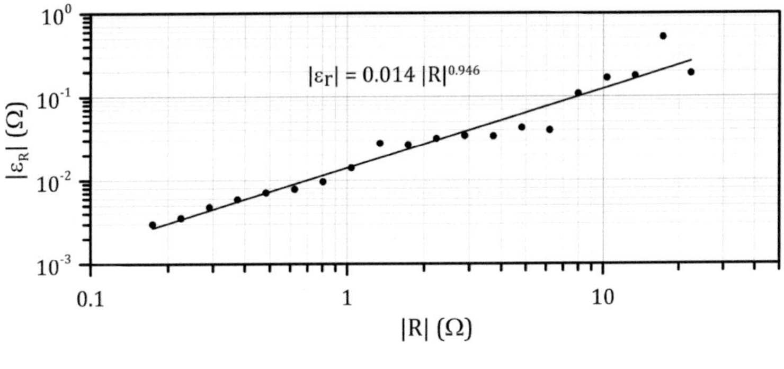

Figure 6.3 Variation in reciprocal error with transfer resistance for cross-borehole resistivity survey reported in Crook et al. (2008). The symbols represent average reciprocal errors for binned samples following the procedure described in Section 4.2.2.1. The solid line is the best fit to the trend in error with transfer resistance.

ratio of 2:1 (see Section 4.2.2.7) although the array of electrodes crossing the stream allows some enhancement of sensitivity beneath the riverbed. Error analysis (see Section 4.2.2.1) of the dataset revealed relatively low reciprocal errors (Figure 6.3). The inverted model

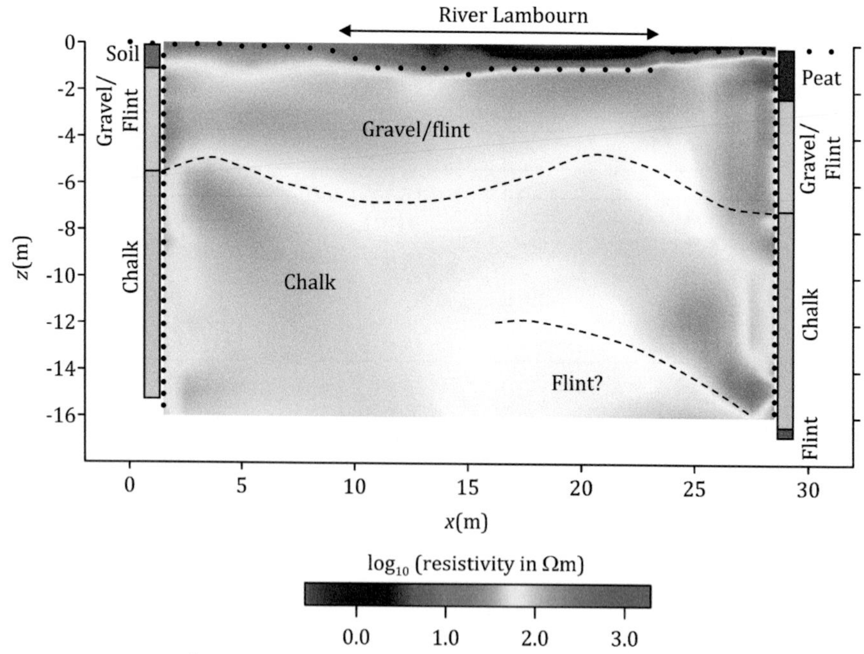

Figure 6.4 Cross-borehole resistivity imaging beneath the river Lambourn using data reported in Crook et al. (2008). The cumulative sensitivity map (Equation 5.41) has been used to highlight opaqueness in the resistivity image in areas of low sensitivity. Electrode positions are shown by solid circles. Borehole logs are shown for comparison. (A black and white version of this figure will appear in some formats. For the colour version, please refer to the plate section.)

(Figure 6.4), again using *R2*, shows a clearer definition of the gravel-Chalk interface due to the use of borehole electrode measurements. Note also that the gravel unit is much thicker at this location compared to the upstream Great Shefford site. The results in Figure 6.4 are also consistent with a larger scale survey carried out by Chambers et al., 2014).

The examples shown above focus on examination of subsurface structure. Resistivity imaging has also been used to identify seepage zones in lakes (e.g. Mitchell et al., 2008; Nyquist et al., 2009). Nyquist et al. (2009) inferred zones of groundwater seepage in a riverbed from a comparison of resistivity surveys conducted 4 months apart. Time-lapse resistivity surveys have also been used to monitor the migration of electrically conductive tracers injected in stream water upstream of a survey site (e.g. Nyquist et al., 2010; Ward et al., 2010, 2012; Toran et al., 2012). Such surveys appear effective in determining the movement of solutes between the river and bed sediments (i.e. the hyporheic zone) and the effectiveness of river restoration schemes. However, the injection of saline tracers in rivers is often prohibited to preserve water quality and ecological health: often we wish to conduct investigations in water bodies that are already under threat. Finally, it should also be noted that induced polarization (IP) methods, although rarely used on aquatic surveys, also show great promise for characterizing stream and riverbed sediments (e.g. Slater et al., 2010; Benoit et al., 2019).

6.1.4 Hydrogeology: Time-Lapse 3D Imaging of Solute Migration in the Unsaturated Zone

Understanding how solutes move within the unsaturated zone (the region of partial saturation above the water table) can improve our ability to assess the vulnerability of aquifers to contaminants originating at or near the ground surface. Conventional hydrological methods for sampling the unsaturated zone are limited since it is difficult to sample pore water above the water table: traditional suction samplers do not abstract water from the full range of pore sizes within the porous media, and are also limited in their measurement support volume (the volume over which the measurement relates). In fractured systems even greater challenges exist.

Geophysical methods offer an alternative approach for the monitoring of fluid flow and solute transport in unsaturated systems. Time-lapse resistivity imaging is particularly attractive in situations where changes in pore water content or solute concentration changes the electrical conductivity. Park (1998) demonstrated how measurements from an array of surface electrodes can be used to track movement of fluids injected into the unsaturated zone. In this study, a significant volume of water (397 m^3) was injected, with monitoring carried out over a 200 m by 200 m region. As shown in a later case study (Section 6.1.6), surface electrodes can be effective for monitoring more subtle changes in soil water content at shallow depths. For investigations at greater depths, surveys using borehole electrode arrays are more suitable given the required sensitivity at depth. Asch and Morrison (1989) is an early illustration of the use of a surface-borehole configuration for tracking fluids in the unsaturated zone; Daily et al. (1992) is the first reported example of the use of cross-borehole resistivity imaging for such a purpose.

This case study illustrates results from one of several tracer experiments carried out in the unsaturated zone of the UK Sherwood Sandstone aquifer. The field experiments were conducted in order to assess likely travel times of solutes moving from the surface to the water table, in part driven by concerns of elevated levels of solutes in nearby public supply wells. In 1998 four boreholes were installed to a depth of 13 m at the Hatfield site in Yorkshire (UK), each equipped with 16 electrodes. Additional boreholes were installed, as documented by Binley et al. (2002b). Within the centre of the plot a shallow borehole was installed to permit injection of tracers at a depth of 3.5 m (see Figure 6.5), directly into the sandstone. The first injection experiment was carried out in October 1998 using a water tracer (Binley et al., 2002b). Cross-borehole resistivity was used to monitor the migration of the water tracer. A similar experiment, also using a water tracer, was carried out in February 1999. These experiments revealed relatively rapid migration of water through the unsaturated zone. In 2003 a final experiment was conducted with a moderately saline tracer (Winship et al., 2006). The purpose of the 2003 campaign was to assess whether changes in resistivity observed at depth were likely to be due to rapid movement of tracer (new) water or remobilized (old) water. To assess this, Winship et al. (2006) adopted a combination of cross-borehole resistivity and ground penetrating radar surveys, using inferred changes in permittivity at depth (from transmission radar) to assess changes in water content, which could then be used to determine changes in pore water salinity (i.e. due to the tracer) from time-lapse images of resistivity.

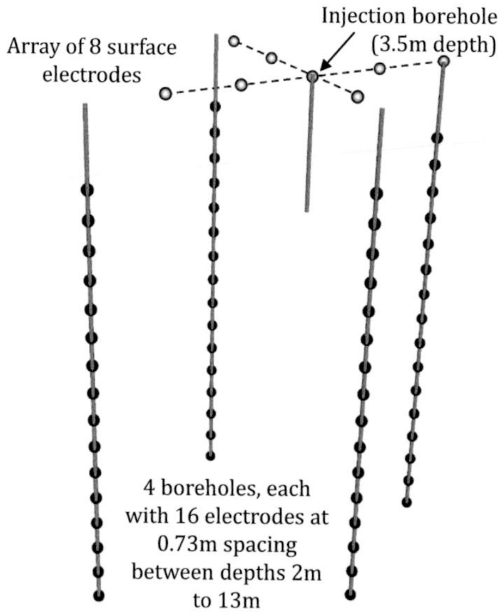

Figure 6.5 Electrode configuration for the Hatfield experiments.

Winship et al. (2006) carried out cross-borehole and surface-borehole measurements using the electrode arrangement shown in Figure 6.5. A six-channel Geoserve RESECS instrument was used, allowing the collection of 6,372 (including a full set of reciprocal) measurements in about 2.5 hours. A four electrode configuration (for cross-borehole measurements) was designed so that current and potential dipoles were horizontal and not separated more than 4.4 m vertically, in order to improve sensitivity to the tracer (injected in the centre of the borehole array) and maintain a good signal-to-noise ratio. The tracer (1200 L NaCl solution; electrical conductivity 2200 μS/cm – approximately three times the conductivity of native pore water) was injected over three days (14 to 17 March 2003) at a steady rate. Two background (pre-tracer) surveys were carried out; the tracer was then monitored for a period of one month. Winship et al. (2006) analysed the resistivity data using an earlier version of *R3t* (see Appendix A.2) based on hexahedral (brick type) elements.

Figure 6.6 shows examples of inferred changes in resistivity during the experiment. The growth of the tracer plume during injection is clearly seen in the first two images; the other images show the gradual downward movement. After 7 days the tracer appears to be impeded vertically and some lateral spreading is evident at approximately 8 to 9 m depth, which is consistent with observed layering of fine sandstone at the site (see Binley et al., 2001). As noted by Winship et al. (2006), despite relatively rapid vertical movement of the tracer, a significant reduction in resistivity remained at shallow depth. By analysing their radar and resistivity data jointly, Winship et al. (2006) were able to demonstrate that some of the reduction in resistivity at depth can be attributed solely to increases in water content, not

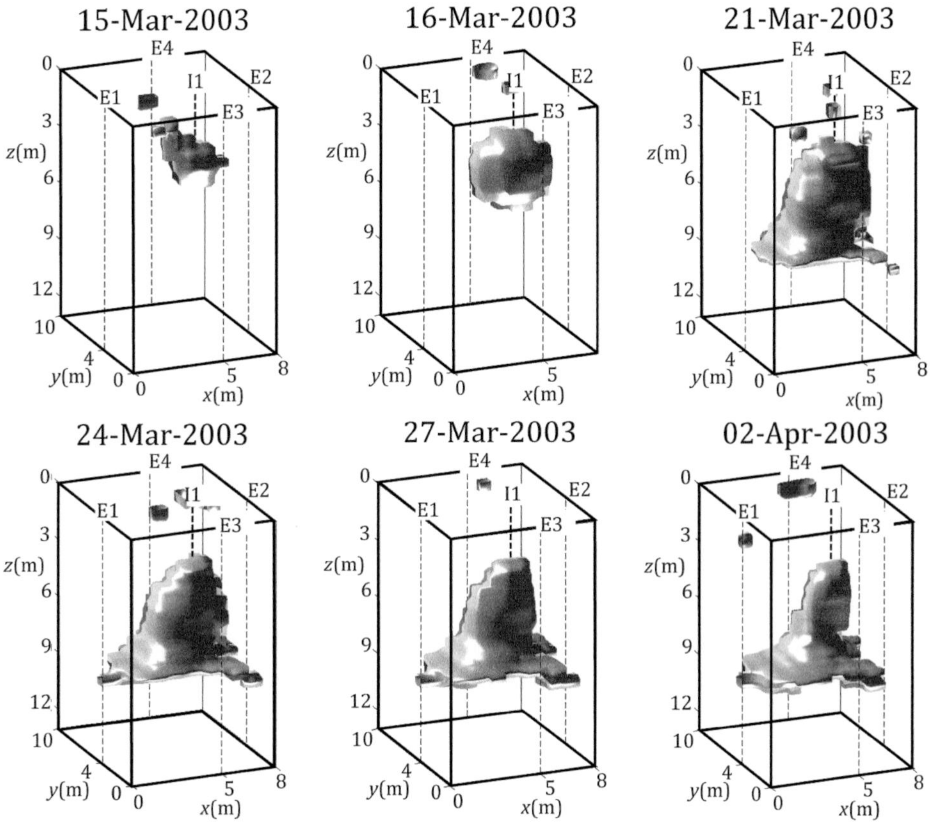

Figure 6.6 Changes in resistivity due to tracer injection at the Hatfield 2003 experiment. Tracer is injected between 14 and 17 March 2003. Boreholes E1, E2, E3 and E4 are equipped with electrodes. Borehole I1 is the injection site. The images show an isosurface of 7.5% change in resistivity from the pre-tracer image. Modified after Winship et al. (2006).

salinity, i.e. the displacement of old water, thus offering new insight into mechanisms controlling the migration of potential contaminants in the unsaturated zone.

A further novel aspect of the work of the Hatfield experiments was the use of hydrological models to interpret the geophysical signals. In fact, Binley et al. (2002b) calibrated an unsaturated flow model using the tracer geophysics dataset. They noted that the change in water content computed from inverted resistivity data underestimated the true (known) change by 50%, attributing this to sensitivity of the measurements, particularly in the centre of the imaged domain. Other investigators subsequently noted similar mass balance errors (e.g. Singha and Gorelick, 2005). Whilst several similar attempts to integrate electrical geophysical data and hydrological models in unsaturated zone studies are reported (e.g. Looms et al., 2008), most have been focussed on synthetic experiments.

Field experiments, like those at Hatfield, are useful demonstrators but because of their plot scale construction are limited, to a degree, in generalizable findings. Such experiments can, however, be used to test hypotheses about processes and can also be effective for investigating local scale problems, e.g. monitoring remediation technologies. Recently, Tso et al. (2019) reanalysed data from the Hatfield experiment of Winship et al. (2006) in order to assess the impact of uncertainty in petrophysical models on inferred changes in water content from time-lapse resistivity. They highlight the significance of such uncertainty – a factor that has been widely overlooked in field and model studies using time-lapse data.

6.1.5 Soil Science: Imaging Solute Transport in Soil Cores

Soil and rock cores are widely used in the laboratory to analyse the transport of solutes, in particular to examine dispersion mechanisms. Many natural soils are structured in fabric, which leads to dual flow behaviour, i.e. slow and fast pathways. Traditionally, this behaviour is analysed by subjecting a core to steady fluid flow and the introduction of a solute at the inflow end of the core. The breakthrough of a solute at the outflow is then measured and modelled to determine bulk properties of the core. In an ideal porous medium such breakthrough should follow a Gaussian type rise and fall of the solute concentration; in a structured medium more complex behaviour will be evident. Although such measurements can be modelled (e.g. Beven et al., 1993) there is a need to gain a greater understanding of the processes operating within a sample. Dye staining of a core has been widely used to gain such insight, however, such methods are destructive and consequently do not permit the investigation of solute transport mechanisms under, for example, different flow conditions.

Recognizing the above constraints of traditional approaches, Binley et al. (1996a) proposed the use of electrical resistivity imaging to analyse the movement of an electrically conductive solute. They documented the results of a laboratory experiment, which we summarize here and provide additional analysis. More comprehensive details of their experimentation are documented in Henry-Poulter (1996) and a companion study is reported in Binley et al. (1996b).

Binley et al. (1996a) extracted an undisturbed soil core, 34 cm in diameter and 46 cm long by excavating around a soil monolith, which was then wrapped in fibreglass before removal from the field. In the laboratory, four circular planes of 16 electrodes were installed around the core (Figure 6.7), which was seated on a porous plate, allowing flow to be injected from the base. Binley et al. (1996a) carried out DC resistivity measurements using two quadrupole geometries (Figure 6.7) resulting in 200 measurements per plane, plus a full set of reciprocal measurements for error analysis. The experiment was conducted in 1995, at which time multiplexed DC resistivity instruments were becoming available, although they suffered from slow data acquisition. Most instruments at the time were able to measure around 400 quadrupoles per hour. With four planes of electrodes, this would equate to a 4-hour data collection period, which is clearly limited if one is interested in relatively fast processes. To address this, Binley et al. (1996a) used a high speed UMIST Mk1b data acquisition system developed for process tomography by Wang et al. (1993). The system was slowed down to enable more stacking of measurements and improve accuracy, but still allowed a data acquisition time of 2 minutes per plane. The challenge with the instrument was that it could not measure effectively in a resistive environment. To address this as an electrically conductive fluid (2.5 mS/cm) was injected into the column at the base at a steady flow rate until the same conductivity was observed in the outflow. Then, a higher conductivity (6 mS/cm) tracer was injected (at the same flow rate) for 97 hours, after which time the flow was returned to the lower conductive solute.

During a monitoring period of 212 hours, 278 sets of data for the four planes were collected and analysed in 2D using a ratio inversion scheme (Section 5.2.2.10) using *R2* (see Appendix A.2). Figure 6.8 shows example time lapse images of the experiment. The non-uniform behaviour of the solute is clearly visible (and verified at the end of the experiment by dye staining; see Binley et al., 1996a).

Although time-lapse resistivity imaging experiments had been demonstrated a few years before this experiment (Daily et al., 1992), the volume of data collected was unprecedented. Furthermore, Binley et al. (1996a) introduced the concept of pixel breakthrough curves, using a measure of relative solute concentration in each pixel as

$$C_r = \left(\frac{\rho_0}{\rho_t} - 1\right) / \left(\frac{EC_0}{EC_1} - 1\right), \tag{6.1}$$

where ρ_0/ρ_t is the ratio of resistivity at background to that at time t, EC_0 is the electrical conductivity of the background solute EC_1 is the electrical conductivity of the tracer solute. These are illustrated in Figure 6.9 for the upper plane of the measurement setup. Figure 6.9b shows clearly contrasting behaviour of different pixels. Note, for example, that pixel 33 shows a decay in C_r immediately after the tracer is stopped, whereas the other two example responses show a later peak.

Binley et al. (1996a) did not model the pixel breakthrough response, but in a following study by Olsen et al. (1999) a transfer function approach was applied to a 3D imaging experiment. The experiment of Binley et al. (1996a) focussed on saturated conditions, avoiding complexity in interpretation due to changes in soil moisture Koestel et al. (2008) took this further by 3D imaging of the transient behaviour of a solute in an unsaturated soil core, deploying time domain reflectometry sensors in the core to allow accounting for variation in soil moisture. Koestel et al. (2009a, 2009b) analysed the data using solute transport models; Koestel et al. (2009c) provide dye staining verification of the analysis. Wehrer et al. (2016) developed the approach further by analysing reactive (nitrate) transport behaviour using resistivity imaging.

Resistivity imaging of solute transport in soil cores offers a means of examining the (sometimes) complex spatio-temporal behaviour of solutes in natural systems. Compared to other imaging technology (e.g. X-ray computed tomography) the resolution is inevitably inferior, however, the ease at which dynamic processes can be monitored and the scalability of experiments (small cores to large laboratory tanks, e.g. Slater et al., 2002; Fernandez et al., 2019) makes them effective in many studies. For time-lapse studies, careful design of experiments is necessary to ensure that each dataset represents a single snapshot of the system. Many modern multiplexed field resistivity instruments offer the capability of high-speed data collection, although few are able to compete with the sampling rate in the original experiment of Binley et al. (1996a).

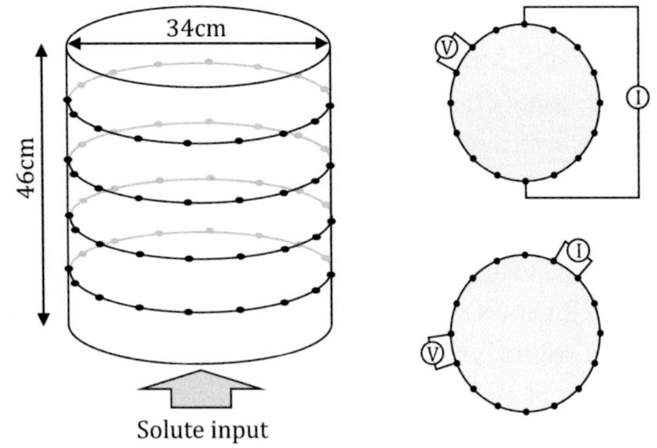

Figure 6.7 Soil column setup and measurement sequence used by Binley et al. (1996a). Four rings of 16 electrodes are installed on the column wall. A flow source at the base of the column allows solute injection. The two schematics on the right show examples of the measurement configuration used for each of the four planes.

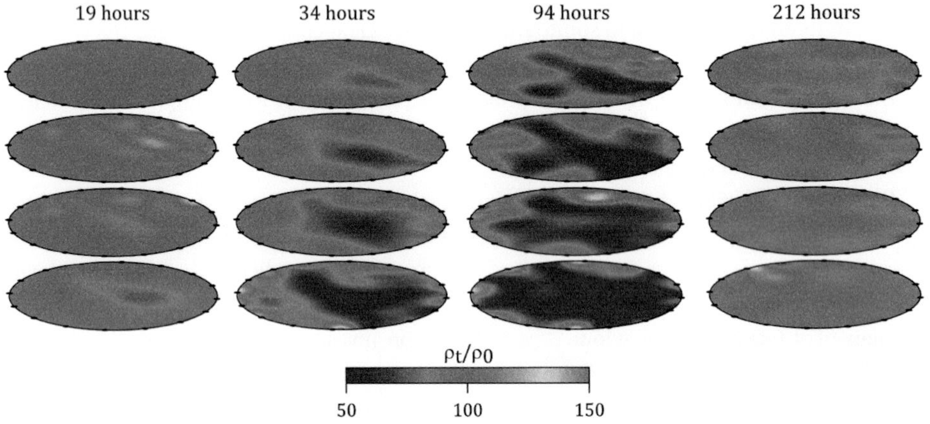

Figure 6.8 Time-lapse inversion of example datasets from the study of Binley et al. (1996a). The four planes are shown in the same layout as in Figure 6.7. The solute concentration is reduced 97 hours after the experiment starts. (A black and white version of this figure will appear in some formats. For the colour version, please refer to the plate section.)

6.1.6 Agriculture: Imaging Crop Water Uptake

One of the major global challenges is food security. Developing crop breeds that can better access water and nutrients in the soil can lead to enhanced yield. Crop breeders can assess plant performance above ground but non-invasive methods for assessment of root function are lacking. As noted by Meister et al. (2014), there is a need for high-throughput phenotyping techniques (characterization of observable traits) that can be applied across

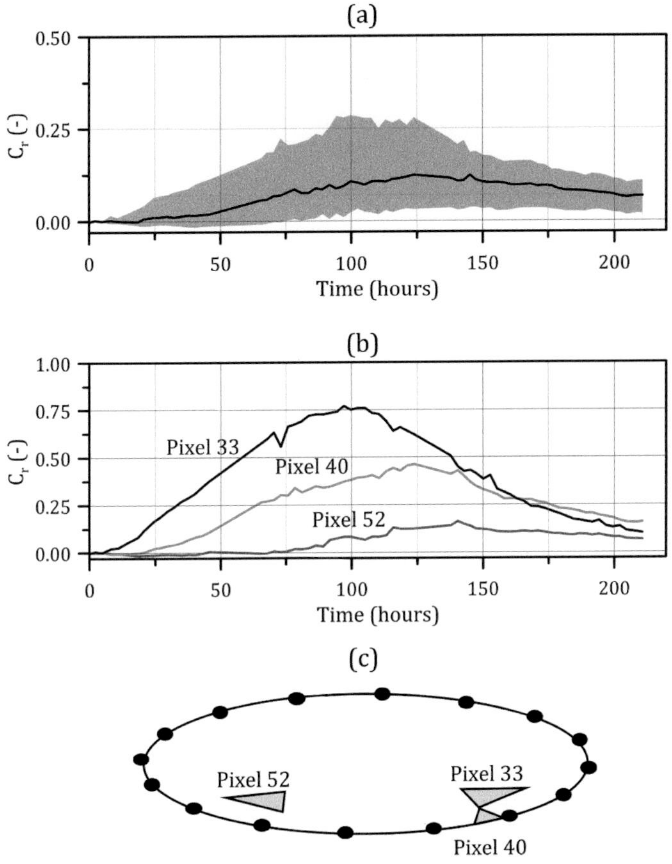

Figure 6.9 Pixel breakthrough curves for upper image plane (36 cm from injection source). (a) Median response for all 104 pixels (solid line) and 1st and 3rd quartile range (shaded area). (b) Example pixel breakthrough curves. (c) Location in image plane of example pixels. Note that the tracer is switched off at 97 hours. Analysis carried out using data from the study of Binley et al. (1996a).

large breeding platforms. One possible measure of root function is the soil water taken up by the plant during growth. For a given crop type, an increase in water uptake is likely to lead to higher yield. In addition, plant breeders may wish to examine the depth in the soil over which a crop extracts water, particularly in developing drought tolerant species. Conventional soil water measurement techniques (e.g. time domain reflectometry) can be used to monitor changes in soil water content beneath a plant; however, such methods are not scalable to large breeding platforms and suffer from a small measurement support volume. Given such constraints, there has been recent interest in the use of surface deployed geophysical methods to estimate soil moisture profiles in agricultural environments.

Jayawickreme et al. (2008) used resistivity imaging to monitor the influence of climate and vegetation on soil moisture in the root zone. Davidson et al. (2011) also used resistivity

Figure 6.10 Field plot layout for the winter wheat study reported in Whalley et al. (2017). Along each transect are up to 13.7 m by 1.8 m plots.

imaging for their investigation of deep (>11 m) rooting soil-water uptake in Amazonian woodlands during periods of drought. Garre et al. (2011) demonstrated the value of 3D resistivity imaging for improved understanding of root water uptake. Time-lapse resistivity imaging has also been shown to be effective for monitoring irrigation of crops (e.g. Michot et al., 2003).

It is clear from the studies highlighted above that the sensing of changes in soil water beneath a plant is possible; however, few attempts have been made to apply geophysical methods for comparing different breeds of a given plant. If such methods are able to differentiate changes in soil water uptake for different crop breeds, then the potential exists for them to be utilized as a phenotyping tool.

Building on the earlier work of Shanahan et al. (2015), Whalley et al. (2017) studied a range of field-based methods for phenotyping root function in breeds of winter wheat. Figure 6.10 shows an example of the field plot layout for the study of Whalley et al. (2017). Arrays were constructed with 0.3 m spaced electrodes to span groups of 7 m by 1.8 m plots of different breeds of wheat.

To assess drying patterns a reference (background) resistivity dataset was collected; then using time-lapse inversion techniques presented in Chapter 5 (Section 5.2.2.10), changes in resistivity were assessed. Note that changes in soil temperature were accounted for in the interpretation of resistivity changes.

Whalley et al. (2017) describe example time-lapse resistivity imaging for a clay rich soil. Figure 6.11 shows an example result for a sand rich soil, examined in the same study. The background resistivity distribution is shown in Figure 6.11, along with images of changes in resistivity, computed using the difference inversion method of LaBrecque and Yang (2001b). All inversions were computed using the *R2* code (Appendix A.2). The time-lapse results show clearly the evolution of a drying front during crop development. Note also the contrast to the fallow plot in the middle of the transect.

As reported by Whalley et al. (2017), resistivity imaging can be effective for differentiating genotypic differences in the depth of water extraction in the soil. However, in order to achieve sufficient sensitivity, semi-permanent electrode arrays need to be installed, which may limit the use of such methods to relatively small breeding platforms. Furthermore, the temporal changes in resistivity can only be attributed to soil drying if pore water electrical conductivity remains stable (which may be problematic if the crop is irrigated or fertilizer

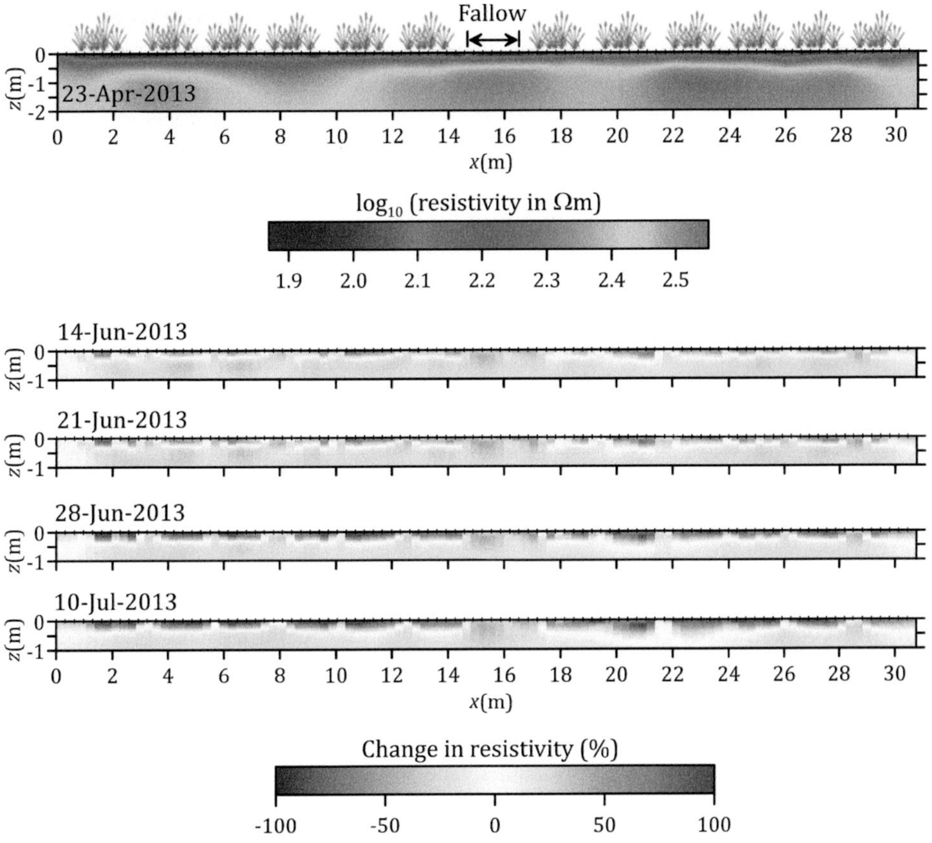

Figure 6.11 Monitoring changes in resistivity due to water uptake by winter wheat. Twelve different breeds of wheat are grown along the 30 m transect. The upper image shows the background resistivity image at early stages of crop growth. The lower images show changes in resistivity during crop growth The position of the fallow plot is shown in the upper image.. (A black and white version of this figure will appear in some formats. For the colour version, please refer to the plate section.)

addition changes the pore water conductivity significantly). Finally, it is also worth noting the study of Macleod et al. (2013), who used resistivity imaging to help in developing breeds of grass to enhance soil structure, with a view to alleviating flooding risk in grassland environments. We believe that more use will be made of electrical geophysical techniques in crop breeding studies.

6.1.7 Geotechnical Engineering: Time-Lapse 3D Imaging of Moisture-Induced Landslides

Landslides can have immense impacts on society, e.g. disruption of infrastructure (buildings, transportation networks, etc.) and, in extreme cases, fatalities. Landslides occur globally, although different triggering mechanisms (e.g. seismic events) can be regional. Moisture-induced landslides are common; they are typically caused by excessive infiltration of rainwater, which changes pore water pressure within the soil to such an extent that the shear strength reduces (e.g. Terzaghi, 1943). Monitoring of landslide prone areas requires measurements of the soil water status and given that significant heterogeneity can exist in hillslopes, conventional point source sensors generally provide insufficient spatial resolution. Geophysical methods are ideally suited for landslide monitoring of localized vulnerable slopes. Bogolovsky and Ogilvy (1977) first reviewed geophysical methods for landslides studies; several other reviews subsequently followed (Jongmans and Garambois, 2007; Perrone et al., 2014; Whiteley et al., 2019). These reviews highlight the potential value of electrical methods for landslide assessment. Typically, 2D electrical resistivity surveys are conducted along a slope in order to assess the internal hydrological and geological structure (e.g. Lapenna et al., 2005). However, given the potential ambiguity of static measures of resistivity, time-lapse resistivity imaging can be more effective, allowing the tracking over time of changes in moisture content on a slope. Given advances in instrumentation and modelling tools, time-lapse 3D approaches are now feasible (e.g. Chambers et al., 2014; Uhlemann et al., 2017).

The Hollin Hill Landslide Observatory (Yorkshire, UK) was established in 2005 by the British Geology Survey (BGS) as a field laboratory for monitoring landslide processes, primarily using terrestrial light detection and ranging (LiDAR) and electrical geophysics. The slope consists of Redcar Mudstone at the base, with an outcrop of the Staithes Sandstone Formation running across the middle of the slope. This is overlain by the Whitby Mudstone and Dogger Formations. The Whitby Mudstone is an actively failing unit at the site. Multiple failure regimes can be observed within the mudstone, towards the top of the slope a rotational failure can be seen, and towards the middle of the slope a progressive deformation associated with translational movements of the flow lobes can be observed.

In 2008, BGS installed their ALERT system (Kuras et al., 2009) for monitoring 3D resistivity. The system was decommissioned in 2018; BGS plans to install their PRIME system (Huntley et al., 2019) for future monitoring. The electrode array originally installed was arranged as 5 parallel lines (9.5 m apart), each with 32 electrodes spaced at 4.75 m intervals. The area of the slope covered by the electrode array is approximately 150 m by 40 m, designed to cover two actively moving zones. The ALERT system allows automatic data collection, which was carried out every two days using a dipole–dipole configuration, giving 2,580 measurements (and their reciprocals) for each survey. Uhlemann et al. (2017) provides details of the time-lapse inversion procedure adopted. A particular novelty (and essential for this kind of study) was the use of an adaptive finite element mesh that allows the repositioning of electrodes over time to be accounted for in the forward modelling (since the position of several electrodes changes as the active landslide progresses). The electrode positions were derived from periodic surveying of GPS markers (Uhlemann et al., 2015), although it is also possible to estimate them just using the time-lapse electrical data (Wilkinson et al., 2016).

Figure 6.12a shows the layout of the electrode array at the site along with the GPS markers; Figure 6.12b shows an example static image of resistivity during winter 2012. This inversion was made using the *R3t* code (Appendix A.2), although Uhlemann et al. (2017) used *E4D* (Johnson et al., 2010; see also Appendix A.1). The resistivity model in Figure 6.12b shows clearly the contrast between the Staithes Sandstone and Whitby Mudstone Formations. Uhlemann et al. (2017) adopted a Waxman-Smits model (Waxman and Smits, 1968) (Equation 2.38) to convert changes in resistivity over time to changes in water content. Figure 6.12c is an example result of using such an approach, showing clearly the contrast in wetting and drying zones of the site. By analysing the long record of DC

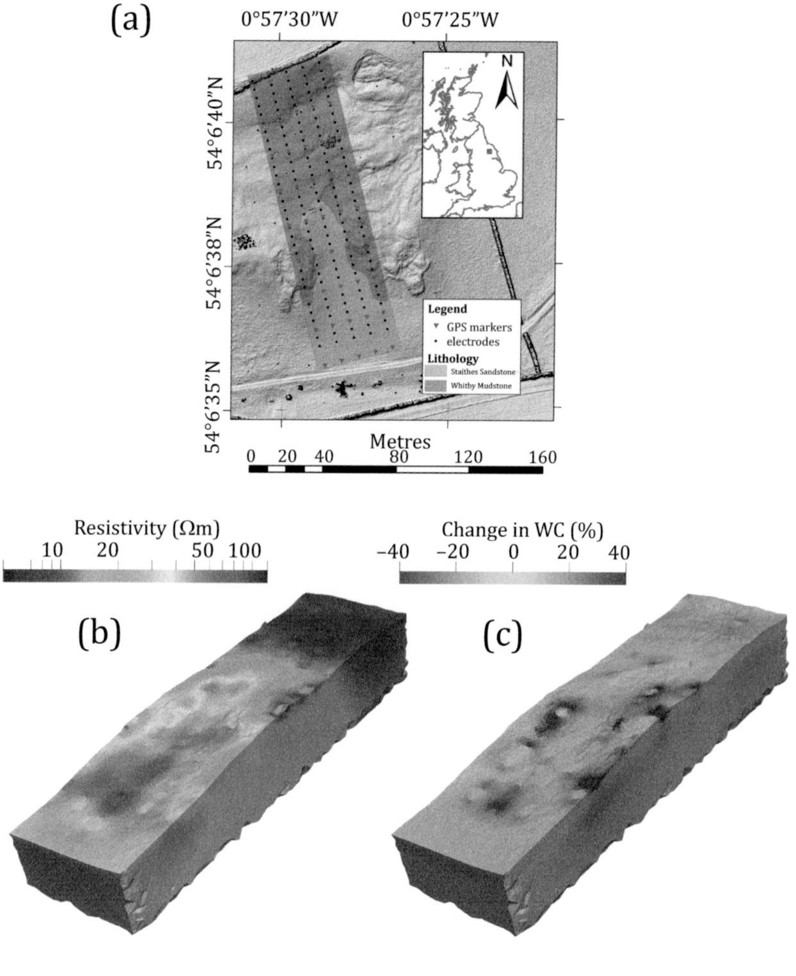

Figure 6.12 Example results from the Hollin Hill Landslide Observatory. (a) LiDAR image overlain by monitoring array positions and geological information. (b) 3D resistivity image from data collected 12 December 2012. (c) Example 3D image of change in gravimetric moisture content (between 12 March 2010 and 12 December 2012) inferred from temporal changes in resistivity. Inversions conducted by Jimmy Boyd (BGS, Lancaster University). (A black and white version of this figure will appear in some formats. For the colour version, please refer to the plate section.)

resistivity data, Uhlemann et al. (2017) analysed data prior and post activation of the slope. They were able to identify normal seasonal variation in soil water content (dry summers, wet winters) and revealed a perturbation to this seasonal trend over a period of several months prior to reactivation of the slope. The focus to date has been on inferring changes in water content, although recent work by Neyamadpour (2019) highlights possible links between changes in resistivity and geotechnical properties more directly related to soil strength.

The study of Uhlemann et al. (2017) highlights the potential value of time-lapse resistivity measurements for monitoring moisture-induced landslides. Given recent developments in low-power automated monitoring systems (most notably the BGS PRIME system (Figure 3.10)), such an approach can be applied to studies up to 1 or 2 km in size. Furthermore, with appropriate installation of buried cables/electrodes and secure enclosures BGS have successfully deployed such systems in public access areas. A vast number of sites may benefit from such installations, e.g. railway embankments, flood barriers, etc. and the expected rate of deployments will no doubt rise in the future.

6.1.8 Emerging Applications: Imaging Deep CO_2 Injection

Over the past few decades carbon capture and storage (CCS) technologies have been developed in an attempt to address increasing levels of greenhouse gas generation and mitigate the impacts of climate change. Part of the process involves securely storing carbon dioxide (CO_2) underground in depleted oil and gas fields or deep saline aquifer formations. Understanding the migration of injected CO_2 in the subsurface is essential for safe storage. Complex pathways may result due to heterogeneity of the formation and reactions that take place as CO_2 interacts with native pore waters. Geophysical techniques offer great potential for monitoring such injection and subsequent migration. In particular, electrical methods are ideally suited given likely contrasts in bulk resistivity when saline fluids are displaced by resistive CO_2, or when native saline pore water is transported to fresh groundwater due to the CO_2 injection process. For deep injection (up to several km below ground level), borehole-based methods are essential.

Field sites have been established to help improve our understanding of subsurface processes related to carbon storage, and to develop suitable monitoring technologies. At the Cranfield, Mississippi, US site (Yang et al., 2014) and Ketzin, Germany site (Schmidt-Hattenberger et al., 2014) permanently installed borehole electrode arrays have been successfully used to monitor deep CO_2 injection. Several electrical investigations of shallow analogue sites have also been carried out in the context of CO_2 injection and migration to shallow groundwater (e.g. Dafflon et al., 2012; Sauer et al., 2014; Yang et al., 2015). A number of technical challenges exist with the application of resistivity imaging of deep injection sites. We focus here on a study of Wagner et al. (2015), which is based on the Ketzin site (although their analysis is directly applicable to the other studies).

The storage zone at the Ketzin site is a sandstone formation 630–650 m below ground level. The injection borehole was installed in 2007 and includes a permanent electrode array containing 15 electrodes between approximately 590–730 m below ground level. Two monitoring boreholes were installed at some distance from the injection array with similar electrode arrays. As the borehole casing used is steel, electrodes (consisting of steel rings) were designed to be isolated from the borehole casing (e.g. Bergmann et al., 2017). Bergmann et al. (2012) illustrate the use of surface-borehole measurements for monitoring injection at the site. Bergmann et al. (2017) compares such an approach with cross-borehole imaging.

In the study of Wagner et al. (2015) an attempt is made to assess the impact of a number of factors on the performance of cross-borehole imaging at the Ketzin site, focussing on a 2D panel between the injection borehole Ktzi201 and a monitoring well Ktzi200, approximately 50 m away from the injection well. Using static and time-lapse DC resistivity data, Wagner et al. (2015) examined the effects of (i) finite electrode size, (ii) borehole deviation and (iii) borehole completion. Such effects are common in most cross-borehole electrical imaging applications. Wagner et al. (2015) also offers methods to mitigate against such effects.

The effect of finite electrodes was studied by comparing modelled measurements with the complete electrode model of Rücker and Günther (2011) (see Section 4.2.2.8) against those using the standard point electrode assumption. At the Ketzin site, electrodes are 10 cm high and electrode spacing is 10 m. Wagner et al. (2015) observed a deviation of up to 2.3% for such a setup and concluded that such effects are sufficiently smaller than the 5% to 10% reciprocity observed in measurements at the site, and thus the finite electrode effect can be ignored.

Borehole deviation (e.g. non-verticality) can have a significant impact on the electrical inversion if they are not accounted for within the model (e.g. Oldenborger et al., 2005; Wilkinson et al., 2008). Wagner et al. (2015) illustrated the impact of borehole deviation at the Ketzin site by comparing inverse model results for different borehole geometry scenarios. Borehole deviation can be measured using an inclinometer. For applications to field sites where such surveys are not carried out (or practical), Wagner et al. (2015) suggests including the electrode coordinates within the parameter set, thus re-meshing the geometry as the inversion progresses.

Borehole completion effects have been studied by other researchers (e.g. Nimmer et al., 2008; Doetsch et al., 2010a). At the Ketzin site Wagner et al. (2015) examined such effects by comparing inverse models from a 2.5D inversion (3D current flow, 2D resistivity) and a 3D inversion that includes the finite borehole within the finite element mesh, which is decoupled from the formation model. Figure 6.13 shows an example result from the analysis of Wagner et al. (2015) based on Ketzin field data. The effect is striking and highlights the need for appropriate modelling of borehole effects. Even greater effects are realized when time-lapse data are analysed (Figure 6.14), showing how results may be incorrectly interpreted if such effects are not accounted for.

Electrical imaging of CO_2 injection offers great potential for improved understanding of subsurface processes that impact on the security and effectiveness of carbon storage. The results of Wagner et al. (2015) and others highlight the need to quantify the impact of a number of geometrical effects and, where appropriate, mitigate against them in order to provide reliable interpretation of subsurface processes that can impact on the success of CCS.

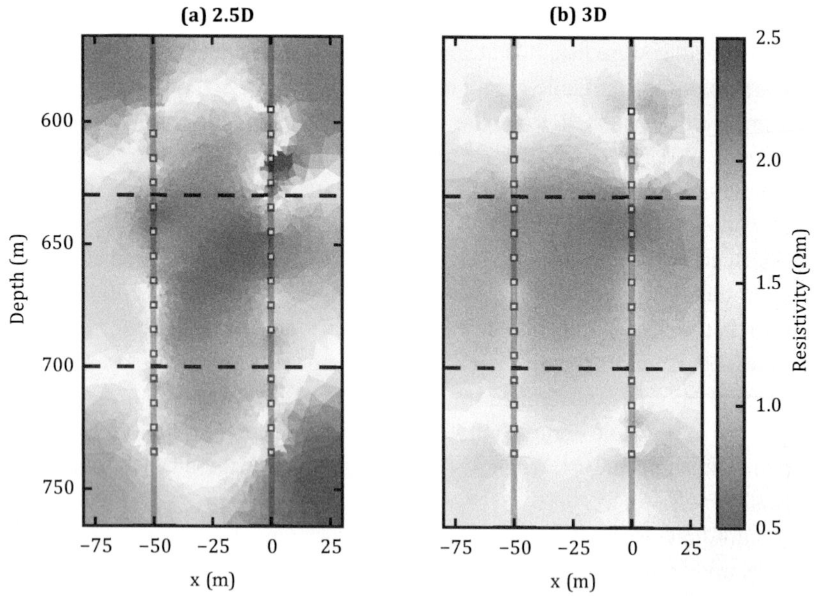

Figure 6.13 Inversion of Ketzin 21-June-2008 data. (a) 2.5D inversion. (b) 3D inversion accounting for borehole geometry. The horizontal broken lines show the formation into which CO_2 is injected. Figure based on Wagner et al. (2015). (A black and white version of this figure will appear in some formats. For the colour version, please refer to the plate section.)

6.1.9 Emerging Applications: Imaging Permafrost Distribution and Properties

Climate warming has focused much scientific attention on the Arctic region, which is highly sensitive to warming and contains a vast reservoir of carbon locked up in permafrost (earth material with a temperature continuously below 0°C for two or more consecutive years). Shallow permafrost properties including ice content, salinity of unfrozen water and soil cryostructure regulate rates of active layer (the periodically unfrozen layer above permafrost) melting, soil settlement and ice wedge growth or retreat. Shallow permafrost properties can exert a significant control on active (unfrozen) layer thickness and ecosystem functioning, in particular through a control on surface microtopography. These properties can also have a strong impact on civil infrastructure in developed regions of the Arctic. New methods are therefore needed to image shallow permafrost in the Arctic to improve understanding of the link between changing permafrost and changing geomorphic features of the Arctic region observed at the land surface.

Strong contrasts in electrical resistivity of frozen permafrost and unfrozen, active layer sediments have encouraged applications of the method for characterization of permafrost and active layer thickness (Hilbich et al., 2008; Lewkowicz et al., 2011). More recently, attention has focused on using resistivity to investigate changes in the physical properties (e.g. unfrozen water content and salinity) of permafrost (see Section 2.2.4.3). Dafflon et al.,

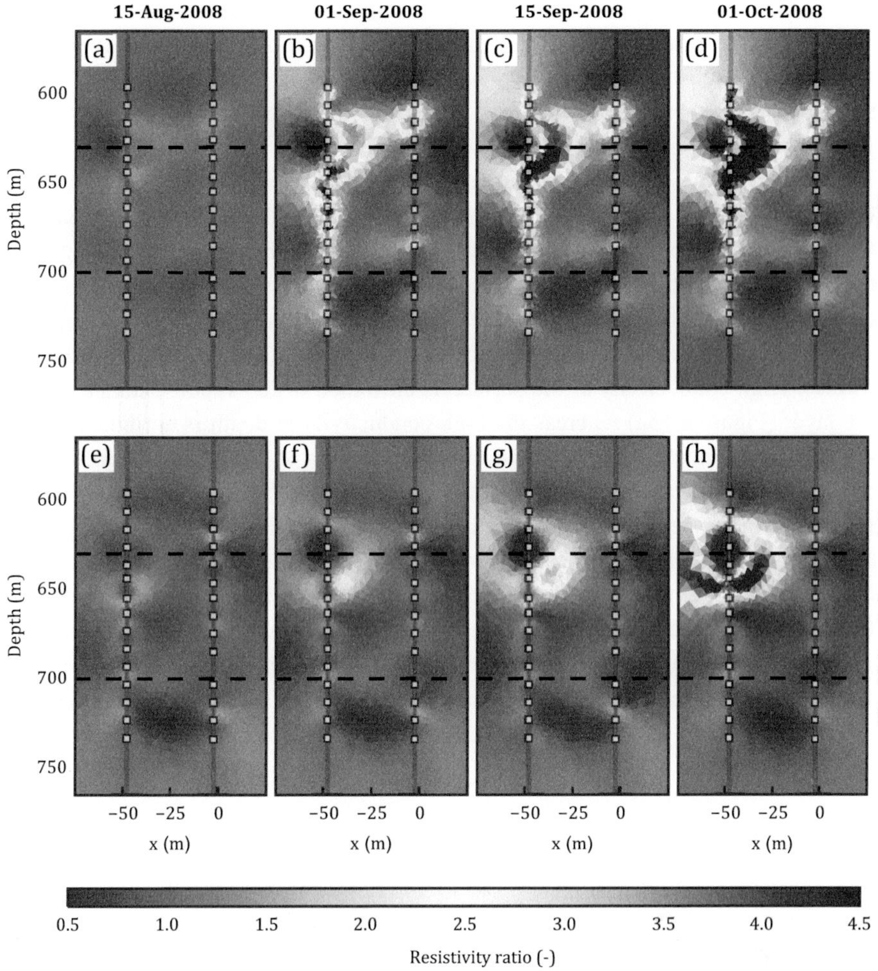

Figure 6.14 Time-lapse inversion of Ketzin data. (a–d) 2.5D inversion. (e–h) 3D inversion accounting for borehole geometry. The horizontal broken lines show the formation into which CO_2 is injected. The results are shown as a ratio of resistivity relative to the 21-June-2008 model. Figure based on Wagner et al. (2015). (A black and white version of this figure will appear in some formats. For the colour version, please refer to the plate section.)

(2016) used resistivity imaging to investigate variations in active layer thickness, shallow permafrost ice content and the distribution of salinity at an experimental research site located near Barrow, Alaska (USA). The surface of the site exhibits a characteristic polygon geomorphology that results from the formation of cracks that develop into ice wedges and that grow with repeated freezing and expansion. Resistivity data were acquired on two 500 m transects, and on 9 parallel 27.5 m long lines forming a high-resolution grid for examining the upper 3 m of the subsurface at the site. High-quality data were acquired and a temperature correction applied. Petrophysical relationships based on those introduced in

Section 2.2.4.3 were used to estimate variations in unfrozen water content of the permafrost as well as either the initial salinity of the unfrozen soil or its porosity. Dafflon et al. (2016) acknowledged the substantial uncertainty in the spatial estimates of variations in unfrozen water content and other parameters given the assumptions in the petrophysical models applied, as well as the limitations of the inversion.

Figure 6.15b shows a 3D reconstruction of electrical conductivity below the high-resolution grid (Figure 6.15a) determined from interpolating the 2D inversion results for the nine lines making up the grid. The resolved subsurface conductivity structure highlights how the subsurface distribution of permafrost correlates with visible surface geomorphological features (Figure 6.15a) as well as the microtopography (Figure 6.15b, top slice). The low conductivity values at depth are interpreted as ice wedges existing beneath troughs in the surface. High conductivity at the surface is attributed to the visible saturation of the surface layer (Figure 6.15a) whereas the high conductivity at depth is attributed to more saline permafrost. Figure 6.16 shows the full interpretation of one of the 500 m transects, where the inverted 2D resistivity image is compared with the depth of the unfrozen active layer and physical properties of the permafrost estimated from the (uncertain) petrophysical relationships are presented. Figure 6.16f provides an interpretation of the resistivity image constrained by supporting information at the site. The interpretation highlights the predicted location of ice wedges, and regions of permafrost with higher salinity in the unfrozen water.

This case study highlights the potential utility of resistivity imaging for investigating variations in active layer thickness, physical characteristics of permafrost and the linkage

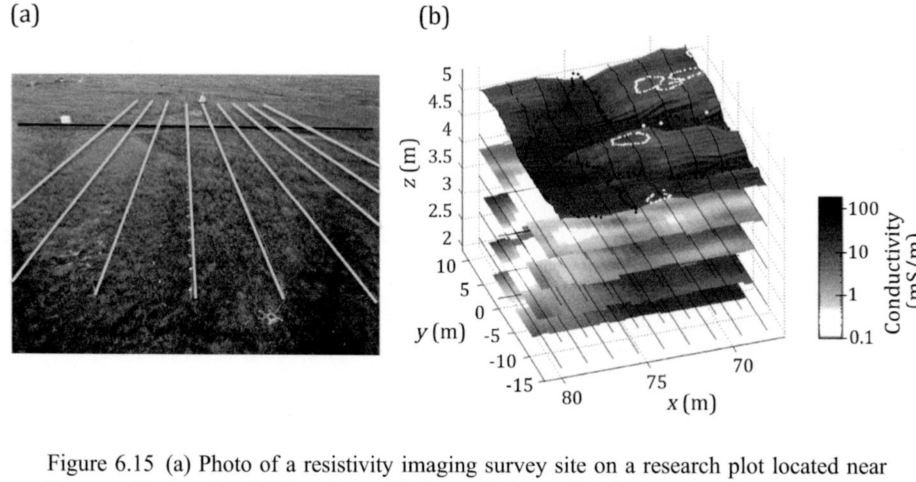

Figure 6.15 (a) Photo of a resistivity imaging survey site on a research plot located near Barrow, Alaska, showing location of high-resolution resistivity imaging lines and surface water ponding in depressions caused by microtopography related to subsurface permafrost structures; (b) horizontal slices of 3D resistivity distribution generated from interpolation of 2D inversion results, highlighting high resistivity associated with ice wedges below depressions. The low resistivity surface layer is due to the saturated surface layer. The low resistivity at depth is interpreted as evidence of high-salinity unfrozen water. After Dafflon et al. (2016).

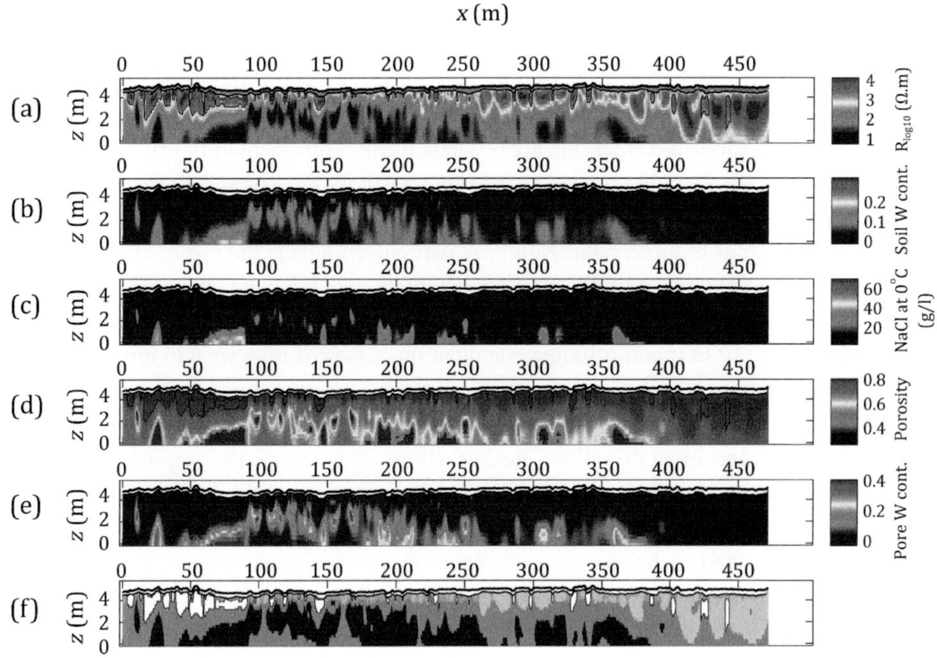

Figure 6.16 Results of a 500 m 2D ERI survey at an experimental research site near Barrow, Alaska. Black line shows depth of active layer from direct probing. (a) Resistivity distribution. Petrophysical interpretation: (b) soil water content, (c) initial salinity of thawed soil, (d) porosity, (e) fraction unfrozen water content and (f) interpretation including predicted location of ice wedges (white), as well as regions of permafrost with higher salinity (blue). After Dafflon et al. (2016). (A black and white version of this figure will appear in some formats. For the colour version, please refer to the plate section.)

between shallow subsurface structures and surface geomorphology. Such applications will increase with further focus on understanding the effects of climate change in this sensitive region of the Earth. Strong opportunities exist to use resistivity imaging to monitor changes in physical properties of permafrost over short to medium timescales.

6.2 IP Case Studies

6.2.1 Introduction

We have repeatedly emphasized how IP can be considered an extension of the electrical resistivity method. When applied and interpreted appropriately, IP can reduce ambiguity in the interpretation of subsurface structures as, in the common case of rocks containing a low electron conducting mineral concentration, the IP measurement is a direct measure of the surface conductivity (Chapter 2). IP measurements are also very useful for investigating structures and processes associated with electron conducting minerals. In the last few decades, interest in this application of IP has gone beyond the classical use of the method for mineral deposit location to novel environmental applications involving electron conducting mineral transformations. Although the number of published IP case studies is undoubtedly much smaller than resistivity case studies, it is still impossible to showcase the full range of application areas in a single textbook. Here, we highlight the breadth of use of the method and showcase emerging areas where the measurement of IP brings additional information over what would be recovered using resistivity measurements alone. The case studies again add support to material appearing in previous chapters, showing specific context for the use of induced polarization. A particular focus is the use of IP to improve lithological discrimination and estimate permeability by resolving the inherent ambiguity in the interpretation of resistivity measurements alone. We also showcase the tantalizing opportunity to use IP to monitor biogeochemical processes of relevance to environmental remediation.

Similar to the resistivity case studies, some context is provided and for most of the cases references are cited to allow further study. Details of the approach used to measure and model data are given, with specific attention paid to the challenge of acquiring reliable measurements. As noted in Chapter 3, IP measurements are inherently more challenging than resistivity measurements due to the much higher signal-to-noise ratio. Model files are again provided in online supplementary information (www.cambridge.org/binley) to allow the interested reader to work with the data using the inversion codes provided with this text (see Appendix A).

6.2.2 Hydrogeology: Characterization of a Hydrogeological Framework

This case study reported in Slater et al. (2010) and in Mwakanyamale et al. (2012) was conducted as part of a multi-method effort to improve understanding of the hydrogeological framework regulating exchange of groundwater with surface water at the US Department of Energy Hanford 300 Facility, Richland, WA, USA. The basic hydrogeological framework at this site consists of a coarse-grained aquifer (the Hanford Formation) underlain by a lower permeability, fine-grained confining unit (the Ringold Formation). Due to a legacy of nuclear waste processing and disposal at the site, the potential for radionuclide contaminated groundwater to discharge into the adjacent Columbia River exists. Of particular concern is the possible presence of relict paleochannels incised into the Ringold Formation that could act as permeable pathways for preferential rapid transport of contaminants from the aquifer into the river. The risks of radionuclide contamination result in a very high cost of drilling at this site.

Two-dimensional IP surveys were performed at this site with the specific objectives of (1) improving estimates in the spatial variability in the depth to the Hanford–Ringold contact across the site, and (2) identifying evidence for incisions in the Ringold formation that might represent paleochannels. The rationale for the application of IP was that a strong contrast in surface conductivity should be expected between the coarse-grained Hanford sediments (low surface conductivity) and the fine-grained Ringold sediments (high surface conductivity). The acquisition of resistivity measurements alone was not considered a good strategy due to the potential for variations in the conductivity of the groundwater driven by surface water–groundwater exchange at the site. Such variations could significantly impact the resistivity measurements through the electrolytic conductivity and mask the variations associated with the surface conductivity contrast between the formations. As noted in Section 2.3.3, the imaginary conductivity measured with induced polarization is directly proportional to the surface conductivity in the absence of electron conducting minerals.

The survey included land-based imaging and waterborne imaging along a much larger portion of the river corridor. On land, a series of parallel 2D induced polarization lines was acquired on 400–500 m long transects using a 5 m and 10 m electrode spacing. A data acquisition scheme that included Wenner and dipole–dipole arrays with n values equal to or less than 3 was used to ensure high recorded primary voltages and thus proportionally high secondary voltages used to record IP. Essentially the same scheme was measured with a waterborne pulled array along a 3 km stretch of the river corridor approximately centred on these land-based survey locations. IP measurements were made in the time domain, with normalized chargeabilities translated to equivalent phase angles based on a laboratory calibration of the time domain receiver readings versus a frequency domain SIP instrument (Section 3.3.3).

On land, the resulting σ'' images, with an approximately 40 m investigation depth, readily resolved the Hanford–Ringold contact as supported at two locations by observations from boreholes (Figure 6.17), and identified locations close to the river where the resistivity structure suggests the presence of paleochannels. A strong similarity in the σ' and σ'' images from IP datasets (Mwakanyamale et al., 2012) indicates that surface conductivity dominates over electrolytic conductivity (Equation 2.30) in controlling conductivity magnitude. With resistivity measurements alone, it would not be possible to reach this conclusion. As

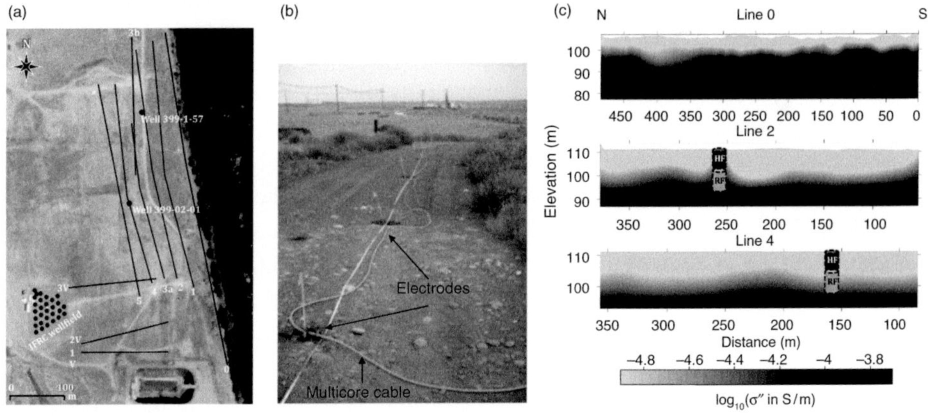

Figure 6.17 IP survey at the DOE Hanford 300 Area, Richland, WA described in Mwakanyamale et al. (2012): (a) 2D survey line locations, (b) example of line showing site conditions and multicore cable/electrodes, and (c) example 2D images of imaginary conductivity (σ'') for three of the ten lines showing a strong contrast between the Hanford Formation (HF) and Ringold Formation (RF), with location of two boreholes shown.

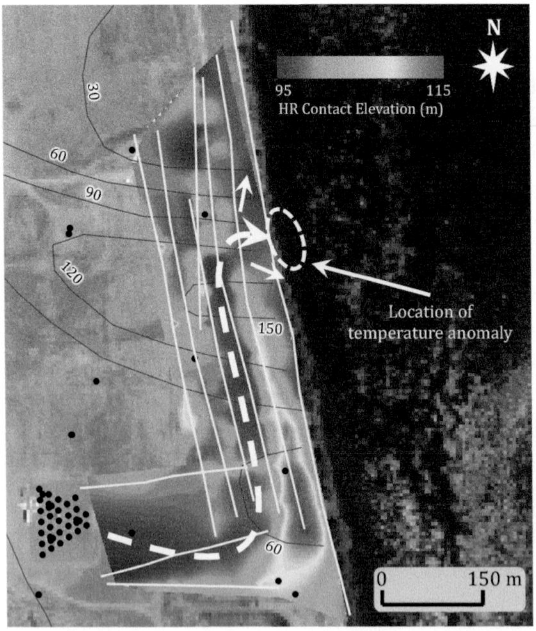

Figure 6.18 Elevation of the Hanford–Ringold contact determined from imaginary conductivity images compared to a prominent temperature anomaly (from distributed temperature sensing [DTS]) showing (1) groundwater–surface water exchange locations, and (2) contours of uranium concentrations (mg/L) in aquifer from boreholes (after Mwakanyamale et al., 2012). (A black and white version of this figure will appear in some formats. For the colour version, please refer to the plate section.)

discussed in Section 2.3.3., the surface conductivity directly sensed with imaginary conductivity measurements is foremost controlled by surface area (or grain size). An interpolated plan view of the elevation of the sharp contrast in imaginary conductivity, representing the Hanford–Ringold contact (constructed using the ten 2D transects shown in Figure 6.17a) is shown in Figure 6.18. It highlights (1) the path of a key paleochannel that coincides with the location of distributed temperature sensing (DTS) anomalies indicative of pronounced groundwater–surface water exchange (Slater et al., 2010), and (2) high uranium concentrations sampled in the river sediments (Williams et al., 2007).

One limitation of this study is that it relied on 2D inversions that were interpolated between survey lines to infer a pseudo 3D distribution of lithology. In this setting, the 2D assumption used in the inversion of the IP dataset is reasonable for the lines parallel to the river, as evident by the strong similarity in structure between these lines. One exception might be the lines closest to the river, due to 3D effects from structure (e.g. the river water and the river channel) perpendicular to the 2D image planes. Logs from two boreholes that were drilled at the site confirm that the imaged depth to the

Hanford–Ringold contact is reasonable, despite the inherent smoothing and limited resolution in the images at depth.

6.2.3 Hydrogeology: Imaging Hydrostratigraphy

Cross-borehole IP imaging can further delineate lithological units in the subsurface, for example, to improve hydrogeological conceptual models for groundwater management. The greater resolution at depth compared to surface-based methods makes cross-borehole imaging effective for delineating fine-scale vertical variations in lithology that could not be obtained from surface measurements such as those presented in Section 6.2.2. However, lateral sensitivity is limited by the aspect ratio (borehole electrode array length divided by borehole separation) as discussed in Section 4.2.2.7. In this case study we show one of these examples from the study of Kemna et al. (2004), using induced polarization in cross-borehole mode.

The study was carried out at the UK's Low Level Waste Repository, situated on the west Cumbria coast, and formerly operated by British Nuclear Fuels plc (LLWR Ltd are current operators). The shallow geology at the site is a complex, heterogeneous sequence of Quaternary sediments, overlying sandstone (Sears, 1998). The overburden is up to 60 m thick in places and difficult to characterize hydrogeologically because of the high spatial variability at the site. Hydrogeological characterization is necessary in order to assist in long-term management of the site, which included maintenance of a satisfactory radiological risk assessment and identification of any potential subsurface flow paths.

Several geophysical investigations have been carried out at the site; here we focus on a field study conducted in 2000, which was part of an experiment to map flow pathways at a specific location of the site. Two boreholes (labelled BH6124 and BH6125), approximately 15 m apart, were drilled to a depth of 41 m using shell and auger technique. The boreholes were geologically and geophysically logged. Geological logging revealed a sequence of sand/clay/gravel sediments (Figure 6.19). Forty-five electrodes were installed in each borehole at 0.8 m intervals over the depth range 5.3 m to 40.5 m. The electrodes were strapped to a permanent plastic casing during final completion of each borehole.

A cross-borehole IP survey was carried out on 26 January 2000 using the RESECS (GeoServe, Germany) time domain instrument. Measurements were made using a square-wave signal with 2.048 s pulse length, sampled at 1 ms. Fourier analysis of the injected current and received voltage waveforms, following Kemna (2000), allowed conversion to an impedance magnitude and phase angle for each measurement. Using the 90 borehole electrodes 2,984 dipole–dipole combinations, with a separation of 3.2 m, were measured in both normal and reciprocal configurations. Measurements with poor reciprocity (transfer impedance or phase angle), often due to low observed voltages, were removed, leaving 1,365 measurements for inversion. Inversion was carried out using *cR2* (see Appendix A.2) with an L_2 norm regularization. An unstructured triangular finite element mesh was used, consisting of 38,230 elements and 19,145 nodes. Anisotropic regularization (30:1 horizontal: vertical) was applied, as justified by the strong consistency between geological logs for the two boreholes (Figure 6.19). Convergence was achieved in three iterations, with one further iteration for final-phase improvement (see Section 5.3.5).

Figure 6.19 shows the cross-borehole electrical images together with the lithological logs for the two boreholes for comparison; natural gamma logs recorded in BH6125 are also shown in the figure. The image of real conductivity (σ') in Figure 6.19 is consistent with the geological logs, showing lateral connectivity of major units. The most noticeable contrast in conductivity occurs between the near-surface clayey units and the underlying sand/gravel. Contrasts in other units can also be seen, for example, the subtle increase in conductivity at about 25 m depth in the transition from silt/sand to sand/gravel. The image of imaginary conductivity (σ'') in Figure 6.19 shows a similar overall pattern to that for σ' except that the contrast between units is greater (note that the real and imaginary conductivity images are shown with an identical range of scale), allowing more lithological discrimination. This example again highlights the major benefit of acquiring induced polarization measurements, which is the ability to separate out surface conduction from electrolytic conduction. As discussed in Section 2.3.3, the imaginary conductivity is directly proportional to the surface conductivity that cannot be uniquely resolved in resistivity measurements alone. The fact that the real conductivity image has the same structure as the imaginary conductivity image is because the real conductivity is, in this case, primarily sensing changes in surface conductivity due to lithological variations (primarily surface area differences) at the site. The stronger contrasts in the imaginary conductivity between units results from the more direct relationship to the lithology, with the real conductivity being more influenced by variations in groundwater fluid conductivity.

The different IP responses reflect the textural and mineralogical variations between the units at the site, due to differences in origin and nature of deposition. The sand/silt unit shows higher polarizability than the sand/gravel unit, despite the higher clay content of the latter (interpreted from the natural gamma log). We can attribute such an effect to be a result of greater pore volume normalized surface area (S_{por}) in the sand/silt unit. Note also that the sandstone and sand/silt unit have similar values of σ'', which may be a result of similar grain size characteristics. The image of phase angle (Figure 6.19) can assist in lithological discrimination but is ambiguous if not combined with the conductivity (or resistivity) magnitude (e.g. as σ'' in the example here).

Images, such as those shown in Figure 6.19, can help in differentiating units and assist in building a hydrogeological conceptual model for a site. In the example presented here, subdivision of the lower sand/gravel unit was of interest in order to assess whether multiple subsurface flow pathways could exist. Addition of IP data proved to be effective in removing some of the ambiguity in interpretation of resistivity data alone, although more quantitative interpretation would require access to core samples for petrophysical analysis. In this example, high resolution was achievable using a cross-borehole configuration; furthermore, a surface array would not have been practical because of access limitations at an active site.

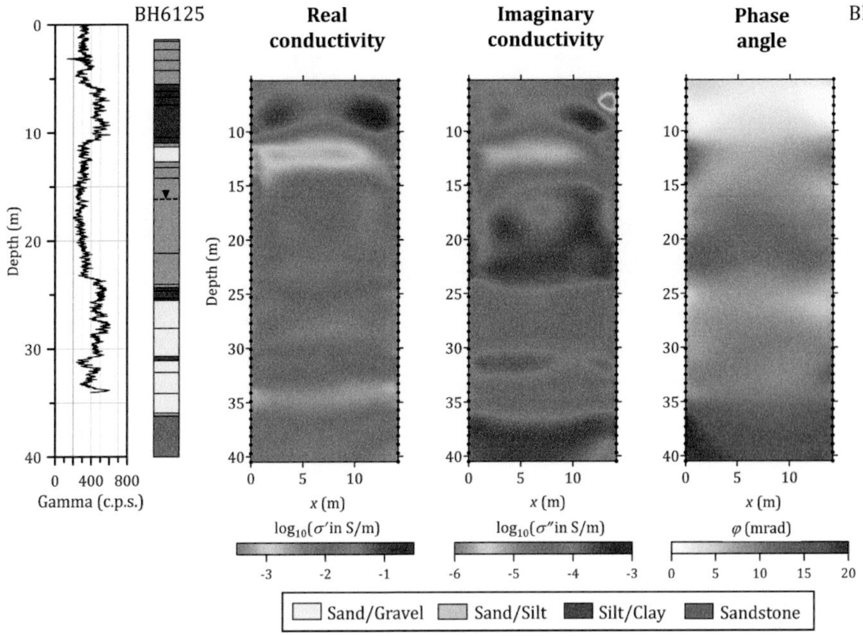

Figure 6.19 Cross-borehole IP imaging at the Low Level Waste Repository site, Cumbria, UK. Each image is between boreholes BH6125 and BH6124. Natural gamma log for BH6125 is shown, along with geological logs for both boreholes. Images based on data reported in Kemna (2004). (A black and white version of this figure will appear in some formats. For the colour version, please refer to the plate section.)

6.2.4 Hydrogeology: Imaging Permeability Distributions in Unconsolidated Sediments

Understanding and modelling groundwater flow and transport is critically dependent on good information on the permeability (or hydraulic conductivity) structure of the subsurface (Section 2.3.6). Conventional hydrological testing methods provide direct measurements of permeability (k) either at localized scales (e.g. from slug tests) or average values representing large portions of an aquifer (e.g. from pumping tests). Consequently, there has long been an interest in the possibility that electrical measurements could be used to estimate spatial variations in k. As described in Chapter 2, the electrical properties of soils and rocks are closely related to the pore geometrical properties (porosity, surface area, grain size distribution, pore size distribution) that control fluid flow. Section 2.3.6 specifically introduced the models for permeability estimation that are based upon substitution of the pore geometric terms in established models for permeability estimation (e.g. capillary bundle, percolation threshold) with proxies of these terms that are measured with IP and SIP techniques.

Although most work on permeability estimation from IP has been conducted at the laboratory scale, there have been some field-scale proof of concept experiments done to demonstrate the opportunity to image permeability at the field scale (Slater and Glaser, 2003; Hördt et al.,

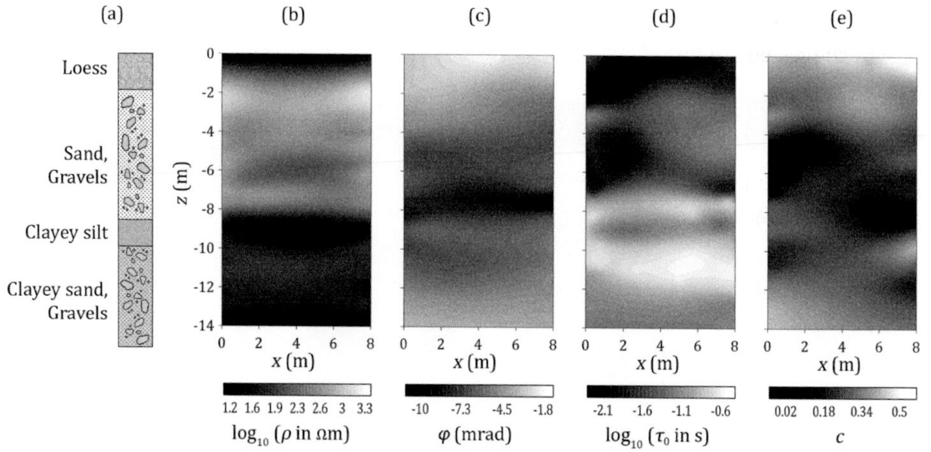

Figure 6.20 Comparison of lithological log (left) with (a) log of resistivity magnitude, (b) phase angle ($-\varphi$ plotted in resistivity space), (c) Cole–Cole time constant and (d) Cole–Cole shape parameter obtained from a 2D inversion of cross-borehole induced polarization data at a site composed of alluvial sediments impacted by jet fuel contamination. Modified from (Kemna et al., 2004).

2007). An early example of the approach is presented in Kemna (2000). This study was directed at using IP to improve understanding of the permeability structure controlling the transport of jet fuels spilled at a military tank storage site. The site geology consists of a sequence of alluvial deposits, with an uppermost loess layer, sands/gravels to approximately 9 m depth and an approximately 1 m thick clayey-silt layer near the water table.

Kemna (2000) performed 2D cross-borehole, frequency domain induced polarization measurements between four approximately 13 m deep boreholes spaced approximately 8 m apart. This maintained an image ratio of about 1.5, which is appropriate for cross-borehole image resolution as discussed in Section 4.2.2.7. Each borehole contained 16 electrodes spaced 0.75 m apart with 10 additional electrodes spanning the surface between the boreholes. A total of 350 measurements were acquired, with full reciprocals measurements acquired for error analysis (Section 4.2.2.1). The scope of the work described in Kemna (2000) was forward thinking and ahead of its time, involving the acquisition of spectral induced polarization measurements from 0.1 to 10,000 Hz, where high-frequency measurements were adversely impacted by electromagnetic (EM) coupling errors (Section 3.3.2.3). Kemna (2000) explored the effectiveness of various methods of EM coupling removal (Section 5.3.3.1) aimed at obtaining reliable field-scale measurements of the polarization relaxation time distribution. The result was effective, with reasonable images of the Cole–Cole time constant (τ_0) and the shape parameter (c) being obtained and related, along with the resistivity and phase images, to the alluvial sequence of sediments at the site (Figure 6.20).

Permeability prediction was based on a single low-frequency measurement of the IP magnitude using an approach proposed by Börner et al. (1996), whereby the PaRiS model derived by Pape et al. (1987) (Equation 2.106) was adapted with the imaginary conductivity being used as a proxy for the pore normalized surface area (S_{por}). The data from Börner et al. (1996) suggested the linear relation,

$$S_{por} = 86\sigma''_{1Hz}, \tag{6.2}$$

providing a proxy for S_{por} in Equation 2.106. Kemna (2000) recognized that the formation factor needed in Equation 2.106 could be determined from σ''_{1Hz}, an estimate of the fluid conductivity (σ_w) and the coefficient l describing an assumed linear relationship between the real (σ'_{surf}) and imaginary (σ''_{surf}) parts of the surface conductivity (Equation 2.71) as first proposed in Börner (1992) and demonstrated for a wide range of sediments by Weller et al. (2013). Kemna (2000) also had to account for the variation in saturation above the water table, which was done based on a simple modelled moisture content profile and an assumed representative value of the saturation exponent $n = 2$.

Kemna (2000) generated a 2D image of the permeability structure between the boreholes at this site (Figure 6.21). Despite the significant assumptions made and considerable uncertainty (e.g. in the value of the formation factor in the unsaturated zone), the resulting image of the permeability distribution is generally consistent with the lithologic structure identified in logs of the composition of the sediments extracted from the boreholes. The upper loess layer and the clayey silt above the water table are imaged as low permeability layers. More importantly, the estimated permeability

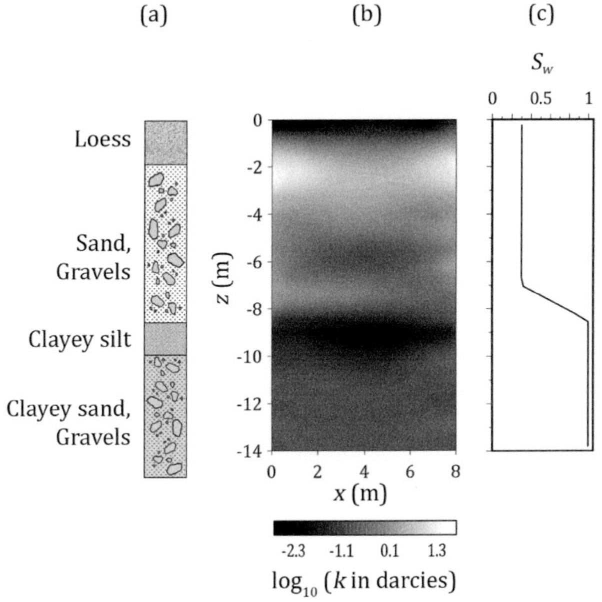

Figure 6.21 Resulting image of permeability obtained from a modification of the PaRiS equation (Equation 2.106) using imaginary conductivity as a proxy of surface area to pore volume (S_{por}). (a) Lithologic log, (b) recovered image of permeability structure and (c) assumed saturation profile for computing a saturation-dependent formation factor. Modified from Kemna et al. (2004).

values are reasonable for the sediment types identified in the borehole logs. Given recent advances in the petrophysical relationships linking permeability to IP methods (Section 2.3.6), this approach has future potential especially when performed in three dimensions (Binley et al., 2016).

6.2.5 Hydrogeology: Relationships between Spectral Induced Polarization and Permeability

Section 2.3.5 describes how more information on the pore geometrical properties controlling fluid flow might be determined by analysing the shape of SIP spectra rather than using a single measure of the polarization magnitude alone. Scott and Barker (2003) showed that the phase spectra of a database of consolidated sandstones contained a distinct peak that was related to the diameter of the pores. As the pore diameter is a critical parameter controlling fluid flow (as demonstrated by the Hagen–Poiseuille law), strong correlations between the position of this phase peak (or, equivalently, a characteristic time constant) and permeability might be expected. Binley et al. (2005) explored these concepts for a set of samples taken from a site on the UK Triassic sandstone aquifer. These experiments were part of a larger study to address issues of aquifer vulnerability associated with nutrient pollution from agricultural activities.

The sandstone at the field sites is of fluvial origin, being composed of medium to fine-grained sandstone with occasional siltstone bands. Individual core samples were extracted from a 17 m long drill core, with additional samples extracted from two nearby quarries. An extensive physical properties analysis was performed on the cores, including mercury injection capillary pressure (MICP) tests to determine pore throat size distribution and gas permeametry to obtain permeability. For the SIP measurements, a sample holder was designed to obtain reliable measurements on samples over a range of saturation states (Section 3.3.1.1). This was achieved by filling end caps with a 4 per cent agar-synthetic groundwater mix that ensured electrical contact with the rock core and also prevented ingress of the fluid into the sample under capillary suction for the case of unsaturated samples. SIP measurements were made from 0.01 to 1,000 Hz with all cores placed in an environmental chamber to ensure strict temperature control.

Section 2.3.6 introduced the theoretical arguments for defining equivalent length scales of a porous medium from IP and SIP datasets. One of these length scales is the characteristic time constant multiplied by the diffusion coefficient of the ions in the Stern Layer (D_+) (Equation 2.112), suggesting that τ should be linearly proportional to the square of a pore diameter (for a constant D_+). Binley et al. (2005) used the time constant (τ_0) found by fitting the Pelton et al. (1978) model to complex conductivity data (Box 2.8). Figure 6.22a shows the empirical relationship between this τ_0 and the square of the characteristic pore-throat size determined from MICP (being the pore radius that relates to an equivalent pressure for maximum intrusion). In this study, τ_0 is best described by a near linear dependence on the pore-throat size. Equation 2.112 shows that τ_0 should be directly proportional to k over a limited range of formation factor (again assuming constant D_+). However, Binley et al. (2005) found that τ_0 showed a much weaker than linear dependence on hydraulic

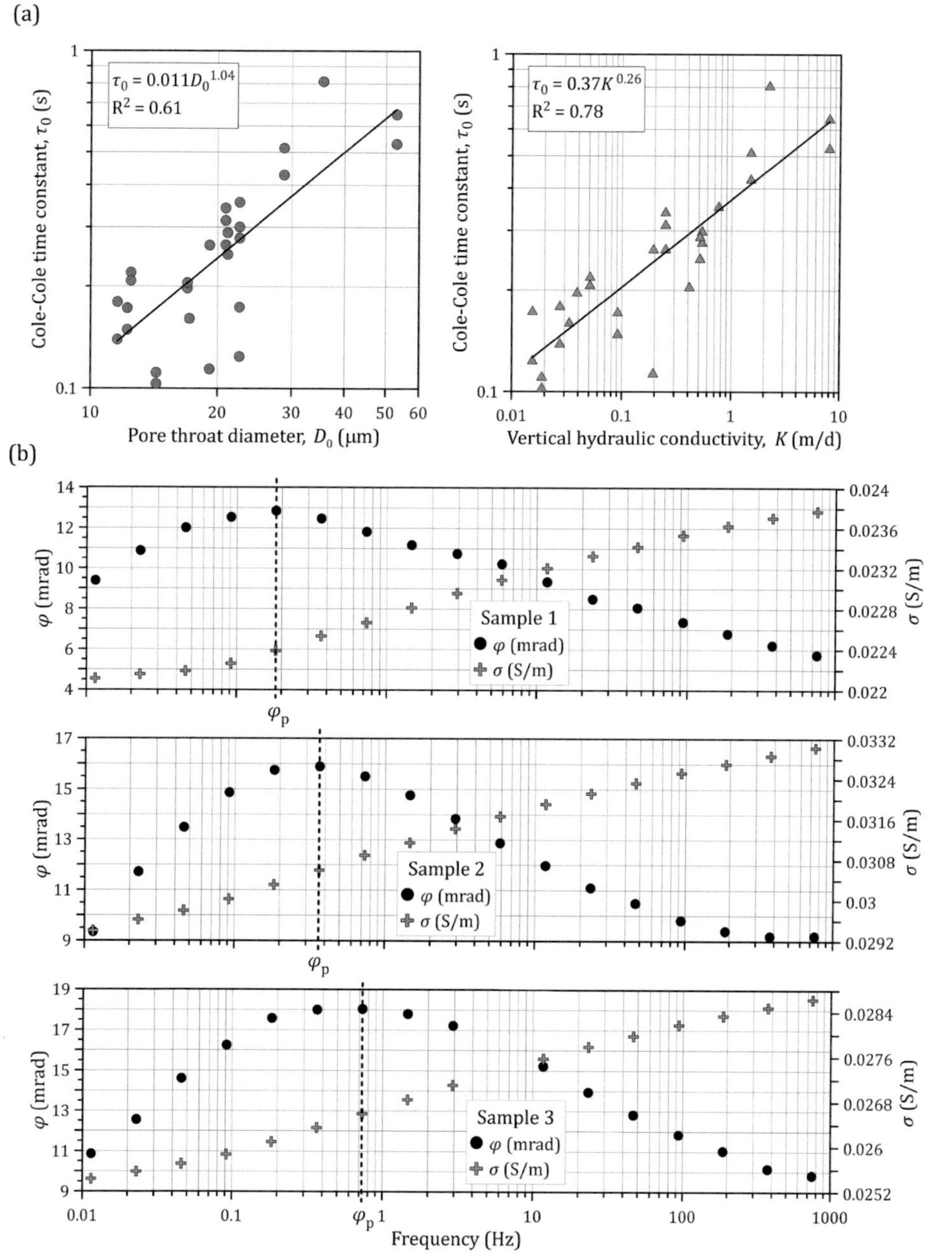

Figure 6.22 Variations in the time constant (τ_0) from the Pelton et al. (1978) modification of the Cole–Cole model (a) with pore throat diameter and with hydraulic conductivity observed for the sandstone database of Binley et al. (2005). (b) Three example datasets showing the shift in the phase peak (φ_p) between samples ($+\varphi$ plotted for conductivity space).

conductivity, with $\tau_0 \propto K^{0.26}$ (Figure 6.22b). Zisser et al. (2010b) explored the empirical relationship between a median relaxation time (τ_{50}) from Debye Decomposition (Section 2.3.4) and k in low permeability sandstones, finding $\tau_{50} \propto k^{1.56}$. Zisser et al. (2010b) also compiled datasets from all available studies at the time, which showed that the exponent on k in the $\tau - k$ relation varied from a minimum of 0.26 to a maximum of 1.56 between sandstones and unconsolidated sands from four independent studies. One reason that the empirical relationships do not agree or match with theory is likely the assumption of a single value of the diffusion coefficient (D_+). Numerous studies have shown that the diffusion coefficient may vary by orders of magnitude with variations in sample mineralogy (Kruschwitz et al., 2010; Titov et al., 2010b; Weller et al., 2016).

Since Binley et al. (2005), additional studies have further explored the relationship between a relaxation time constant (or the frequency of the peak in the phase spectrum) and permeability (Revil and Florsch, 2010; Titov et al., 2010b; Revil et al., 2015b). Although study-specific empirical relationships are found, reconciling all the datasets to satisfy Equation 2.112 proves challenging. In addition, the accurate measurement of broad frequency range SIP datasets at the field-scale remains a setback to applying such concepts beyond the laboratory.

6.2.6 Engineering: Imaging Engineered Permeable Reactive Barriers for Remediating Groundwater

Permeable reactive barriers are often used as an active remediation strategy for addressing groundwater contamination (Blowes et al., 2000). An engineering challenge is to construct a barrier in situ that conforms to design expectations and performs to expected remediation standards. Electrical imaging is a promising technology for investigating the post construction geometry and integrity of engineered structures. One of the most successful technologies for *in situ* remediation of contaminated groundwater is the zero valent iron (Fe^0) barrier, a highly oxidizing material that accelerates contaminant removal from groundwater (Puls et al., 1999). The case study reported in Slater and Binley (2003) and extended in Slater and Binley (2006) was conducted to assess the value of electrical imaging for evaluating whether an engineered Fe^0 barrier was installed per design specifications. The high metallic content of the iron wall barrier made resistivity and IP imaging excellent candidate technologies for investigating issues relating to construction and performance.

The investigated Fe^0 barrier was installed at a US Department of Energy (DOE) facility located in Kansas City, Missouri, USA. The design specifications were precisely known, but some failure to meet expected clean-up targets for contaminants in groundwater raised a concern that the barrier was compromized in locations. The barrier design, composed of a thinner upper section and thicker lower section, is summarized in Figure 6.23a. The thicker lower section of the barrier was created to compensate for higher permeability sediments (thus higher groundwater flow rates) in this section of the barrier. The length of the barrier relative to its width and height made it an excellent target for 2D imaging of cross-sectional structure. Indeed, this is a rare example of a study where the often employed

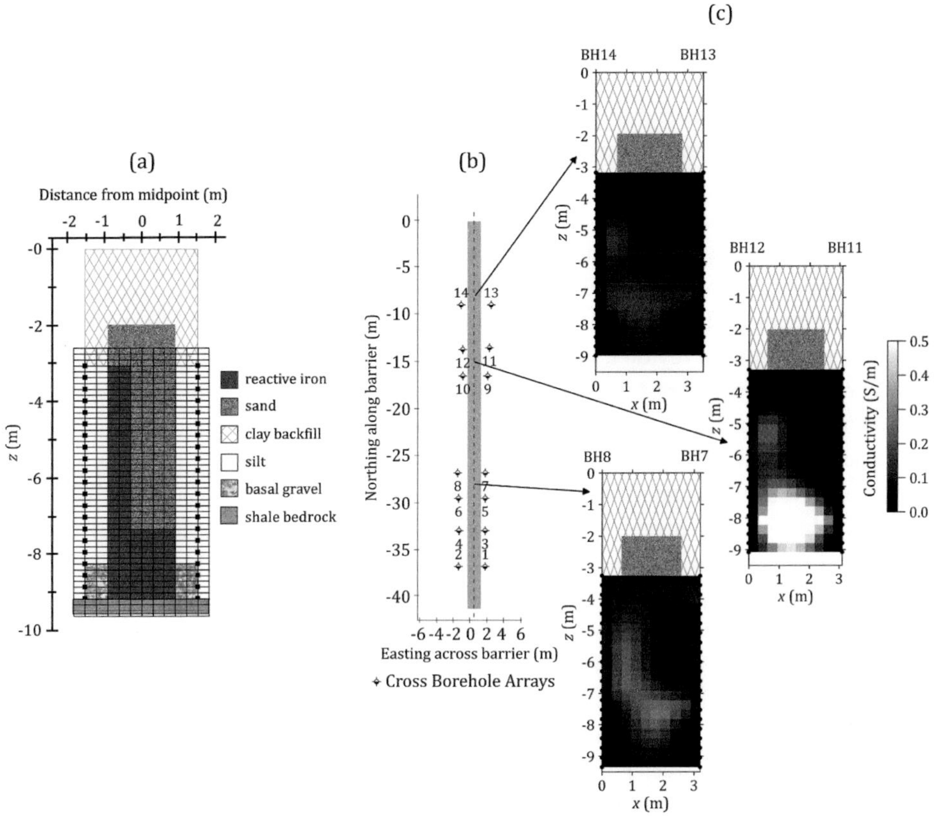

Figure 6.23 (a) Summary of idealized Fe^0 barrier cross-section at the Kansas City site, with position of electrodes and finite element modelling mesh superimposed; (b) example cross-sectional electrical images from cross-borehole measurements showing the estimated barrier configuration at different locations along the barrier investigated in Slater and Binley (2003).

2D assumption is valid: geology very rarely provides a scenario where this assumption is truly reasonable.

Slater and Binley (2003) performed 2D resistivity surveys at specific locations along the barrier in order to contrast locations where the barrier was performing well versus locations where barrier performance was apparently compromized. Initial surface surveys showed that cross-borehole imaging would be necessary to reliably resolve the barrier cross-section throughout its entire depth. Vertical electrode arrays were installed either side of the barrier wall to image 2D cross-sectional panels at discrete locations along the barrier. The measurement scheme consisted of 770 measurements in which the current and voltage dipoles were split between boreholes. Figure 6.23b shows example images of the electrical structure of the barrier at different locations along the wall. As expected due to electron conduction through the iron particles, the Fe^0 barrier is imaged as a high conductivity material relative to the background sediments. The electrical images effectively resolve variations in the integrity of the barrier along its length. The electrical evidence for poor construction towards the northern end of the barrier is consistent with groundwater sampling that showed compromized barrier performance in remediating chlorinated solvents towards the northern end.

The concept of electrical imaging of Fe^0 barriers has subsequently been extended to consider the effectiveness of new technologies that inject Fe^0 particles into the subsurface (Flores Orozco et al., 2015). The idea is that the injected particles flow with the injection fluids through high permeability sediments. However, obtaining a uniform distribution of particles is extremely challenging and uncertain relative to an engineered construction based on an excavation. Electrical imaging surveys can again assist by helping to evaluate the uniformity of the Fe^0 injected into the subsurface. Figure 6.24 shows results of a cross-borehole resistivity survey on a Fe^0 wall that was created by such an injection procedure. The survey was performed because of failure of the barrier to effectively remediate contaminants at the site. The nominal design specifications for this barrier called for the construction of a continuous 7.5 to 10 cm wide barrier from 6 to 18 m below surface. Relative to the barrier created from the excavation (Figure 6.23), this injected Fe^0 barrier is clearly highly heterogeneous, showing evidence for incomplete distribution of the Fe^0 throughout the targeted barrier. In some places the imaging suggests that the injected Fe^0 approaches 90 cm. The 2D images highlight the obvious fact that the injection procedure was ineffective in delivering a uniform barrier at this location, contributing to the poor performance of the injected barrier at this site.

One problem with the PRB technology is that the effectiveness of the barrier to remediate contaminants in groundwater decreases with ageing. Iron corrosion and mineral precipitation due to reactions between Fe^0 and groundwater constituents are widely recognized as the primary causes of Fe^0 barrier performance reduction. Laboratory research with the IP method suggests that future opportunities might exist to monitor this degradation of PRB performance with time. The IP method may have some sensitivity to changes in the mineral surface chemistry and iron mineralogy, which has been investigated for applications of the method to studies of mineral deposits (Bérubé et al., 2018), mine tailings (Placencia-Gómez et al., 2015) and precipitation of metals associated with bioremediation processes (Ntarlagiannis et al., 2005b). Wu et al. (2006) extracted horizontal cores from the Kansas City barrier. They found that the reacted zone on the up-gradient edge of the barrier had normalized chargeability (m_n) and conductivity magnitude σ_0 from the modified Cole–Cole model of Pelton et al. (1978) (Box 2.8) between 2 and 10 times higher than unreacted Fe^0 further inside the barrier (Figure 6.25a). Wu et al. (2006) analysed the solid phase components to show that these increases corresponded with thick corrosion rinds on the Fe^0 minerals in the reacted zone (Figure 6.25b). This increased the surface area of the iron minerals and led to enhanced mineral complexity as confirmed by the precipitation of iron oxides/hydroxides and other iron minerals in the reacted zone.

Slater et al. (2006) explored the possibility of detecting such mineralogical alterations at the field scale. Returning to the Kansas City barrier, they analysed IP images for evidence of changes in complex conductivity with time related to corrosion of the up-gradient reacted zone. In order to more reliably resolve subtle changes in the internal complex conductivity structure of the PRB, they employed a disconnect in the regularization (discussed in Section

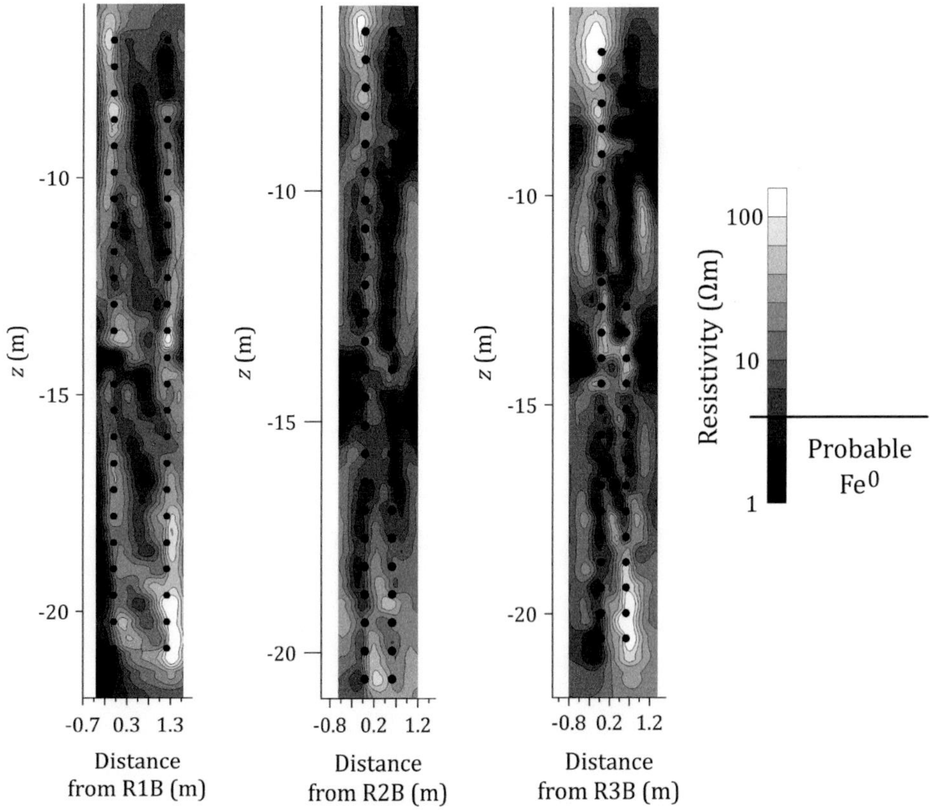

Figure 6.24 Example cross-borehole images of a Fe^0 barrier installation performed via an injection process. The design specifications called for a uniform 7.5 to 10 cm wide barrier from 6 to 18 m below surface. The electrical imaging indicates that the injection failed to provide a uniform barrier, likely contributing to the ineffectiveness of this barrier in remediating contaminated groundwater at the site.

5.2.2.8) based on the design specifications of the barrier to define a sharp boundary in the finite element mesh across which smoothing was not expected. Engineered barriers are an example of a target where such disconnects can be employed with relative confidence. The result of this analysis at one section of the barrier is shown in Figure 6.26. The real and imaginary conductivity structure shows some variability within the barrier, probably representing variable Fe^0 concentration, although the thin reaction front is not directly resolvable. However, a time-lapse inversion of changes in the complex conductivity over a 14-month time span does reveal an approximately 10% increase in conductivity focused on the up-gradient side of the barrier. Such results encourage future installation of resistivity and IP infrastructure during the construction of such barriers in support of long-term monitoring of these corrosion processes. Similar opportunities may exist for monitoring corrosion processes in mine tailings (Placencia-Gomez et al., 2015) and, importantly, in ageing infrastructure (e.g. bridges, tunnels, buildings) that are reinforced with reinforcing bar (rebar).

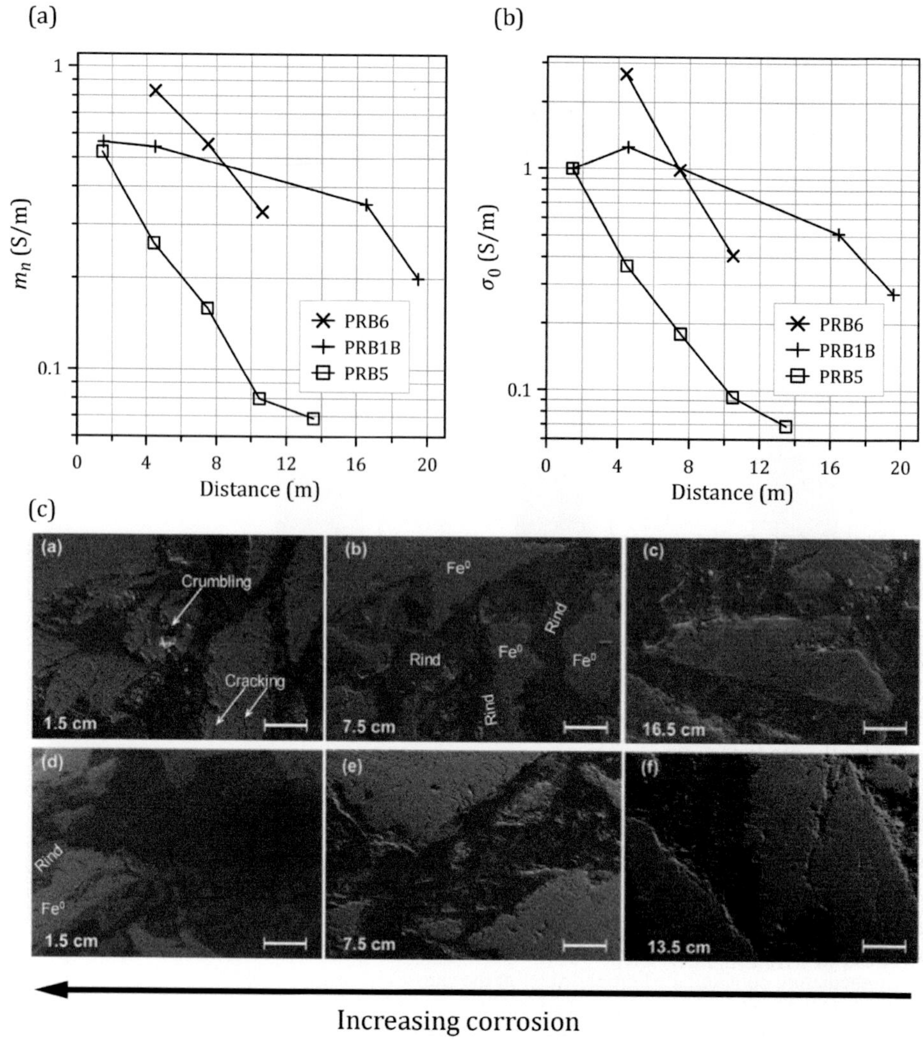

Figure 6.25 Profiles of the normalized chargeability (m_n) and DC conductivity (σ_0) from the modified Cole–Cole model of Pelton et al. (1978) recorded on horizontal cores drilled into the up-gradient edge of the Kansas City PRB at three locations, showing increases in conduction and polarization associated with the reacted zone; (b) Scanning Electron Microscope (SEM) images of the reacted zone showing thick corrosion rinds on Fe^0 a few cm into the up-gradient side of the barrier relative to unaltered Fe^0 deeper inside the barrier. Modified from Wu et al. (2006).

6.2.7 Engineering: Imaging of Soil Strengthening

Microbial induced calcite precipitation (MICP) is increasingly being implemented to stabilize soils and concrete (Achal et al., 2011). The approach offers some advantages over mechanical methods (compaction) that are not always feasible to implement, e.g. in

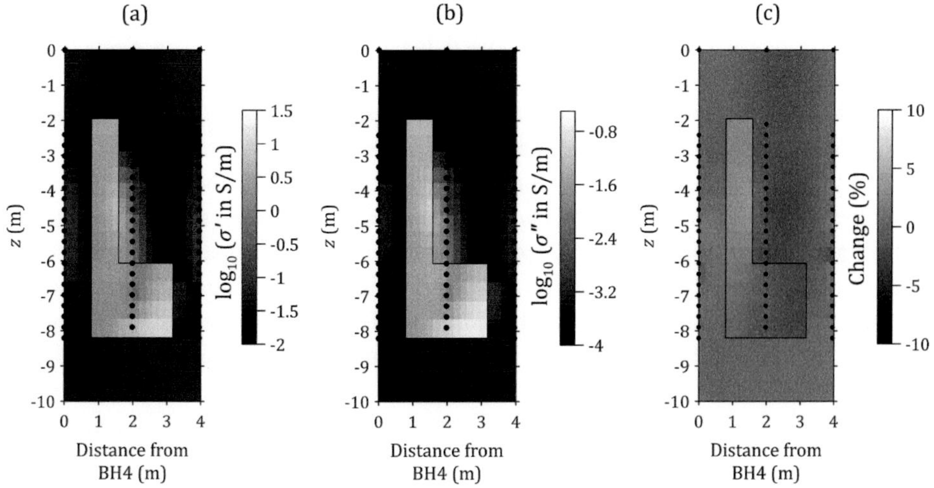

Figure 6.26 Complex conductivity inversion of the Kansas City Fe^0 barrier with a regularization disconnect defined based on the design specifications: (a) real conductivity, (b) imaginary conductivity and (c) changes in real conductivity over a 14-month period after Slater and Binley (2006).

heavily urbanized areas. MICP also avoids groundwater and soil contamination issues commonly associated with the chemical injections of grouts. MICP relies on promoting microbially mediated production of CO_2 in addition to products of urea decomposition (ureolysis). During this process, calcite (and other carbonate minerals) is readily precipitated throughout the pore space of Ca^{2+}-rich aquifers (Fujita et al., 2000). This biomineralization of carbonates serves as a strong cementation agent that strengthens the grain to grain contacts in soils. As with the implementation of groundwater remediation methods described previously, there is a pressing need for effective methods for monitoring the progress of MICP in strengthening/stabilizing soils.

Saneiyan et al. (2019) hypothesized that changes to the pore geometry and the soil mineralogy resulting from biomineralization of carbonates would result in readily detectable IP signals. Indeed, IP signatures resulting from calcite precipitation have been recorded in laboratory experiments (Wu et al., 2010) with subsequent efforts to develop models to explain these signals (Leroy et al., 2017). Building on prior laboratory measurements, Saneiyan et al. (2019) performed a pilot MICP study in a calcium rich aquifer where naturally occurring microbes had been shown to promote ureolysis. The pilot experiment involved the injection of nutrients as well as urea to stimulate the growth and activity of naturally occurring bacteria, resulting in calcite precipitation throughout the pore space.

Time domain surface resistivity and IP measurements were collected during this pilot MICP study. Careful data acquisition procedures were also followed, with tests performed to show that IP errors at the site were reduced when using graphite electrodes. One important component of this experiment relates to the careful efforts made to characterize the measurement errors and develop appropriate error models to apply in the inversion of these IP datasets. Reciprocal error modelling followed the procedures outlined in Section 4.3.2 resulted in well-behaved IP inversions using the code *cR2* (see Appendix A.2). Informative error plots describing the patterning of the reciprocal errors (Figure 6.27b) and the effects of the filtering out of high error measurements were presented by Saneiyan et al. (2019).

Saneiyan et al. (2019) showed that this MICP pilot test resulted in the clear development of a time-varying polarization anomaly that was interpreted to represent the progressive formation of calcite (Figure 6.27c). Supporting evidence for the formation of calcite came from solid phase geochemical analysis of sediments incubated in the treatment zone and measurements of hydraulic conductivity reduction. The MICP process did not result in a detectable change in the real part of the conductivity, presumably because the precipitation of dispersed calcite minerals at low volumetric concentrations did not measurably alter the current conduction pathways. However, the precipitation of a dispersed new mineral phase did sufficiently alter the interfacial polarization processes recorded with IP. Although the phase anomaly is small (the isocontour in Figure 6.27c is for phase between −6 and −4.5 mrad), the careful attention to error analysis makes the imaging of these relatively small polarization processes possible. This case study therefore highlights the emerging possibilities for IP monitoring of subsurface mineral formation (and dissolution) processes that cannot be detected with conventional resistivity monitoring alone.

6.2.8 Emerging Applications: Tracking Biomineralization Processes during Remediation

Industrial activities, military operations and acid mine drainage operations all result in contamination of groundwater by heavy metals. Microorganisms may enhance remediation of groundwater contaminated with heavy metals by sequestering metals as insoluble precipitates. As with most active techniques for remediation of contaminated groundwater, verifying the effectiveness of such bioremediation strategies is hard to do because of the lack of reliable methods for monitoring both the delivery of the treatment (where and when treatment is occurring) and the long-term stability of the remediation process.

There is growing recognition that the alteration of the solid and liquid phases of soils and rocks by microbial activity can be detected with geophysical methods (Sauck, 2000). The term ‘biogeophysics’ has been used to define a new field of geophysics focused on studying the geophysical signatures that result from microbial interactions with geologic media (Atekwana and Slater, 2009). Resistivity and IP have to date proven the most valuable geophysical techniques for monitoring microbial processes. The precipitation of disseminated metals associated with microbial processes within the interconnected pores of a soil or rock can result in a strong IP signal (Williams et al., 2005; Slater et al., 2007). Only when the precipitation is extensive enough that continuous mineral veins are produced will a resistivity signal be observed. Figure 6.28 shows the results of laboratory experiments where SIP measurements were made on sand columns where FeS biomineralization (Figure 6.28a) associated with sulphate reducing bacteria occurred during an anaerobic transition,

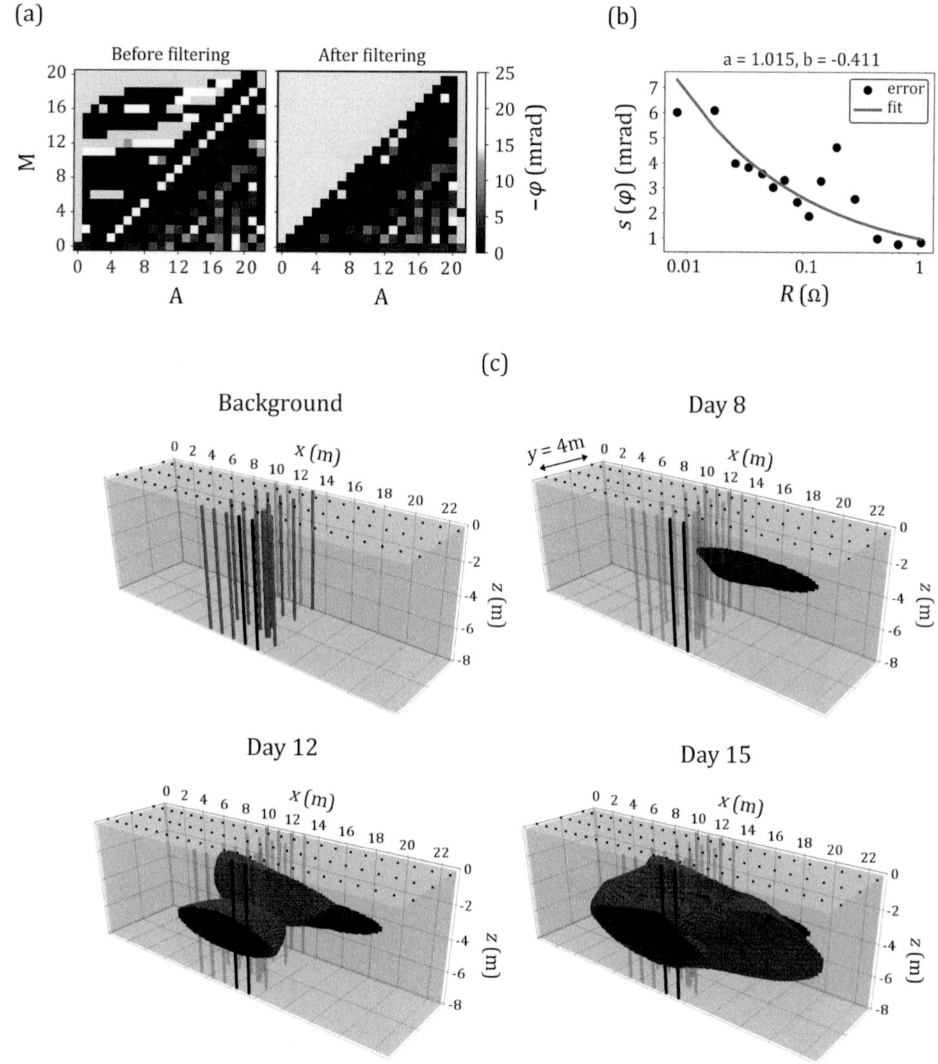

Figure 6.27 Selected results from Saneiyan et al. (2019) where IP imaging was used to monitor a pilot MICP process: (a) example plot of analysis of reciprocal errors for phase datasets acquired; (b) phase ($+\varphi$ plotted for conductivity space) error model developed for reliable inversion of the IP datasets; (c) images of the evolution of the phase anomaly associated with the microbial induced precipitation of calcite. Although the phase anomaly is small (the isocontour is for phase between 4.5 and 6 mrad), the careful attention to error analysis made this a successful application of the IP method. Vertical lines represent PVC cased groundwater sampling wells.

with subsequent biomineral dissolution as the column was returned to an aerobic state. A pronounced increase in the phase response was associated with the formation of the biominerals (Figure 6.28b); the phase change reversed and decreased with subsequent

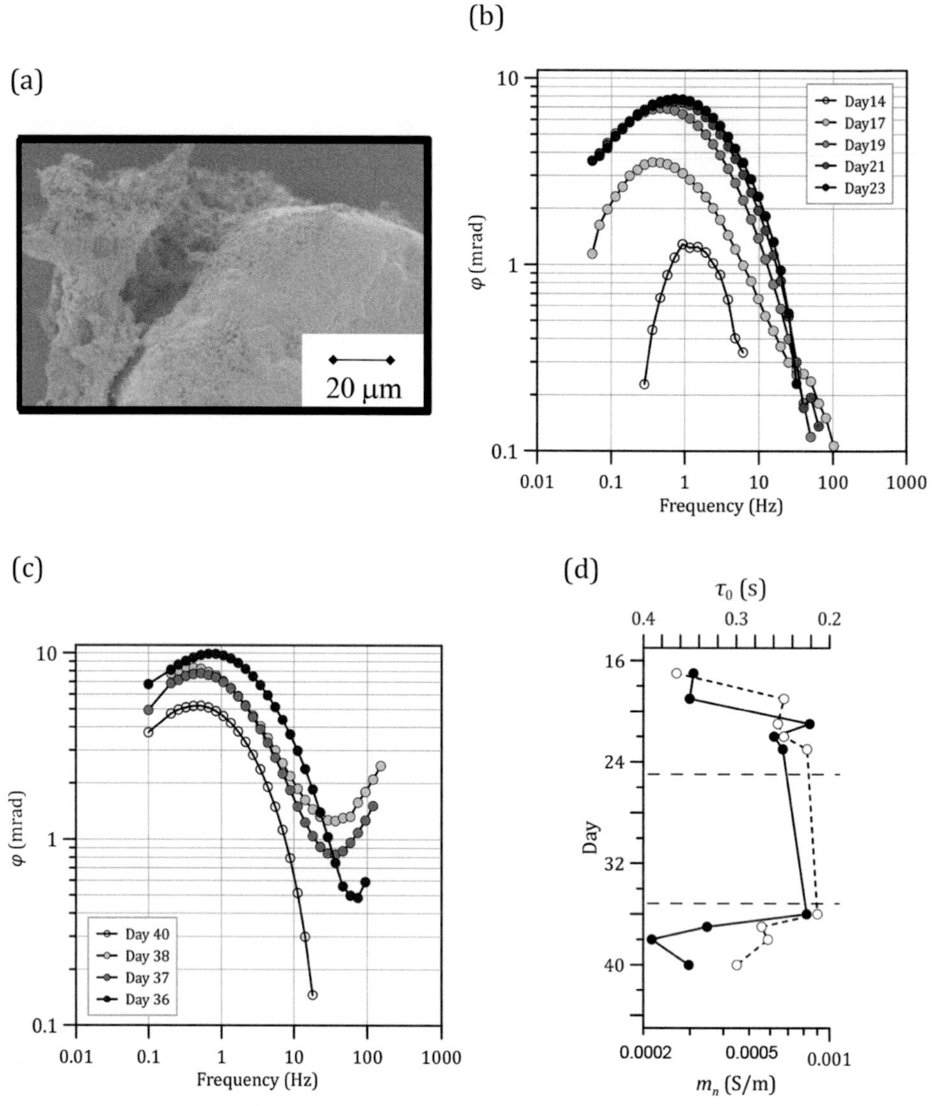

Figure 6.28 Results from an experiment where sulphate reducing bacteria *Desulfovibrio vulgaris* cause the precipitation of biominerals during an aerobic-anaerobic transition: (a) Scanning electron microscope image showing the FeS biominerals forming a crust on sand grains; Measured phase ($+\varphi$ plotted for conductivity space) as a function of time resulting from (b) precipitation of FeS biominerals, and (c) FeS dissolution associated with a subsequent aerobic transition (d) Variation of modelled Cole–Cole parameters τ_0 (filled circles) and m_n (open circles) as a function of time. Modified from Slater et al. (2007).

dissolution of the biominerals (Figure 6.28c). A pronounced change in the Cole–Cole relaxation time constant (τ_0) is observed across the aerobic–anaerobic transition (Figure 6.28d).

Williams et al. (2009) adopted these concepts to perform a field-scale demonstration of how IP could be used to monitor stimulated activity of sulphate reducing bacteria in a heavy metal contaminated aquifer. They set up an amendment delivery experiment at the Rifle field site located in north-western Colorado (USA). Groundwater was extracted from the aquifer upgradient from the study site, amended and then injected back into the aquifer through two sets of galleries (used for experiments conducted in 2004 and 2005). The amendment included sodium acetate (a source of carbon and nutrients) and targeted both sulphate reduction and iron reduction. Time-lapse 2D surface IP measurements were acquired on two orthogonal lines traversing the injection site. Anomalous phase signals developed down hydraulic gradient from the injection galleries (Figure 6.29), showing a clear correlation with microbial-induced sulphate and iron reduction as inferred from geochemical sampling. Subsequent retrieval of sediments from locations where the anomalous phase signal was recorded directly confirmed the presence of extensive FeS precipitation. No anomalous phase signals were observed in the absence of microbial

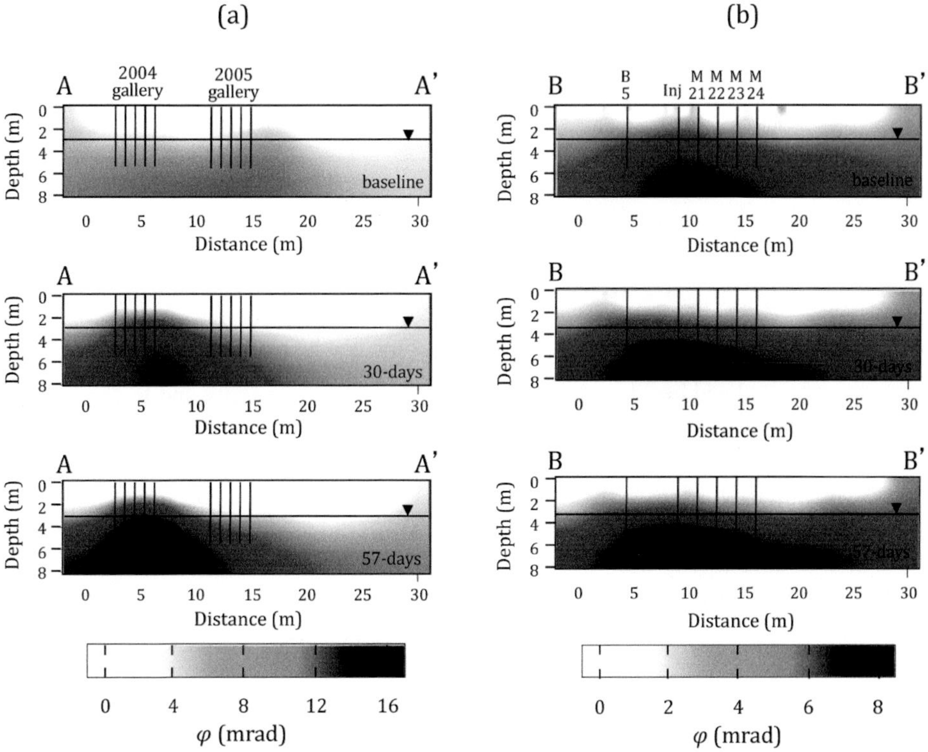

Figure 6.29 Phase inversion ($+\varphi$ plotted for conductivity space) results for surface IP datasets (0.125 Hz) acquired at three time intervals along two roughly orthogonal lines traversing the locations where sodium acetate was injected to promote microbially induced sulphate and iron reduction leading to the precipitation of iron sulphide minerals. Vertical black lines represent the galleries used to inject and extract amendment. After Williams et al. (2009).

activity. The authors attributed the substantial phase anomalies in part to the increasing concentrations of Fe(II) in solution, serving as a redox active ion and reducing the charge transfer resistance across the fluid–mineral interface (see discussion of the Wong model in Section 2.3.7). Precipitation of the new iron minerals also resulted in a weak late time increase in the electrical conductivity, suggesting that mineral precipitation was eventually extensive enough to support some electron conduction through connected minerals.

Without detracting from this very innovative application of IP, it is worth noting that the generation of a strong IP signal associated with this microbial process is perhaps unsurprising. As discussed in Chapter 1, the IP method was foremost developed as a technology for mineral detection and discrimination. The IP measurement is highly sensitive to the presence of disseminated mineralization (where the conductors or semi-conductors are dispersed in relatively low concentrations) throughout the rock. The microbial processes monitored here are essentially forming new disseminated mineral deposits in the sediments, probably being the best possible target for time-lapse biogeophysical monitoring with the IP method. Given the success of such demonstrations, IP should see increasing use for monitoring such microbial-based remediation strategies and be valuable for evaluating the long-term stability of sequestered heavy metals.

6.2.9 Emerging Applications: Characterization and Monitoring of Trees

Much like medical imaging of the human body, minimally invasive technologies are needed to evaluate structural variations in the wood of trees due to diseases that are not visibly apparent on the bark. Wood is a porous, capillary material with a large porosity and a high internal surface area. Although dry wood is resistive, the resistivity of wood decreases with the moisture content. Most trees have a moisture content greater than 60%. This contrast between dry wood and healthy wood encourages the use of resistivity and IP for imaging trees, although other challenges, e.g. ensuring electrical contact, must be overcome. Basic electrical resistivity measurements have long been used to detect variations in resistivity associated with the transformations of the physical structure of wood due to decay and rot leading to tree failure (Stamm, 1930; Skutt et al., 1972; Shigo and Shigo, 1974). However, these studies have mostly been invasive, with the electrodes inserted into the tree trunk, being somewhat detrimental to the health of the trees.

Electrical resistivity and IP imaging of trees has been relatively recently adopted to non-invasively assess structural changes in wood. Such structural changes can result from increases in the dissolved cation concentrations of pore fluids as cell walls break down, commonly caused by fungal infections (Schwarze et al., 1999). These increased cation concentrations provide a readily detectable electrical conductivity increase. Non-invasive imaging of the trunks of trees is typically performed using a radial array of electrodes to acquire 2D cross-sectional conductivity images. A tree trunk represents a case where the 2D structure approximation is valid as the trunk extends a long way in the direction perpendicular to the image plane. Al Hagrey (2006) utilized resistivity monitoring to observe sap flow in the outer living sapwood of trunks. Subsequent work has mapped conductivity variations in an effort to differentiate between sapwood and heartwood (the aged wood found in the interior of trees) (Guyot et al., 2013). The work of Guyot et al. (2013) highlighted the potential of resistivity imaging to resolve sapwood from heartwood. However, comparison of the images with direct visualization of the trunk cross-section highlighted the limited predictive capability of electrical imaging for detecting the boundary between sapwood and heartwood. Guyot et al. (2013) identified a need to improve accuracy in the detection of this boundary, e.g. by increasing the resolution by collecting more data using a larger number of electrodes.

More recently, complex resistivity measurements have been related to changes in surface area to pore volume, interconnected porosity and even surface sorption in trees (Martin, 2012; Martin and Günther, 2013). Martin and Günther (2013) demonstrated how IP imaging could be used to discriminate between healthy and fungus infected oak (Figure 6.30). They showed that 2D cross-sectional images of a healthy oak tree are characterized by a ring-like structure, where the rings differ between summer and winter seasons. Figure 6.30a shows that the studied healthy oak tree in winter is characterized by two rings, with an outer ring of high resistivity and an inner ring of low resistivity. The phase image for the healthy oak also shows

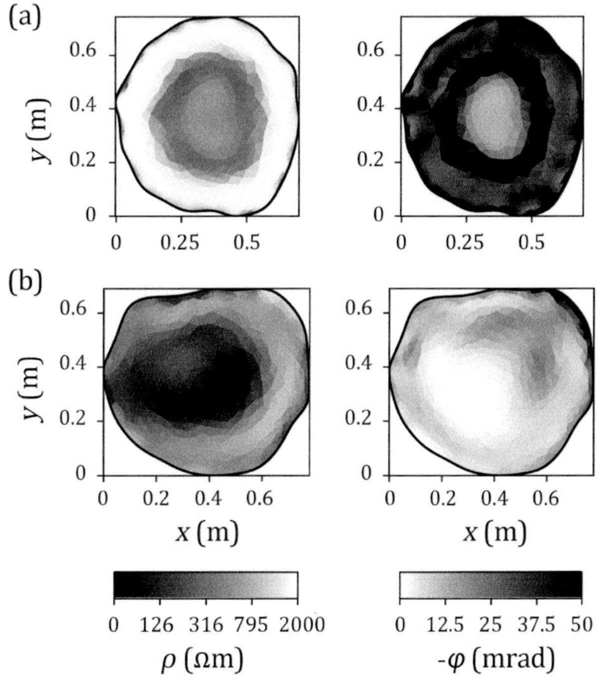

Figure 6.30 Comparison of 2D cross-sectional electrical images of the trunks of (a) a healthy oak tree and (b) a fungi-infected oak tree described in Martin and Günther (2013). The healthy tree shows a characteristic ring-like resistivity and phase ($-\varphi$ plotted for resistivity space) pattern. This pattern is disrupted in the case of fungi infection, with the level of disruption being particularly pronounced in the phase images.

a ring-like structure, although three rings can be distinguished. In contrast, the image for the oak tree damaged by fungal rot is characterized by disruption of the ring like structure, with the disruption being particularly pronounced for the image of the phase angle (Figure 6.30b).

Future developments in non-invasive electrical investigations of trees might include capacitance-based tomography (Carcangiu et al., 2019), which offers the significant advantage of avoiding the contact resistance problem inherent with galvanic methods. Furthermore, it also avoids the possibility of damaging a tree as a result of inserting electrodes into the trunk. Improved understanding of the basic electrical properties of wood, e.g. the temperature dependence (Luo et al., 2019), will improve implementation of the technology for estimating moisture content variations in trees and monitoring moisture dynamics with time.

7
Future Developments

Resistivity methods were conceived a century ago and have evolved to become one of the most popular geophysical methods for near-surface application. Mineral and petroleum exploration were the driving force of early developments, fuelled by the onset of two world wars. Following initial work on resistivity methods, induced polarization, as a technique, emerged given the realization that electrical polarization of Earth materials can offer additional insight into their physical and chemical characteristics (particularly from a mineral exploration perspective). At the start of this book we charted key stages in the development of resistivity and IP methods, developments that have resulted in remarkably flexible and scalable tools to tackle a wide range of scientific areas, including hydrology, soil science, ecology, forensic science and many more, not to mention parallel activities in archaeology, civil engineering, biomedicine and process engineering. Exactly 100 years ago, in his seminal report, Schlumberger (1920) noted '. . . *the great adaptability and wide application* [of electrical methods] *should be noted. There is every reason to suppose that, among the innumerable problems presented to the mining industry, hydraulic researchers and studies of the Earth's crust, certain ones will present themselves in a favorable manner to be solved conveniently*'. He then continued: '*The present study constitutes merely a beginning*'. Conrad Schlumberger clearly recognized great opportunities for electrical methods, but could not have foreseen the true potential. Time-lapse imaging, in particular, has opened up numerous avenues of application that could have not been foreseen 100 years ago. Schlumberger also appreciated the weaknesses, stating '*Electrical processes will probably never give certain and precise results. Their domain is to furnish more or less clear indications, serving as a guide* . . . .' Clearly, the context of Schlumberger's remarks was exploration, but the same can be said for many applications that have emerged over the past 100 years. Like all geophysical methods, resistivity and IP techniques should not be used in isolation. They can support additional investigations, but also in some cases reveal new insights into subsurface processes that could not have been appreciated with conventional methods.

Throughout the book we have attempted to illustrate the current state of knowledge regarding resistivity and IP methods, covering petrophysical relationships (that underpin any interpretation of inferred variation in electrical properties), instrumentation (for field- and laboratory-scale mapping and imaging, and also measurement of bulk electrical properties), field acquisition (including the selection of electrode configurations and their sensitivity to variation in properties), forward and inverse modelling (to allow improved survey design in addition to transforming measured data to electrical models of the subsurface). We have also tried to capture a range of application areas throughout the book, including a targeted set of case studies in Chapter 6. We have cited, what we believe are, key references to the development of the methods. The literature is, however, immense and some contributions have inevitably been overlooked in our reviewing process.

The methods continue to evolve and the past few decades have seen significant development in our understanding of the relationships between electrical properties and the physical–chemical–biological characteristics of Earth materials. More notable is the advancement of methods for modelling the relationships between observed signals and underlying subsurface models, facilitated by the advances in computer technology. Improvements in instrumentation have also been realized, although these have tended to lag behind modelling developments and the emergence of new application areas. Given that the subject area continues to grow and develop, to close the book we offer here some final remarks on areas of potential future development, targeting: petrophysics, instrumentation and modelling.

7.1 Developments in Petrophysical Relationships

Decades of geophysical research have passed since the publication of Archie's classic paper quantifying the relationship between Earth resistivity, pore fluid resistivity, pore geometry and saturation (Archie, 1942). It is testimony to the painstakingly careful observations made by Archie that his laws remain in routine use for interpreting resistivity measurements today. Although Archie's laws were empirically derived, they have been theoretically proven for specific conditions (e.g. Sen et al., 1981). However, a full understanding of the relationship between the cementation exponent (m) in Archie's law and the complex geometry of the pore space remains incomplete. Recent reinterpretations of the cementation exponent emphasize the link to the 'connectedness' of the pore space (Glover, 2009) and provide a more holistic view of the power law exponent relative to a concept mostly associated with sandstone petroleum reservoirs. This viewpoint simplifies the application of these laws to a wider range of fields, including hydrogeophysics. Increasing computer power has enabled pore-scale numerical modelling of soils and rocks to better explore the dependence of m on texture (Niu and Zhang, 2018). Extensions of Archie's law to multiple porosity domains containing different conductivity fluids (Glover, 2010) has the potential to advance use of resistivity for understanding multi-phase systems and transport of mass (e.g. ions) between more mobile and less mobile domains (Singha et al., 2007).

Starting with the observations recorded by Schlumberger (1920), the underlying causes of the induced polarization phenomenon have now intrigued geophysicists for over a century. Although much progress was made through laboratory and field observations driven by the mineral exploration community, the phenomenon still intrigues and perplexes geophysicists to this day. The complexity of the measurement is partly attributed to the electrochemical origin of the signal, depending on both the geometry of the mineral–fluid interface and the electrical double layer chemistry. It has taken a merging of geophysicists, physicists and electrochemists to advance the understanding of these signals. The challenge has perhaps been greater as the IP measurements show a very different relationship to soil/rock properties in the presence of electron conducting minerals versus in their absence.

The development of IP relationships in mineral exploration led to an electrochemical model based on the polarization of dilute suspensions of polarizing particles (Wong, 1979). Although technically a model for colloidal suspensions rather than a porous medium, the major predictions of this model form the basis for using IP to estimate the volume concentration of electron conductive particles (e.g. iron minerals) and the distribution of particle sizes. Although little explored for 30 years since its publication, the Wong model has recently been revisited to advance the use of IP in a broader range of applications beyond mineral exploration, e.g. in environmental studies of mine tailings (Placencia-Gómez et al., 2015). Consensus on the underlying source of the IP response is incomplete, with some relating it to the diffusion of ions in the ionic cloud forming around electron conducting particles and others additionally considering the diffusion of electrons and holes within the particle (Revil et al., 2015a). Numerical models that solve the fundamental electrochemical equations describing polarization of a single particle are yielding new insights into the IP phenomenon in the case of conducting particles (Bücker et al., 2018). Further work in this area will advance the implementation of IP for investigating a growing range of exploration and environmental problems, including those seeking to utilize IP to monitor biomineral transformations (e.g. Williams et al., 2009).

Understanding of the 'background' polarization in the absence of electron conducting particles, first identified by Schlumberger (1920), evolved rapidly after the 1980s following the realization that this signal could be directly related to the surface conductivity that had been so difficult to estimate from resistivity measurements alone (Vinegar and Waxman, 1984). Börner (1992) first showed that the real and imaginary parts of the surface conductivity are linearly proportional. Börner et al. (1996) suggested that the combination of resistivity and IP measurements could lead to a reliable estimation of the electrical formation factor in the presence of surface conduction. Recent petrophysical studies highlight the improvements in the estimation of formation factor using IP measurements (Weller et al., 2013). However, these studies also reveal significant variability in the proportionality between the real and imaginary parts of the surface conductivity, suggesting that secondary factors (such as mineralogy) may play a role. Further studies to explore how mineralogy (e.g. clay minerals) modifies the proportionality constant linking the real and imaginary parts of the surface conductivity are needed.

The link between the surface conductivity (e.g. as measured with IP) and pore geometric properties is well understood, including the key role of the mineral–fluid interfacial area in controlling the magnitude of the surface conductivity (Börner and Schön, 1991; Weller et al., 2010a). However, the non-negligible role of the tortuosity of the pore space results in a dependence of the surface conductivity on the formation factor, which can limit the prediction of surface area from IP particularly in fine-grained, low-permeability rocks (Niu et al., 2016). Furthermore the influence of the EDL on the proportionality constant linking the surface conductivity to pore geometry requires more investigation. Improved understanding of the most representative length-scale controlling the characteristic relaxation time of the polarization is also needed. Recent mechanistic models fluctuate between invoking the pore size, the grain size or some distance between narrow and wide pores. Improvements in such relationships will be necessary to enhance the predictive capabilities of IP-based models for permeability prediction beyond the approximate order of magnitude effectiveness currently demonstrated (Revil et al., 2015b; Weller et al., 2015b). Mechanistic modelling frameworks have led to important new insights into the nature of the IP phenomenon by linking the pore geometry to the EDL chemistry (e.g. Leroy et al., 2008). However, additional empirical studies are needed on a wider range of rock types, e.g. carbonates and mudstones.

7.2 Future Instrument Development Needs

Automated, multi-channel instruments capable of addressing hundreds of electrodes are now standard practice in resistivity and IP surveys. These instruments are versatile, being deployable over a range of scales (i.e. investigation depth and resolution). However, advances in these instruments have been mostly incremental in the last twenty years. One obvious need is for lighter, more transportable instrumentation as conventional instrumentation is heavy and sometimes difficult to deploy in remote areas. Some developments in this arena have shown promise, with higher-sensitivity voltage receivers compensating for lower-power current injection for long-term monitoring (Weller et al., 2014). Reduced data acquisition time is a second need, which would have particularly significant benefits in monitoring surveys. Some opportunities exist to exploit developments made in the biomedical electrical tomography field where code-division multiple access (CDMA) can be used for simultaneous current injections from multiple pairs of electrodes. When only resistivity measurements are needed, it should be possible to acquire measurements using much higher frequency (e.g. 100s of Hz) waveforms similar to those used in process tomography and biomedical imaging. At present, such approaches are being considered for increased data collection speed, however, in other fields of electrical imaging, multiple current electrode excitations have been used to enhance resolution (see e.g. Hua et al., 1991; Li et al., 2004). So far, this concept, first proposed by Lytle and Dines (1978), has not been examined for geoscience applications; studies to assess the potential value of such approaches are warranted.

A largely untapped opportunity still exists to develop autonomous resistivity monitoring systems for long-term monitoring of a wide range of environmental processes that result in resistivity changes. This concept goes back to simple (non-autonomous) systems installed in the early 1990s (e.g. Park and Van, 1991) and is well exemplified by the PRIME system developed by the British Geological Survey (Chambers et al., 2015). Most commercially available geophysical instruments are not optimized for deployment as a stand-alone monitoring system. In contrast, the PRIME system was specifically designed to serve as an off-the-grid monitoring technology. The rapid development of open-source hardware and software provides opportunities for researchers and practitioners to develop their own low-cost solutions for acquiring resistivity monitoring datasets. With such developments, the routine deployment of resistivity for monitoring and detection (e.g. on slopes, levees, embankments) may be realized.

Instrumentation for field-scale induced polarization measurements has developed incrementally since the 1980s. The measurement of an integral of the voltage decay following current termination has been the main way to characterize the IP effect in the field since the technology was first developed. Recent developments have focused on approaches to reduce the capacitive and electromagnetic coupling effects to improve the chances of acquiring meaningful spectral information from field-scale surveys. Some improvements have been realized with the use of fibre optic cables to allow analogue-to-digital signal conversion at the potential electrodes (Radic, 2004). Other approaches have focused on optimizing the geometric layout of the cables connecting electrodes to the transmitter/receiver (Dahlin and Leroux, 2012). Such approaches have not yet transitioned to widespread commercially available instrumentation, which may partly reflect the fact that the market for these instruments is relatively small. Future improvements in the instrumentation to reduce coupling effects may ultimately realize the opportunity to acquire broadband frequency domain measurements at the field scale.

Some improvements in field-scale induced polarization instrumentation have come from implementing data processing strategies to improve the information extractable from time domain waveforms. This includes high-frequency sampling of the full waveform first implemented in the mid 1990s. A number of commercially available instruments now record the full waveform. Other instrumentation, such as the ABEM (Sweden) system, now acquire data during the on-period, which speeds up IP data acquisition (Olsson et al., 2015). Further implementation of such signal processing strategies may further improve IP data acquisition and the information content in the measurements.

One need for resistivity and induced polarization surveys has been recently met by the development of distributed systems that eliminate the requirement for bulky and long cables, connecting electrodes to the receiver unit (Truffert et al., 2019). This development is valuable for surveys over rough terrain, where laying out lines/grids of electrodes connected to cables is impractical. The distributed system uses a network of time-synchronized receivers and transmitters that are placed to acquire fully 3D surveys. The acquisition of true 3D datasets over rough terrain has the potential to significantly improve the utility of resistivity and induced polarization for watershed-scale studies in hydrology, as well as for mineral exploration work. The wider availability and utilization of such technologies may come to represent a paradigm shift in the acquisition of resistivity and induced polarization datasets in the near future.

7.3 Future Modelling Development Needs

As mentioned above, the principles behind forward and inverse modelling approaches have, in general, not changed significantly over the past three decades. What has changed is the availability of computational power (processing speed and core memory) to allow much larger problems to be solved. As discussed in Chapter 5, 3D inverse modelling can now be carried out (for relatively small problems at least) on personal computers. That said, most applications remain based on 2D analysis (e.g. single transects). This will no doubt remain for some time, not because of modelling constraints, but simply because of the cost (labour, cabling, field time, etc.) to carry out 3D surveys. The limitations of 2D inversions should always be remembered in the interpretation of the results: the subsurface is rarely 2D and the 3D resistivity variations will result in false structure in 2D images. 3D applications are likely to grow, however, particularly those based on monitoring solutions.

Solving 3D inverse problems on large parameter grids, with a large number of electrode sites and associated measurements, is currently constrained to high-performance computing. For standard gradient-based methods (e.g. Equation 5.25), the Jacobian matrix contains $N \times M$ values, where N is the number of measurements and M is the number of parameters (e.g. unknown resistivity). To allow an appropriate level of number precision in computation, 16-byte storage is necessary, and so a problem involving 50,000 measurements with a grid of 100,000 parameter cells requires 80 GB storage – exceeding current personal computer specifications. Computer memory costs are likely to continue to decrease over time but the availability of non-specialist computer hardware will always be driven by the market demands in other areas. Video gaming, e.g., has driven some of the availability of computer processing power. We, therefore, wait for similar (bigger market) drivers, which we can benefit from. The use of parallel computing has opened up opportunities for tackling large problems (e.g. Johnson and Wellman, 2015), as discussed in Section 5.2.2.6. Distributed memory architecture is particularly appealing since it is relatively easy to decompose the gradient-based inverse problem into many discrete tasks. Parallel computing is not new, however. Parallel systems utilizing distributed heterogeneous computers (e.g. a network of standard desktop computers) have been available since the 1980s. What is new is the software that makes some of the adaption to the computer architecture much easier, and also the potential for using processing power within 'the cloud'. The advancement of graphical processing capability (again, in part driven by video gaming) has led to the exploitation of graphic processing units (GPUs) for computation. Čuma and Zhdanov (2014) illustrate how arrays of GPUs can be used for inverse modelling of resistivity problems. Other examples include Ma et al. (2015) and Anwar and Kistijantoro (2016). Exploitation of such hardware will no doubt continue to rise in the future.

Another area that is not necessarily new but has received recent attention for modelling geophysical data is machine learning (e.g. Russell, 2019), e.g. using neural networks. As stated in Chapter 1, constrained by computational power to carry out gradient-based inverse modelling, the inversions of Pelton et al. (1978) were conducted by using pre-calculated forward models stored on disk. One could view such approaches as early, albeit crude, machine learning approaches – a search for the optimum model being based on current knowledge of known responses. The avoidance of using a Jacobian matrix and inherent parallel nature of the computations are likely to prove attractive in future applications of electrical problems.

As discussed in Section 5.2.5, non-gradient-based methods have also become more common in tackling the inverse problem in electrical geophysics. What makes some of these methods attractive is the avoidance of local optima in the search for the solution; furthermore, they allow some assessment of uncertainty in the final model. They also permit the inclusion of multiple data sources (e.g. from different geophysical modalities or non-geophysical observations). They come at a cost, however. Methods based on Monte Carlo sampling are currently restricted to 1D or simple 2D problems (see Section 5.2.5). This will no doubt change in the future, but better methods are needed to make such application viable.

Uncertainty estimation in electrical inverse modelling also requires some change to standard practice. Rarely are uncertainty bounds reported or shown in inverse models of electrical properties and yet, if the models are to be used in decision-making, then some assessment of confidence is surely required. Tso et al. (2019) also highlight that the uncertainty in petrophysical models can propagate through the modelling process, leading to high uncertainty in quantitative interpretation of a geophysical model (e.g. the assessment of soil water content from an image of resistivity).

Time-lapse inversion of electrical data is now routinely carried out. Commonly this is done using pairs of datasets (e.g. following the difference inversion method of LaBrecque and Yang (2001b)), although in some cases multiple datasets are used (e.g. Kim et al., 2009; Karaoulis et al., 2014), i.e. potentially as a true 4D solution. It is not clear to what extent all datasets are required for such inversion, i.e. a quasi-4D solution may be equally good. Some further work on this is required in order to understand the performance of such techniques, allowing improvement in the computational viability. Just as space + time inversions can be conducted, there is no reason why similar regularization in the frequency domain cannot be applied to allow space + frequency inversion of complex resistivity data. In fact, a logical extension is 5D inversion (3D space + time + frequency). Such an approach is technically feasible, although the computational demands will limit the approach to very specialized applications.

Recognizing the value of additional information in the inverse problem, joint and coupled inverse modelling applications will also continue to grow in popularity, but again are likely to be constrained to specialist applications. To date, most uses have focused on research sites, serving as a demonstration of their merit.

Static grids (for forward and inverse modelling) are clearly the norm, but in some cases adaptive meshing has been exploited, e.g. to account for changing electrode positions in monitored arrays (e.g. Uhlemann et al., 2017). Similar approaches may be useful for time-lapse inversion of dynamic subsurface processes, e.g. allowing the parameter mesh to transform as the process (e.g. solute plume) evolves. Applications to time-lapse volcanology studies where subsurface properties and geometry of the study region can change may be particularly appealing. Adaptive meshing (e.g. Ren and Tang, 2010) may also be effective for modelling potential fields using the emerging distributed technology discussed in Section 7.2.

7.4 Closing Remarks

The previous sections offer some perspective on future developments of the DC resistivity and IP methods. There will no doubt be application areas that we currently cannot conceive. Instrumentation and modelling approaches will surely advance, hopefully coupled with improved insight into the mechanisms relating electrical properties to characteristics of Earth materials. It should be noted that one of the reasons for the popularity of resistivity methods in near-surface studies is their simplicity (in terms of data acquisition and modelling). There is no doubt that many future applications will continue to exploit the current approaches 'warts and all'. It is often appealing in academia to explore more complex solutions, ignoring the fact that there may be some instances where even more simplified approaches are preferable. In many real problems simple approaches are likely to remain effective.

Appendix A
Modelling Tools

Chapter 5 details the mathematical basis behind forward and inverse modelling of resistivity and IP data. Throughout much of the book we have used the first author's codes (see details later in this appendix), although a number of other codes are available to carry out such modelling. We can classify such software as either commercial (paid for) or non-commercial (free), although some non-commercial software may be available commercially, for profit-making activities. Commercial software is often freely available in a demonstration form, e.g. limiting the size of problem that can be analysed or disabling some functions. Some of the non-commercial software are open source, while some are available in executable form only. Some non-commercial software are so new that it is really only suitable for research activities. There are, therefore, a range of options available that will depend on the user's needs, budgets and intended application.

In this appendix we provide a listing of a number of software tools available. We also provide details of the first author's codes used throughout the book. Finally, we present an open-source graphical interface that can assist an interested reader in using these modelling tools – the interface, *ResIPy*, was developed specifically for this book. Our intention is to provide the reader with a modelling environment to work on many datasets covered in the book (including some of the case studies in Chapter 6) and apply similar methods to their own datasets. By developing *ResIPy* as open source, the reader can also modify and develop, and hopefully share such enhancements with the community.

A.1 Available Modelling Tools

Table A.1 lists a range of modelling tools available. Web links are provided, although we recognize that some of these links may not be permanent. Some software is available through academic journals – for such cases a reference is provided.

A.2 R Family of Codes

The first author has developed a series of codes for inversion of resistivity and IP data. The codes have evolved since the early 1990s and have been available (in executable form) for non-commercial use (see link in Table A.1). The codes were designed to allow users, without access to budgets for commercial software, to carry out forward and inverse modelling. They do not include graphical interfaces (but see Section A.3) but can utilize

Table A.1 Selected modelling tools. Source means that the source code is available

Name	Free*	Source	Brief description	Reference or web link
IPI2WIN	✓		1D resistivity and IP.	http://geophys.geol.msu.ru/ipi2win.htm
Sensing 1D	✓		1D resistivity and IP.	www.harbourdom.de/sensiv1d.htm
VES1dinv	✓	✓	1D resistivity. MATLAB® source.	Ekinci and Demirci (2008)
EarthImager 1D			1D resistivity commercial code.	www.agiusa.com/agi-earthimager-1d-ves
SPIA			1D VES can be used as part of the Aarhus Workbench (see below).	https://hgg.au.dk/software/spia-ves/
ZONDIP1D			1D resistivity and IP.	http://zond-geo.com/english/zond-software/ert-and-ves /zondip1d/
1X1Dv3			1D resistivity and IP.	www.interpex.com/ix1dv3/ix1dv3.htm
CR1Dinv	✓	✓	1D resistivity and IP relaxation modelling. MATLAB® source.	Ghorbani et al. (2009)
SEISRES	✓	✓	1D and 2D resistivity guided by seismic refraction inversion. Visual C++ code.	Nath et al. (2000) See also www.iamg.org/documents/oldftp/VOL26/v26-2–5.zip
FW2_5D	✓	✓	2D resistivity. MATLAB® source.	Pidlidesky and Knight (2008)
R2, cR2	✓		2D resistivity and FDIP.	www.es.lancs.ac.uk/people/amb/Freeware/Freeware.htm See also Section A.2
DCIP2D	✓		2D resistivity and TDIP.	https://dcip2d.readthedocs.io/en/latest/
EarthImager 2D			2D resistivity commercial code.	www.agiusa.com/agi-earthimager-2d
X2IPI			2D resistivity and TDIP.	http://x2ipi.ru/en
ZONDRES2D			2D resistivity and TDIP/FDIP.	http://zond-geo.com/english/zond-software/ert-and-ves /zondres2d/
Res2DInv			Widely used 2D resistivity and IP commercial code.	www.geotomosoft.com/index.php
Aarhus Workbench			Commercial modelling platform for a range of applications including 2D resistivity and TDIP.	https://hgg.au.dk/software/aarhus-workbench/
E4D	✓		3D resistivity and FDIP. Designed for large problems with parallel computation.	https://e4d.pnnl.gov/Pages/Home.aspx
IP4DI	✓	✓	2D and 3D resistivity and TDIP/FDIP. MATLAB® source.	Karaoulis et al. (2013)

Table A.1 *(cont.)*

Name	Free*	Source	Brief description	Reference or web link
RESINVM3D	✓	✓	3D resistivity. MATLAB® source.	Pidlisecky et al. (2007) See also https://software.seg.org/2007/0001/
AIM4RES	✓	✓	2D anisotropic resistivity inverse code. MATLAB® source.	Gernez et al. (2020) See also https://github.com/Simoger/AIM4RES
DCIP3D	✓		3D resistivity and TDIP.	https://dcip3d.readthedocs.io/en/latest/
EarthImager 3D			3D resistivity and TDIP commercial code.	www.agiusa.com/agi-earthimager-3d
BERT	✓	✓	2D and 3D resistivity and FDIP.	https://gitlab.com/resistivity-net/bert See also Günther et al. (2006) and *pyGIMLi* below.
pyGIMLi	✓	✓	Generalized inversion suite allowing multi-dimensional inversion. Python source.	Rucker et al. (2017) See also www.pygimli.org/ and *BERT* above.
Eidors	✓	✓	2D and 3D resistivity targeted for biomedical applications. MATLAB® source.	http://eidors3d.sourceforge.net/
R3t, cR3t	✓		3D resistivity and FDIP.	www.es.lancs.ac.uk/people/amb/Freeware/Freeware.htm See also Section A.2
Res3DInv			Widely used 3D resistivity and IP commercial code.	www.geotomosoft.com/index.php
ERTLab64			3D resistivity and IP commercial code.	http://ertlab64.com/
Geccoinv	✓	✓	Time-lapse SIP relaxation modelling using Debye decomposition. Python source.	Weigand and Kemna (2016a) See also https://github.com/m-weigand/Debye_Decomposition_Tools
BISIP	✓	✓	Markov-chain Monte Carlo SIP relaxation modelling. Python source.	Bérubé et al. (2017) See also https://github.com/clberube/bisip
SimPEG	✓	✓	Simulation- and gradient-based parameter estimation in geophysical applications. Python source.	Cockett et al. (2015). See also https://simpeg.xyz/
ResIPy	✓	✓	Modelling environment for analysis and inversion of 2D and 3D resistivity and IP data. Python source.	Boyd et al. (2019), Blanchy et al. (2020) See also Section A.3.

* Will be subject to licence agreement in some cases.

freely available software for mesh generation (e.g. *Gmsh,* http://gmsh.info) and presentation of 2D and 3D images (e.g. *ParaView,* www.paraview.org). The codes have been used in a wide range of applications. Recent examples include: Kiflai et al. (2019); Ward et al. (2019); Sparacino et al. (2019); Brindt et al. (2019); Cheng et al. (2019a, 2019b); Zarif et al. (2018); Vanella et al. (2018); Perri et al. (2018); Parsekian et al. (2017); Zarif et al. (2017); Raffelli et al. (2017); Mares et al. (2016); Wehrer et al. (2016). The codes form the basis of much of the modelling examples used in this book.

The codes are based on potential field modelling using the finite element method on structured and unstructured meshes (triangles and quadrilaterals in 2D; tetrahedra and triangular prisms in 3D). 2D analysis is done assuming 3D current flow (Equation 5.7). The modelled region is defined by the user, and semi-infinite conditions are not imposed, allowing the analysis of bounded regions (e.g. Figure 5.14). Electrodes can be placed anywhere in the modelled region, e.g. on a line or plane representing the ground surface, or at depth (e.g. representing borehole arrays). Inversion is based on the L_2 norm regularization, although disconnected regularization (e.g. Figure 5.20) can be applied. In all codes robust inversion (see Section 5.2.3.1) can be applied. Resistivity data are input as transfer resistances; IP data are input as transfer impedances (magnitude and phase angle). IP modelling is carried out in the frequency domain, i.e. as complex resistivity (see Section 5.3.5). Output from any inversion is accompanied with the cumulative sensitivity matrix (Equation 5.41), allowing mapping of data coverage (e.g. Figure 5.23).

The principle code is *R2*. This is a 2D resistivity forward and inverse modelling program. *cR2* is the complex resistivity equivalent of *R2* for IP modelling. *R3t* is a 3D resistivity forward and inverse modelling program (the ‘t’ suffix was added to differentiate *R3t* from an earlier (now-obsolete) hexahedral element-based code *R3*). *cR3t* is a complex resistivity version of *R3t* for 3D IP modelling. All codes have similar data structure requirements. *R2*, *cR2*, *R3t* and *cR3t* form the R family of codes. Links to access the codes, along with documentation, are provided in the accompanying online material for the book (see www.cambridge.org/binley).

A.3 *ResIPy*

The codes described in the previous section can be executed independently as stand-alone software. In order to provide a more user-friendly tool for the reader to apply these codes, the *ResIPy* interface (https://gitlab.com/hkex/resipy) was written. Blanchy et al. (2020) describe the code structure and provide examples, illustrating its use; Boyd et al. (2019) illustrate the use of *ResIPy* for 3D inverse modelling. The code allows the user to carry out forward and inverse modelling of resistivity and IP data. In both cases, the user is able to define the geometry of the problem and then design a modelling mesh (2D or 3D). Unstructured meshing is done by *ResIPy* with calls to *Gmsh* (http://gmsh.info). Forward modelling can assist in survey design and allow the user to understand the sensitivity of different measurement configurations. Inverse modelling in *ResIPy* includes data quality checking and the construction of an error model (e.g. Figure 4.15) if reciprocal measurements are available. *ResIPy* includes graphical output of inversion results, although for 3D modelling the user is recommended to use the powerful (and freely available) *ParaView* environment (www.paraview.org) for visualization. *ResIPy* is available in source code form or as a stand-alone executable. The former allows the user to customize the interface, should they so wish. A stand-alone executable of *ResIPy* (for Windows, macOS or Linux) includes the R family of codes (*R2, cR2, R3t, cR3t*).

ResIPy is made up of three layers. The bottom layer contains the executables (*R2*, *cR2*, *R3t*, *cR3t*, along with *Gmsh*). A central layer is composed of the Python application programming interface (API). This interface contains a set of functions allowing the writing of input files to the executables along with reading of their outputs. The Python API also contains specific processing routines, e.g. for data filtering and error model construction. It can be independently downloaded from pypi (https://pypi.org/project/resipy/) and used in Jupyter Notebooks, for instance, which can be useful for automated operations. The upper layer of *ResIPy* is composed of a set of visualization tools that provide a graphical environment to the user.

ResIPy allows forward (2D) and inverse (2D and 3D) modelling of resistivity and IP data. The first stage is to define the geometry of the problem (topography, electrode positions) and design a mesh. In 2D mode, the mesh can be a structured (quadrilateral element) or unstructured (triangular element) mesh. In 3D mode an unstructured (tetrahedral element) mesh is used. Refinement of an unstructured mesh is achieved by defining characteristic lengths (spacing near electrodes) and growth factors (rate of increase in mesh size away from electrodes). A custom mesh can also be imported for closed or more complex geometry.

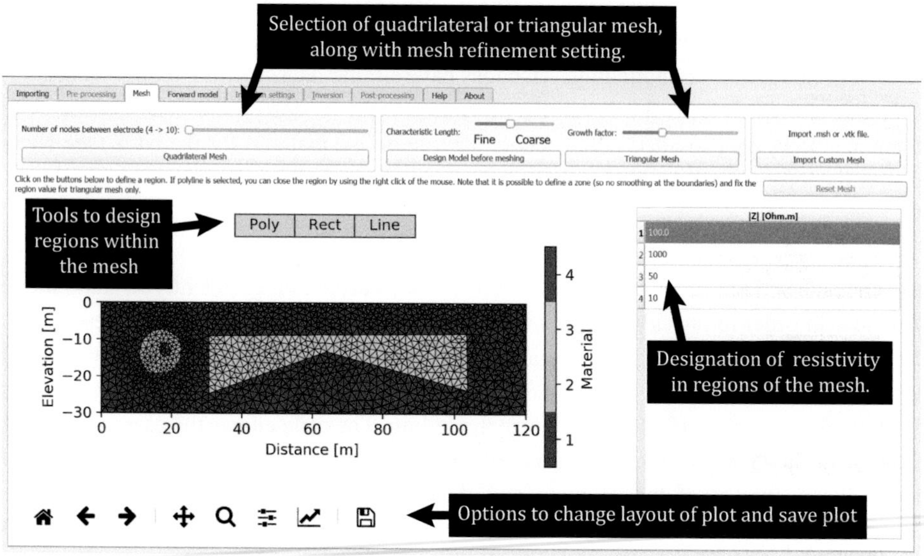

Figure A.1 Screenshot from *ResIPy* showing meshing options and designation of resistivity structure for forward modelling. (A black and white version of this figure will appear in some formats. For the colour version, please refer to the plate section.)

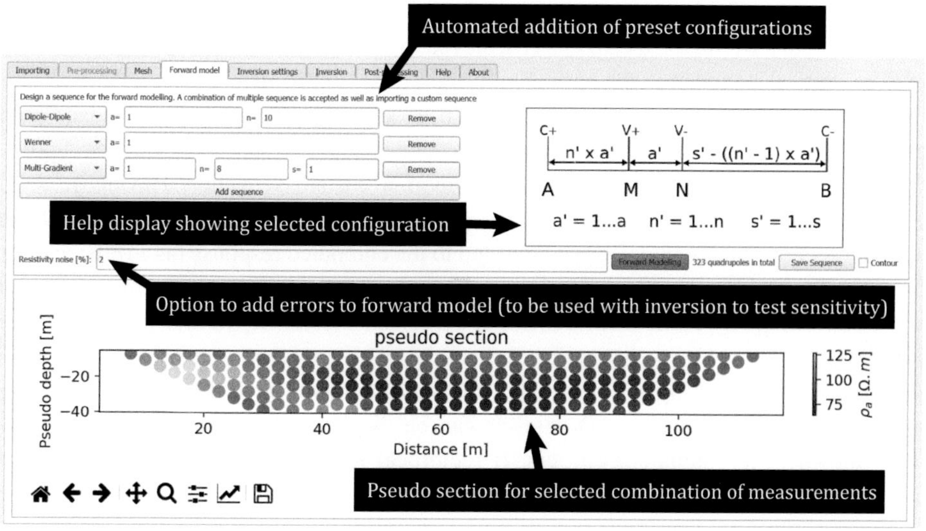

Figure A.2 Screenshot from *ResIPy* showing design of measurement sequence and pseudosection plotting following forward modelling stage. (A black and white version of this figure will appear in some formats. For the colour version, please refer to the plate section.)

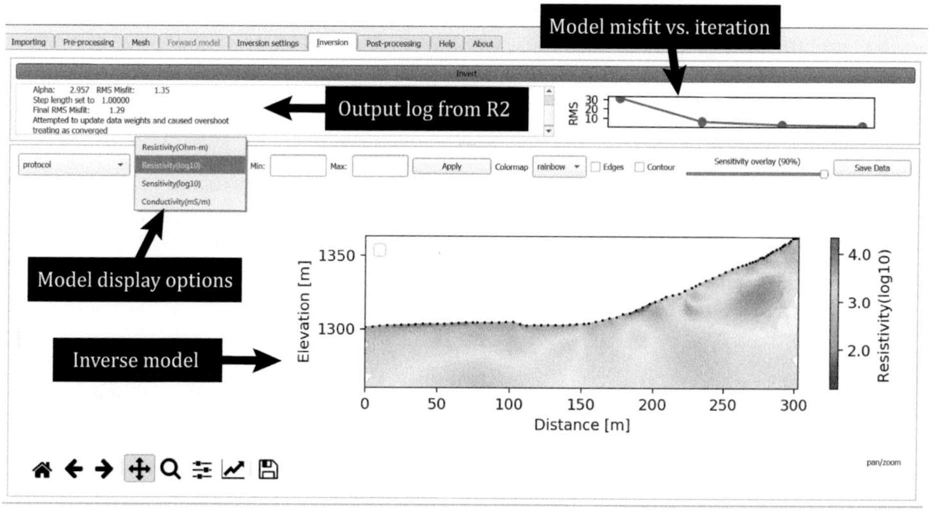

Figure A.3 Screenshot from *ResIPy* showing inverse model for a field dataset.

For forward modelling (2D) the user can design a geometrical arrangement of resistivity (and IP) regions within the mesh; this is achieved by selecting rectangles, polygons or lines (e.g. as vertical boundaries for horizontally infinite layers) to define such geometry (see Figure A.1). Next, the electrode configuration is selected; full flexibility of quadrupole geometry is permitted: the user can choose predefined standard schemes (Wenner, dipole–dipole, Schlumberger), combine them or import a custom sequence. Having selected a measurement sequence, the forward model is computed, which is shown graphically as a pseudosection. Gaussian noise can be added to the computed response (as a relative error for resistivity and absolute error for phase angle), allowing the user to run an inverse model and assess sensitivity of the measurement to the selected resistivity (and IP) geometry (see Figure A.2).

For inverse modelling (2D and 3D), the measurement set is first imported along with the electrode geometry. 2D datasets can be displayed as a pseudosection, allowing removal of any apparent outliers. If reciprocal measurements are available, these can be analysed to produce an error model (e.g. Figure 4.15 for DC resistivity or Figure 4.41 for IP). The mesh is then designed, as in a forward modelling phase (see Figure A.1), and after defining inversion parameters such as choice of regularization options, the inversion is run. Figure A.3 shows an example screenshot from an inversion of a field dataset. Post-processing of the model is then possible, e.g. to investigate the distribution of individual model misfits (see Figure 5.15). For both forward and inverse modelling, results can be stored, e.g. in vtk format, allowing presentation of model results in *ParaView*. This is particularly useful for 3D inverse modelling.

The accompanying online material for the book (see www.cambridge.org/binley) provides a version of *ResIPy*, example datasets used in this book, Jupyter Notebooks to illustrate specific use, along with a more detailed guide and tutorials on the use of *ResIPy*. *ResIPy* continues to be developed and so links are provided in the online set of materials to direct the interested reader to up-to-date versions. Once installed, *ResIPy* will automatically be updated, provided the user has internet access.

References

Abdulsamad, F., Florsch, N. and Camerlynck, C. (2017) 'Spectral induced polarization in a sandy medium containing semiconductor materials: Experimental results and numerical modelling of the polarization mechanism', *Near Surface Geophysics*, 15 (6), pp. 669–683. DOI: 10.3997/1873-0604.2017052

Abdulsamad, F., Revil, A., Soueid Ahmed, A., Coperey, A., Karaoulis, M., Nicaise, S. and Peyras, L. (2019) 'Induced polarization tomography applied to the detection and the monitoring of leaks in embankments', *Engineering Geology*, 254, pp. 89–101. DOI: 10.1016/j.enggeo.2019.04.001

Accerboni, E. (1970) 'Sur la correlation existant entre porosite et fagteur de formation dans les sediments non consolides', *Geophysical Prospecting*, 18(4), pp. 505–515.

Achal, V., Pan, X. and Özyurt, N. (2011) 'Improved strength and durability of fly ash-amended concrete by microbial calcite precipitation', *Ecological Engineering*, 37(4), pp. 554–559.

Ahmed, A. S., Revil, A. and Gross, L. (2019) 'Multiscale induced polarization tomography in hydrogeophysics: A new approach', *Advances in Water Resources*, 134, p. 103451. DOI: 10.1016/j.advwatres.2019.103451

Ahmed, A. S., Revil, A., Byrdina, S., Coperey, A., Gailler, L., Grobbe, N., Viveiros, F., Silva, C., Jougnot, D. and Ghorbani, A. (2018) '3D electrical conductivity tomography of volcanoes', *Journal of Volcanology and Geothermal Research*, 356, pp. 243–263.

Aines, R., Nitao, J., Newmark, R., Carle, S., Ramirez, A., Harris, D., Johnson, J., Johnson, V., Ermak, D. and Sugiyama, G. (2002) The stochastic engine initiative: Improving prediction of behavior in geologic environments we cannot directly observe. Lawrence Livermore National Lab., CA (US). UCRL-ID-148221. DOI: 10.2172/15002143

Aizebeokhai, A. P., Olayinka, A. I., Singh, V. S. and Uhuegbu, C. C. (2011) 'Effectiveness of 3D geoelectrical resistivity imaging using parallel 2D profiles', *Current Science*, 101(8), pp. 1036–1052.

Al Hagrey, S. A. (2006) 'Electrical resistivity imaging of tree trunks', *Near Surface Geophysics*, 4(3), pp. 179–187.

Alfano, L. (1962) 'Geoelectrical prospecting with underground electrodes', *Geophysical Prospecting*, 10(3), pp. 290–303.

Alle, I. C., Descloitres, M., Vouillamoz, J.-M., Yalo, N., Lawson, F. M. A. and Adihou, A. C. (2018) 'Why 1D electrical resistivity techniques can result in inaccurate siting of boreholes in hard rock aquifers and why electrical resistivity tomography must be preferred: The example of Benin, West Africa', *Journal of African Earth Sciences*, 139, pp. 341–353.

Allen, D. A. and Merrick, N. P. (2005) 'Imaging of Aquifers beneath watercourses', in Where Waters Meet. New Zealand Society of Soil Science, pp. 1–8.

Allred, B., Daniels, J. J. and Ehsani, M. R. (2008) *Handbook of Agricultural Geophysics*. CRC Press, Boca Raton, FL, p. 432.

Alumbaugh, D. L. and Newman, G. A. (2000) 'Image appraisal for 2-D and 3-D electromagnetic inversion', *Geophysics*, 65(5), pp. 1455–1467.

Ambegaokar, V., Halperin, B. I. and Langer, J. S. (1971) 'Hopping conductivity in disordered systems', *Physical Review B*, 4(8), p. 2612.

Andersen, K. E., Brooks, S. P. and Hansen, M. B. (2003) 'Bayesian inversion of geoelectrical resistivity data', *Journal of the Royal Statistical Society: Series B (Statistical Methodology)*, 65(3), pp. 619–642.

Anderson, R. (1981) 'Nonlinear induced polarization spectra'. PhD dissertation thesis, Department of Geology and Geophysics, University of Utah.

Anderson, W. L. (1979) 'Numerical integration of related Hankel transforms of orders 0 and 1 by adaptive digital filtering', *Geophysics*, 44(7), pp. 1287–1305.

André, F., van Leeuwen, C., Saussez, S., Van Durmen, R., Bogaert, P., Moghadas, D., de Rességuier, L., Delvaux, B., Vereecken, H. and Lambot, S. (2012) 'High-resolution imaging of a vineyard in south of France using ground-penetrating radar, electromagnetic induction and electrical resistivity tomography', *Journal of Applied Geophysics*, 78, pp. 113–122.

Angoran, Y. and Madden, T. R. (1977) 'Induced polarization: A preliminary of its chemical basis study', *Geophysics*, 42(4), pp. 788–803.

Anwar, H. and Kistijantoro, A. I. (2016) 'Acceleration of finite element method for 3D DC resistivity modeling using multi-GPU', in 2016 International Conference on Information Technology Systems and Innovation (ICITSI). IEEE, pp. 1–5.

Archie, G. E. (1942) 'The electrical resistivity log as an aid in determining some reservoir characteristics', *Transactions of the American Institute of Mining, Metallurgical, and Petroleum Engineers*, 146, pp. 54–62.

Archie, G. E. (1950) 'Introduction to petrophysics of reservoir rocks', *AAPG Bulletin. American Association of Petroleum Geologists*, 34(5), pp. 943–961.

Asami, K. (2002) 'Characterization of heterogeneous systems by dielectric spectroscopy', *Progress in Polymer Science*, 27(8), pp. 1617–1659.

Asch, T. and Morrison, H. F. (1989) 'Mapping and monitoring electrical resistivity with surface and subsurface electrode arrays', *Geophysics*, 54(2), pp. 235–244.

Aster, R. C., Borchers, B. and Thurber, C. H. (2018) *Parameter Estimation and Inverse Problems*. Elsevier, p. 383.

Atekwana, E. and Slater, L. D. (2009) 'Biogeophysics: A New Frontier in Earth Science Research', *Reviews of Geophysics*, 47(RG4004/2009), pp. 1–30. DOI: 10.1029/2009RG000285

Atekwana, E. A., Sauck, W. A. and Werkema, D. D. (2000) 'Investigations of geoelectrical signatures at a hydrocarbon contaminated site', *Journal of Applied Geophysics*, 44(2–3), pp. 167–180. DOI: 10.1016/S0926-9851(98)00033-0

Atekwana, Estella A. and Atekwana, Eliot A. (2009) 'Geophysical signatures of microbial activity at hydrocarbon contaminated sites: A review', *Surveys in Geophysics*, 31(2), pp. 247–283. DOI: 10.1007/s10712-009-9089-8

Atkinson, R. J. C. (1953) *Field Archaeology*. Methuen, pp. 1–233.

Auken, E. and Christiansen, A. V. (2004) 'Layered and laterally constrained 2D inversion of resistivity data', *Geophysics*, 69(3), pp. 752–761.

Auken, E., Christiansen, A. V., Jacobsen, B. H., Foged, N. and Sørensen, K. I. (2005) 'Piecewise 1D laterally constrained inversion of resistivity data', *Geophysical Prospecting*, 53(4), pp. 497–506.

Backus, G. and Gilbert, F. (1970) 'Uniqueness in the inversion of inaccurate gross earth data', *Philosophical Transactions of the Royal Society of London. Series A, Mathematical and Physical Sciences*, 266(1173), pp. 123–192.

Bairlein, K., Bücker, M., Hördt, A. and Hinze, B. (2016) 'Temperature dependence of spectral induced polarization data: Experimental results and membrane polarization theory', *Geophysical Journal International*, 205(1), pp. 440–453. DOI: 10.1093/gji/ggw027

Barber, D. C. (1989) 'A review of image reconstruction techniques for electrical impedance tomography', *Medical Physics*, 16(2), pp. 162–169.

Barber, D. C. and Brown, B. H. (1984) 'Applied potential tomography', *Journal of Physics E: Scientific Instruments*, 17(9), p. 723.

Barboza, F. M., Medeiros, W. E. and Santana, J. M. (2019) 'A user-driven feedback approach for 2D direct current resistivity inversion based on particle swarm optimization', *Geophysics*, 84(2), pp. E105–E124.

Barker, R. D. (1979) 'Signal contribution sections and their use in resistivity studies', *Geophysical Journal International*, 59(1), pp. 123–129.

Barker, R. D. (1981) 'The offset system of electrical resistivity sounding and its use with a multicore cable', *Geophysical Prospecting*, 29(1), pp. 128–143. DOI: 10.1111/j.1365-2478.1981.tb01015.x

Barton, D. C. (1927) 'Applied geophysical methods in America', *Economic Geology*, 22(7), pp. 649–668.

Barus, C. (1882) 'On the electrical activity of ore bodies', in *Geology of the Comstock Lode and the Washoe District* (Becker, G. F. ed.). U.S. Gological Survey, pp. 309–367.

Baumgartner, F. and Christensen, N. B. (1998) 'Analysis and application of a non-conventional underwater geoelectrical method in Lake Geneva, Switzerland', *Geophysical Prospecting*, 46(5), pp. 527–541.

Bayford, R. H. (2006) 'Bioimpedance tomography (electrical impedance tomography)', *Annual Review of Biomedical Engineering*, 8, pp. 63–91.

Bear, J. (1972) *Dynamics of Fluids in Porous Media*, Elsevier Publishing Co., p 764.

Beard, L. P., Hohmann, G. W. and Tripp, A. C. (1996) 'Fast resistivity/IP inversion using a low-contrast approximation', *Geophysics*, 61(1), pp. 169–179.

Beasley, C. W. and Ward, S. H. (1986) 'Three-dimensional mise-a-la-masse modeling applied to mapping fracture zones', *Geophysics*, 51(1), pp. 98–113.

Benoit, S., Ghysels, G., Gommers, K., Hermans, T., Nguyen, F. and Huysmans, M. (2019) 'Characterization of spatially variable riverbed hydraulic conductivity using electrical resistivity tomography and induced polarization', *Hydrogeology Journal*, 27(1), pp. 395–407.

Bergmann, P., Schmidt-Hattenberger, C., Kiessling, D., Rücker, C., Labitzke, T., Henninges, J., Baumann, G. and Schütt, H. (2012) 'Surface-downhole electrical resistivity tomography applied to monitoring of CO2 storage at Ketzin, Germany', *Geophysics*, 77(6), pp. B253–B267.

Bergmann, P., Schmidt-Hattenberger, C., Labitzke, T., Wagner, F. M., Just, A., Flechsig, C. and Rippe, D. (2017) 'Fluid injection monitoring using electrical resistivity tomography: Five years of CO2 injection at Ketzin, Germany', *Geophysical Prospecting*, 65(3), pp. 859–875.

Bernabé, Y. (1995) 'The transport properties of networks of cracks and pores', *Journal of Geophysical Research*, 100(B3), pp. 4231–4241. DOI: 10.1029/94JB02986

Bernabé, Y., Li, M. and Maineult, A. (2010) 'Permeability and pore connectivity: A new model based on network simulations', *Journal of Geophysical Research: Solid Earth*, 115 (10), pp. 1–14. DOI: 10.1029/2010JB007444

Bernabé, Y., Zamora, M., Li, M., Maineult, A. and Tang, Y. B. (2011) 'Pore connectivity, permeability, and electrical formation factor: A new model and comparison to experimental data', *Journal of Geophysical Research: Solid Earth*, 116 (11), pp. 1–15. DOI: 10.1029/2011JB008543

Bernard, J. and Valla, P. (1991) 'Groundwater exploration in fissured media with electrical and VLF methods', *Geoexploration*, 27(1–2), pp. 81–91.

Berryman, J. G. (1995) 'Mixture theories for rock properties', *Rock Physics and Phase Relations: A Handbook of Physical Constants*, 3, pp. 205–228, DOI: 10.1029/RF003p0205

Bertin, J. and Loeb, J. (1976) *Experimental and Theoretical Aspects of Induced Polarization, Vol 1*, Presentation and application of the IP method case histories. Gebrüder Borntraeger, p. 250.

Bérubé, C. L., Chouteau, M., Shamsipour, P., Enkin, R. J. and Olivo, G. R. (2017) 'Bayesian inference of spectral induced polarization parameters for laboratory complex resistivity measurements of rocks and soils', *Computers & Geosciences*, 105, pp. 51–64.

Bérubé, C. L., Olivo, G. R., Chouteau, M. and Perrouty, S. (2018) 'Mineralogical and textural controls on spectral induced polarization signatures of the Canadian Malartic gold deposit: Applications to mineral exploration', *Geophysics*, 84(2), pp. 1–83. DOI: 10.1190/geo2018-0404.1

Bevc, D. and Morrison, H. F. (1991) 'Borehole-to-surface electrical resistivity monitoring of a salt water injection experiment', *Geophysics*, 56(6), pp. 769–777.

Beven, K. J., Henderson, D. E. and Reeves, A. D. (1993) 'Dispersion parameters for undisturbed partially saturated soil', *Journal of Hydrology*, 143(1–2), pp. 19–43.

Bhattacharya, P. K. and Patra, H. P. (1968) *Direct Current Geoelectric, Sounding Methods in Geochemistry and Geophysics*. Elsevier, Amsterdam, p. 135.

Bhattacharya, B. B., Gupta, D., Banerjee, B. and Shalivahan (2001) 'Mise-a-la-masse survey for an auriferous sulfide deposit', *Geophysics*, 66(1), pp. 70–77.

Bibby, H. M. (1977) 'The apparent resistivity tensor', *Geophysics*, 42(6), pp. 1258–1261.

Bigelow, E. (1992) *Introduction to Wireline Log Analysis*. Baker Hughes Inc., Western Atlas International, p. 312

Bing, Z. and Greenhalgh, S. A. (1997) 'A synthetic study on crosshole resistivity imaging using different electrode arrays', *Exploration Geophysics*, 28(1–2), pp. 1–5.

Bing, Z. and Greenhalgh, S. A. (2000) 'Cross-hole resistivity tomography using different electrode configurations', *Geophysical Prospecting*, 48(5), pp. 887–912.

Bing, Z. and Greenhalgh, S. A. (2001) 'Finite element three dimensional direct current resistivity modelling: Accuracy and efficiency considerations', *Geophysical Journal International*, 145(3), pp. 679–688.

Binley, A. and Daily, W. (2003) 'The performance of electrical methods for assessing the integrity of geomembrane liners in landfill caps and waste storage ponds', *Journal of Environmental & Engineering Geophysics*, 8(4), pp. 227–237.

Binley, A., Ramirez, A. and Daily, W. (1995) 'Regularised image reconstruction of noisy electrical resistance tomography data', in *Process Tomography – 1995, Proceedings of the 4th Workshop of the European Concerted Action on Process Tomography.* Bergen, pp. 6–8.

Binley, A., Henry-Poulter, S. and Shaw, B. (1996a) 'Examination of solute transport in an undisturbed soil column using electrical resistance tomography', *Water Resources Research*, 32(4), pp. 763–769.

Binley, A., Shaw, B. and Henry-Poulter, S. (1996b) 'Flow pathways in porous media: Electrical resistance tomography and dye staining image verification', *Measurement Science and Technology*, 7(3), pp. 384–390. DOI: 10.1088/0957-0233/7/3/020

Binley, A., Pinheiro, P. and Dickin, F. (1996c) 'Finite element based three-dimensional forward and inverse solvers for electrical impedance tomography', in *Colloquium on Advances in Electrical Tomography, Computing and Control Division, IEE, Digest No. 96/143*. Manchester, UK, pp. 6/1–6/3.

Binley, A., Daily, W. and Ramirez, A. (1997) 'Detecting leaks from environmental barriers using electrical current imaging', *Journal of Environmental and Engineering Geophysics*, 2(1), pp. 11–19.

Binley, A., Slater, L. D., Fukes, M. and Cassiani, G. (2005) 'Relationship between spectral induced polarization and hydraulic properties of saturated and unsaturated sandstone', *Water Resources Research*, 41 (12), DOI: 10.1029/2005WR004202

Binley, A., Winship, P., Middleton, R., Pokar, M. and West, J. (2001) 'High-resolution characterization of vadose zone dynamics using cross-borehole radar', *Water Resources Research*, 37(11), pp. 2639–2652.

Binley, A., Winship, P., West, L. J., Pokar, M. and Middleton, R. (2002a) 'Seasonal variation of moisture content in unsaturated sandstone inferred from borehole radar and resistivity profiles', *Journal of Hydrology*, 267(3–4), pp. 160–172.

Binley, A., Cassiani, G., Middleton, R. and Winship, P. (2002b) 'Vadose zone flow model parameterisation using cross-borehole radar and resistivity imaging', *Journal of Hydrology*, 267 (3–4), pp. 147–159. DOIDOI: 10.1016/S0022-1694(02)00146-4

Binley, A., Hubbard, S. S., Huisman, J. A., Revil, A., Robinson, D. A., Singha, K. and Slater, L. D. (2015) 'The emergence of hydrogeophysics for improved understanding of subsurface processes over multiple scales', *Water Resources Research*, 51, pp. 3837–3866. DOI: 10.1002/2015WR017016

Binley, A., Keery, J., Slater, L., Barrash, W. and Cardiff, M. (2016) 'The hydrogeologic information in cross-borehole complex conductivity data from an unconsolidated conglomeratic sedimentary aquifer', *Geophysics*, 81(6), pp. E409–E421. DOI: 10.1190/geo2015-0608.1

Blanchy, G., Saneiyan, S., Boyd, J., McLachlan, P., Binley, A. (2020) 'ResIPy, an intuitive open source software for complex geoelectrical inversion/modeling in 2D space', *Computers and Geosciences*, 137. DOI: 10.1016/j.cageo.2020.104423

Blaschek, R., Hördt, A. and Kemna, A. (2008) 'A new sensitivity-controlled focusing regularization scheme for the inversion of induced polarization data based on the minimum gradient support', *Geophysics*, 73(2). DOI: 10.1190/1.2824820

Bleil, D. F. (1953) 'Induced polarization: A method of geophysical prospecting', *Geophysics*, 18, pp. 605–635. DOI: 10.1190/1.1437917

Blome, M., Maurer, H. and Greenhalgh, S. (2011) 'Geoelectric experimental design: Efficient acquisition and exploitation of complete pole-bipole data sets', *Geophysics*, 76(1), pp. F15–F26.

Blowes, D. W., Ptacek, C. J., Benner, S. G., McRae, C. W. T., Bennett, T. A. and Puls, R. W. (2000) 'Treatment of inorganic contaminants using permeable reactive barriers', *Journal of Contaminant Hydrology*, 45(1–2), pp. 123–137.

Bobachev A. A. (2003) Reshenie pryamyh i obratnyh zadach elektrorazvedki metodom soprotivlenij dlya slozhno-postroennyh sred (Direct and inverse problems of electrical prospecting by the resistivity method for difficult-built environments, in Russian). PhD dissertation. Moscow University, Russia.

Bodin, T. and Sambridge, M. (2009) 'Seismic tomography with the reversible jump algorithm', *Geophysical Journal International*, 178(3), pp. 1411–1436.

Bodmer, R., Ward, S. H. and Morrison, H. F. (1968) 'On induced electrical polarization and groundwater', *Geophysics*, 33(5), pp. 805–821.

Bogoslovsky, V. A. and Ogilvy, A. A. (1977) 'Geophysical methods for the investigation of landslides', *Geophysics*, 42(3), pp. 562–571.

Bording, T. S., Fiandaca, G., Maurya, P. K., Auken, E., Christiansen, A. V., Tuxen, N., Klint, K. E. S. and Larsen, T. H. (2019) 'Cross-borehole tomography with full-decay spectral time-domain induced polarization for mapping of potential contaminant flow-paths', *Journal of Contaminant Hydrology*, 226. DOI: 10.1016/j.jconhyd.2019.103523

Börner, F. (1992) *Complex conductivity measurements of reservoir properties. Advances in Core Evaluation III (Reservoir Management)*. Gordon and Breach Science Publishers, London.

Börner, F. D. and Schön, J. H. (1991) 'A relation between the quadrature component of electrical conductivity and the specific surface area of sedimentary rocks', *The Log Analyst*, 32, pp. 612–613.

Börner, F. Gruhne, M. and Schön, J. (1993) 'Contamination indications derived from electrical properties in the low frequency range 1', *Geophysical Prospecting*, 41(1), pp. 83–98.

Börner, F. D., Schopper, J. R. and Weller, A. (1996) 'Evaluation of transport and storage properties in the soil and groundwater zone from induced polarization measurements', *Geophysical Prospecting*, 44(4), pp. 583–601. DOI: 10.1111/j.1365-2478.1996.tb00167.x

Boryta, D. A. and Nabighian, M. N. (1985) 'Method for determining a leak in a pond liner of electrically insulating sheet material', U.S. Patent 4,543,525.

Bouchedda, A., Chouteau, M., Binley, A. and Giroux, B. (2012) '2-D joint structural inversion of cross-hole electrical resistance and ground penetrating radar data', *Journal of Applied Geophysics*, 78, pp. 52–67. DOI: 10.1016/j.jappgeo.2011.10.009

Boyd, J., Blanchy, G., Saneiyan, S., McLachlan, P., Binley, A. (2019) '3D geoelectrical problems with ResiPy, an open source graphical user interface for geoelectrical data processing', *Fast Times*, 24(4), pp. 85–92.

Bradbury, K. R. and Taylor, R. W. (1984) 'Determination of the hydrogeologic properties of lakebeds using offshore geophysical surveys', *Groundwater*, 22(6), pp. 690–695.

Brindt, N., Rahav, M. and Wallach, R. (2019) 'ERT and salinity: A method to determine whether ERT-detected preferential pathways in brackish water-irrigated soils are water-induced or an artifact of salinity', *Journal of Hydrology*, 574, pp. 35–45.

Brown, B. H. (2001) 'Medical impedance tomography and process impedance tomography: A brief review', *Measurement Science and Technology*, 12(8), p. 991.

Brown, B. H., Barber, D. C., Wang, W., Lu, L., Leathard, A. D., Smallwood, R. H., Hampshire, A. R., Mackay, R. and Hatzigalanis, K. (1994) 'Multi-frequency imaging and modelling of respiratory related electrical impedance changes', *Physiological Measurement*, 15(2A), p. A1.

Brown, F. H. (1900) 'Process of locating metallic minerals or buried treasures', U.S. Patent 645,910.

Brown, F. H. (1901) 'Process of locating metallic minerals', U.S. Patent 672,309.

Brown, S. R., Lesmes, D., Fourkas, J. and Sorenson, J. R. (2003) *Complex Electrical Resistivity for Monitoring DNAPL Contamination*. New England Research Inc. (US), p. 29.

Bruggeman, V. D. A. G. (1935) 'Berechnung verschiedener physikalischer Konstanten von heterogenen Substanzen. I. Dielektrizitätskonstanten und Leitfähigkeiten der Mischkörper aus isotropen Substanzen', *Annalen der physik*, 416(7), pp. 636–664.

Brunauer, S., Emmett, P. H. and Teller, E. (1938) 'Adsorption of gases in multimolecular layers', *Journal of the American Chemical Society*, 60(2), pp. 309–319.

Bücker, M. and Hördt, A. (2013a) 'Analytical modelling of membrane polarization with explicit parametrization of pore radii and the electrical double layer', *Geophysical Journal International*, 194 (2), pp. 804–813. DOI: 10.1093/gji/ggt136

Bücker, M. and Hördt, A. (2013b) 'Long and short narrow pore models for membrane polarization', *Geophysics*, 78(6), pp. E299–E314. DOI: 10.1190/geo2012-0548.1

Bücker, M., Bairlein, K., Bielefeld, A., Kuhn, E., Nordsiek, S. and Stebner, H. (2016) 'The dependence of induced polarization on fluid salinity and pH, studied with an extended model of membrane polarization', *Journal of Applied Geophysics*, 135, pp. 408–417. DOI: 10.1016/j.jappgeo.2016.02.007

Bücker, M., Orozco, A. F. and Kemna, A. (2018) 'Electrochemical polarization around metallic particles – Part 1: The role of diffuse-layer and volume-diffusion relaxation', *Geophysics*, 83(4), pp. E203–E217. DOI: 10.1190/geo2017-0401.1

Bücker, M., Flores Orozco, A., Undorf, S. and Kemna, A. (2019) 'On the role of Stern- and diffuse-layer polarization mechanisms in porous media', *Journal of Geophysical Research: Solid Earth*, 124 (6), pp. 5656–5677. DOI: 10.1029/2019JB017679

Butler, K. E. (2009) 'Trends in waterborne electrical and EM induction methods for high resolution sub-bottom imaging', *Near Surface Geophysics*, 7(4), pp. 241–246.

Cai, J., Wei, W., Hu, X. and Wood, D. A. (2017) 'Electrical conductivity models in saturated porous media: A review', Earth-Science Reviews. pp. 419–433. DOI: 10.1016/j.earscirev.2017.06.013

Calderón-Macías, C., Sen, M. K. and Stoffa, P. L. (2000) 'Artificial neural networks for parameter estimation in geophysics', *Geophysical Prospecting*, 48(1), pp. 21–47.

Candansayar, M. E. (2008) 'Two-dimensional individual and joint inversion of three-and four-electrode array dc resistivity data', *Journal of Geophysics and Engineering*, 5(3), pp. 290–300.

Carcangiu, S., Fanni, A. and Montisci, A. (2019) 'Electric capacitance tomography for nondestructive testing of standing trees', *International Journal of Numerical Modelling: Electronic Networks, Devices and Fields*, 32 (4), pp. 1–10. DOI: 10.1002/jnm.2252

Carman, P. C. (1939) 'Permeability of saturated sands, soils and clays', *The Journal of Agricultural Science*, 29(2), pp. 262–273.

Carpenter, E. W. (1955) 'Some notes concerning the Wenner configuration', *Geophysical Prospecting*, 3(4), pp. 388–402.

Carpenter, E. W. and Habberjam, G. M. (1956) 'A tri-potential method of resistivity prospecting', *Geophysics*, 21(2), pp. 455–469.

Cassiani, G., Binley, A., Kemna, A., Wehrer, M., Orozco, A. F., Deiana, R., Boaga, J., Rossi, M., Dietrich, P., Werban, U., Zschornack, L., Godio, A., JafarGandomi, A. and Deidda, G. P. (2014) 'Noninvasive characterization of the Trecate (Italy) crude-oil contaminated site: Links between contamination and geophysical signals', *Environmental Science and Pollution Research*, 21(15), pp. 8914–8931. DOI: 10.1007/s11356-014-2494-7

Caterina, D., Beaujean, J., Robert, T. and Nguyen, F. (2013) 'A comparison study of different image appraisal tools for electrical resistivity tomography', *Near Surface Geophysics*, 11(6), pp. 639–657. DOI: 10.3997/1873-0604.2013022

Chambers, J., Ogilvy, R., Kuras, O., Cripps, J. and Meldrum, P. (2002) '3D electrical imaging of known targets at a controlled environmental test site', *Environmental Geology*, 41(6), pp. 690–704.

Chambers, J. E., Meldrum, P. I., Ogilvy, R. D. and Wilkinson, P. B. (2005) 'Characterisation of a NAPL-contaminated former quarry site using electrical impedance tomography', *Near Surface Geophysics*, 3(2), pp. 81–92. DOI: 10.3997/1873-0604.2005003

Chambers, J. E., Wilkinson, P. B., Uhlemann, S., Sorensen, J. P. R., Roberts, C., Newell, A. J., Ward, W. O. C., Binley, A., Williams, P. J., Gooddy, D. C., Old, G. and Bai, L. (2014) 'Derivation of lowland riparian wetland deposit architecture using geophysical image analysis and interface detection', *Water Resources Research*, 50 (7), pp. 5886–5905. DOI: 10.1002/2014WR015643

Chambers, J. E., Wilkinson, P. B., Wardrop, D., Hameed, A., Hill, I., Jeffrey, C., Loke, M. H., Meldrum, P. I., Kuras, O., Cave, M. and Gunn, D. A. (2012) 'Bedrock detection beneath river terrace deposits using three-dimensional electrical resistivity tomography', *Geomorphology*, 177–178, pp. 17–25. DOI: 10.1016/j.geomorph .2012.03.034

Chambers, J. E., Gunn, D. A., Wilkinson, P. B., Meldrum, P. I., Haslam, E., Holyoake, S., Kirkham, M., Kuras, O., Merritt, A. and Wragg, J. (2014) '4D electrical resistivity tomography monitoring of soil moisture dynamics in an operational railway embankment', *Near Surface Geophysics*, 12(1), pp. 61–72.

Chambers, J., Meldrum, P., Gunn, D., Wilkinson, P., Uhlemann, S., Kuras, O. and Swift, R. (2015) 'Proactive infrastructure monitoring and evaluation (PRIME): A new electrical resistivity tomography system for remotely monitoring the internal condition of geotechnical infrastructure assets', in 3rd International Workshop on Geoelectrical Monitoring (GELMON).

Chave, A. D., Constable, S. C. and Edwards, R. N. (1991) 'Electrical exploration methods for the seafloor', in *Electromagnetic Methods in Applied Geophysics: Volume 2, Application, Parts A and B*. Society of Exploration Geophysicists, pp. 931–966.

Chelidze, T. L. and Gueguen, Y. (1999) 'Electrical spectroscopy of porous rocks: A review-I. Theoretical models', *Geophysical Journal International*, 137(1), pp. 1–15. DOI: 10.1046/j.1365-246X.1999.00799.x

Chelidze, T. L., Guéguen, Y. and Ruffet, C. (1999) 'Electrical spectroscopy of porous rocks: A review- II. Experimental results and interpretation', *Geophysical Journal International*, 137, pp. 16–34.

Chen, J., Kemna, A. and Hubbard, S. S. (2008) 'A comparison between Gauss-Newton and Markov-chain Monte Carlo-based methods for inverting spectral induced-polarization data for Cole–Cole parameters', *Geophysics*, 73(6). DOI: 10.1190/1.2976115

Chen, Q., Pardo, D., Li, H. and Wang, F. (2011) 'New post-processing method for interpretation of through casing resistivity (TCR) measurements', *Journal of Applied Geophysics*, 74(1), pp. 19–25.

Cheng, Q., Chen, X., Tao, M. and Binley, A. (2019a) 'Characterization of karst structures using quasi-3D electrical resistivity tomography', *Environmental Earth Sciences*, 78 (9). DOI: 10.1007/s12665-019-8284-2

Cheng, Q., Tao, M., Chen, X. and Binley, A. (2019b) 'Evaluation of electrical resistivity tomography (ERT) for mapping the soil–rock interface in karstic environments', *Environmental Earth Sciences*, 78(15), p. 439.

Cho, Y., Sudduth, K. A. and Chung, S.-O. (2016) 'Soil physical property estimation from soil strength and apparent electrical conductivity sensor data', *Biosystems Engineering*, 152, pp. 68–78.

Christensen, N. B. (1990) 'Optimized fast Hankel transform filters 1', *Geophysical Prospecting*, 38(5), pp. 545–568.

Christensen, N. B. and Sørensen, K. (2001) 'Pulled array continuous electrical sounding with an additional inductive source: An experimental design study', *Geophysical Prospecting*, 49(2), pp. 241–254.

Chuprinko, D. and Titov, K. (2017) 'Influence of mineral composition on spectral induced polarization in sediments', *Geophysical Journal International*, pp. 186–191. DOI: 10.1093/GJI/GGX018

Claerbout, J. F. and Muir, F. (1973) 'Robust modeling with erratic data', *Geophysics*, 38(5), pp. 826–844.

Clark, A. (1990) *Seeing Beneath the Soil: Prospecting Methods in Archaeology*. Routledge, p. 176.

Clark, A. R. and Salt, D. J. (1951) 'The investigation of earth resistivities in the vicinity of a diamond drill hole', *Geophysics*, 16(4), pp. 659–665.

Clavier, C., Coates, G. and Dumanoir, J. (1984) 'Theoretical and experimental bases for the dual-water model for interpretation of shaly sands', *Society of Petroleum Engineers Journal*, 24(02), pp. 153–168.

Cockett, R., Kang, S., Heagy, L. J., Pidlisecky, A. and Oldenburg, D. W. (2015) 'SimPEG: An open source framework for simulation and gradient based parameter estimation in geophysical applications', *Computers & Geosciences*, 85, pp. 142–154.

Coggon, J. H. (1971) 'Electromagnetic and electrical modeling by the finite element method', *Geophysics*, 36(1), pp. 132–155.

Cole, K. S. and Cole, R. H. (1941) 'Dispersion and absorption in dielectrics I. Alternating current characteristics', *The Journal of Chemical Physics*, 9(4), pp. 341–351.

Colton, D. L. and Kress, R. (1992) *Inverse Acoustic and Electromagnetic Scattering Theory*. Springer-Verlag (Applied Mathematical Sciences), p. 334.

Commer, M., Newman, G. A., Williams, K. H. and Hubbard, S. S. (2011) '3D induced-polarization data inversion for complex resistivity', *Geophysics*, 76(3). DOI: 10.1190/1.3560156

Constable, S. C., Parker, R. L. and Constable, C. G. (1987) 'Occam's inversion: A practical algorithm for generating smooth models from electromagnetic sounding data', *Geophysics*, 52(3), pp. 289–300.

Crestani, E., Camporese, M. and Salandin, P. (2015) 'Assessment of hydraulic conductivity distributions through assimilation of travel time data from ERT-monitored tracer tests', *Advances in Water Resources*, 84, pp. 23–36.

Crook, N., Musgrave, H. and Binley, A. (2006) 'Geophysical characterisation of the riparian zone in groundwater fed catchments', in *19th Symposium on the Application of Geophysics to Engineering and Environmental Problems, SAGEEP 2006*. Geophysical Applications for Environmental and Engineering Hazards – Advances and Constraints.

Crook, N., Binley, A., Knight, R., Robinson, D. A., Zarnetske, J. and Haggerty, R. (2008) 'Electrical resistivity imaging of the architecture of substream sediments', *Water Resources Research*, 44, W00D13, DOI:10.1029/2008WR006968

Čuma, M. and Zhdanov, M. S. (2014) 'Massively parallel regularized 3D inversion of potential fields on CPUs and GPUs', *Computers & Geosciences*, 62, pp. 80–87.

Dafflon, B., Wu, Y., Hubbard, S. S., Birkholzer, J. T., Daley, T. M., Pugh, J. D., Peterson, J. E. and Trautz, R. C. (2012) 'Monitoring CO2 intrusion and associated geochemical transformations in a shallow groundwater system using complex electrical methods', *Environmental Science & Technology*, 47(1), pp. 314–321.

Dafflon, B., Hubbard, S., Wainwright, H., Kneafsey, T. J., Ulrich, C., Peterson, J. and Wu, Y. (2016) 'Geophysical estimation of shallow permafrost distribution and properties in an ice-wedge polygon-dominated Arctic tundra region', *Geophysics*, 81 (1), pp. WA247–WA263. DOI: 10.1190/geo2015-0175.1

Daft, L. and Williams, A. (1906) 'Apparatus for detecting and localizing mineral deposits', U.S. Patent 817,736.

Dahlin, T. (1995) 'On the automation of 2D resistivity surveying for engineering and environmental applications'. PhD thesis, Lund University, Sweden, p. 187.
Dahlin, T. (2000) 'Short note on electrode charge-up effects in DC resistivity data acquisition using multi-electrode arrays', *Geophysical Prospecting*, 48(1), pp. 181–187.
Dahlin, T. and Loke, M. H. (1997) 'Quasi-3D resistivity imaging-mapping of three dimensional structures using two dimensional DC resistivity techniques', in 3rd EEGS Meeting, European Association of Geoscientists & Engineers (EAGE). DOI: 10.3997/2214-4609.201407298
Dahlin, T. and Leroux, V. (2012) 'Improvement in time-domain induced polarization data quality with multi-electrode systems by separating current and potential cables', *Near Surface Geophysics*, 10(6), pp. 545–565. DOI: 10.3997/1873-0604.2012028
Dahlin, T. and Loke, M. H. (2015) 'Negative apparent chargeability in time-domain induced polarisation data', *Journal of Applied Geophysics*, 123, pp. 322–332.
Dahlin, T. and Zhou, B. (2006) 'Multiple-gradient array measurements for multichannel 2D resistivity imaging', *Near Surface Geophysics*, 4(2), pp. 113–123.
Dahlin, T., Bernstone, C. and Loke, M. H. (2002a) 'A 3-D resistivity investigation of a contaminated site at Lernacken, Sweden', *Geophysics*, 67(6), pp. 1692–1700.
Dahlin, T., Leroux, V. and Nissen, J. (2002b) 'Measuring techniques in induced polarisation imaging', *Journal of Applied Geophysics*, 50(3), pp. 279–298. DOI: 10.1016/S0926-9851(02)00148-9
Daily, W. and Owen, E. (1991) 'Cross-borehole resistivity tomography', *Geophysics*, 56(8), pp. 1228–1235.
Daily, W. and Ramirez, A. (1995) 'Electrical resistance tomography during in-situ trichloroethylene remediation at the Savannah River Site', *Journal of Applied Geophysics*, 33 (4), pp. 239–249.
Daily, W. D. and Ramirez, A. L. (1999) 'Electrical resistance tomography using steel cased boreholes as electrodes', U.S. Patent 5,914,603.
Daily, W. and Ramirez, A. L. (2000) 'Electrical imaging of engineered hydraulic barriers', *Geophysics*, 65(1), pp. 83–94.
Daily, W., Ramirez, A., LaBrecque, D. and Nitao, J. (1992) 'Electrical resistivity tomography of vadose water movement', *Water Resources Research*, 28(5), pp. 1429–1442.
Daily, W., Ramirez, A. and Binley, A. (2004) 'Remote monitoring of leaks in storage tanks using electrical resistance tomography: Application at the Hanford Site', *Journal of Environmental and Engineering Geophysics*, 9(1), pp. 11–24.
Daniels, J. J. (1977) 'Three-dimensional resistivity and induced-polarization modeling using buried electrodes', *Geophysics*, 42(5), pp. 1006–1019.
Davidson, D. W. and Cole, R. H. (1951) 'Dielectric relaxation in glycerol, propylene glycol, and n-propanol', *The Journal of Chemical Physics*, 19(12), pp. 1484–1490. DOI: 10.1063/1.1748105
Davidson, E., Lefebvre, P. A., Brando, P. M., Ray, D. M., Trumbore, S. E., Solorzano, L. A., Ferreira, J. N., Bustamante, M. M. da C. and Nepstad, D. C. (2011) 'Carbon inputs and water uptake in deep soils of an eastern Amazon forest', *Forest Science*, 57(1), pp. 51–58.
Davis, J. A., James, R. O. and Leckie, J. O. (1978) 'Surface ionization and complexation at the oxide/water interface: I. Computation of electrical double layer properties in simple electrolytes', *Journal of Colloid and Interface Science*, 63(3), pp. 480–499. DOI: https://DOI.org/10.1016/S0021-9797(78)80009-5
Day-Lewis, F. D., Singha, K. and Binley, A. M. (2005) 'Applying petrophysical models to radar travel time and electrical resistivity tomograms: Resolution-dependent

limitations', *Journal of Geophysical Research: Solid Earth*, 110(8), pp. 1–17. DOI: 10.1029/2004JB003569

Day-Lewis, F. D., White, E. A., Johnson, C. D., Lane Jr, J. W. and Belaval, M. (2006) 'Continuous resistivity profiling to delineate submarine groundwater discharge: Examples and limitations', *The Leading Edge*, 25(6), pp. 724–728.

Day-Lewis, F. D., Linde, N., Haggerty, R., Singha, K. and Briggs, M. A. (2017) 'Pore network modeling of the electrical signature of solute transport in dual-domain media', *Geophysical Research Letters*, 44(10), pp. 4908–4916. DOI: 10.1002/2017GL073326

De Donno, G. and Cardarelli, E. (2011) 'Assessment of errors from different electrode materials and configurations for electrical resistivity and time-domain IP data on laboratory models', *Bollettino di Geofisica Teorica ed Applicata*, 52(2), pp. 211–223.

de Sosa, L. L., Glanville, H. C., Marshall, M. R., Schnepf, A., Cooper, D. M., Hill, P. W., Binley, A. and Jones, D. L. (2018) 'Stoichiometric constraints on the microbial processing of carbon with soil depth along a riparian hillslope', *Biology and Fertility of Soils*. 54(8), pp. 949–963. DOI: 10.1007/s00374-018-1317-2

De Witt, G. W. (1979) 'Parametric studies of induced polarization data'. MS disseration thesis, University of Utah, p. 178.

Deceuster, J. and Kaufmann, O. (2012) 'Improving the delineation of hydrocarbon-impacted soils and water through induced polarization (IP) tomographies: A field study at an industrial waste land', *Journal of Contaminant Hydrology*, 136, pp. 25–42.

Deceuster, J., Etienne, A., Robert, T., Nguyen, F. and Kaufmann, O. (2014) 'A modified DOI-based method to statistically estimate the depth of investigation of dc resistivity surveys', *Journal of Applied Geophysics*, 103, pp. 172–185. DOI: 10.1016/j.jappgeo.2014.01.018

deGroot-Hedlin, C. (1990) 'Occam's inversion to generate smooth, two-dimensional models from magnetotelluric data', *Geophysics*, 55 (12), pp. 1613–1624. DOI: 10.1190/1.1442813

Demirel, C. and Candansayar, M. E. (2017) 'Two-dimensional joint inversions of cross-hole resistivity data and resolution analysis of combined arrays', *Geophysical Prospecting*, 65(3), pp. 876–890. DOI: 10.1111/1365-2478.12432

Dey, A. and Morrison, H. F. (1973) 'Electromagnetic coupling in frequency and time-domain induced-polarization surveys over a multilayered earth', *Geophysics*, 38(2), pp. 380–405.

Dey, A. and Morrison, H. F. (1979) 'Resistivity modeling for arbitrarily shaped three-dimensional structures', *Geophysics*, 44(4), pp. 753–780.

Dias, C. A. (1972) 'Analytical model for a polarizable medium at radio and lower frequencies', *Journal of Geophysical Research*, 77(26), pp. 4945–4956. DOI: 10.1029/jb077i026p04945

Dias, C. A. (2000) 'Developments in a model to describe low-frequency electrical polarization of rocks', *Geophysics*, 65(2), pp. 437–451. DOI: 10.1190/1.1444738

Dickin, F. and Wang, M. (1996) 'Electrical resistance tomography for process applications', *Measurement Science and Technology*, 7(3), p. 247.

Dissado, L. A. and Hill, R. M. (1984) 'Anomalous low-frequency dispersion. Near direct current conductivity in disordered low-dimensional materials', *Journal of the Chemical Society, Faraday Transactions 2: Molecular and Chemical Physics*, 80 (3), pp. 291–319.

Doetsch, J. A., Coscia, I., Greenhalgh, S., Linde, N., Green, A. and Günther, T. (2010a) 'The borehole-fluid effect in electrical resistivity imaging', *Geophysics*, 75 (4), pp. F107–F114.

Doetsch, J., Linde, N., Coscia, I., Greenhalgh, S. A. and Green, A. G. (2010b) 'Zonation for 3D aquifer characterization based on joint inversions of multimethod crosshole geophysical data', *Geophysics*, 75(6). DOI: 10.1190/1.3496476

Doetsch, J., Linde, N., Vogt, T., Binley, A. and Green, A. G. (2012a) 'Imaging and quantifying salt-tracer transport in a riparian groundwater system by means of 3D ERT monitoring', *Geophysics*, 77(5). DOI: 10.1190/geo2012-0046.1

Doetsch, J., Linde, N., Pessognelli, M., Green, A. G. and Günther, T. (2012b) 'Constraining 3-D electrical resistance tomography with GPR reflection data for improved aquifer characterization', *Journal of Applied Geophysics*, 78, pp. 68–76.

Dosso, S. E. and Oldenburg, D. W. (1989) 'Linear and non-linear appraisal using extremal models of bounded variation', *Geophysical Journal International*, 99(3), pp. 483–495.

Draskovits, P. and Fejes, I. (1994) 'Geophysical methods in drinkwater protection of near-surface reservoirs', *Journal of Applied Geophysics*, 31(1–4), pp. 53–63.

Draskovits, P., Hobot, J., Vero, L. and Smith, B. D. (1990) 'Induced-polarization surveys applied to evaluation of groundwater resources, Pannonian Basin, Hungary', *Induced Polarization: Applications and Case Histories, Investigations in Geophysics*, 4, pp. 379–410.

Duckworth, K. and Calvert, H. T. (1995) 'An examination of the relationship between time-domain integral chargeability and the Cole–Cole impedance model', *Geophysics*, 60(4), pp. 1249–1252.

Dukhin, S. S. and Shilov, V. N. (1974) *Dielectric Phenomena and the Double Layer in Disperse Systems and Polyelectrolytes*. Naukova Duma, Kiev, p. 206.

Dunlap, H. F. and Hawthorne, R. R. (1951) 'The calculation of water resistivities from chemical analyses', *Journal of Petroleum Technology*, 3(03), p. 17.

Edwards, L. S. (1977) 'A modified pseudosection for resistivity and IP', *Geophysics*, 42(5), pp. 1020–1036.

Efron, B. and Tibshirani, R. J. (1994) *An Introduction to the Bootstrap*. CRC Press, p. 456.

Ekinci, Y. L. and Demirci, A. (2008) 'A damped least-squares inversion program for the interpretation of Schlumberger sounding curves', *Journal of Applied Sciences*, 8(22), pp. 4070–4078.

Ellis, D. V. and Singer, J. M. (2008) *Well Logging for Earth Scientists*. Springer Science and Business Media, p. 708.

Everett, M. E. (2013) *Near-Surface Applied Geophysics*. Cambridge University Press, p. 400.

Evjen, H. M. (1938) 'Depth factors and resolving power of electrical measurements', *Geophysics*, 3(2), pp. 78–95.

Farias, V. J. da C., Maranhão, C. H. de M., da Rocha, B. R. P. and de Andrade, N. de P. O. (2010) 'Induced polarization forward modelling using finite element method and the fractal model', *Applied Mathematical Modelling*, 34(7), pp. 1849–1860.

Farquharson, C. G. and Oldenburg, D. W. (1998) 'Non-linear inversion using general measures of data misfit and model structure', *Geophysical Journal International*, 134(1), pp. 213–227.

Fatt, I. (1956) 'The network model of porous media'. *Petroleum Transactions, AIME*, Volume 207, pp. 144–181.

Fernandez, P. M., Bloem, E., Binley, A., Philippe, R. S. B. A. and French, H. K. (2019) 'Monitoring redox sensitive conditions at the groundwater interface using electrical resistivity and self-potential', *Journal of Contaminant Hydrology*, 226, p. 103517.

Fernández-Muñiz, Z., Khaniani, H. and Fernández-Martínez, J. L. (2019) 'Data kit inversion and uncertainty analysis', *Journal of Applied Geophysics*, 161, pp. 228–238. DOI: 10.1016/j.jappgeo.2018.12.022

Ferré, T., Bentley, L., Binley, A., Linde, N., Kemna, A., Singha, K., Holliger, K., Huisman, J. A. and Minsley, B. (2009) 'Critical steps for the continuing advancement of hydrogeophysics', *Eos, Transactions American Geophysical Union*, 90(23), p. 200.

Fiandaca, G., Auken, E., Christiansen, A. V. and Gazoty, A. (2012) 'Time-domain-induced polarization: Full-decay forward modeling and 1D laterally constrained inversion of Cole-Cole parameters', *Geophysics*, 77(3), pp. E213–E225. DOI: 10.1190/geo2011-0217.1

Fiandaca, G., Ramm, J., Binley, A., Gazoty, A., Christiansen, A. V. and Auken, E. (2013) 'Resolving spectral information from time domain induced polarization data through 2-D inversion', *Geophysical Journal International*, 192(2), pp. 631–646. DOI: 10.1093/gji/ggs060

Fiandaca, G., Madsen, L. M. and Maurya, P. K. (2018a) 'Re-parameterisations of the Cole–Cole model for improved spectral inversion of induced polarization data', *Near Surface Geophysics*, 16(4), pp. 385–399. DOI: 10.3997/1873-0604.2017065

Fiandaca, G., Maurya, P. K., Balbarini, N., Hördt, A., Christiansen, A. V., Foged, N., Bjerg, P. L. and Auken, E. (2018b) 'Permeability estimation directly from logging-while-drilling induced polarization data', *Water Resources Research*, 54(4), pp. 2851–2870. DOI: 10.1002/2017WR022411

Fixman, M. (1980) 'Charged macromolecules in external fields. I. The sphere', *The Journal of Chemical Physics*, 72(9), pp. 5177–5186. DOI: 10.1063/1.439753

Flathe, H. (1955) 'A practical method of calculating geoelectrical model graphs for horizontally stratified media', *Geophysical Prospecting*, 3(3), pp. 268–294. DOI: 10.1111/j.1365-2478.1955.tb01377.x

Flis, M. F., Newman, G. A. and Hohmann, G. W. (1989) 'Induced-polarization effects in time-domain electromagnetic measurements', *Geophysics*, 54(4), pp. 514–523.

Flores Orozco, A., Kemna, A. and Zimmermann, E. (2012) 'Data error quantification in spectral induced polarization imaging', *Geophysics*, 77(3). DOI: 10.1190/geo2010-0194.1

Flores Orozco, A. F., Williams, K. H. and Kemna, A. (2013) 'Time-lapse spectral induced polarization imaging of stimulated uranium bioremediation', *Near Surface Geophysics*, 11(5), pp. 531–544. DOI: 10.3997/1873-0604.2013020

Flores Orozco, A., Velimirovic, M., Tosco, T., Kemna, A., Sapion, H., Klaas, N., Sethi, R. and Bastiaens, L. (2015) 'Monitoring the injection of microscale zerovalent iron particles for groundwater remediation by means of complex electrical conductivity imaging', *Environmental Science and Technology*, 49(9), pp. 5593–5600. DOI: 10.1021/acs.est.5b00208

Flores Orozco, A., Micić, V., Bücker, M., Gallistl, J., Hofmann, T. and Nguyen, F. (2019a) 'Complex-conductivity monitoring to delineate aquifer pore clogging during nano-particles injection', *Geophysical Journal International*, 218(3), pp. 1838–1852. DOI: 10.1093/gji/ggz255

Flores Orozco, A. F., Kemna, A., Binley, A. and Cassiani, G. (2019b) 'Analysis of time-lapse data error in complex conductivity imaging to alleviate anthropogenic noise for site characterization', *Geophysics*, 84(2), pp. B181–B193. DOI: 10.1190/GEO2017-0755.1

Florsch, N. and Muhlach, F. (2017) *Everyday Applied Geophysics 1: Electrical Methods*. Elsevier, p. 202.

Frangos, W. (1997) 'Electrical detection of leaks in lined waste disposal ponds', *Geophysics*, 62(6), pp. 1737–1744.

Freedman, R. and Vogiatzis, J. P. (1986) 'Theory of induced-polarization logging in a borehole', *Geophysics*, 51(9), pp. 1830–1849.

Fujita, Y., Ferris, F. G., Lawson, R. D., Colwell, F. S. and Smith, R. W. (2000) 'Subscribed content calcium carbonate precipitation by ureolytic subsurface bacteria', *Geomicrobiology Journal*, 17(4), pp. 305–318.

Fuller, B. D. and Ward, S. H. (1970) 'Linear system description of the electrical parameters of rocks', *IEEE Transactions on Geoscience Electronics*, 8(1), pp. 7–18.

Fuoss, R. M. and Kirkwood, J. G. (1941) 'Electrical properties of solids. VIII. Dipole moments in polyvinyl chloride-diphenyl systems*', *Journal of the American Chemical Society*, 63(2), pp. 385–394. DOI: 10.1021/ja01847a013

Furman, A., Ferré, T. P. and Heath, G. L. (2007) 'Spatial focusing of electrical resistivity surveys considering geologic and hydrologic layering', *Geophysics*, 72(2), pp. F65–F73.

Gaffney, C. F. and Gater, J. (2003) *Revealing the Buried Past: Geophysics for Archaeologists*. Tempus, p. 192.

Gaffney, C., Harris, C., Pope-Carter, F., Bonsall, J., Fry, R. and Parkyn, A. (2015) 'Still searching for graves: An analytical strategy for interpreting geophysical data used in the search for "unmarked" graves', *Near Surface Geophysics*, 13(6), pp. 557–569.

Galetti, E. and Curtis, A. (2018) 'Transdimensional electrical resistivity tomography', *Journal of Geophysical Research: Solid Earth*, 123(8), pp. 6347–6377.

Gallardo, L. A. and Meju, M. A. (2003) 'Characterization of heterogeneous near-surface materials by joint 2D inversion of dc resistivity and seismic data', *Geophysical Research Letters*, 30(13). DOI: 10.1029/2003GL017370

Gallardo, L. A. and Meju, M. A. (2011) 'Structure-coupled multiphysics imaging in geophysical sciences', *Reviews of Geophysics*, 49(1). DOI: 10.1029/2010RG000330

Gan, F., Han, K., Lan, F., Chen, Y. and Zhang, W. (2017) 'Multi-geophysical approaches to detect karst channels underground: A case study in Mengzi of Yunnan Province, China', *Journal of Applied Geophysics*, 136, pp. 91–98.

Gardner, F. D. (1897) 'The electrical method of moisture determination in soils: Results and modifications in 1897', in *Bulletin No. 12*. U.S. Department of Agriculture, Division of Soils, Washington, D.C., p. 24.

Gardner, F. D. (1898) *The Electrical Method of Moisture Determination in Soils, Results and Modifications in 1897*. US Government Printing Office, p. 38.

Garré, S., Javaux, M., Vanderborght, J. and Vereecken, H. (2011) 'Three-dimensional electrical resistivity tomography to monitor root zone water dynamics', *Vadose Zone Journal*, 10(1), pp. 412–424.

Gazoty, A., Fiandaca, G., Pedersen, J., Auken, E., Christiansen, A. V. and Pedersen, J. K. (2012) 'Application of time domain induced polarization to the mapping of lithotypes in a landfill site', *Hydrology and Earth System Sciences*, 16(6), pp. 1793–1804. DOI: 10.5194/hess-16-1793-2012

Gazoty, A., Fiandaca, G., Pedersen, J., Auken, E. and Christiansen, A. V. (2013) 'Data repeatability and acquisition techniques for Time-Domain spectral Induced Polarization', *Near Surface Geophysics*, 11(1983), pp. 391–406. DOI: 10.3997/1873-0604.2013013

Gebbers, R., Lück, E., Dabas, M. and Domsch, H. (2009) 'Comparison of instruments for geoelectrical soil mapping at the field scale', *Near Surface Geophysics*, 7(3), pp. 179–190.

Gelman, A. and Rubin, D. B. (1992) 'Inference from iterative simulation using multiple sequences', *Statistical Science*, 7(4), pp. 457–472.

Geometrics (2001) *OhmMapper TR1 29005–01 REV.F Operation Manual*. Geometrics Inc.

Gernez, S., Bouchedda, A., Gloaguen, E. and Paradis, D. (2020) 'AIM4RES, an open-source 2.5D finite differences MATLAB library for anisotropic electrical resistivity modeling', *Computers and Geosciences*, 135, p. 104401. DOI: 10.1016/j.cageo.2019.104401

Geselowitz, D. B. (1971) 'An application of electrocardiographic lead theory to impedance plethysmography', *IEEE Transactions on Biomedical Engineering*, 1, pp. 38–41.

Geuzaine, C. and Remacle, J. (2009) 'Gmsh: A 3-D finite element mesh generator with built-in pre-and post-processing facilities', *International Journal for Numerical Methods in Engineering*, 79(11), pp. 1309–1331.

Ghorbani, A., Camerlynck, C., Florsch, N., Cosenza, P. and Revil, A. (2007) 'Bayesian inference of the Cole–Cole parameters from time-and frequency-domain induced polarization', *Geophysical Prospecting*, 55(4), pp. 589–605.

Ghorbani, A., Camerlynck, C. and Florsch, N. (2009) 'CR1Dinv: A Matlab program to invert 1D spectral induced polarization data for the Cole–Cole model including electromagnetic effects', *Computers & Geosciences*, 35(2), pp. 255–266.

Ghosh, D. P. (1971a) 'Inverse filter coefficients for the computation of apparent resistivity standard curves for a horizontally stratified earth', *Geophysical Prospecting*, 19(4), pp. 769–775.

Ghosh, D. P. (1971b) 'The application of linear filter theory to the direct interpretation of geoelectrical resistivity sounding measurements', *Geophysical Prospecting*, 19(2), pp. 192–217.

Gish, O. H. and Rooney, W. J. (1925) 'Measurement of resistivity of large masses of undisturbed earth', *Terrestrial Magnetism and Atmospheric Electricity*, 30(4), pp. 161–188.

Gisser, D. G., Isaacson, D. and Newell, J. C. (1987) 'Current topics in impedance imaging', *Clinical Physics and Physiological Measurement*, 8(4A), pp. 39–46. DOI: 10.1088/0143-0815/8/4A/005

Glover, P. W. J. (2009) 'What is the cementation exponent? A new interpretation', *The Leading Edge*, 28(1), pp. 82–85.

Glover, P. W. J. (2010) 'A generalized Archie's law for n phases', *Geophysics*, 75(6), pp. E247–E265.

Glover, P. W. J. (2015) 'Geophysical properties of the near surface earth: Electrical properties', in *Treatise on Geophysics* (G. Schubert ed.). Elsevier B.V. DOI: 10.1016/B978-0-444-53802-4.00189-5

Glover, P. W. J. (2016) 'Archie's law: A reappraisal', *Solid Earth*, 7(4), pp. 1157–1169. DOI: 10.5194/se-7-1157-2016

Gómez-Treviño, E. and Esparza, F. J. (2014) 'What is the depth of investigation of a resistivity measurement?', *Geophysics*, 79(2), pp. W1–W10.

Grahame, D. C. (1952) 'Mathematical theory of the faradaic admittance', *Journal of the Electrochemical Society*, 99(12), pp. 370–385.

Green, P. J. (1995) 'Reversible jump Markov chain Monte Carlo computation and Bayesian model determination', *Biometrika*, 82(4), pp. 711–732.

Greenberg, R. J. and Brace, W. F. (1969) 'Archie's law for rocks modeled by simple networks', *Journal of Geophysical Research*, 74(8), pp. 2099–2102. DOI: 10.1029/JB074i008p02099

Greenhalgh, S. A., Zhou, B., Greenhalgh, M., Marescot, L. and Wiese, T. (2009) 'Explicit expressions for the Fréchet derivatives in 3D anisotropic resistivity inversion', *Geophysics*, 74(3), pp. F31–F43.

Greenhalgh, S., Wiese, T. and Marescot, L. (2010) 'Comparison of DC sensitivity patterns for anisotropic and isotropic media', *Journal of Applied Geophysics*, 70(2), pp. 103–112.

Griffiths, D. H. and Turnbull, J. (1985) 'A multi-electrode array for resistivity surveying', *First Break*, 3(7), pp. 16–20.

Griffiths, D. H., Turnbull, J. and Olayinka, A. I. (1990) 'Two-dimensional resistivity mapping with a computer-controlled array', *First Break*, 8(4), pp. 121–129.

Guillemoteau, J., Lück, E. and Tronicke, J. (2017) '1D inversion of direct current data acquired with a rolling electrode system', *Journal of Applied Geophysics*, 146, pp. 167–177.

Günther, T., Rücker, C. and Spitzer, K. (2006) 'Three-dimensional modelling and inversion of dc resistivity data incorporating topography – II. Inversion', *Geophysical Journal International*, 166(2), pp. 506–517. DOI: 10.1111/j.1365-246X.2006.03011.x

Gupta, P. K., Niwas, S. and Gaur, V. K. (1997) 'Straightforward inversion of vertical electrical sounding data', *Geophysics*, 62(3), pp. 775–785.

Guptasarma, D. (1982) 'Optimization of short digital linear filters for increased accuracy', *Geophysical Prospecting*, 30(4), pp. 501–514.

Gurin, G., Ilyin, Y., Nilov, S., Ivanov, D., Kozlov, E. and Titov, K. (2018) 'Induced polarization of rocks containing pyrite: Interpretation based on X-ray computed tomography', *Journal of Applied Geophysics*, 154, pp. 50–63. DOI: 10.1016/j.jappgeo.2018.04.019

Gurin, G., Tarasov, A., Ilyin, Y. and Titov, K. (2013) 'Time domain spectral induced polarization of disseminated electronic conductors: Laboratory data analysis through the Debye decomposition approach', *Journal of Applied Geophysics*, 98, pp. 44–53. DOI: 10.1016/j.jappgeo.2013.07.008

Gurin, G., Titov, K., Ilyin, Y. and Tarasov, A. (2015) 'Induced polarization of disseminated electronically conductive minerals: A semi-empirical model', *Geophysical Journal International*, 200, pp. 1555–1565. DOI: 10.1093/gji/ggu490

Guyot, A., Ostergaard, K. T., Lenkopane, M., Fan, J. and Lockington, D. A. (2013) 'Using electrical resistivity tomography to differentiate sapwood from heartwood: Application to conifers', *Tree Physiology*, 33(2), pp. 187–194. DOI: 10.1093/treephys/tps128

Habberjam, G. M. (1972) 'The effects of anisotropy on square array resistivity measurements', *Geophysical Prospecting*, 20(2), pp. 249–266.

Hallbauer-Zadorozhnaya, V., Santarato, G. and Abu Zeid, N. (2015) 'Non-linear behaviour of electrical parameters in porous, water-saturated rocks: A model to predict pore size distribution', *Geophysical Journal International*, 202(2), pp. 871–886. DOI: 10.1093/gji/ggv161

Hallof, P. G. (1957) 'On the interpretation of resistivity and induced polarization field measurements'. PhD dissertation thesis, Massachusetts Institute of Technology, p. 200.

Hamdan, H. A. and Vafidis, A. (2013) 'Joint inversion of 2D resistivity and seismic travel time data to image saltwater intrusion over karstic areas', *Environmental Earth Sciences*, 68(7), pp. 1877–1885. DOI: 10.1007/s12665-012-1875-9

Hanai, T. (1960) 'Theory of the dielectric dispersion due to the interfacial polarization and its application to emulsions', *Kolloid-Zeitschrift*, 171(1), pp. 23–31.

Hanai, T. (1968) 'Electrical properties of emulsions', in *Emulsion Science* (P. Sherman ed.). Academic Press, New York.

Hansen, P. C. (1992) 'Analysis of discrete ill-posed problems by means of the L-curve', *SIAM Review*, 34(4), pp. 561–580.

Hao, N., Moysey, S. M. J., Powell, B. A. and Ntarlagiannis, D. (2015) 'Evaluation of surface sorption processes using spectral induced pPolarization and a 22 Na tracer', *Environmental Science & Technology*, 49(16), pp. 9866–9873. DOI: 10.1021/acs.est.5b01327

Hauck, C. and Kneisel, C. (2006) 'Application of capacitively-coupled and DC electrical resistivity imaging for mountain permafrost studies', *Permafrost and Periglacial Processes*, 17(2), pp. 169–177. DOI: 10.1002/ppp.555

Hayley, K., Pidlisecky, A. and Bentley, L. R. (2011) 'Simultaneous time-lapse electrical resistivity inversion', *Journal of Applied Geophysics*, 75(2), pp. 401–411.

Heenan, J., Slater, L., Ntarlagiannis, D., Atekwana, E. A., Fathepure, B. Z., Dalvi, S., Ross, C., Werkema, D. D. and Atekwana, Estella, E. A. (2014) 'Electrical resistivity imaging for long-term autonomous monitoring of hydrocarbon degradation: Lessons from the Deepwater Horizon oil spill', *Geophysics*, 80(1), pp. B1–B11. DOI: 10.1190/geo2013-0468.1

Henderson, R. D., Day-Lewis, F. D., Abarca, E., Harvey, C. F., Karam, H. N., Liu, L. and Lane, J. W. (2010) 'Marine electrical resistivity imaging of submarine groundwater discharge: Sensitivity analysis and application in Waquoit Bay, Massachusetts, USA', *Hydrogeology Journal*, 18(1), pp. 173–185.

Hennig, T., Weller, A. and Möller, M. (2008) 'Object orientated focussing of geoelectrical multielectrode measurements', *Journal of Applied Geophysics*, 65(2), pp. 57–64.

Henry-Poulter, S. (1996) 'An investigation of transport properties in natural soils using electrical resistance tomography'. PhD thesis, Lancaster University, UK, p. 237.

Hering, A., Misiek, R., Gyulai, A., Ormos, T., Dobroka, M. and Dresen, L. (1995) 'A joint inversion algorithm to process geoelectric and sutface wave seismic data. Part I: Basic ideas1', *Geophysical Prospecting*, 43(2), pp. 135–156. DOI: 10.1111/j.1365-2478.1995.tb00128.x

Herwanger, J. V., Worthington, M. H., Lubbe, R., Binley, A. and Khazanehdari, J. (2004a) 'A comparison of cross-hole electrical and seismic data in fractured rock', *Geophysical Prospecting*, 52(2), pp. 109–121.

Herwanger, J. V., Pain, C. C., Binley, A., De Oliveira, C. R. E. and Worthington, M. H. (2004b) 'Anisotropic resistivity tomography', *Geophysical Journal International*, 158(2), pp. 409–425.

Hilbich, C., Hauck, C., Hoelzle, M., Scherler, M., Schudel, L., Völksch, I., Vonder Mühll, D. and Mäusbacher, R. (2008) 'Monitoring mountain permafrost evolution using electrical resistivity tomography: A 7-year study of seasonal, annual, and long-term variations at Schilthorn, Swiss Alps', *Journal of Geophysical Research: Earth Surface*, 113 (1), pp. 1–12. DOI: 10.1029/2007JF000799

Hilbich, C., Marescot, L., Hauck, C., Loke, M. H. and Mausbacher, R. (2009) 'Applicability of electrical resistivity tomography monitoring to coarse blocky and ice-rich permafrost landforms', *Permafrost and Periglacial Processes*, 20(3), pp. 269–284. DOI: 10.1002/ppp.652

Hilchie, D. W. (1984) 'A new water resistivity versus temperature equation; Technical notes', *The Log Analyst*, 25 (04).SPWLA-1984-vXXVn4a3.

Hill, H. J. and Milburn, J. D. (1956) 'Effect of clay and water salinity on electrochemical behavior of reservoir rocks', *Transactions, AIME*, 207, pp. 65–72.

Hinnell, A. C., Ferré, T. P. A., Vrugt, J. A., Huisman, J. A., Moysey, S., Rings, J. and Kowalsky, M. B. (2010) 'Improved extraction of hydrologic information from geophysical data through coupled hydrogeophysical inversion', *Water Resources Research*, 46, W00D40, DOI:10.1029/2008WR007060

Hohmann, G. W. (1973) 'Electromagnetic coupling between grounded wires at the surface of a two-layer earth', *Geophysics*, 38(5), pp. 854–863.

Hohmann, G. W. (1975) 'Three-dimensional induced polarization and electromagnetic modeling', *Geophysics*, 40(2), pp. 309–324.

Hohmann, G. W. (1988) 'Numerical modeling for electromagnetic methods of geophysics', *Electromagnetic Methods in Applied Geophysics*, 1, pp. 313–363.

Hördt, A., Hanstein, T., Hönig, M. and Neubauer, F. M. (2006) 'Efficient spectral IP-modelling in the time domain', *Journal of Applied Geophysics*, 59(2), pp. 152–161. DOI: 10.1016/j.jappgeo.2005.09.003

Hördt, A., Blaschek, R., Kemna, A. and Zisser, N. (2007) 'Hydraulic conductivity estimation from induced polarisation data at the field scale: The Krauthausen case history', *Journal of Applied Geophysics*, 62(1), pp. 33–46.

Hua, P., Woo, E. J., Webster, J. G. and Tompkins, W. J. (1991) 'Iterative reconstruction methods using regularization and optimal current patterns in electrical impedance tomography', *IEEE Transactions on Medical Imaging*, 10(4), pp. 621–628.

Huisman, J. A., Rings, J., Vrugt, J. A., Sorg, J. and Vereecken, H. (2010) 'Hydraulic properties of a model dike from coupled Bayesian and multi-criteria hydrogeophysical inversion', *Journal of Hydrology*, 380(1–2), pp. 62–73. DOI: 10.1016/j.jhydrol.2009.10.023

Hunkel, H. (1924) 'Verfahren zur Feststellung und Lokalisierung von Koerpern im Untergrunde', German Patent 442,832.

Huntley, D. (1986) 'Relations between permeability and electrical resistivity in granular aquifers', *Groundwater*, 24(4), pp. 466–474.

Huntley, D., Bobrowsky, P., Hendry, M., Macciotta, R., Elwood, D., Sattler, K., Best, M., Chambers, J. and Meldrum, P. (2019) 'Application of multi-dimensional electrical resistivity tomography datasets to investigate a very slow-moving landslide near Ashcroft, British Columbia, Canada', *Landslides*, 16, pp. 1033–1042.

Hupfer, S., Martin, T., Weller, A., Günther, T., Kuhn, K., Djotsa Nguimeya Ngninjio, V. and Noell, U. (2016) 'Polarization effects of unconsolidated sulphide-sand-mixtures', *Journal of Applied Geophysics*, 135, pp. 456–465. DOI: 10.1016/j.jappgeo.2015.12.003

Ingeman-Nielsen, T. and Baumgartner, F. (2006) 'CR1Dmod: A Matlab program to model 1D complex resistivity effects in electrical and electromagnetic surveys', *Computers & Geosciences*, 32(9), pp. 1411–1419.

Inman Jr, J. R., Ryu, J. and Ward, S. H. (1973) 'Resistivity inversion', *Geophysics*, 38 (6), pp. 1088–1108.

Inman, J. R. (1975) 'Resistivity inversion with ridge regression', *Geophysics*, 40(5), pp. 798–817.

Iseki, S. and Shima, H. (1992) 'Induced-polarization tomography: A crosshole imaging technique using chargeability and resistivity', in *SEG Technical Program Expanded Abstracts 1992*. Society of Exploration Geophysicists, pp. 439–442.

Ishizu, K., Goto, T., Ohta, Y., Kasaya, T., Iwamoto, H., Vachiratienchai, C., Siripunvaraporn, W., Tsuji, T., Kumagai, H. and Koike, K. (2019) 'Internal structure of a seafloor massive sulfide deposit by electrical resistivity tomography, Okinawa Trough', *Geophysical Research Letters*, 46(20), pp. 11025–11034.

Jackson, P., Smith, D. and Stanford, P. (1978) 'Resistivity-porosity-particle shape relationships for marine sands', *Geophysics*, 43(6), pp. 1250–1268. DOI: 10.1190/1.1440891

JafarGandomi, A. and Binley, A. (2013) 'A Bayesian trans-dimensional approach for the fusion of multiple geophysical datasets', *Journal of Applied Geophysics*, 96, pp. 38–54. DOI: 10.1016/j.jappgeo.2013.06.004

Jayawickreme, D. H., Van Dam, R. L. and Hyndman, D. W. (2008) 'Subsurface imaging of vegetation, climate, and root-zone moisture interactions', *Geophysical Research Letters*, L18404, DOI:10.1029/2008GL034690

Jha, M. K., Kumar, S. and Chowdhury, A. (2008) 'Vertical electrical sounding survey and resistivity inversion using genetic algorithm optimization technique', *Journal of Hydrology*, 359(1–2), pp. 71–87.

Johansen, H. K. (1975) 'An interactive computer/graphic-display-terminal system for interpretation of resistivity soundings', *Geophysical Prospecting*, 23(3), pp. 449–458.

Johansen, H. K. (1977) 'A man/computer interpretation system for resistivity soundings over a horizontally strafified earth', *Geophysical Prospecting*, 25(4), pp. 667–691.

Johansen, H. K. and Sørensen, K. (1979) 'Fast hankel transforms', *Geophysical Prospecting*, 27(4), pp. 876–901.

Johnson, D. L., Koplik, J. and Schwartz, L. M. (1986) 'New pore-size parameter characterizing transport in porous media', *Physical Review Letters*, 57(20), pp. 2564–2567. DOI: 10.1103/PhysRevLett.57.2564

Johnson, H. M. (1962) 'A history of well logging', *Geophysics*, 27(4), pp. 507–527.

Johnson, I. M. (1984) 'Spectral induced polarization parameters as determined through time-domain measurements', *Geophysics*, 49(11), pp. 1993–2003.

Johnson, T. C. and Wellman, D. (2015) 'Accurate modelling and inversion of electrical resistivity data in the presence of metallic infrastructure with known location and dimension', *Geophysical Journal International*, 202(2), pp. 1096–1108. DOI: 10.1093/gji/ggv206

Johnson, T. C. and Thomle, J. (2018) '3-D decoupled inversion of complex conductivity data in the real number domain', *Geophysical Journal International*, 212(1), pp. 284–296. DOI: 10.1093/gji/ggx416

Johnson, T. C., Versteeg, R. J., Huang, H. and Routh, P. S. (2009) 'Data-domain correlation approach for joint hydrogeologic inversion of time-lapse hydrogeologic and geophysical data', *Geophysics*, 74(6). DOI: 10.1190/1.3237087

Johnson, T. C., Versteeg, R. J., Ward, A., Day-Lewis, F. D. and Revil, A. (2010) 'Improved hydrogeophysical characterization and monitoring through parallel modeling and inversion of time-domain resistivity and induced-polarization data', *Geophysics*, 75 (4), pp. WA27–WA41.

Johnson, T. C., Slater, L. D., Ntarlagiannis, D., Day-Lewis, F. D. and Elwaseif, M. (2012a) 'Monitoring groundwater-surface water interaction using time-series and time-frequency analysis of transient three-dimensional electrical resistivity changes', *Water Resources Research*, 48(7), pp. 1–13. DOI: 10.1029/2012WR011893

Johnson, T. C., Versteeg, R. J., Rockhold, M., Slater, L. D., Ntarlagiannis, D., Greenwood, W. J. and Zachara, J. (2012b) 'Characterization of a contaminated wellfield using 3D electrical resistivity tomography implemented with geostatistical, discontinuous boundary, and known conductivity constraints', *Geophysics*, 77 (6), pp. EN85–EN96.

Johnson, T. C., Versteeg, R. J., Day-Lewis, F. D., Major, W. and Lane, J. W. (2015). 'Time-lapse electrical geophysical monitoring of amendment-based biostimulation', *Groundwater*, 53(6), pp. 920–932.

Johnson, T. C., Hammond, G. E. and Chen, X. (2017) 'PFLOTRAN-E4D: A parallel open source PFLOTRAN module for simulating time-lapse electrical resistivity data', *Computers and Geosciences*, 99, pp. 72–80. DOI: 10.1016/j.cageo.2016.09.006

Jol, H. M. (2008) *Ground Penetrating Radar Theory and Applications*. Elsevier, p. 544.

Jongmans, D. and Garambois, S. (2007) 'Geophysical investigation of landslides: A review', *Bulletin de la Société géologique de France*, 178(2), pp. 101–112.

Kaipio, J. P., Kolehmainen, V., Somersalo, E. and Vauhkonen, M. (2000) 'Statistical inversion and Monte Carlo sampling methods in electrical impedance tomography', *Inverse Problems*, 16(5), p. 1487. DOI: 10.1088/0266-5611/16/5/321

Kang, X., Shi, X., Revil, A., Cao, Z., Li, L., Lan, T. and Wu, J. (2019) 'Coupled hydrogeophysical inversion to identify non-Gaussian hydraulic conductivity field by jointly assimilating geochemical and time-lapse geophysical data', *Journal of Hydrology*, 578, p. 124092. DOI: /10.1016/j.jhydrol.2019.124092

Karaoulis, M. C., Kim, J.-H. and Tsourlos, P. I. (2011) '4D active time constrained resistivity inversion', *Journal of Applied Geophysics*, 73(1), pp. 25–34.

Karaoulis, M., Revil, A., Zhang, J. and Werkema, D. D. (2012) 'Time-lapse joint inversion of crosswell DC resistivity and seismic data: A numerical investigation', *Geophysics*, 77(4). DOI: 10.1190/geo2012-0011.1

Karaoulis, M., Revil, A., Tsourlos, P., Werkema, D. D. and Minsley, B. J. (2013) 'IP4DI: A software for time-lapse 2D/3D DC-resistivity and induced polarization tomography', *Computers and Geosciences*, 54, pp. 164–170. DOI: 10.1016/j.cageo.2013.01.008

Karaoulis, M., Tsourlos, P., Kim, J.-H. and Revil, A. (2014) '4D time-lapse ERT inversion: introducing combined time and space constraints', *Near Surface Geophysics*, 12(1), pp. 25–34.

Karhunen, K., Seppänen, A., Lehikoinen, A., Monteiro, P. J. M. and Kaipio, J. P. (2010) 'Electrical resistance tomography imaging of concrete', *Cement and Concrete Research*, 40(1), pp. 137–145.

Katz, A. J. and Thompson, A. H. (1986) 'Quantitative prediction of permeability in porous rock', *Physical Review B*, 34(11), pp. 8179–8181. DOI: 10.1103/PhysRevB.34.8179

Katz, A. J. and Thompson, A. H. (1987) 'Prediction of rock electrical-conductivity from mercury injection measurements', *Journal of Geophysical Research-Solid Earth and Planets*, 92(B1), pp. 599–607. DOI: 10.1029/JB092iB01p00599

Kauahikaua, J., Mattice, M. and Jackson, D. (1980) 'Mise-a-la-masse mapping of the HGP-A geothermal reservoir, Hawaii', Proc. Geothermal Resources Council 1980 Annual Meeting, September 9–11,1980, Salt Lake City, Utah. Vol. 4. pp. 65–68.

Kaufman, A. A. and Wightman, W. E. (1993) 'A transmission-line model for electrical logging through casing', *Geophysics*, 58(12), pp. 1739–1747.

Kaufman, A. A., Alekseev, D. and Oristaglio, M. (2014) *Principles of Electromagnetic Methods in Surface Geophysics*. Newnes, p. 794.

Keery, J., Binley, A., Elshenawy, A. and Clifford, J. (2012) 'Markov-chain Monte Carlo estimation of distributed Debye relaxations in spectral induced polarization', *Geophysics*, 77(2). DOI: 10.1190/geo2011-0244.1

Keller, C. (2012) 'Hydro-geologic spatial resolution using flexible Liners', *The Professional Geologist*, 49(3), pp. 45–51.

Keller, G. V. and Frischknecht, F. C. (1966) *Electrical Methods in Geophysical Prospecting*. Pergamon Press, Oxford, p. 517.

Kelter, M., Huisman, J. A., Zimmermann, E. and Vereecken, H. (2018) 'Field evaluation of broadband spectral electrical imaging for soil and aquifer characterization', *Journal of Applied Geophysics*, 159, pp. 484–496.

Kelter, M., Huisman, J. A., Zimmermann, E., Kemna, A. and Vereecken, H. (2015) 'Quantitative imaging of spectral electrical properties of variably saturated soil columns', *Journal of Applied Geophysics*, 123, pp. 333–344. DOI: 10.1016/j.jappgeo.2015.09.001

Kemna, A. (2000) *Tomographic Inversion of Complex Resistivity: Theory and Application*. Der Andere Verlag Osnabrück, p. 196.

Kemna, A., Rakers, E. and Binley, A. (1997) 'Application of complex resistivity tomography to field data from a kerosene-contaminated site', in *Environmental and Engineering Geophysics (EEGS)*. European Section, pp. 151–154.

Kemna, A., Binley, A., Ramirez, A. and Daily, W. (2000) 'Complex resistivity tomography for environmental applications', *Chemical Engineering Journal*, 77(1–2), pp. 11–18.

Kemna, A., Vanderborght, J., Kulessa, B. and Vereecken, H. (2002) 'Imaging and characterisation of subsurface solute transport using electrical resistivity tomography (ERT) and equivalent transport models', *Journal of Hydrology*, 267(3–4), pp. 125–146.

Kemna, A., Binley, A. and Slater, L. (2004) 'Crosshole IP imaging for engineering and environmental applications', *Geophysics*, 69(1), pp. 97–107. DOI: 10.1190/1.1649379

Kemna, A., Binley, A., Cassiani, G., Niederleithinger, E., Revil, A., Slater, L., Williams, K. H., Orozco, A. F., Haegel, F. H., Hördt, A., Kruschwitz, S., Leroux, V., Titov, K. and Zimmermann, E. (2012) 'An overview of the spectral induced polarization method for near-surface applications', *Near Surface Geophysics*, 10(6), pp. 453–468. DOI: 10.3997/1873-0604.2012027

Kenkel, J., Hördt, A. and Kemna, A. (2012) '2D modelling of induced polarization data with anisotropic complex conductivities', *Near Surface Geophysics*, 10(6), pp. 533–544.

Ketola, M. (1972) 'Some points of view concerning mise-a-la-masse measurements', *Geoexploration*, 10(1), pp. 1–21.

Key, K. T. (1977) *Nuclear Waste Tank and Pipeline External Leak Detection Systems*. Atlantic Richfield Hanford Co., Richland, WA, p. 144.

Keys, W. S. (1989) *Borehole Geophysics Applied to Ground-Water Investigations*. National Water Well Association Dublin, OH, p. 150.

Kiessling, D., Schmidt-Hattenberger, C., Schuett, H., Schilling, F., Krueger, K., Schoebel, B., Danckwardt, E., Kummerow, J. and Group, C. (2010) 'Geoelectrical methods for monitoring geological CO2 storage: First results from cross-hole and surface–downhole measurements from the CO2SINK test site at Ketzin (Germany)', *International Journal of Greenhouse Gas Control*, 4(5), pp. 816–826.

Kiflai, M. E., Whitman, D., Ogurcak, D. E. and Ross, M. (2019) 'The effect of Hurricane Irma storm surge on the freshwater lens in Big Pine Key, Florida, using electrical resistivity tomography', *Estuaries and Coasts*, DOI: 10.1007/s12237-019-00666-3

Kim, J.-H., Yi, M.-J., Cho, S.-J., Son, J.-S. and Song, W.-K. (2006) 'Anisotropic crosshole resistivity tomography for ground safety analysis of a high-storied building over an abandoned mine', *Journal of Environmental & Engineering Geophysics*, 11(4), pp. 225–235.

Kim, J.-H., Yi, M.-J., Park, S.-G. and Kim, J. G. (2009) '4-D inversion of DC resistivity monitoring data acquired over a dynamically changing earth model', *Journal of Applied Geophysics*, 68(4), pp. 522–532.

Kim, B., Nam, M. J. and Kim, H. J. (2018) 'Inversion of time-domain induced polarization data based on time-lapse concept', *Journal of Applied Geophysics*, 152, pp. 26–37. DOI: 10.1016/j.jappgeo.2018.03.010

King, M. S., Zimmerman, R. W. and Corwin, R. F. (1988) 'Seismic and electrical properties of unconsolidated permafrost', *Geophysical Prospecting*, 36(4), pp. 349–364. DOI: 10.1111/j.1365-2478.1988.tb02168.x

Kirkpatrick, S., Gelatt, C. D. and Vecchi, M. P. (1983) 'Optimization by simulated annealing', *Science*, 220(4598), pp. 671–680.

Klein, J. D. and Sill, W. R. (1982) 'Electrical properties of artificial clay-bearing sandstone', *Geophysics*, 47(11), pp. 1593–1605. DOI: 10.1190/1.1441310

Klein, J. D., Biegler, T. and Horne, M. D. (1984) 'Mineral interfacial processes in the method of induced polarization', *Geophysics*, 49(7), pp. 1105–1114.

Knight, R. J. and Nur, A. (1987) 'The dielectric constant of sandstones, 60 kHz to 4 MHz', *Geophysics*, 52(5), pp. 644–654.

Koefoed, O. (1970) 'A fast method for determining the layer distribution from the raised kernel function in geoelegtrical sounding', *Geophysical Prospecting*, 18(4), pp. 564–570.

Koefoed, O. (1979) *Geosounding Principles, 1, Resistivity Sounding Measurements*. Elsevier, p. 276.

Koestel, J., Kemna, A., Javaux, M., Binley, A. and Vereecken, H. (2008) 'Quantitative imaging of solute transport in an unsaturated and undisturbed soil monolith with 3-D ERT and TDR', *Water Resources Research*, 44(12), pp. 1–17. DOI: 10.1029/2007WR006755

Koestel, J., Vanderborght, J., Javaux, M., Kemna, A., Binley, A. and Vereecken, H. (2009a) 'Noninvasive 3-D transport characterization in a sandy soil using ERT: 1. Investigating the validity of ERT-derived transport parameters', *Vadose Zone Journal*, 8(3), pp. 711–722.

Koestel, J., Vanderborght, J., Javaux, M., Kemna, A., Binley, A. and Vereecken, H. (2009b) 'Noninvasive 3-D transport characterization in a sandy soil using ERT: 2. Transport process inference', *Vadose Zone Journal*, 8(3), pp. 723–734.

Koestel, J., Kasteel, R., Kemna, A., Esser, O., Javaux, M., Binley, A. and Vereecken, H. (2009c) 'Imaging brilliant blue stained soil by means of electrical resistivity tomography', *Vadose Zone Journal*, 8(4), pp. 963–975.

Komarov, V. A. (1980) *Electrical Prospecting with the Induced Polarization Method*. Nedra, Leningrad, p. 391.

Kormiltsev, V. V. (1963) 'O vozbuzdenii i spade vyzvannoi polarizatsii v kapillarnoi srede (On excitation and decay of induced polarization in capillary medium). Izvestia AN SSSR', *Seria Geofizicheskaya*, 11, pp. 1658–1666.

Kosinski, W. K. and Kelly, W. E. (1981) 'Geoelectric soundings for predicting aquifer properties', *Groundwater*, 19(2), pp. 163–171.

Kowalsky, M. B., Finsterle, S. and Rubin, Y. (2004) 'Estimating flow parameter distributions using ground-penetrating radar and hydrological measurements during transient flow in the vadose zone', *Advances in Water Resources*, 27(6), pp. 583–599. DOI: 10.1016/j.advwatres.2004.03.003

Kruschwitz, S. and Yaramanci, U. (2004) 'Detection and characterization of the disturbed rock zone in claystone with the complex resistivity method', *Journal of Applied Geophysics*, 57(1), pp. 63–79.

Kruschwitz, S., Binley, A., Lesmes, D. and Elshenawy, A. (2010) 'Textural controls on low-frequency electrical spectra of porous media', *Geophysics*, 75(4), pp. WA113-WA123.

Kuras, O., Meldrum, P. I., Beamish, D., Ogilvy, R. D. and Lala, D. (2007) 'Capacitive resistivity imaging with towed arrays', *Journal of Environmental and Engineering Geophysics*, 12(3), pp. 267–279. DOI: 10.2113/JEEG12.3.267

Kuras, O., Pritchard, J. D., Meldrum, P. I., Chambers, J. E., Wilkinson, P. B., Ogilvy, R. D. and Wealthall, G. P. (2009) 'Monitoring hydraulic processes with automated time-lapse electrical resistivity tomography (ALERT)', *Comptes Rendus Geoscience*, 341(10–11), pp. 868–885.

LaBrecque, D. J. (1991) 'IP tomography', in *SEG Technical Program Expanded Abstracts 1991*. Society of Exploration Geophysicists, pp. 413–416.

LaBrecque, D. and Daily, W. (2008) 'Assessment of measurement errors for galvanic-resistivity electrodes of different composition', *Geophysics*, 73(2), p. F55. DOI: 10.1190/1.2823457

LaBrecque, D. J. and Ward, S. H. (1990) 'Two-dimensional cross-borehole resistivity model fitting', in *Geotechnical and Environmental Geophysics*. Society of Exploration Geophysicists: Tulsa, OK, 1, pp. 51–57.

LaBrecque, D. and Yang, X. (2001a), 'The effects of anisotropy on ERT images for Vadose Zone monitoring', in *Symposium on the Application of Geophysics to Engineering and Environmental* Problems 2001. Society of Exploration Geophysicists. pp. VZC2–VZC2.

LaBrecque, D. J. and Yang, X. (2001b) 'Difference inversion of ERT data: A fast inversion method for 3-D in situ monitoring', *Journal of Environmental & Engineering Geophysics*, 6(2), pp. 83–89.

LaBrecque, D. J., Morelli, G., Daily, W., Ramirez, A. and Lundegard, P. (1999) 'Occam's inversion of 3-D electrical resistivity tomography', in *Three-Dimensional Electromagnetics*. Society of Exploration Geophysicists, pp. 575–590.

LaBrecque, D. J., Ramirez, A. L., Daily, W. D., Binley, A. M. and Schima, S. A. (1996a) 'ERT monitoring of environmental remediation processes', *Measurement Science and Technology*, 7(3), p. 375.

LaBrecque, D. J., Miletto, M., Daily, W., Ramirez, A. and Owen, E. (1996b) 'The effects of noise on Occam's inversion of resistivity tomography data', *Geophysics*, 61(2), pp. 538–548.

LaBrecque, D. J., Heath, G., Sharpe, R. and Versteeg, R. (2004) 'Autonomous monitoring of fluid movement using 3-D electrical resistivity tomography', *Journal of Environmental & Engineering Geophysics*, 9(3), pp. 167–176.

Lagabrielle, R. (1983) 'The effect of water on direct current resistivity measurement from the sea, river or lake floor', *Geoexploration*, 21(2), pp. 165–170.

Laloy, E., Hérault, R., Jacques, D. and Linde, N. (2018) 'Training-image based geostatistical inversion using a spatial generative adversarial neural network', *Water Resources Research*, 54(1), pp. 381–406.

Landauer, R. (1952) 'The electrical resistance of binary metallic mixtures', *Journal of Applied Physics*, 23(7), pp. 779–784.

Lane Jr, J. W., Haeni, F. P. and Watson, W. M. (1995) 'Use of a square-array direct-current resistivity method to detect fractures in crystalline bedrock in New Hampshire', *Groundwater*, 33(3), pp. 476–485.

Lapenna, V., Lorenzo, P., Perrone, A., Piscitelli, S., Rizzo, E. and Sdao, F. (2005) '2D electrical resistivity imaging of some complex landslides in Lucanian Apennine chain, southern Italy', *Geophysics*, 70(3), pp. B11–B18.

Leroy, P. and Revil, A. (2004) 'A triple-layer model of the surface electrochemical properties of clay minerals', *Journal of Colloid and Interface Science*, 270(2), pp. 371–380. DOI: 10.1016/j.jcis.2003.08.007

Leroy, P. and Revil, A. (2009) 'A mechanistic model for the spectral induced polarization of clay materials', *Journal of Geophysical Research: Solid Earth*, 114 (10), pp. 1–21. DOI: 10.1029/2008JB006114

Leroy, P., Revil, A., Kemna, A., Cosenza, P. and Ghorbani, A. (2008) 'Complex conductivity of water-saturated packs of glass beads', *Journal of Colloid and Interface Science*, 321(1), pp. 103–17. DOI: 10.1016/j.jcis.2007.12.031

Leroy, P., Li, S., Jougnot, D., Revil, A. and Wu, Y. (2017) 'Modelling the evolution of complex conductivity during calcite precipitation on glass beads', *Geophysical Journal International*, 209(1), pp. 123–140. DOI: 10.1093/gji/ggx001

Leroy, P., Hördt, A., Gaboreau, S., Zimmermann, E., Claret, F., Bücker, M., Stebner, H. and Huisman, J. A. (2019) 'Spectral induced polarization of low-pH cement and concrete',

Cement and Concrete Composites, p. 103397. DOI: 10.1016/j.cemconcomp.2019.103397

Lesmes, D. P. (1993) 'Electrical impedance spectroscopy of sedimentary rocks'. PhD dissertation thesis, Texas A&M University, p. 168.

Lesmes, D. P. and Morgan, F. D. (2001) 'Dielectric spectroscopy of sedimentary rocks', *Journal of Geophysical Research-Solid Earth*, 106(B7), 13329–13346, DOI:10.1029/2000JB900402

Lesmes, P. and Frye, M. (2001) 'Influence of pore fluid chemistry on the complex conductivity and induced polarization responses of Berea sandstone', *Journal of Geophysical Research*, 106(2000), pp. 4079–4090.

Lesparre, N., Nguyen, F., Kemna, A., Robert, T., Hermans, T., Daoudi, M. and Flores-Orozco, A. (2017) 'A new approach for time-lapse data weighting in electrical resistivity tomography', *Geophysics*, 82(6), pp. E325–E333.

Lesur, V., Cuer, M. and Straub, A. (1999) '2-D and 3-D interpretation of electrical tomography measurements, Part 1: The forward problem', *Geophysics*, 64(2), pp. 386–395.

Lévy, L., Gibert, B., Sigmundsson, F., Flóvenz, Ó. G., Hersir, G. P., Briole, P. and Pezard, P. A. (2018) 'The role of smectites in the electrical conductivity of active hydrothermal systems: Electrical properties of core samples from Krafla volcano, Iceland', *Geophysical Journal International*, 215(3), pp. 1558–1582. DOI: 10.1093/gji/ggy342

Lewkowicz, A. G., Etzelmüller, B. and Smith, S. L. (2011) 'Characteristics of discontinuous permafrost based on ground temperature measurements and electrical resistivity tomography, Southern Yukon, Canada', *Permafrost and Periglacial Processes*, 22 (4), pp. 320–342. DOI: 10.1002/ppp.703

Li, T., Isaacson, D., Newell, J. C. and Saulnier, G. J. (2014) 'Adaptive techniques in electrical impedance tomography reconstruction', *Physiological Measurement*, 35 (6), pp. 1111–1124.

Li, Y. and Oldenburg, D. W. (1999) '3-D inversion of DC resistivity data using an L-curve criterion', in *SEG Technical Program Expanded Abstracts 1999*. Society of Exploration Geophysicists, pp. 251–254.

Li, Y. and Oldenburg, D. W. (2000) '3-D inversion of induced polarization data', *Geophysics*, 65(6), pp. 1931–1945.

Lichtenecker, K. and Rother, K. (1931) 'Die Herleitung des logarithmischen Mischungsgesetzes aus allgemeinen Prinzipien der stationaren Stromung', *Physikalische Zeitschrift*, 32, pp. 255–260.

Lima, O. A. L. De and Sharma, M. M. (1992) 'A generalized Maxwell-Wagner theory for membrane polarization in shaly sands', *Geophysics*, 57(3), pp. 431–440.

Linde, N., Binley, A., Tryggvason, A., Pedersen, L. B. and Revil, A. (2006) 'Improved hydrogeophysical characterization using joint inversion of cross-hole electrical resistance and ground-penetrating radar traveltime data', *Water Resources Research*, 42, W12404, DOI:10.1029/2006WR005131

Lionheart, W. R. B. (2004) 'EIT reconstruction algorithms: Pitfalls, challenges and recent developments', *Physiological Measurement*, 25(1), p. 125. DOI: 10.1088/0967-3334/25/1/021

Lippmann, R. P. (1987) 'An introduction to computing with neural nets', *IEEE ASSP Magazine*, 4(2), pp. 4–22.

Liu, B., Li, S. C., Nie, L. C., Wang, J., L. X. and Zhang, Q. S. (2012) '3D resistivity inversion using an improved Genetic Algorithm based on control method of mutation direction', *Journal of Applied Geophysics*, 87, pp. 1–8. DOI: 10.1016/j.jappgeo.2012.08.002

LoCoco, J. (2018) 'Advances in slimline borehole geophysical logging', in *Symposium on the Application of Geophysics to Engineering and Environmental Problems 2018*. Society of Exploration Geophysicists and Environment and Engineering, pp. 221–222.

Loke, M. H. and Barker, R. D. (1995) 'Least-squares deconvolution of apparent resistivity pseudosections', *Geophysics*, 60(6), pp. 1682–1690.

Loke, M. H. and Barker, R. D. (1996a) 'Practical techniques for 3D resistivity surveys and data inversion1', *Geophysical Prospecting*, 44(3), pp. 499–523.

Loke, M. H. and Barker, R. D. (1996b) 'Rapid least-squares inversion of apparent resistivity pseudosections by a quasi-Newton method', *Geophysical Prospecting*, 44(1), pp. 131–152.

Loke, M. H., Acworth, I. and Dahlin, T. (2003) 'A comparison of smooth and blocky inversion methods in 2D electrical imaging surveys', *Exploration Geophysics*, 34(3), pp. 182–187.

Loke, M. H., Wilkinson, P. B. and Chambers, J. E. (2010) 'Fast computation of optimized electrode arrays for 2D resistivity surveys', *Computers & Geosciences*, 36(11), pp. 1414–1426.

Loke, M. H., Wilkinson, P. B., Uhlemann, S. S., Chambers, J. E. and Oxby, L. S. (2014a) 'Computation of optimized arrays for 3-D electrical imaging surveys', *Geophysical Journal International*, 199(3), pp. 1751–1764.

Loke, M. H., Wilkinson, P. B., Chambers, J. E. and Strutt, M. (2014b) 'Optimized arrays for 2D cross-borehole electrical tomography surveys', *Geophysical Prospecting*, 62(1), pp. 172–189.

Looms, M. C., Binley, A., Jensen, K. H., Nielsen, L. and Hansen, T. M. (2008) 'Identifying unsaturated hydraulic parameters using an integrated data fusion approach on cross-borehole geophysical data', *Vadose Zone Journal*, 7(1), p. 238. DOI: 10.2136/vzj2007.0087

Lowry, T., Allen, M. B. and Shive, P. N. (1989) 'Singularity removal: A refinement of resistivity modeling techniques', *Geophysics*, 54(6), pp. 766–774.

Lück, E. and Rühlmann, J. (2013) 'Resistivity mapping with GEOPHILUS ELECTRICUS: Information about lateral and vertical soil heterogeneity', *Geoderma*, 199, pp. 2–11.

Lund, E. D., Christy, C. D. and Drummond, P. E. (1999) 'Practical applications of soil electrical conductivity mapping', *Precision Agriculture*, 99, pp. 771–779.

Lundegard, P. D. and LaBrecque, D. (1995) 'Air sparging in a sandy aquifer (Florence, Oregon, USA): Actual and apparent radius of influence', *Journal of Contaminant Hydrology*, 19(1), pp. 1–27.

Luo, Z., Guan, H. and Zhang, X. (2019) 'The temperature effect and correction models for using electrical resistivity to estimate wood moisture variations', *Journal of Hydrology*, p. 124022. DOI: 10.1016/j.jhydrol.2019.124022

Lytle, R. J. and Dines, K. A. (1978) Impedance camera: A system for determining the spatial variation of electrical conductivity, Report No. UCRL-52413. Lawrence Livermore Lab., 1978. p. 11.

Ma, H., Tan, H. and Guo, Y. (2015) 'Three-dimensional induced polarization parallel inversion using nonlinear conjugate gradients method', in *Mathematical Problems in Engineering*. Hindawi, 2015. DOI: 10.1155/2015/464793

MacDonald, A. M., Davies, J. and Peart, R. J. (2001) 'Geophysical methods for locating groundwater in low permeability sedimentary rocks: Examples from southeast Nigeria', *Journal of African Earth Sciences*, 32(1), pp. 115–131.

Macleod, C. J. A., Humphreys, M. W., Whalley, W. R., Turner, L., Binley, A., Watts, C. W., Skøt, L., Joynes, A., Hawkins, S. and King, I. P. (2013) 'A novel grass hybrid to reduce flood generation in temperate regions'. *Scientific Reports*, 3, DOI: 10.1038/srep01683

Macnae, J. (2016) 'Quantifying Airborne Induced Polarization effects in helicopter time domain electromagnetics', *Journal of Applied Geophysics*, 135, pp. 495–502. DOI: 10.1016/j.jappgeo.2015.10.016

Madden, T. R. (1972) Transmission systems and network analogies to geophysical forward and inverse problems, Department of Defense report, p. 52.

Madsen, L. M., Fiandaca, G., Auken, E. and Christiansen, A. V. (2017) 'Time-domain induced polarization: An analysis of Cole–Cole parameter resolution and correlation using Markov ChainMonte Carlo inversion', *Geophysical Journal International*, 211 (3), pp. 1341–1353. DOI: 10.1093/gji/ggx355

Maineult, A., Revil, A., Camerlynck, C., Florsch, N. and Titov, K. (2017a) 'Upscaling of spectral induced polarization response using random tube networks', *Geophysical Journal International*, 209(2), pp. 948–960. DOI: 10.1093/gji/ggx066

Maineult, A., Jougnot, D. and Revil, A. (2017b) 'Variations of petrophysical properties and spectral induced polarization in response to drainage and imbibition: A study on a correlated random tube network', *Geophysical Journal International*, pp. 1398–1411. DOI: 10.1093/gji/ggx474

Major, J. and Silic, J. (1981) 'Restrictions on the use of Cole–Cole dispersion models in complex resistivity interpretation', *Geophysics*, 46(6), pp. 916–931.

Mansinha, L. and Mwenifumbo, C. J. (1983) 'A mise-a-la-masse study of the Cavendish geophysical test site', *Geophysics*, 48(9), pp. 1252–1257.

Mansoor, N. and Slater, L. (2007) 'Aquatic electrical resistivity imaging of shallow-water wetlands', *Geophysics*, 72(5), p. F211. DOI: 10.1190/1.2750667

Mansoor, N., Slater, L., Artigas, F. and Auken, E. (2006) 'High-resolution geophysical characterization of shallow-water wetlands', *Geophysics*, 71(4). DOI: 10.1190/1.2210307

Mares, R., Barnard, H. R., Mao, D., Revil, A. and Singha, K. (2016) 'Examining diel patterns of soil and xylem moisture using electrical resistivity imaging', *Journal of Hydrology*. 536, pp. 327–338.

Marescot, L., Lopes, S. P., Lagabrielle, R. and Chapellier, D. (2002) 'Designing surface-to-borehole electrical resisitivity tomography surveys using the frechet derivative', Proceedings of 8th Meeting of the Environmental and Engineering Geophysical Society, European Section, pp. 289–292.

Marescot, L., Loke, M. H., Chapellier, D., Delaloye, R., Lambiel, C. and Reynard, E. (2003) 'Assessing reliability of 2D resistivity imaging in mountain permafrost studies using the depth of investigation index method', *Near Surface Geophysics*, 1(2), pp. 57–67.

Marshall, D. J. and Madden, T. R. (1959) 'Induced polarization, a study of its causes', *Geophysics*, 24(4), pp. 790–816. DOI: 10.1190/1.1438659

Martin, T. (2012) 'Complex resistivity measurements on oak', *European Journal of Wood and Wood Products*, 70(1–3), pp. 45–53. DOI: 10.1007/s00107-010-0493-z

Martin, T. and Günther, T. (2013) 'Complex resistivity tomography (CRT) for fungus detection on standing oak trees', *European Journal of Forest Research*, 132(5–6), pp. 765–776. DOI: 10.1007/s10342-013-0711-4

Mary, B., Peruzzo, L., Boaga, J., Schmutz, M., Wu, Y., Hubbard, S. S. and Cassiani, G. (2018) 'Small-scale characterization of vine plant root water uptake via 3-D electrical resistivity tomography and mise-à-la-masse method', *Hydrology and Earth System Sciences*, 22(10), pp. 5427–5444. DOI: 10.5194/hess-22-5427-2018

Maurya, P. K., Fiandaca, G., Weigand, M., Kemna, A., Christiansen, A. V. and Auken, E. (2017) 'Comparison of frequency-domain and time-domain spectral induced polarization methods at field scale', 23rd European Meeting of Environmental and Engineering Geophysics, (September 2017). DOI: 10.3997/2214-4609.201701977

Maurya, P. K., Balbarini, N., Møller, I., Rønde, V., Christiansen, A. V., Bjerg, P. L., Auken, E. and Fiandaca, G. (2018) 'Subsurface imaging of water electrical conductivity, hydraulic permeability and lithology at contaminated sites by induced polarization', *Geophysical Journal International*, 213(2), pp. 770–785. DOI: 10.1093/gji/ggy018

Mboh, C. M., Huisman, J. A., Van Gaelen, N., Rings, J. and Vereecken, H. (2012) 'Coupled hydrogeophysical inversion of electrical resistances and inflow measurements for topsoil hydraulic properties under constant head infiltration', *Near Surface Geophysics*, 10(5), pp. 413–426. DOI: 10.3997/1873-0604.2012009

McClatchey, A. (1901a) 'Apparatus for locating metals, minerals, ores, etc.,' U.S. Patent 681,654.

McClatchey, A. (1901b) 'Electric prospecting apparatus', U.S. Patent 681,654.

McCollum, B. and Logan, K. H. (1915) Earth resistance and its relation to electrolysis of underground structures. Technical Papers of the Bureau of Standards, U.S. Dept. Commerse, p. 48.

McCulloch, W. S. and Pitts, W. (1943) 'A logical calculus of the ideas immanent in nervous activity', *The Bulletin of Mathematical Biophysics*, 5(4), pp. 115–133.

McLachlan, P. J. (2020) 'Geophysical characterisation of the groundwater-surface water interface'. PhD thesis, Lancaster University, UK.

McLachlan, P. J., Chambers, J. E., Uhlemann, S. S. and Binley, A. (2017) 'Geophysical characterisation of the groundwater–surface water interface', *Advances in Water Resources*, 109. DOI: 10.1016/j.advwatres.2017.09.016

Macnae, J. (2015) 'Comment on: Tarasov, A. & Titov, K., 2013, On the use of the Cole–Cole equations in spectral induced polarization, Geophys. J. Int., 195, 352–356', *Geophysical Journal International*, 202(1), pp. 529–532.

McNeil, J. D. (1980) *Electromagnetic terrain conductivity measurement at low induction numbers: Technical Note TN-6*. GEONICS Limited, Ontario, Canada, p. 15.

Meister, R., Rajani, M. S., Ruzicka, D. and Schachtman, D. P. (2014) 'Challenges of modifying root traits in crops for agriculture', *Trends in Plant Science*, 19(12), pp. 779–788.

Mejus, L. (2015) 'Using multiple geophysical techniques for improved assessment of aquifer vulnerability'. PhD thesis, Lancaster University, UK, p. 307.

Melo, A. and Li, Y. (2016) 'Geological characterization applying k-means clustering to 3D magnetic, gravity gradient, and DC resistivity inversions: A case study at an iron oxide copper gold (IOCG) deposit', SEG Technical Program Expanded Abstracts. September 2016, pp. 2180–2184.

Mendelson, K. S. and Cohen, M. H. (1982) 'The effect of grain anisotropy on the electrical properties of sedimentary rocks', *Geophysics*, 47(2), pp. 257–263.

Mendonça, C. A., Doherty, R., Amaral, N. D., McPolin, B., Larkin, M. J. and Ustra, A. (2015) 'Resistivity and induced polarization monitoring of biogas combined with microbial ecology at a brownfield site', *Interpretation*, 3(4), pp. SAB43–SAB56.

Menke, W. (2015) 'Review of the generalized least squares method', *Surveys in Geophysics*, 36(1), pp. 1–25.

Mester, A., van der Kruk, J., Zimmermann, E. and Vereecken, H. (2011) 'Quantitative two-layer conductivity inversion of multi-configuration electromagnetic induction measurements', *Vadose Zone Journal*, 10(4), pp. 1319–1330.

Metherall, P., Barber, D. C., Smallwood, R. H. and Brown, B. H. (1996) 'Three-dimensional electrical impedance tomography', *Nature*, 380(6574), p. 509.

Metropolis, N., Rosenbluth, A. W., Rosenbluth, M. N., Teller, A. H. and Teller, E. (1953) 'Equation of state calculations by fast computing machines', *The Journal of Chemical Physics*. 21(6), pp. 1087–1092.

Michot, D., Benderitter, Y., Dorigny, A., Nicoullaud, B., King, D. and Tabbagh, A. (2003) 'Spatial and temporal monitoring of soil water content with an irrigated corn crop cover using surface electrical resistivity tomography', *Water Resources Research*, 39, 1138, DOI:10.1029/2002WR001581

Miller, C. R. and Routh, P. S. (2007) 'Resolution analysis of geophysical images: Comparison between point spread function and region of data influence measures', *Geophysical Prospecting*, 55(6), pp. 835–852.

Millett F. B., Jr. (1967) 'Electromagnetic coupling of collinear dipoles on a uniform half-space', in *Mining Geophysics* Vol. II (Hansen, D. A. et al. eds.). Society of Exploration Geophysicists, pp. 401–419.

Minsley, B. J., Sogade, J. and Morgan, F. D. (2007) 'Three-dimensional source inversion of self-potential data', *Journal of Geophysical Research: Solid Earth*, 112(2), B02202. DOI: 10.1029/2006JB004262

Misiek, R., Liebig, A., Gyulai, A., Ormos, T., Dobroka, M. and Dresen, L. (1997) 'A joint inversion algorithm to process geoelectric and surface wave seismic data. Part II: Applications', *Geophysical Prospecting*, 43(2), pp. 135–156.

Misra, S., Torres-Verdín, C., Revil, A., Rasmus, J. and Homan, D. (2016) 'Interfacial polarization of disseminated conductive minerals in absence of redox-active species – Part 1: Mechanistic model and validation', *Geophysics*, 81(2), pp. E139–E157. DOI: 10.1190/geo2015-0346.1

Mitchell, N., Nyquist, J. E., Toran, L., Rosenberry, D. O. and Mikochik, J. S. (2008) 'Electrical resistivity as a tool for identifying geologic heterogeneities which control seepage at Mirror Lake, NH', in *Symposium on the Application of Geophysics to Engineering and Environmental Problems 2008*. Society of Exploration Geophysicists, pp. 749–759.

Monteiro Santos, F. A., Andrade Afonso, A. R. and Dupis, A. (2007) '2D joint inversion of dc and scalar audio-magnetotelluric data in the evaluation of low enthalpy geothermal fields', *Journal of Geophysics and Engineering*, 4(1), pp. 53–62. DOI: 10.1088/1742-2132/4/1/007

Morelli, G. and LaBrecque, D. J. (1996) 'Advances in ERT inverse modelling', *European Journal of Environmental and Engineering Geophysics*, 1(2), pp. 171–186.

Morris, G., Binley, A. M. and Ogilvy, R. D. (2004) 'Comparison of different electrode materials for induced polarization measurements', in *Proceedings of the 2004 Symposium on the Application of Geophysics to Engineering and Environmental Problems*. Environmental and Engineering Geophysical Society (EEGS), p. 4.

Mosteller, F. and Tukey, J. W. (1977) 'Data analysis and regression: A second course in statistics', Addison-Wesley Series in Behavioral Science: Quantitative Methods, p. 588.

Mualem, Y. and Friedman, S. P. (1991) 'Theoretical prediction of electrical conductivity in saturated and unsaturated soil', *Water Resources Research*, 27(10), pp. 2771–2777. DOI: 10.1029/91WR01095

Mudler, J., Hördt, A., Przyklenk, A., Fiandaca, G., Kumar Maurya, P. and Hauck, C. (2019) 'Two-dimensional inversion of wideband spectral data from the capacitively coupled resistivity method: First applications in periglacial environments', *Cryosphere*, 13(9), pp. 2439–2456. DOI: 10.5194/tc-13-2439-2019

Musgrave, H. and Binley, A. (2011) 'Revealing the temporal dynamics of subsurface temperature in a wetland using time-lapse geophysics', *Journal of Hydrology*, 396 (3–4), pp. 258–266. DOI: 10.1016/j.jhydrol.2010.11.008

Mustopa, E. J., Srigutomo, W. and Sutarno, D. (2011) 'Resistivity imaging of mataloko geothermal field by Mise-Á-La-Masse method', *Indonesian Journal of Physics*, 22 (2), pp. 45–51.

Mwakanyamale, K., Slater, L., Binley, A. and Ntarlagiannis, D. (2012) 'Lithologic imaging using complex conductivity: Lessons learned from the Hanford 300 Area', *Geophysics*, 77(6), p. E397. DOI: 10.1190/geo2011-0407.1

Nabighian, M. N. and Elliot, C. L. (1976) 'Negative induced-polarization effects from layered media', *Geophysics*, 41(6), pp. 1236–1255.

Nabighian, M. N. and Macnae, J. C. (1991) 'Time domain electromagnetic prospecting methods', *Electromagnetic Methods in Applied Geophysics*, 2 (Part A), pp. 427–509.

Nadler, A. and Frenkel, H. (1980) 'Determination of soil solution electrical conductivity from bulk soil electrical conductivity measurements by the four-electrode method1', *Soil Science Society of America Journal*, 44, pp. 1216–1221. DOI: 10.2136/sssaj1980.03615995004400060017x

Nagy, V., Milics, G., Smuk, N., Kovács, A. J., Balla, I., Jolánkai, M., Deákvári, J., Szalay, K. D., Fenyvesi, L. and Štekauerová, V. (2013) 'Continuous field soil moisture content mapping by means of apparent electrical conductivity (ECa) measurement', *Journal of Hydrology and Hydromechanics*, 61(4), pp. 305–312.

Nath, S. K., Shahid, S. and Dewangan, P. (2000) 'SEISRES: A visual C++ program for the sequential inversion of seismic refraction and geoelectric data', *Computers & Geosciences*, 26(2), pp. 177–200.

Newman, G. A. and Alumbaugh, D. L. (1997) 'Three-dimensional massively parallel electromagnetic inversion – I. Theory', *Geophysical Journal International*, 128(2), pp. 345–354.

Neyamadpour, A. (2019) '3D electrical resistivity tomography as an aid in investigating gravimetric water content and shear strength parameters', *Environmental Earth Sciences*, Springer, 78(19), p. 583.

Neyamadpour, A., Abdullah, W. A. T. W., Taib, S. and Niamadpour, D. (2010) '3D inversion of DC data using artificial neural networks', *Studia Geophysica et Geodaetica*, 54(3), pp. 465–485.

Nguyen, F., Garambois, S., Jongmans, D., Pirard, E. and Loke, M. H. (2005) 'Image processing of 2D resistivity data for imaging faults', *Journal of Applied Geophysics*, 57(4), pp. 260–277.

Nguyen, F., Kemna, A., Robert, T. and Hermans, T. (2016) 'Data-driven selection of the minimum-gradient support parameter in time-lapse focused electric imaging', *Geophysics*, 81(1), pp. A1–A5.

Nimmer, R. E. and Osiensky, J. L. (2002) 'Using mise-a-la-masse to delineate the migration of a conductive tracer in partially saturated basalt', *Environmental Geosciences*, 9(2), pp. 81–87.

Nimmer, R. E., Osiensky, J. L., Binley, A. M. and Williams, B. C. (2008) 'Three-dimensional effects causing artifacts in two-dimensional, cross-borehole, electrical imaging', *Journal of Hydrology*, 359(1–2), pp. 59–70. DOI: 10.1016/j.jhydrol.2008.06.022

Niu, Q. and Revil, A. (2016) 'Connecting complex conductivity spectra to mercury porosimetry of sedimentary rocks', *Geophysics*, 81(1), pp. E17–E32. DOI: 10.1190/geo2015-0072.1

Niu, Q. and Zhang, C. (2018) 'Physical explanation of Archie's porosity exponent in granular materials: A process-based, pore-scale numerical study', *Geophysical Research Letters*, 45 (4), pp. 1870–1877. DOI: 10.1002/2017GL076751

Niu, Q., Prasad, M., Revil, A. and Saidian, M. (2016a) 'Textural control on the quadrature conductivity of porous media', *Geophysics*, 81(5), pp. E297–E309. DOI: 10.1190/geo2015-0715.1

Niu, Q., Revil, A. and Saidian, M. (2016b) 'Salinity dependence of the complex surface conductivity of the Portland sandstone', *Geophysics*, 81 (2). DOI: 10.1190/geo2015-0426.1

Niwas, S. and Israil, M. (1986) 'Computation of apparent resistivities using an exponential approximation of kernel functions', *Geophysics*, 51(8), pp. 1594–1602.

Nordsiek, S. and Weller, A. (2008) 'A new approach to fitting induced-polarization spectra', *Geophysics*, 73(6), pp. F235–F245. DOI: 10.1190/1.2987412

Ntarlagiannis, D., Yee, N. and Slater, L. (2005a) 'On the low-frequency electrical polarization of bacterial cells in sands', *Geophysical Research Letters*, 32(24), pp. 1–4. DOI: 10.1029/2005GL024751

Ntarlagiannis, D., Williams, K. H., Slater, L. and Hubbard, S. (2005b) 'Low-frequency electrical response to microbial induced sulfide precipitation', *Journal of Geophysical Research*, 110(G2), pp. 1–12. DOI: 10.1029/2005JG000024

Nunn, K. R., Barker, R. D. and Bamford, D. (1983) 'In situ seismic and electrical measurements of fracture anisotropy in the Lincolnshire Chalk', *Quarterly Journal of Engineering Geology and Hydrogeology*, 16(3), pp. 187–195.

Nyquist, J. E. and Roth, M. J. S. (2005) 'Improved 3D pole-dipole resistivity surveys using radial measurement pairs', *Geophysical Research Letters*, 32, L21416, DOI:10.1029/2005GL024153

Nyquist, J. E., Heaney, M. J. and Toran, L. (2009) 'Characterizing lakebed seepage and geologic heterogeneity using resistivity imaging and temperature measurements', *Near Surface Geophysics*, 7(5–6), pp. 487–498.

Nyquist, J. E., Toran, L., Fang, A. C., Ryan, R. J. and Rosenberry, D. O. (2010) 'Tracking tracer breakthrough in the hyporheic zone using time-lapse DC resistivity, Crabby Creek, Pennsylvania', in 23rd EEGS Symposium on the Application of Geophysics to Engineering and Environmental Problems.

O'Neill, D. J. (1975) 'Improved linear filter coefficients for application in apparent resistivity computations', *Exploration Geophysics*, 6(4), pp. 104–109.

Ochs, J. and Klitzsch, N. (2020) 'Considerations regarding small-scale surface and borehole-to-surface electrical resistivity tomography', *Journal of Applied Geophysics*, 172, p. 103862. DOI: 10.1016/j.jappgeo.2019.103862

Okay, G., Leroy, P., Ghorbani, A., Cosenza, P., Camerlynck, C., Cabrera, J., Florsch, N. and Revil, A. (2014) 'Spectral induced polarization of clay-sand mixtures: Experiments and modeling', *Geophysics*, 79(6), pp. E353-375.

Oldenborger, G. A. and LeBlanc, A. M. (2018) 'Monitoring changes in unfrozen water content with electrical resistivity surveys in cold continuous permafrost', *Geophysical Journal International*, 215(2), pp. 965–977. DOI: 10.1093/GJI/GGY321

Oldenborger, G. A. and Routh, P. S. (2009) 'The point-spread function measure of resolution for the 3-D electrical resistivity experiment', *Geophysical Journal International*, 176(2), pp. 405–414. DOI: 10.1111/j.1365-246X.2008.04003.x

Oldenborger, G. A., Routh, P. S. and Knoll, M. D. (2005) 'Sensitivity of electrical resistivity tomography data to electrode position errors', *Geophysical Journal International, UK*, 163(1), pp. 1–9.

Oldenborger, G. A., Routh, P. S. and Knoll, M. D. (2007) 'Model reliability for 3D electrical resistivity tomography: Application of the volume of investigation index to a time-lapse monitoring experiment', *Geophysics*, 72(4). DOI: 10.1190/1.2732550

Oldenburg, D. W. and Li, Y. (1994) 'Inversion of induced polarization data', *Geophysics*, 59 (9), pp. 1327–1341.

Oldenburg, D. W. and Li, Y. (1999) 'Estimating depth of investigation in dc resistivity and IP surveys', *Geophysics*, 64(2), pp. 403–416.

Oldenburg, D. W., Mcgillivray, P. R. and Ellis, R. G. (1993) 'Generalized subspace methods for large-scale inverse problems', *Geophysical Journal International*, 114(1), pp. 12–20. DOI: 10.1111/j.1365-246X.1993.tb01462.x

Olhoeft, G. R. (1974) 'Electrical properties of rocks', *Physical Properties of Rocks and Minerals*, 2, pp. 257–297.

Olhoeft, G. R. (1979) 'Nonlinear electrical properties', in *Nonlinear Behavior of Molecules, Atoms and Ions in Electric, Magnetic, or Electromagnetic Fields* (Neel, L. ed.). Elsevier Science Publishing Co, pp. 395–410.

Olhoeft, G. R. (1985) 'Low-frequency electrical properties', *Geophysics*, 50(12), pp. 2492–2503. DOI: 10.1190/1.1441880

Olsen, P. A., Binley, A., Henry-Poulter, S. and Tych, W. (1999) 'Characterizing solute transport in undisturbed soil cores using electrical and X-ray tomographic methods', *Hydrological Processes*, 13(2), pp. 211–221.

Olsson, P. I., Fiandaca, G., Larsen, J. J., Dahlin, T. and Auken, E. (2016) 'Doubling the spectrum of time-domain induced polarization by harmonic de-noising, drift correction, spike removal, tapered gating and data uncertainty estimation', *Geophysical Journal International*, 207(2), pp. 774–784. DOI: 10.1093/gji/ggw260

Olsson, P.-I. I., Dahlin, T., Fiandaca, G. and Auken, E. (2015) 'Measuring time-domain spectral induced polarization in the on-time: Decreasing acquisition time and increasing signal-to-noise ratio', *Journal of Applied Geophysics*, 123, pp. 6–11. DOI: 10.1016/j.jappgeo.2015.08.009

Orlando, L. (2013) 'Some considerations on electrical resistivity imaging for characterization of waterbed sediments', *Journal of Applied Geophysics*, 95, pp. 77–89.

Osiensky, J. L., Nimmer, R. and Binley, A. M. (2004) 'Borehole cylindrical noise during hole-surface and hole-hole resistivity measurements', *Journal of Hydrology*, 289(1–4), pp. 78–94. DOI: 10.1016/j.jhydrol.2003.11.003

Osterman, G., Keating, K., Binley, A. and Slater, L. (2016) 'A laboratory study to estimate pore geometric parameters of sandstones using complex conductivity and nuclear magnetic resonance for permeability prediction', *Water Resources Research*, 52(6), pp. 4321–4337. DOI: 10.1002/2015WR018472

Osterman, G., Sugand, M., Keating, K., Binley, A. and Slater, L. (2019) 'Effect of clay content and distribution on hydraulic and geophysical properties of synthetic sand-clay mixtures', *Geophysics*, 84(4), pp. E239–E253.

Oware, E. K., Moysey, S. M. J. and Khan, T. (2013) 'Physically based regularization of hydrogeophysical inverse problems for improved imaging of process-driven systems', *Water Resources Research*, 49(10), pp. 6238–6247. DOI: 10.1002/wrcr.20462

Pandit, B. I. and King, M. S. (1979) 'A study of the effects of pore-water salinity on some physical properties of sedimentary rocks at permafrost temperatures', *Canadian Journal of Earth Sciences*, 16(8), pp. 1566–1580. DOI: 10.1139/e79-143

Panissod, C., Dabas, M., Hesse, A., Jolivet, A., Tabbagh, J. and Tabbagh, A. (1998) 'Recent developments in shallow-depth electrical and electrostatic prospecting using mobile arrays', *Geophysics*, 63(5), pp. 1542–1550.

Pape, H., Clauser, C. and Iffland, J. (1999) 'Permeability prediction based on fractal pore-space geometry', *Geophysics*, 64(5), pp. 1447–1460.

Pape, H., Riepe, L. and Schopper, J. R. (1987) 'Theory of self-similar network structures in sedimentary and igneous rocks and their investigation with microscopical and physical methods', *Journal of Microscopy*, 148(2), pp. 121–147. DOI: 10.1111/j.1365-2818.1987.tb02861.x

Parasnis, D. (1988) 'Reciprocity theorems in geoelectric and geoelectromagnetic work', *Geoexploration*, 25(3), pp. 177–198.

Parasnis, D. S. (1967) 'Three-dimensional electric mise-a-la-masse survey of an irregular lead-zinc-copper deposit in central Sweden', *Geophysical Prospecting*, 15(3), pp. 407–437.

Park, S. (1998) 'Fluid migration in the vadose zone from 3-D inversion of resistivity monitoring data', *Geophysics*, 63(1), pp. 41–51.

Park, S. K. and Fitterman, D. V. (1990) 'Sensitivity of the telluric monitoring array in Parkfield, California, to changes of resistivity', *Journal of Geophysical Research: Solid Earth*, 95(B10), pp. 15557–15571.

Park, S. K. and Van, G. P. (1991) 'Inversion of pole-pole data for 3-D resistivity structure beneath arrays of electrodes', *Geophysics*, 56(7), pp. 951–960.

Parra, J. O. (1988) 'Electrical response of a leak in a geomembrane liner', *Geophysics*, 53 (11), pp. 1445–1452.

Parsekian, A. D., Claes, N., Singha, K., Minsley, B. J., Carr, B., Voytek, E., Harmon, R., Kass, A., Carey, A. and Thayer, D. (2017) 'Comparing measurement response and inverted results of electrical resistivity tomography instruments', *Journal of Environmental and Engineering Geophysics*, 22(3), pp. 249–266.

Passaro, S. (2010) 'Marine electrical resistivity tomography for shipwreck detection in very shallow water: A case study from Agropoli (Salerno, southern Italy)', *Journal of Archaeological Science*, 37(8), pp. 1989–1998.

Patella, D. (1972) 'An interpretation theory for induced polarization vertical soundings (time-domain)', *Geophysical Prospecting*, 20(3), pp. 561–579.

Pelton, W. H., Rijo, L. and Swift Jr, C. M. (1978) 'Inversion of two-dimensional resistivity and induced-polarization data', *Geophysics*, 43(4), pp. 788–803.

Pelton, W. H., Ward, S. H., Hallof, P. G., Sill, W. R. and Nelson, P. H. (1978) 'Mineral discrimination and removal of inductive coupling with multifrequency IP', *Geophysics*, 43(3), pp. 588–609.

Perri, M. T., De Vita, P., Masciale, R., Portoghese, I., Chirico, G. B. and Cassiani, G. (2018) 'Time-lapse Mise-à-la-Masse measurements and modeling for tracer test monitoring in a shallow aquifer', *Journal of Hydrology*, 561, pp. 461–477.

Perrone, A., Lapenna, V. and Piscitelli, S. (2014) 'Electrical resistivity tomography technique for landslide investigation: A review', *Earth-Science Reviews*, 135, pp. 65–82.

Pessel, M. and Gibert, D. (2003) 'Multiscale electrical impedance tomography', *Journal of Geophysical Research*, 108, p. 2054, DOI:10.1029/2001JB000233, B1

Petiau, G. (2000) 'Second generation of lead-lead chloride electrodes for geophysical applications', *Pure and Applied Geophysics*, 157(3), pp. 357–382. DOI: 10.1007/s000240050004

Phuong Tran, A., Dafflon, B., Hubbard, S. S., Kowalsky, M. B., Long, P., Tokunaga, T. K. and Williams, K. H. (2016) 'Quantifying shallow subsurface water and heat dynamics using coupled hydrological-thermal-geophysical inversion', *Hydrology and Earth System Sciences*, 20(9), pp. 3477–3491.

Pidlisecky, A. and Knight, R. (2008) 'FW2_5D: A MATLAB 2.5-D electrical resistivity modeling code', *Computers & Geosciences*, 34(12), pp. 1645–1654.

Pidlisecky, A., Haber, E. and Knight, R. (2007) 'RESINVM3D: A 3D resistivity inversion package', *Geophysics*, 72(2), pp. H1–H10.

Pidlisecky, A., Rowan Cockett, A. and Knight, R. (2013) 'Electrical conductivity probes for studying vadose zone processes: Advances in data acquisition and analysis', *Vadose Zone Journal*, 12(1), pp. 1–12. DOI: 10.2136/vzj2012.0073

Pinheiro, P. A. T., Loh, W. W. and Dickin, F. J. (1998) 'Optimal sized electrodes for electrical resistance tomography', *Electronics Letters*, 34(1), pp. 69–70.

Placencia-Gómez, E., Parviainen, A., Slater, L. and Leveinen, J. (2015) 'Spectral induced polarization (SIP) response of mine tailings', *Journal of Contaminant Hydrology*, 173, pp. 8–24. DOI: 10.1016/j.jconhyd.2014.12.002

Portniaguine, O. and Zhdanov, M. S. (1999) 'Focusing geophysical inversion images', *Geophysics*, 64(3), pp. 874–887.

Potapenko, G. (1940) 'Method of determining the presence of oil', U.S. Patent 2,190,320.

Powell, H. M., Barber, D. C. and Freeston, I. L. (1987) 'Impedance imaging using linear electrode arrays', *Clinical Physics and Physiological Measurement*, 8(4A), p. 109.

Power, C., Tsourlos, P., Ramasamy, M., Nivorlis, A. and Mkandawire, M. (2018), 'Combined DC resistivity and induced polarization (DC-IP) for mapping the internal composition of a mine waste rock pile in Nova Scotia, Canada', *Journal of Applied Geophysics*, 150, pp. 40–51.

Pride, S. (1994) 'Governing equations for the coupled electromagnetics and acoustics of porous media', *Physical Review B*, 50(21), p. 15678.

Pridmore, D. F., Hohmann, G. W., Ward, S. H. and Sill, W. R. (1981) 'An investigation of finite-element modeling for electrical and electromagnetic data in three dimensions', *Geophysics*, 46(7), pp. 1009–1024.

Prodan, C. and Bot, C. (2009) 'Correcting the polarization effect in very low frequency dielectric spectroscopy', *Journal of Physics D: Applied Physics*, 42(17), p. 175505.

Pullen, M. W. (1929) *Tentative Method for Making Resistivity Measurements of Drill Cores and Hand Specimens of Rocks and Ores*. US Department of Commerce, Bureau of Mines.

Puls, R. W., Paul, C. J. and Powell, R. M. (1999) 'The application of in situ permeable reactive (zero-valent iron) barrier technology for the remediation of chromate-contaminated groundwater: A field test', *Applied Geochemistry*, 14(8), pp. 989–1000.

Purvance, D. T. and Andricevic, R. (2000) 'On the electrical-hydraulic conductivity correlation in aquifers', *Water Resources Research*, 36(10), pp. 2905–2913.

Qing, C., Pardo, D., Hong-bin, L. and Fu-rong, W. (2017) 'New post-processing method for interpretation of through casing resistivity (TCR) measurements', *Journal of Applied Geophysics*, 74, pp. 19–25.

Radic, T. (2004) 'Elimination of cable effects while multi-channel SIP measurements', Near Surface 2004 – 10th EAGE European Meeting of Environmental and Engineering Geophysics, pp. 1–4.

Radic, T. and Klitzsch, N. (2012) 'Compensation technique to minimize capacitive cable coupling effects in multi-channel IP systems', Near Surface Geoscience 2012, (September 2012), pp. 3–5. DOI: 10.3997/2214-4609.20143487

Radic, T., Kretzschmar, D. and Niederleithinger, E. (1998) 'Improved characterization of unconsolidated sediments under field conditions based on complex resistivity measurements', in Proceedings of the 4th Environmental and Engineering Geophysical Society (EEGS) Meeting.

Raffelli, G., Previati, M., Canone, D., Gisolo, D., Bevilacqua, I., Capello, G., Biddoccu, M., Cavallo, E., Deiana, R. and Cassiani, G. (2017) 'Local-and plot-scale measurements of soil moisture: Time and spatially resolved field techniques in plain, hill and mountain sites', *Water*, 9(9), p. 706.

Ramirez, A. and Daily, W. (2001) 'Electrical imaging at the large block test: Yucca Mountain, Nevada', *Journal of Applied Geophysics*, 46(2), pp. 85–100.

Ramirez, A. L., Nitao, J. J., Hanley, W. G., Aines, R., Glaser, R. E., Sengupta, S. K., Dyer, K. M., Hickling, T. L. and Daily, W. D. (2005) 'Stochastic inversion of electrical resistivity changes using a Markov Chain Monte Carlo approach', *Journal of Geophysical Research*, 110, B02101, DOI:10.1029/2004JB003449

Ramirez, A., Daily, W., Binley, A., LaBrecque, D. and Roelant, D. (1996) 'Detection of leaks in underground storage tanks using electrical resistance methods', *Journal of Environmental and Engineering Geophysics*, 1(3), pp. 189–203.

Ramirez, A., Daily, W., LaBrecque, D., Owen, E. and Chesnut, D. (1993) 'Monitoring an underground steam injection process using electrical resistance tomography', *Water Resources Research*, 29(1), pp. 73–87.

Randles, J. E. B. (1947) 'Kinetics of rapid electrode reactions', *Discussions of the Faraday Society*, 1, pp. 11–19.

Ray, A. and Myer, D. (2019) 'Bayesian geophysical inversion with trans-dimensional Gaussian process machine learning', *Geophysical Journal International*, 217(3), pp. 1706–1726.

Razavirad, F., Schmutz, M. and Binley, A. (2018) 'Estimation of the permeability of hydrocarbon reservoir samples using induced polarization (ip) and nuclear magnetic resonance (nmr) methods', *Geophysics*, pp. 1–76. DOI: 10.1190/geo2017-0745.1

Ren, Z. and Tang, J. (2010) '3D direct current resistivity modeling with unstructured mesh by adaptive finite-element method', *Geophysics*, 75(1), pp H1-H17, DOI: 10.1190/1.3298690

Revil, A. (2012) 'Spectral induced polarization of shaly sands: Influence of the electrical double layer', *Water Resources Research*, 48(2). DOI: 10.1029/2011WR011260

Revil, A. (2013) 'On charge accumulation in heterogeneous porous rocks under the influence of an external electric field', *Geophysics*, 78(4), pp. D271–D291.

Revil, A. and Florsch, N. (2010) 'Determination of permeability from spectral induced polarization in granular media', *Geophysical Journal International*, 181(3), pp. 1480–1498. DOI: 10.1111/j.1365-246X.2010.04573.x

Revil, A. and Glover, P. W. J. (1997) 'Theory of ionic-surface electrical conduction in porous media', *Physical Review B*, 55(3), p. 1757.

Revil, A. and Glover, P. W. J. (1998) 'Nature of surface electrical conductivity in natural sands, sandstones, and clays', *Geophysical Research Letters*, 25(5), pp. 691–694.

Revil, A. and Jardani, A. (2013) *The Self-Potential Method: Theory and Applications in Environmental Geosciences*. Cambridge University Press, p. 369.

Revil, A. and Skold, M. (2011) 'Salinity dependence of spectral induced polarization in sands and sandstones', *Geophysical Journal International*, 187(2), pp. 813–824. DOI: 10.1111/j.1365-246X.2011.05181.x

Revil, A., Abdel Aal, G. Z., Atekwana, E. A., Mao, D. and Florsch, N. (2015c) 'Induced polarization response of porous media with metallic particles – Part 2: Comparison with a broad database of experimental data', *Geophysics*, 80(5), pp. D539–D552. DOI: 10.1190/geo2014-0578.1

Revil, A., Binley, A., Mejus, L. and Kessouri, P. (2015b) 'Predicting permeability from the characteristic relaxation time and intrinsic formation factor of complex conductivity spectra', *Water Resources Research*, 51(8). DOI: 10.1002/ 2015WR017074

Revil, A., Coperey, A., Mao, D., Abdulsamad, F., Ghorbani, A., Rossi, M. and Gasquet, D. (2018a) 'Induced polarization response of porous media with metallic particles – Part 8. Influence of temperature and salinity', *Geophysics*, 83(6), pp. 1–78. DOI: 10.1190/geo2018-0089.1

Revil, A., Coperey, A., Shao, Z., Florsch, N., Fabricius, I. L., Deng, Y., Delsman, J. R., Pauw, P. S., Karaoulis, M., de Louw, P. G. B., van Baaren, E. S., Dabekaussen, W., Menkovic, A. and Gunnink, J. L. (2017a) 'Complex conductivity of soils', *Water Resources Research*, 53(8), pp. 7121–7147. DOI: 10.1002/2017WR020655

Revil, A., Florsch, N. and Camerlynck, C. (2014) 'Spectral induced piorosimetry', *Geophysical Journal International*, pp. 1016–1033. DOI: 10.1093/gji/ggu180

Revil, A., Florsch, N. and Mao, D. (2015a) 'Induced polarization response of porous media with metallic particles – Part 1: A theory for disseminated semiconductors', *Geophysics*, 80(5), D525–D538.

Revil, A., Johnson, T. C. and Finizola, A. (2010) 'Three-dimensional resistivity tomography of Vulcan's forge, Vulcano Island, southern Italy', *Geophysical Research Letters*, 37(15).

Revil, A., Murugesu, M., Prasad, M. and Le Breton, M. (2017b) 'Alteration of volcanic rocks: A new non-intrusive indicator based on induced polarization measurements', *Journal of Volcanology and Geothermal Research*, 341, pp. 351–362. DOI: 10.1016/j.jvolgeores.2017.06.016

Revil, A., Qi, Y., Ghorbani, A., Soueid Ahmed, A., Ricci, T. and Labazuy, P. (2018b) 'Electrical conductivity and induced polarization investigations at Krafla volcano, Iceland', *Journal of Volcanology and Geothermal Research*, 368, pp. 73–90. DOI: 10.1016/j.jvolgeores.2018.11.008

Rhoades, J. D., Manthegi, N. A., Shouse, P. J. and Alves, W. J. (1989) 'Soil electrical conductivity and soil salinity: New formulations and falibrations', *Soil Science Society of America Journal*, 53(2), pp. 433–439.

Rhoades, J. D., Ratts, P. A. C. and Prather, R. J. (1976) 'Effects of liquid-phase electrical conductivity, water content and surface conductivity on bulk electrical conductivity', *Soil Science Society of America Journal*, 40, pp. 651–655.

Rink, M. and Schopper, J. R. (1974) 'Interface conductivity and its implications to electric logging', *Transactions of the SPWLA 15th Annual Logging Symposium*, 15, p. 15.

Robinson, J., Slater, L., Johnson, T., Shapiro, A., Tiedeman, C., Ntarlagiannis, D., Johnson, C., Day-Lewis, F., Lacombe, P., Imbrigiotta, T. and Lane, J. (2016) 'Imaging pathways in fractured rock using three-dimensional electrical resistivity tomography', *Groundwater*, 54(2). DOI: 10.1111/gwat.12356

Robinson, J., Slater, L., Weller, A., Keating, K., Robinson, T., Rose, C. and Parker, B. (2018) 'On permeability prediction from complex conductivity measurements using polarization magnitude and relaxation time', *Water Resources Research*, 54(5), pp. 3436–3452. DOI: 10.1002/2017WR022034

Rödder, A. and Junge, A. (2016) 'The influence of anisotropy on the apparent resistivity tensor: A model study', *Journal of Applied Geophysics*, 135, pp. 270–280.

Routh, P. S. and Oldenburg, D. W. (2001) 'Electromagnetic coupling in frequency-domain induced polarization data: A method for removal', *Geophysical Journal International*, 145(1), pp. 59–76.

Roy, A. and Apparao, A. (1971) 'Depth of investigation in direct current methods', *Geophysics*, 36(5), pp. 943–959.

Roy, K. K. and Elliott, H. M. (1980) 'Resistivity and IP survey for delineating saline water and fresh water zones', *Geoexploration*, 18(2), pp. 145–162.

Rücker, C. and Günther, T. (2011) 'The simulation of finite ERT electrodes using the complete electrode model', *Geophysics*, 76(4), pp. F227–F238.

Rücker, C., Günther, T. and Spitzer, K. (2006) 'Three-dimensional modelling and inversion of dc resistivity data incorporating topography – I. Modelling', *Geophysical Journal International*, 166(2), pp. 495–505. DOI: 10.1111/j.1365-246X.2006.03010.x

Rücker, C., Günther, T. and Wagner, F. M. (2017) 'pyGIMLi: An open-source library for modelling and inversion in geophysics', *Computers & Geosciences*, 109, pp. 106–123.

Rucker, D. F., Fink, J. B. and Loke, M. H. (2011) 'Environmental monitoring of leaks using time-lapsed long electrode electrical resistivity', *Journal of Applied Geophysics*, 74 (4), pp. 242–254.

Rucker, D. F., Loke, M. H., Levitt, M. T. and Noonan, G. E. (2010) 'Electrical-resistivity characterization of an industrial site using long electrodes', *Geophysics*, 75(4), pp. WA95–WA104.

Rucker, D. F., Noonan, G. E. and Greenwood, W. J. (2011) 'Electrical resistivity in support of geological mapping along the Panama Canal', *Engineering Geology*, 117(1–2), pp. 121–133.

Russell, B. (2019) 'Machine learning and geophysical inversion: A numerical study', *Leading Edge*, 38(7), pp. 512–519. DOI: 10.1190/tle38070512.1

Rust Jr, W. M. (1938) 'A historical review of electrical prospecting methods', *Geophysics*, 3 (1), pp. 1–6.

Sambuelli, L., Comina, C., Bava, S. and Piatti, C. (2011) 'Magnetic, electrical, and GPR waterborne surveys of moraine deposits beneath a lake: A case history from Turin, Italy', *Geophysics*, 76(6), pp. B213–B224.

Samouëlian, A., Richard, G., Cousin, I., Guerin, R., Bruand, A. and Tabbagh, A. (2004) 'Three-dimensional crack monitoring by electrical resistivity measurement', *European Journal of Soil Science*, 55(4), pp. 751–762.

Saneiyan, S., Ntarlagiannis, D., Ohan, J., Lee, J., Colwell, F. and Burns, S. (2019) 'Induced polarization as a monitoring tool for in-situ microbial induced carbonate precipitation (MICP) processes', *Ecological Engineering*, 127, pp. 36–47. DOI: 10.1016/j.ecoleng.2018.11.010

Santini, R. and Zambrano, R. (1981) 'A numerical method of calculating the kernel function from Schlumberger apparent resistivity data', *Geophysical Prospecting*, 29(1), pp. 108–127. DOI: 10.1111/j.1365-2478.1981.tb01014.x

Sasaki, Y. (1989) 'Two-dimensional joint inversion of magnetotelluric and dipole-dipole resistivity data', *Geophysics*, 54(2), pp. 254–262.

Sasaki, Y. (1992) 'Resolution of resistivity tomography inferred from numercial simulation', *Geophysical Prospecting*, 40(4), pp. 453–463.

Sasaki, Y. (1993) 'Surface-to-tunnel resistivity tomography at the Kamaishi Mine', *Butsuri-Tansa, Geophysics*, 46, pp. 128–134.

Sasaki, Y. (1994) '3-D resistivity inversion using the finite-element method', *Geophysics*, 59(12), pp. 1839–1848.

Sauck, W. A. (2000) 'A model for the resistivity structure of LNAPL plumes and their environs in sandy sediments', *Journal of Applied Geophysics*, 44(2–3), pp. 151–165. DOI: 10.1016/S0926-9851(99)00021-X

Sauer, U., Watanabe, N., Singh, A., Dietrich, P., Kolditz, O. and Schütze, C. (2014) 'Joint interpretation of geoelectrical and soil-gas measurements for monitoring CO_2 releases at a natural analogue', *Near Surface Geophysics*, 12(1), pp. 165–187.

Schenkel, C. J. (1991) 'The electrical resistivity method in cased boreholes'. PhD thesis, University of California.

Schenkel, C. J. (1994) 'DC resistivity imaging using a steel cased well', in *SEG Technical Program Expanded Abstracts 1994*. Society of Exploration Geophysicists, pp. 403–406.

Schenkel, C. J. and Morrison, H. F. (1990) 'Effects of well casing on potential field measurements using downhole current sources', *Geophysical Prospecting*, 38(6), pp. 663–686.

Schima, S., LaBrecque, D. J. and Lundegard, P. D. (1993) 'Monitoring air sparging using resistivity tomography', *Groundwater Monitoring & Remediation*, 16(2), pp. 131–138.

Schlumberger, C. (1912) 'Verfahren zur Bestimmung der Beshaffenheit des Erbodens mittels Elektrizität', German Patent 269,928.

Schlumberger, C. (1915) 'Process for determining the nature of the subsoil by the aid of electricity', U.S. Patent 1,163,468.

Schlumberger, C. (1920) *Etude sur la prospection electrique du sous-sol*. Gauthier-Villars.

Schlumberger, C. (1926) 'Method for the location of oil bearing formation', U.S. Patent 1,719,786.

Schlumberger, C. (1933) 'Electrical process for the geological investigation of the porous strata traversed by drill holes', U.S. Patent 1,913,293.

Schlumberger, C. (1939) 'Method and apparatus for identifying the nature of the formations in a borehole', U.S. Patent 2,165,013.

Schlumberger, C., Schlumberger, M. and Leonardon, E. G. (1934) 'Electrical exploration of water-covered areas', *Transactions of the American Institute of Mining and Metallurgical Engineers*, 110, pp. 122–134.

Schmidt-Hattenberger, C., Bergmann, P., Labitzke, T. and Wagner, F. (2014) 'CO2 migration monitoring by means of electrical resistivity tomography (ERT)–Review on five years of operation of a permanent ERT system at the Ketzin pilot site', *Energy Procedia*, 63, pp. 4366–4373.

Schmutz, M., Revil, A., Vaudelet, P., Batzle, M., Viñao, P. F. and Werkema, D. D. (2010) 'Influence of oil saturation upon spectral induced polarization of oil-bearing sands', *Geophysical Journal International*, 183(1), pp. 211–224. DOI: 10.1111/j.1365-246X.2010.04751.x

Schnaidt, S. and Heinson, G. (2015) 'Bootstrap resampling as a tool for uncertainty analysis in 2-D magnetotelluric inversion modelling', *Geophysical Journal International*, 203 (1), pp. 92–106. DOI: 10.1093/gji/ggv264

Schön, J. H. (2011) *Physical Properties of Rocks: A Workbook*. Elsevier B.V., p. 494.

Schulmeister, M. K., Butler Jr, J. J., Healey, J. M., Zheng, L., Wysocki, D. A. and McCall, G. W. (2003) 'Direct-push electrical conductivity logging for high-resolution hydrostratigraphic characterization', *Groundwater Monitoring & Remediation*, 23(3), pp. 52–62.

Schurr, J. M. (1964) 'On the theory of the dielectric dispersion of spherical colloidal particles in electrolyte solution', *Journal of Physical Chemistry*, 68(9), pp. 2407–2413. DOI: 10.1021/j100791a004

Schwarz, G. (1962) 'A theory of the low-frequency dielectric dispersion of colloidal particles in electrolyte solution1,2', *The Journal of Physical Chemistry*, 66(12), pp. 2636–2642. DOI: 10.1021/j100818a067

Schwarz, H. R. (1991) *FORTRAN-Programme zur Methode der finiten Elemente*. Springer, p. 224.

Schwarzbach, C., Börner, R.-U. and Spitzer, K. (2005) 'Two-dimensional inversion of direct current resistivity data using a parallel, multi-objective genetic algorithm', *Geophysical Journal International*, 162(3), pp. 685–695.

Schwarze, F., Engels, J., Mattheck, C. and others (1999) *Holzzersetzende Pilze in Baumen – Strategien der Holzzersetzung*. Rombach Verlag.

Scott, J. B. T. and Barker, R. D. (2003) 'Determining pore-throat size in Permo-Triassic sandstones from low-frequency electrical spectroscopy', *Geophysical Research Letters*, 30(9), p. 1450. DOI: 10.1029/2003GL016951

Scott, W. J., Sellmann, P. and Hunter, J. A. (1990) 'Geophysics in the study of permafrost', in *Geotechnical and Environmental Geophysics* (Ward, S. ed.). Society of Exploration Geophysicists, 1, pp. 355–384.

Searle, G. F. C. (1911) 'On resistances with current and potential terminals', *Electrician*, 66, p. 999.

Sears, R. (1998) 'The British Nuclear Fuels Drigg low-level waste site characterization programme', *Geological Society, London, Special Publications*, 130(1), pp. 37–46.

Segesman, F. F. (1980) 'Well-logging method', *Geophysics*, 45(11), pp. 1667–1684.

Seigel, H. et al. (2007) 'The early history of the induced polarization method', *The Leading Edge*, 26(3), pp. 312–321.

Seigel, H. O. (1949) 'Theoretical and experimental investigations into the applications of the phenomenon of overvoltage to geophysical prospecting,' Toronto, Unpublished doctoral dissertation, University of Toronto.

Seigel, H. O. (1959) 'Mathematical formulation and type curves for induced polarization', *Geophysics*, 24(3), pp. 547–565.

Seigel, H., Nabighian, M., Parasnis, D. S. and Vozoff, K. (2007) 'The early history of the induced polarization method', *The Leading Edge*, 26(3), pp. 312–321.

Sen, M. K. and Stoffa, P. L. (2013) *Global Optimization Methods in Geophysical Inversion*. Cambridge University Press, p. 279.

Sen, M. K., Bhattacharya, B. B. and Stoffa, P. L. (1993) 'Nonlinear inversion of resistivity sounding data', *Geophysics*, 58(4), pp. 496–507.

Sen, P. N. (1984) 'Grain shape effects on dielectric and electrical properties of rocks', *Geophysics*, 49(5), pp. 586–587.

Sen, P. N. and Goode, P. A. (1992) 'Influence of temperature on electrical conductivity on shaly sands', *Geophysics*, 57(1), pp. 89–96.

Sen, P. N., Scala, C. and Cohen, M. H. (1981) 'A self-similar model for sedimentary rocks with application to the dielectric constant of fused glass beads', *Geophysics*, 46(5), pp. 781–795.

Shah, P. H. and Singh, D. N. (2005) 'Generalized Archie's law for estimation of soil electrical conductivity', *Journal of ASTM International*, 2(5), pp. 1–20.

Shanahan, P. W., Binley, A., Whalley, W. R. and Watts, C. W. (2015) 'The use of electromagnetic induction to monitor changes in soil moisture profiles beneath different wheat genotypes', *Soil Science Society of America Journal*, 79(2). DOI: 10.2136/sssaj2014.09.0360

Shaw, R. and Srivastava, S. (2007) 'Particle swarm optimization: A new tool to invert geophysical data', *Geophysics*, 72(2). DOI: 10.1190/1.2432481

Sheriff, S. D. (1992) 'Spreadsheet modeling of electrical sounding experiments', *Groundwater*, 30(6), pp. 971–974.

Sherrod, L., Sauck, W. and Werkema, D. D. J. (2012) 'A low-cost, in situ resistivity and temperature monitoring system', *Groundwater Monitoring and Remediation*, 32(2), pp. 31–39. DOI: 10.1111/j1745

Shigo, A. L. and Shigo, A. (1974) 'Detection of discoloration and decay in living trees and utility poles', Res. Pap. NE-294. Upper Darby, PA: US Department of Agriculture, Forest Service, Northeastern Forest Experiment Station. 11p.

Shima, H. (1990) 'Two-dimensional automatic resistivity inversion technique using alpha centers', *Geophysics*, 55(6), pp. 682–694.

Shima, H. (1992) '2-D and 3-D resistivity image reconstruction using crosshole data', *Geophysics*, 57(10), pp. 1270–1281.

Simms, J. E. and Morgan, F. D. (1992) 'Comparison of four least-squares inversion schemes for studying equivalence in one-dimensional resistivity interpretation', *Geophysics*, 57(10), pp. 1282–1293.

Simpson, D., Van Meirvenne, M., Lück, E., Bourgeois, J. and Rühlmann, J. (2010) 'Prospection of two circular Bronze Age ditches with multi-receiver electrical conductivity sensors (North Belgium)', *Journal of Archaeological Science*, 37(9), pp. 2198–2206.

Simyrdanis, K., Tsourlos, P., Soupios, P. and Tsokas, G. (2016) 'Simulation of ERT surface-to-tunnel measurements', *Bulletin of the Geological Society of Greece*, 47 (3), p. 1251. DOI: 10.12681/bgsg.10981

Singha, K. and Gorelick, S. M. (2005) 'Saline tracer visualized with three-dimensional electrical resistivity tomography: Field-scale spatial moment analysis', *Water Resources Research*, 41(5).

Singha, K., Day-Lewis, F. D. and Lane, J. W. (2007) 'Geoelectrical evidence of bicontinuum transport in groundwater', *Geophysical Research Letters*, 34(12), pp. 1–5. DOI: 10.1029/2007GL030019

Skutt, H. R., Shigo, A. L. and Lessard, R. A. (1972) 'Detection of discolored and decayed wood in living trees using a pulsed electric current', *Canadian Journal of Forest Research*, 2(1), pp. 54–56.

Slater, L. and Binley, A. (2003) 'Evaluation of permeable reactive barrier (PRB) integrity using electrical imaging methods', *Geophysics*, 68(3), pp. 911–921.

Slater, L. and Binley, A. M. (2006) 'Synthetic and field-based electrical imaging of a zerovalent iron barrier: Implications for monitoring long-term barrier performance', *Geophysics*, 71(5). DOI: 10.1190/1.2235931

Slater, L. and Lesmes, D. (2002a) 'IP interpretation in environmental investigations', *Geophysics*, 67(1), pp. 77–88. DOI: 10.1190/1.1451353

Slater, L. and Lesmes, D. P. (2002b) 'Electrical-hydraulic relationships observed for unconsolidated sediments', *Water Resources Research*, 38(10), pp. 1–13. DOI: 10.1029/2001WR001075

Slater, L. D. and Glaser, D. R. (2003) 'Controls on induced polarization in sandy unconsolidated sediments and application to aquifer characterization', *Geophysics*, 68(5), pp. 1547–1558. DOI: 10.1190/1.1620628

Slater, L. D., Binley, A. and Brown, D. (1997a) 'Electrical imaging of fractures using ground-water salinity change', *Ground Water*, 35(3), pp. 436–442. DOI: 10.1111/j.1745-6584.1997.tb00103.x

Slater, L. D., Choi, J. and Wu, Y. (2005) 'Electrical properties of iron-sand columns: Implications for induced polarization investigation and performance monitoring of iron-wall barriers', *Geophysics*, 70(4), p. G87. DOI: 10.1190/1.1990218

Slater, L. D., Ntarlagiannis, D., Day-Lewis, F. D., Mwakanyamale, K., Versteeg, R. J., Ward, A., Strickland, C., Johnson, C. D., Lane Jr., J. W. and Lane, J. W. (2010) 'Use of electrical imaging and distributed temperature sensing methods to characterize surface water-groundwater exchange regulating uranium transport at the Hanford 300 Area, Washington', *Water Resources Research*, 46(10), pp. 1–13. DOI: 10.1029/2010WR009110

Slater, L., Barrash, W., Montrey, J. and Binley, A. (2014) 'Electrical-hydraulic relationships observed for unconsolidated sediments in the presence of a cobble framework', *Water Resources Research*, 50(7), pp. 5721–5742. DOI: 10.1002/2013WR014631

Slater, L., Binley, A., Versteeg, R., Cassiani, G., Birken, R. and Sandberg, S. (2002) 'A 3D ERT study of solute transport in a large experimental tank', *Journal of Applied Geophysics*, 49(4), pp. 211–229. DOI: 10.1016/S0926-9851(02)00124-6

Slater, L., Binley, A. M., Daily, W. and Johnson, R. (2000) 'Cross-hole electrical imaging of a controlled saline tracer injection', *Journal of Applied Geophysics*, 44(2–3), pp. 85–102.

Slater, L., Brown, D. and Binley, A. (1996) 'Determination of hydraulically conductive pathways in fractured limestone using cross-borehole electrical resistivity tomography', *European Journal of Environmental and Engineering Geophysics*, 1 (1), pp. 35–52.

Slater, L., Ntarlagiannis, D., Personna, Y. R. and Hubbard, S. (2007) 'Pore-scale spectral induced polarization signatures associated with FeS biomineral transformations', *Geophysical Research Letters*, 34(21), pp. 3–7. DOI: 10.1029/2007GL031840

Slater, L., Ntarlagiannis, D., Yee, N., O'Brien, M., Zhang, C. and Williams, K. H. (2008) 'Electrodic voltages in the presence of dissolved sulfide: Implications for monitoring natural microbial activity', *Geophysics*, 73(2). DOI: 10.1190/1.2828977

Slater, L., Zaidman, M. D., Binley, A. M. and West, L. J. (1997b) 'Electrical imaging of saline tracer migration for the investigation of unsaturated zone transport mechanisms', *Hydrology and Earth System Sciences*, 1(2), pp. 291–302.

Slichter, L. B. (1933) 'The interpretation of the resistivity prospecting method for horizontal structures', *Physics*, 4(9), pp. 307–322.

Smith, R. S. and Klein, J. (1996) 'A special circumstance of airborne induced-polarization measurements', *Geophysics*, 61(1), pp. 66–73.

Snyder, D. D. and Merkel, R. M. (1973) 'Analytic models for the interpretation of electrical surveys using buried current electrodes', *Geophysics*, 38(3), pp. 513–529.

Song, L. (1984) 'A new IP decoupling scheme', *Exploration Geophysics*, 15(2), pp. 99–112.

Sørensen, K. (1996) 'Pulled array continuous electrical profiling', *First Break*, 14(3), pp. 85–90. DOI: 10.3997/1365-2397.1996005

Sparacino, M. S., Rathburn, S. L., Covino, T. P., Singha, K. and Ronayne, M. J. (2019) 'Form-based river restoration decreases wetland hyporheic exchange: Lessons learned from the Upper Colorado River', *Earth Surface Processes and Landforms*, 44(1), pp. 191–203.

Sparrenbom, C. J., Åkesson, S., Johansson, S., Hagerberg, D. and Dahlin, T. (2017) 'Investigation of chlorinated solvent pollution with resistivity and induced polarization', *Science of the Total Environment*, 575, pp. 767–778. DOI: 10.1016/j .scitotenv.2016.09.117

Stamm, A. J. (1930) 'An electrical conductivity method for determining the moisture content of wood', *Industrial & Engineering Chemistry Analytical Edition*, 2(3), pp. 240–244.

Stefanesco, S., Schlumberger, C. and Schlumberger, M. (1930) 'Sur la distribution électrique potentielle autour d'une prise de terre ponctuelle dans un terrain à couches horizontales, homogènes et isotropes', *Journal de Physique et le Radium. Société Française de Physique*, 1(4), pp. 132–140.

Stummer, P., Maurer, H. and Green, A. G. (2004) 'Experimental design: Electrical resistivity data sets that provide optimum subsurface information', *Geophysics*, 69(1), pp. 120–139.

Stummer, P., Maurer, H., Horstmeyer, H. and Green, A. G. (2002) 'Optimization of DC resistivity data acquisition: Real-time experimental design and a new multielectrode system', *IEEE Transactions on Geoscience and Remote Sensing*, 40(12), pp. 2727–2735.

Sudduth, K. A., Kitchen, N. R., Bollero, G. A., Bullock, D. G. and Wiebold, W. J. (2003) 'Comparison of electromagnetic induction and direct sensing of soil electrical conductivity', *Agronomy Journal*, 95(3), pp. 472–482.

Suman, R. J. and Knight, R. J. (1997) 'Effects of pore structure and wettability on the electrical resistivity of partially saturated rocks: A network study', *Geophysics*, 62(4), pp. 1151–1162. DOI: 10.1190/1.1444216

Sumner, J. S. (1976) *Principles of Induced Polarization for Geophysical Exploration*. Elsevier Scientific Publishing Company, p. 277.

Sundberg, K. (1932) 'Effect of impregnating waters on electrical conductivity of soil and rocks', *Trans. Am. Inst. Mining Metall. Petrol. Eng.*, 97, pp. 367–391.

Swift Jr, C. M. (1973) 'The L/M parameter of time-domain IP measurements: A computational analysis', *Geophysics*, 38(1), pp. 61–67.

Szalai, S. and Szarka, L. (2008) 'On the classification of surface geoelectric arrays', *Geophysical Prospecting*, 56(2), pp. 159–175.

Tarasov, A. and Titov, K. (2007) 'Relaxation time distribution from time domain induced polarization measurements', *Geophysical Journal International*, 170(1), pp. 31–43.

Tarasov, A. and Titov, K. (2013) 'On the use of the Cole–Cole equations in spectral induced: polarization', *Geophysical Journal International*, 195(1), pp. 352–356. DOI: 10.1093/gji/ggt251

Taylor, R. W. and Fleming, A. H. (1988) 'Characterizing jointed systems by azimuthal resistivity surveys', *Groundwater*, 26(4), pp. 464–474.

Taylor, S. and Barker, R. (2002) 'Resistivity of partially saturated Triassic sandstone', *Geophysical Prospecting*, 50(6), pp. 603–613.

Telford, W. M., Geldart, L. P. and Sheriff, R. E. (1990) *Applied Geophysics*. 2nd edn. Cambridge: Cambridge University Press,

Terrón, J. M., Mayoral, V., Salgado, J. Á., Galea, F. A., Pérez, V. H., Odriozola, C., Mateos, P. and Pizzo, A. (2015) 'Use of soil apparent electrical resistivity contact sensors for the extensive study of archaeological sites', *Archaeological Prospection*, 22(4), pp. 269–281.

Terzaghi, K. (1943) *Theoretical Soil Mechanics*. John Wiley & Sons, New York, pp. 11–15.

Thomas, E. C. (1992) '50th anniversary of the Archie equation: Archie left more than just an equation', *The Log Analyst* (May–June 1992), 199.

Thomsen, R., Søndergaard, V. H. and Sørensen, K. I. (2004) 'Hydrogeological mapping as a basis for establishing site-specific groundwater protection zones in Denmark', *Hydrogeology Journal*, 12(5), pp. 550–562.

Tikhonov, A. N. and Arsenin, V. I. (1977) *Solutions of Ill-Posed Problems*. Winston, Washington, DC, p. 258.

Titov, K., Kemna, A., Tarasov, A. and Vereecken, H. (2010a) 'Induced polarization of unsaturated sands determined through time domain measurements', *Vadose Zone Journal*, 3(4), p. 1160. DOI: 10.2136/vzj2004.1160

Titov, K., Komarov, V., Tarasov, V. and Levitski, A. (2002) 'Theoretical and experimental study of time domain-induced polarization in water-saturated sands', *Journal of Applied Geophysics*, 50(4), pp. 417–433. DOI: 10.1016/S0926-9851(02)00168-4

Titov, K., Tarasov, A., Ilyin, Y., Seleznev, N. and Boyd, A. (2010b) 'Relationships between induced polarization relaxation time and hydraulic properties of sandstone', *Geophysical Journal International*, 180(3), pp. 1095–1106. DOI: 10.1111/j.1365-246X.2009.04465.x

Tombs, J. M. C. (1981) 'The feasibility of making spectral IP measurements in the time domain', *Geoexploration*, 19(2), pp. 91–102. DOI: 10.1016/0016-7142(81)90022-3

Tong, M., Li, L., Wang, W. and Jiang, Y. (2006) 'Determining capillary-pressure curve, pore-size distribution, and permeability from induced polarization of shaley sand', *Geophysics*, 71(3), pp. N33–N40.

Toran, L., Hughes, B., Nyquist, J. and Ryan, R. (2012) 'Using hydrogeophysics to monitor change in hyporheic flow around stream restoration structures', *Environmental & Engineering Geoscience*, 18(1), pp. 83–97.

Truffert, C., Gance, J., Leite, O. and Texier, B. (2019) 'New instrumentation for large 3D electrical resistivity tomography and induced polarization surveys', pp. 124–127. DOI: 10.1190/gem2019-032.1

Ts, M.-E., Lee, E., Zhou, L., Lee, K. H. and Seo, J. K. (2016) 'Remote real time monitoring for underground contamination in Mongolia using electrical impedance tomography', *Journal of Nondestructive Evaluation*, 35(1), p. 8.

Tso, C. H. M., Kuras, O. and Binley, A. (2019) 'On the field estimation of moisture content using electrical geophysics: The impact of petrophysical model uncertainty', *Water Resources Research*, 55(8), pp. 7196–7211. DOI: 10.1029/2019WR024964

Tso, C. H. M., Kuras, O., Wilkinson, P. B., Uhlemann, S., Chambers, J. E., Meldrum, P. I., Graham, J., Sherlock, E. F. and Binley, A. (2017) 'Improved characterisation and modelling of measurement errors in electrical resistivity tomography (ERT) surveys', *Journal of Applied Geophysics*, 146, pp. 103–119. DOI: 10.1016/j.jappgeo.2017.09.009

Tsokas, G. N., Tsourlos, P. I. and Szymanski, J. E. (1997) 'Square array resistivity anomalies and inhomogeneity ratio calculated by the finite-element method', *Geophysics*, 62(2), pp. 426–435.

Tsourlos, P., Ogilvy, R., Meldrum, P. and Williams, G. (2003) 'Time-lapse monitoring in single boreholes using electrical resistivity tomography', *Journal of Environmental & Engineering Geophysics*, 8(1), pp. 1–14.

Tsourlos, P., Ogilvy, R., Papazachos, C. and Meldrum, P. (2011) 'Measurement and inversion schemes for single borehole-to-surface electrical resistivity tomography surveys', *Journal of Geophysics and Engineering*, 8(4), pp. 487–497.

Udphuay, S., Günther, T., Everett, M. E., Warden, R. R. and Briaud, J.-L. (2011) 'Three-dimensional resistivity tomography in extreme coastal terrain amidst dense cultural signals: Application to cliff stability assessment at the historic D-Day site', *Geophysical Journal International, UK*, 185(1), pp. 201–220.

Uhlemann, S. (2018) 'Geoelectrical monitoring of moisture driven processes in natural and engineered slopes'. PhD thesis, ETH Zurich, Switzerland. p. 446.

Uhlemann, S., Chambers, J., Wilkinson, P., Maurer, H., Merritt, A., Meldrum, P., Kuras, O., Gunn, D., Smith, A. and Dijkstra, T. (2017) 'Four-dimensional imaging of moisture dynamics during landslide reactivation', *Journal of Geophysical Research: Earth Surface*, 122 (1), pp. 398–418. DOI: 10.1002/2016JF003983

Uhlemann, S., Wilkinson, P. B., Chambers, J. E., Maurer, H., Merritt, A. J., Gunn, D. A. and Meldrum, P. I. (2015) 'Interpolation of landslide movements to improve the accuracy of 4D geoelectrical monitoring', *Journal of Applied Geophysics*, 121, pp. 93–105.

Ulrich, C. and Slater, L. (2004) 'Induced polarization measurements on unsaturated, unconsolidated sands', *Geophysics*, 69(3), p. 762. DOI: 10.1190/1.1759462

Ulrych, T. J., Sacchi, M. D. and Woodbury, A. (2001) 'A Bayes tour of inversion: A tutorial', *Geophysics*, 66(1), pp. 55–69.

Ustra, A., Mendonça, Carlos Alberto, Ntarlagiannis, D. and Slater, L. D. (2015) 'Relaxation time distribution obtained from a Debye decomposition of spectral induced polarization data', *Geophysics*, 81(2). DOI: 10.1190/GEO2015-0095.1

Vacquier, V. et al. (1957) 'Prospecting for ground water by induced electrical polarization', *Geophysics*, 22(3), pp. 660–687.

Van der Baan, M. and Jutten, C. (2000) 'Neural networks in geophysical applications', *Geophysics*, 65(4), pp. 1032–1047.

Vanella, D., Consoli, S., Cassiani, G., Busato, L., Boaga, J., Barbagallo, S. and Binley, A. (2018) 'The use of small scale electrical resistivity tomography to identify trees root water uptake patterns', *Journal of Hydrology*, 556, pp 310–324.

Van Nostrand, R. G. and Cook, K. L. (1966) 'Interpretation of resistivity data', *Geological Survey Professional Paper*, 499, p. 310.

Van Schoor, M. and Binley, A. (2010) 'In-mine (tunnel-to-tunnel) electrical resistance tomography in South African platinum mines', *Near Surface Geophysics*, 8(6), pp. 563–574.

Van, G. P., Park, S. K. and Hamilton, P. (1991) 'Monitoring leaks from storage ponds using resistivity methods', *Geophysics*, 56(8), pp. 1267–1270.

Vanhala, H. and Soininen, H. (1995) 'Laboratory technique for measurement of spectral induced polarization (SIP) response of soil samples', *Geophysical Prospecting*, 43, pp. 655–676.

Vaudelet, P., Revil, A., Schmutz, M., Franceschi, M. and Bégassat, P. (2011) 'Induced polarization signatures of cations exhibiting differential sorption behaviors in saturated sands', *Water Resources Research*, 47(2), pp. 1–21. DOI: 10.1029/2010WR009310

Veeken P. C., Legeydo P. J., Davidenko Y. A., Kudryavceva E. O., Ivanov S. A., Chuvaev A. (2009) 'Benefits of the induced polarization geoelectric method to hydrocarbon exploration', *Geophysics*, 74(2), pp. B47–B59.

Verdet, C., Anguy, Y., Sirieix, C., Clément, R. and Gaborieau, C. (2018) 'On the effect of electrode finiteness in small-scale electrical resistivity imaging', *Geophysics*, 83(6), pp. EN39–EN52.

Vernon, R. W. (2008) 'Alfred Williams, Leo Daft and "The Electrical Ore-Finding Company Limited"', *British Mining*, 86, pp. 4–30.

Vinciguerra, A., Aleardi, M. and Costantini, P. (2019) 'Full-waveform inversion of complex resistivity IP spectra: Sensitivity analysis and inversion tests using local and global optimization strategies on synthetic datasets: Sensitivity', *Near Surface Geophysics*, 17(2), pp. 109–125. DOI: 10.1002/nsg.12034

Vinegar, H. and Waxman, M. (1984) 'Induced polarization of shaly sands', *Geophysics*, 49 (8), pp. 1267–1287. DOI: 10.1190/1.1441755

Wagner, F. M., Bergmann, P., Rücker, C., Wiese, B., Labitzke, T., Schmidt-Hattenberger, C. and Maurer, H. (2015) 'Impact and mitigation of borehole related effects in permanent crosshole resistivity imaging: An example from the Ketzin CO2 storage site', *Journal of Applied Geophysics*, 123, pp. 102–111.

Wainwright, H. M., Flores Orozco, A., Bücker, M., Dafflon, B., Chen, J., Hubbard, S. S. and Williams, K. H. (2016) 'Hierarchical Bayesian method for mapping biogeochemical hot spots using induced polarization imaging', *Water Resources Research*, 52(1), pp. 533–551. DOI: 10.1002/2015WR017763

Wait, J. R. and Gruszka, T. P. (1986) 'On electromagnetic coupling "removal" from induced polarization surveys', *Geoexploration*, 24(1), pp. 21–27.

Walker, J. P. and Houser, P. R. (2002) 'Evaluation of the OhmMapper instrument for soil moisture measurement', *Soil Science Society of America Journal*, 66(3), pp. 728–734.

Walker, S. E. (2008) 'Should we care about negative transients in helicopter TEM data?', in *SEG Technical Program Expanded Abstracts 2008*. Society of Exploration Geophysicists, pp. 1103–1107.

Wang, C. and Slater, L. D. (2019) 'Extending accurate spectral induced polarization measurements into the kHz range: Modelling and removal of errors from interactions between the parasitic capacitive coupling and the sample holder', *Geophysical Journal International*, 218(2), pp. 895–912. DOI: 10.1093/gji/ggz199

Wang, M. (2015) 'Electrical impedance tomography', in *Industrial Tomography* (Wang, M. ed.). Woodhead Publishing, pp. 23–59. DOI: https://DOI.org/10.1016/B978-1-78242-118-4.00002-2

Wang, M., Dickin, F. J. and Beck, M. S. (1993) 'Improved electrical impedance tomography data collection system and measurement protocols', in *Tomographic Techniques for Process Design and Operation*. Computational Mechanics, Billerica, MA, pp. 75–88.

Ward, A. S., Gooseff, M. N. and Singha, K. (2010) 'Imaging hyporheic zone solute transport using electrical resistivity', *Hydrological Processes*, 24(7), pp. 948–953.

Ward, A. S., Fitzgerald, M., Gooseff, M. N., Voltz, T. J., Binley, A. M. and Singha, K. (2012) 'Hydrologic and geomorphic controls on hyporheic exchange during baseflow recession in a headwater mountain stream', *Water Resources Research*, 48, W04513.

Ward, A. S., Kurz, M. J., Schmadel, N. M., Knapp, J. L. A., Blaen, P. J., Harman, C. J., Drummond, J. D., Hannah, D. M., Krause, S. and Li, A. (2019) 'Solute transport and transformation in an intermittent, headwater mountain stream with diurnal discharge fluctuations', *Water*, 11(11), p. 2208.

Ward, S. H. (1980) 'Electrical, electromagnetic, and magnetotelluric methods', *Geophysics*, 45(11), pp. 1659–1666. DOI: 10.1190/1.1441056

Ward, S. H. and Fraser, D. C. (1967) 'Part B: Conduction of electricity in rocks', *Mining Geophysics*, 2, pp. 197–223.

Ward, S. H., Sternberg, B. K., LaBrecque, D. J. and Poulton, M. M. (1995) 'Recommendations for IP research', *The Leading Edge*, 14(April), p. 243. DOI: 10.1190/1.1437120

Watson, K. A. and Barker, R. D. (1999) 'Differentiating anisotropy and lateral effects using azimuthal resistivity offset Wenner soundings', *Geophysics*, 64(3), pp. 739–745.

Waxman, M. H. and Smits, L. J. M. (1968) 'Electrical conductivities in oil-bearing shaly sands', *Society of Petroleum Engineers Journal*, 8(02), pp. 107–122.

Webster, J. G. (1990) *Electrical Impedance Tomography*. Taylor & Francis Group, p. 224.

Wehrer, M. and Slater, L. D. (2015) 'Characterization of water content dynamics and tracer breakthrough by 3-D electrical resistivity tomography (ERT) under transient unsaturated conditions', *Water Resources Research*, 51, 97–124, DOI:10.1002/2014WR016131

Wehrer, M., Binley, A. and Slater, L. D. (2016) 'Characterization of reactive transport by 3-D electrical resistivity tomography (ERT) under unsaturated conditions', *Water Resources Research*, 52(10). DOI: 10.1002/2016WR019300

Weigand, M. and Kemna, A. (2016a) 'Debye decomposition of time-lapse spectral induced polarisation data', *Computers and Geosciences*, 86, pp. 34–45. DOI: 10.1016/j.cageo.2015.09.021

Weigand, M. and Kemna, A. (2016b) 'Relationship between Cole–Cole model parameters and spectral decomposition parameters derived from SIP data', *Geophysical Journal International*, 205(3), pp. 1414–1419. DOI: 10.1093/gji/ggw099

Weigand, M. and Kemna, A. (2017) 'Multi-frequency electrical impedance tomography as a non-invasive tool to characterize and monitor crop root systems', *Biogeosciences*, 14 (4), pp. 921–939. DOI: 10.5194/bg-14-921-2017

Weigand, M. and Kemna, A. (2019) 'Imaging and functional characterization of crop root systems using spectroscopic electrical impedance measurements', *Plant and Soil*, 435 (1–2), pp. 201–224. DOI: 10.1007/s11104-018-3867-3

Weigand, M., Orozco, A. F. and Kemna, A. (2017) 'Reconstruction quality of SIP parameters in multi-frequency complex resistivity imaging', *Near Surface Geophysics*, 15 (2), pp. 187–199. DOI: 10.3997/1873-0604.2016050

Weiss, O. (1933) 'The limitations of geophysical methods and the new possibilities opened up by an electrochemical method for determining geological formations at great depths', in *Proceeding of the 1st World Petroleum Congress*. London.

Weller, A. and Slater, L. (2012) 'Salinity dependence of complex conductivity of unconsolidated and consolidated materials: Comparisons with electrical double layer models', *Geophysics*, 77(5), pp. 185–198. DOI: 10.1190/geo2012-0030.1

Weller, A. and Slater, L. (2019) 'Permeability estimation from induced polarization: An evaluation of geophysical length scales using an effective hydraulic radius concept', *Near Surface Geophysics*, pp. 1–14. DOI: 10.1002/nsg.12071

Weller, A., Breede, K., Slater, L. and Nordsiek, S. (2011) 'Effect of changing water salinity on complex conductivity spectra of sandstones', *Geophysics*, 76(5), p. F315. DOI: 10.1190/geo2011-0072.1

Weller, A., Gruhne, M., Seichter, M. and Börner, F. D. (1996b), 'Monitoring hydraulic experiments by complex conductivity tomography,' *European Journal of Environmental and Engineering Geophysics*, 1, pp. 209–228.

Weller, A., Lewis, R., Canh, T., Möller, M. and Scholz, B. (2014) 'Geotechnical and geophysical long-term monitoring at a levee of Red River in Vietnam', *Journal of Environmental and Engineering Geophysics*, 19(3), pp. 183–192.

Weller, A., Nordsiek, S. and Debschütz, W. (2010) 'Estimating permeability of sandstone samples by nuclear magnetic resonance and spectral-induced polarization', *Geophysics*, 75(6), pp. E215–E226. DOI: 10.1190/1.3507304

Weller, A., Seichter, M. and Kampke, A. (1996a) 'Induced-polarization modelling using complex electrical conductivities', *Geophysical Journal International*, 127(2), pp. 387–398. DOI: 10.1111/j.1365-246X.1996.tb04728.x

Weller, A., Slater, L., Binley, A., Nordsiek, S. and Xu, S. (2015b) 'Permeability prediction based on induced polarization: Insights from measurements on sandstone and unconsolidated samples spanning a wide permeability range', *Geophysics*, 80(2), pp. D161–D173. DOI: 10.1190/GEO2014-0368.1

Weller, A., Slater, L., Nordsiek, S. and Ntarlagiannis, D. (2010a) 'On the estimation of specific surface per unit pore volume from induced polarization: A robust empirical relation fits multiple data sets', *Geophysics*, 75 (4), pp. WA105–WA112. DOI: 10.1190/1.3471577

Weller, A., Zhang, Z. and Slater, L. (2015a) 'High-salinity polarization of sandstones', *Geophysics*, 80(3), pp. 1–10. DOI: 10.1190/GEO2014-0483.1

Weller, A., Zhang, Z., Slater, L., Kruschwitz, S. and Halisch, M. (2016) 'Induced polarization and pore radius: A discussion', *Geophysics*, 81(5). DOI: 10.1190/GEO2016-0135.1

Weller, Andreas, Slater, L. and Nordsiek, S. (2013) 'On the relationship between induced polarization and surface conductivity: Implications for petrophysical interpretation of electrical measurements', *Geophysics*, 78(5), pp. 315–325. DOI: 10.1190/geo2013-0076.1

Wenner, F. (1912) 'Characteristics and applications of vibration galvanometers', *Proceedings of the American Institute of Electrical Engineers. IEEE*, 31(6), pp. 1073–1084.

Wenner, F. (1912) 'The four-terminal conductor and the Thomson bridge', *Bulletin of the Bureau of Standards*, 8, pp. 559–610.

Wenner, F. (1915) 'A method for measuring Earth resistivity', *Journal of the Washington Academy of Sciences*, 5(16), pp. 561–563.

West, S. S. (1940) 'Three-layer resistivity curves for the Eltran electrode configuration', *Geophysics*, 5(1), pp. 43–46.

Wetzel, W. W. and McMurry, H. V (1937) 'A set of curves to assist in the interpretation of the three layer resistivity problem', *Geophysics*, 2(4), pp. 329–341.

Wexler, A., Fry, B. and Neuman, M. R. (1985) 'Impedance-computed tomography algorithm and system', *Applied Optics*, 24(23), pp. 3985–3992.

Whalley, W. R., Binley, A., Watts, C. W., Shanahan, P., Dodd, I. C., Ober, E. S., Ashton, R. W., Webster, C. P., White, R. P. and Hawkesford, M. J. (2017) 'Methods to estimate changes in soil water for phenotyping root activity in the field', *Plant and Soil*, 415 (1–2). DOI: 10.1007/s11104-016-3161-1

White, C. C. and Barker, R. D. (1997) 'Electrical leak detection system for landfill liners: A case history', *Groundwater Monitoring & Remediation*, 17(3), pp. 153–159.

Whiteley, J. S., Chambers, J. E., Uhlemann, S., Wilkinson, P. B. and Kendall, J. M. (2019) 'Geophysical monitoring of moisture-induced landslides: A review', *Reviews of Geophysics*, 57(1), pp. 106–145.

Whitney, W., Gardner, F. D. and Briggs, L. J. (1897) *An Electrical Method of Determining the Moisture Content of Arable Soils, Bulletin No.6*. U.S. Department of Agriculture, Washington, D.C.

Wilkinson, P. B., Chambers, J. E., Lelliott, M., Wealthall, G. P. and Ogilvy, R. D. (2008) 'Extreme sensitivity of crosshole electrical resistivity tomography measurements to geometric errors', *Geophysical Journal International*, 173(1), pp. 49–62.

Wilkinson, P. B., Chambers, J. E., Meldrum, P. I., Ogilvy, R. D. and Caunt, S. (2006a) 'Optimization of array configurations and panel combinations for the detection and imaging of abandoned mineshafts using 3D cross-hole electrical resistivity tomography', *Journal of Environmental & Engineering Geophysics*, 11(3), pp. 213–221.

Wilkinson, P. B., Fromhold, T. M., Tench, C. R., Taylor, R. P. and Micolich, A. P. (2001) 'Compact fourth-order finite difference method for solving differential equations', *Physical Review E: Statistical Physics, Plasmas, Fluids, and Related Interdisciplinary Topics*, 64 (4), p. 4. DOI: 10.1103/PhysRevE.64.047701

Wilkinson, P. B., Loke, M. H., Meldrum, P. I., Chambers, J. E., Kuras, O., Gunn, D. A. and Ogilvy, R. D. (2012) 'Practical aspects of applied optimized survey design for electrical resistivity tomography', *Geophysical Journal International*, 189(1), pp. 428–440.

Wilkinson, P. B., Meldrum, P. I., Chambers, J. E., Kuras, O. and Ogilvy, R. D. (2006b) 'Improved strategies for the automatic selection of optimized sets of electrical resistivity tomography measurement configurations', *Geophysical Journal International*, 167(3), pp. 1119–1126.

Wilkinson, P. B., Meldrum, P. I., Kuras, O., Chambers, J. E., Holyoake, S. J. and Ogilvy, R. D. (2010) 'High-resolution electrical resistivity tomography monitoring of a tracer test in a confined aquifer', *Journal of Applied Geophysics*, 70(4), pp. 268–276.

Wilkinson, P. B., Uhlemann, S., Meldrum, P. I., Chambers, J. E., Carrière, S., Oxby, L. S. and Loke, M. H. (2015) 'Adaptive time-lapse optimized survey design for electrical resistivity tomography monitoring', *Geophysical Journal International*, 203(1), pp. 755–766.

Wilkinson, P., Chambers, J., Uhlemann, S., Meldrum, P., Smith, A., Dixon, N. and Loke, M. H. (2016) 'Reconstruction of landslide movements by inversion of 4-D electrical resistivity tomography monitoring data', *Geophysical Research Letters*, 43 (3), pp. 1166–1174.

Williams, B. A., Brown, C. F., Um, W., Nimmons, M. J., Peterson, R. E., Bjornstad, B. N., Lanigan, D. C., Serne, R. J., Spane, F. A. and Rockhold, M. L. (2007) Limited field investigation report for uranium contamination in the 300 Area, 300-FF-5 operable unit, Hanford Site, Washington, Report PNNL-16435.

Williams, K. H, Ntarlagiannis, D., Slater, L. D, Dohnalkova, A., Hubbard, S. S. and Banfield, J. F. (2005) 'Geophysical imaging of stimulated microbial biomineralization', *Environmental Science & Technology*, 39(19), pp. 7592–600. DOI: 10.1021/es0504035

Williams, K. H., Kemna, A., Wilkins, M. J., Druhan, J., Arntzen, E., N'Guessan, A. L., Long, P. E., Hubbard, S. S. and Banfield, J. F. (2009) 'Geophysical monitoring of coupled microbial and geochemical processes during stimulated subsurface bioremediation', *Environmental Science and Technology*, 43(17), pp. 6717–6723. DOI: 10.1021/es900855j

Winsauer, W. O., Shearin Jr, H. M., Masson, P. H. and Williams, M. (1952) 'Resistivity of brine-saturated sands in relation to pore geometry', *AAPG Bulletin*, 36(2), pp. 253–277.

Winship, P., Binley, A. and Gomez, D. (2006) 'Flow and transport in the unsaturated Sherwood sandstone: Characterization using cross-borehole geophysical methods', *Geological Society, London, Special Publications*, 263(1), pp. 219–231.

Wong, J. (1979) 'An electrochemical model of the induced-polarization phenomenon in disseminated sulfide ores', *Geophysics*, 44(7), pp. 1245–1265. DOI: 10.1190/1.1441005

Wong, J. and Strangway, D. W. (1981) 'Induced polarization in disseminated sulfide ores containing elongated mineralization', *Geophysics*, 46(9), pp. 1258–1268.

Wood, J. (2017) 'Roman Lancaster: The archaeology of Castle Hill', *British Archaeology*, Nov–Dec 2017, pp. 38–45.

Worthington, P. F. (1993) 'The uses and abuses of the Archie equations, 1: The formation factor-porosity relationship', *Journal of Applied Geophysics*, 30(3), pp. 215–228. DOI: 10.1016/0926-9851(93)90028-W

Wu, Y., Hubbard, S., Williams, K. H. and Ajo-Franklin, J. (2010) 'On the complex conductivity signatures of calcite precipitation', *Journal of Geophysical Research*, 115, p. G00G04. DOI: 10.1029/2009JG001129

Wu, Y., Slater, L. D. and Korte, N. (2006) 'Low frequency electrical properties of corroded iron barrier cores', *Environmental Science & Technology*, 40(7), pp. 2254–2261

Wyllie, M. R. J. and Rose, W. D. (1950) 'Some theoretical considerations related to the quantitative evaluation of the physical characteristics of reservoir rock from electrical log data', *Journal of Petroleum Technology*, 2(04), pp. 105–118.

Wyllie, M. R. J. and Southwick, P. F. (1954) 'An experimental investigation of the SP and resistivity phenomena in dirty sands', *Journal of Petroleum Technology*, 6(02), pp. 44–57.

Xiang, J., Jones, N. B., Cheng, D. and Schlindwein, F. S. (2001) 'Direct inversion of the apparent complex-resistivity spectrum', *Geophysics*, 66(5), pp. 1399–1404.

Xu, B. and Noel, M. (1993) 'On the completeness of data sets with multielectrode systems for electrical resistivity survey', *Geophysical Prospecting*, 41(6), pp. 791–801.

Yamashita, Y. and Lebert, François (2015a) 'The characteristic and practical issue of resistivity measurement by multiple-current injection based on CDMA technique', in *Proceedings of the 12th SEGJ International Symposium*. Tokyo, Japan, 18–20 November 2015, pp. 9–12. DOI: 10.1190/segj122015-003

Yamashita, Y. and Lebert, Francois (2015b) 'The characteristic of multiple current resistivity profile using Code-Division Multiple-Access technique regarding data quality', in Near-Surface Asia Pacific Conference, Waikoloa, Hawaii, 7–10 July 2015, pp. 367–370.

Yamashita, Y. Y., Kobayashi, T., Saito, H. S., Sugii, T., Kodaka, T., Maeda, K. M. and Cui, Y. C. (2017) '3D ERT monitoring of levee flooding experiment using multi-current transmission technique', 23rd European Meeting of Environmental and Engineering Geophysics, (September 2017), pp. 3–7. DOI: 10.3997/2214-4609.201701980

Yamashita, Y., Lebert, F., Gourry, J. C., Bourgeois, B. and Texier, B. (2014) 'A method to calculate chargeability on multiple-transmission resistivity profile using Code-Division Multiple-Access', *Society of Exploration Geophysicists International Exposition and 84th Annual Meeting SEG 2014*, (2), pp. 3892–3897. DOI: 10.1190/SEG-2014-1270.pdf

Yang, X., Chen, X., Carrigan, C. R. and Ramirez, A. L. (2014) 'Uncertainty quantification of CO2 saturation estimated from electrical resistance tomography data at the Cranfield site', *International Journal of Greenhouse Gas Control*, 27, pp. 59–68. DOI: 10.1016/j.ijggc.2014.05.006

Yang, X., Lassen, R. N., Jensen, K. H. and Looms, M. C. (2015) 'Monitoring CO2 migration in a shallow sand aquifer using 3D crosshole electrical resistivity tomography', *International Journal of Greenhouse Gas Control*, 42, pp. 534–544.

Yerworth, R. J., Bayford, R. H., Brown, B., Milnes, P., Conway, M. and Holder, D. S. (2003) 'Electrical impedance tomography spectroscopy (EITS) for human head imaging', *Physiological Measurement*, 24(2), p. 477.

Yi, M.-J., Kim, J.-H. and Chung, S.-H. (2003) 'Enhancing the resolving power of least-squares inversion with active constraint balancing', *Geophysics*, 68(3), pp. 931–941.

Yukselen, Y. and Kaya, A. (2008) 'Suitability of the methylene blue test for surface area, cation exchange capacity and swell potential determination of clayey soils', *Engineering Geology*, 102(1–2), pp. 38–45. DOI: 10.1016/j.enggeo.2008.07.002

Yuval, D. and Oldenburg, D. W. (1997) 'Computation of Cole–Cole parameters from IP data', *Geophysics*, 62(2), pp. 436–448.

Yuval, D. and Oldenburg, W. (1996) 'DC resistivity and IP methods in acid mine drainage problems: Results from the Copper Cliff mine tailings impoundments', *Journal of Applied Geophysics*, 34(3), pp. 187–198.

Zaidman, M. D., Middleton, R. T., West, L. J. and Binley, A. M. (1999) 'Geophysical investigation of unsaturated zone transport in the Chalk in Yorkshire', *Quarterly Journal of Engineering Geology and Hydrogeology*, 32(2), pp. 185–198.

Zarif, F., Kessouri, P. and Slater, L. (2017) 'Recommendations for field-scale Induced Polarization (IP) data acquisition and interpretation', *Journal of Environmental and Engineering Geophysics*, 22(4). DOI: 10.2113/JEEG22.4.395

Zarif, F., Slater, L., Mabrouk, M., Youssef, A., Al-Temamy, A., Mousa, S., Farag, K. and Robinson, J. (2018) 'Groundwater resources evaluation in calcareous limestone using geoelectrical and VLF-EM surveys (El Salloum Basin, Egypt)', *Hydrogeology Journal*, 26(4), pp. 1169–1185.

Zhang, C., Revil, A., Fujita, Y., Munakata-Marr, J. and Redden, G. (2014) 'Quadrature conductivity: A quantitative indicator of bacterial abundance in porous media', *Geophysics*, 79(6), pp. D363–D375. DOI: 10.1190/geo2014-0107.1

Zhang, J. and Revil, A. (2015) 'Cross-well 4-D resistivity tomography localizes the oil–water encroachment front during water flooding', *Geophysical Journal International*, 201(1), pp. 343–354.

Zhang, J., Mackie, R. L. and Madden, T. R. (1995) '3-D resistivity forward modeling and inversion using conjugate gradients', *Geophysics*, 60(5), pp. 1313–1325.

Zhao, S. and Yedlin, M. J. (1996) 'Some refinements on the finite-difference method for 3-D dc resistivity modeling', *Geophysics*, 61(5), pp. 1301–1307.

Zhao, Y., Zimmermann, E., Huisman, J. A., Treichel, A., Wolters, B., van Waasen, S. and Kemna, A. (2013) 'Broadband EIT borehole measurements with high phase accuracy using numerical corrections of electromagnetic coupling effects', *Measurement Science and Technology*, 24(8), p. 85005.

Zhao, Y., Zimmermann, E., Huisman, J. A., Treichel, A., Wolters, B., van Waasen, S. and Kemna, A. (2014) 'Phase correction of electromagnetic coupling effects in cross-borehole EIT measurements', *Measurement Science and Technology*, 26(1), p. 15801, DOI: 10.1088/0957-0233/26/1/015801

Zhou, B. and Greenhalgh, S. A. (2002) 'Rapid 2-D/3-D crosshole resistivity imaging using the analytic sensitivity function', *Geophysics*, 67(3), pp. 755–765. DOI: 10.1190/1.1484518

Zhou, B., Greenhalgh, M. and Greenhalgh, S. A. (2009) '2.5-D/3-D resistivity modelling in anisotropic media using Gaussian quadrature grids', *Geophysical Journal International*, 176(1), pp. 63–80.

Zhou, J., Revil, A., Karaoulis, M., Hale, D., Doetsch, J. and Cuttler, S. (2014) 'Image-guided inversion of electrical resistivity data', *Geophysical Journal International*, 197(1), pp. 292–309.

Zhou, Q. Y. (2007) 'A sensitivity analysis of DC resistivity prospecting on finite, homogeneous blocks and columns', *Geophysics*, 72(6), pp. F237–F247.

Zhou, X., Bhat, P., Ouyang, H. and Yu, J. (2017) 'Localization of cracks in cementitious materials under uniaxial tension with electrical resistance tomography', *Construction and Building Materials*, 138, pp. 45–55.

Zimmermann, E., Kemna, A., Berwix, J., Glaas, W. and Vereecken, H. (2008b) 'EIT measurement system with high phase accuracy for the imaging of spectral induced polarization properties of soils and sediments', *Measurement Science and Technology*, 19 (9), p. 094010. DOI: 10.1088/0957-0233/19/9/094010

Zimmermann, E., Kemna, a, Berwix, J., Glaas, W., Münch, H. M. and Huisman, J. A. (2008a) 'A high-accuracy impedance spectrometer for measuring sediments with low polarizability', *Measurement Science and Technology*, 19 (10), p. 105603. DOI: 10.1088/0957-0233/19/10/105603

Zisser, N., Kemna, A. and Nover, G. (2010a) 'Dependence of spectral-induced polarization response of sandstone on temperature and its relevance to permeability estimation', *Journal of Geophysical Research: Solid Earth*, 115 (9), pp. 1–15. DOI: 10.1029/2010JB007526

Zisser, N., Kemna, A. and Nover, G. (2010b) 'Relationship between low-frequency electrical properties and hydraulic permeability of low-permeability sandstones', *Geophysics*, 75 (3). DOI: 10.1190/1.3413260

Zohdy, A. A. R. (1975) Automatic interpretation of Schlumberger sounding curves, using modified Dar Zarrouk functions. US Geological Survey Bulletin 1313-E. US Govt. Print. Off., p. 39. DOI: 10.3133/b1313E

Zohdy, A. A. R. (1989) 'A new method for the automatic interpretation of Schlumberger and Wenner sounding curves', *Geophysics*, 54(2), pp. 245–253.

Zonge, K. L. and Wynn, J. C. (1975) 'Recent advances and applications in complex resistivity measurements', *Geophysics*, 40(5), pp. 851–864. DOI: 10.1190/1.1440572

Zonge, K. L., Sauck, W. A. and Sumner, J. S. (1972) 'Comparison of time, frequency, and phase measurements in induced polarization', *Geophysical Prospecting*, 20(3), pp. 626–648. DOI: 10.1111/j.1365-2478.1972.tb00658.x

Zonge, K., Wynn, J. and Urquhart, S. (2005) 'Resistivity, induced polarization, and complex resistivity', in *Near-Surface Geophysics* (Butler, D. K. ed.). Society of Exploration Geophysicists, pp. 265–300.

Index

Printed in the United States
by Baker & Taylor Publisher Services